Optical Properties of Graphene

Optical Properties of Graphene

Editor

Rolf Binder

University of Arizona, USA

World Scientific

NEW JERSEY • LONDON • SINGAPORE • BEIJING • SHANGHAI • HONG KONG • TAIPEI • CHENNAI • TOKYO

Published by

World Scientific Publishing Co. Pte. Ltd.
5 Toh Tuck Link, Singapore 596224
USA office: 27 Warren Street, Suite 401-402, Hackensack, NJ 07601
UK office: 57 Shelton Street, Covent Garden, London WC2H 9HE

Library of Congress Cataloging-in-Publication Data
Names: Binder, Rolf H., editor.
Title: Optical properties of graphene / editor, Rolf Binder (University of Arizona, USA).
Description: Hackensack, NJ : World Scientific Publishing Co. Pte. Ltd., [2016]
Identifiers: LCCN 2016029473| ISBN 9789813148741 (hardcover ; alk. paper) |
ISBN 9813148748 (hardcover ; alk. paper)
Subjects: LCSH: Graphene--Optical properties.
Classification: LCC QD341.H9 O65 2016 | DDC 546/.681--dc23
LC record available at https://lccn.loc.gov/2016029473

British Library Cataloguing-in-Publication Data
A catalogue record for this book is available from the British Library.

Desk Editor: Rhaimie Wahap

Typeset by Stallion Press
Email: enquiries@stallionpress.com

Printed in Singapore

Preface

The importance of graphene can hardly be overstated. Graphene is the thinnest material comprised of atoms. It is only one atomic layer thick and hence an almost ideal representation of a two-dimensional material in our three-dimensional world. It has unique physical properties, including extreme mechanical strength and large electrical and thermal conductivity. Ever since the isolation of graphene with the "sticky-tape method" by Andre Geim and Konstantin Novoselov in 2004, for which they received the Nobel prize in 2010, research on graphene has increased dramatically. According to an article by Andrew Plume in the September 2014 issue of Research Trends (Issue 38, www.researchtrends.com), publications on graphene have increased almost exponentially from about 100 in 2004 to almost 10,000 in 2013.

In addition to technologically-important physical properties in the areas of electrical and thermal transport, the optical properties are among those that have been studied extensively. First of all, understanding of the optical properties of graphene is an interesting research area in its own right. The optical properties are very different from those of III–V semiconductors such as GaAs, which are presently the most important material systems used in opto-electronic device applications. Moreover, the research on graphene has initiated large-scale research on other two-dimensional materials, such as transition-metal dichalcogenides, and understanding graphene helps us to understand many properties of those materials as well. Secondly, in addition to research on its fundamental optical properties, investigations aimed at laying the foundation for graphene to be used in opto-electronic device applications have also dramatically increased.

With the foregoing in mind, it appeared useful to have a review book that exclusively focuses on the optical properties of graphene. The book aims to be accessible to graduate students and researchers that are interested in graphene. In addition to the graduate-level introductory tutorial, 13 review chapters provide an in-depth look at the state-of-the-art of graphene research from a distinctive optics point of view. They cover basic aspects of linear and nonlinear spectroscopy and optical response, including ultrafast pump-and-probe spectroscopy, terahertz spectroscopy, Raman spectroscopy, ultraviolet spectroscopy, with some chapters focusing on experiment and others on theory, and they also cover examples of applications such as saturable absorbers, optical modulators and transparent electrodes.

Due to space limitations, some optics-related topics have been omitted here, with the hope that they be the focus of future review books. This includes the broad area of plasmonics, artificial graphene-like structures in photonic systems, as well as more in-depth comparisons between graphene and other two-dimensional materials.

I would like to thank the contributing authors for the very enjoyable cooperation.

Rolf Binder
Tucson, Arizona
June 2016

Contents

Introductory Tutorial

Chapter 1

Introductory Tutorial

Rolf Binder* and Nai-Hang Kwong
College of Optical Sciences and Department of Physics
University of Arizona
1630 E. University Blvd., Tucson, Arizona 85721, USA
**binder@optics.arizona.edu*

In this Introductory Tutorial, we first present a basic overview of optical processes and optical transitions in graphene and compare them to optical transitions in other material systems, such as semiconductor quantum wells. We discuss basic aspects of the bandstructure of graphene, both close to the Dirac point and away from that point. This is followed by a text-book level introduction to the tight-binding model and its application to graphene. Analytical expressions for the Bloch wave functions and the electronic bandstructure close to the Dirac point will be derived. The Bloch wave functions will then be used as an input to the calculation of optical matrix elements. The derivation of the interband polarization will be performed within the frame work of many-particle systems described in second quantization. Finally, the linear optical absorption spectrum of graphene will be derived, using a transfer matrix method for the approximate solution of Maxwell's equations.

1.1 Introduction and Overview of Linear Optical Properties of Graphene

Graphene is a material system with remarkable properties. Consisting of only one atomic layer of carbon atoms, it is as thin as a system composed of atoms can get. From that physical point of view, it can be viewed as an ideal 2-dimensional system. Yet it has robust mechanical properties and conducts both electricity and heat efficiently, which in the future might make it useful for many real-life applications.

Since 2004, the year Andre Geim and Konstantin Novoselov isolated and characterized graphene samples using the exfoliation or 'Scotch tape' technique (work for which they received the Noble prize in 2010), research on the basic physical properties and potential applications has sky-rocketed. The potential electronic and photonic applications include high-speed field-effect transistors, efficient terahertz sources, ultrafast broadband photodetectors, modulators, plasmonic devices, to name a few. Some of the publications on potential applications include Refs. [1–13].

Since many potential applications of graphene involve its optical properties, we bring together in this book studies that elucidate many different aspects of fundamental optical properties and optical application concepts of graphene. We have asked experts in the field to provide insight into their state-of-the-art research on a variety of properties, which, in addition to the above-mentioned device concepts, include fundamental electronic, vibronic, and optical properties. We present ongoing developments of previously reviewed investigations, for example [14–18], and provide detailed information on a number of topics that are corner stones in the area of graphene physics, such as Raman spectroscopy [19–23], linear and nonlinear spectroscopy close to the Dirac point including THz spectroscopy and generation (e.g. [24–43]), electron-electron interaction (e.g. [44, 45]), phonons (e.g. [46–50]), electron-phonon interaction (e.g. [51–57]), and linear and nonlinear spectroscopy close to the saddle point (e.g. [40, 58–62]). Of course, some aspects of the physics of graphene are closely related to other material systems based on carbon atoms, for example graphite, a 3-dimensional crystal consisting of weakly bound graphene sheets, and commonly used in pencils, and carbon nanotubes (e.g. [63]).

Much of the enthusiasm graphene has found among researchers worldwide is based on its unique electronic bandstructure. The electronic bandstructure does not only yield exotic electronic and electron transport properties, but also makes graphene very different from the materials conventionally used in optical applications. Let us therefore first look at the basic optical properties in relation to the electronic bandstructure. After discussing the basic physical elements, we provide in the following sections of this Introductory Tutorial detailed derivations of the formulas and concepts needed to understand the electronic bandstructure and linear optical properties of graphene.

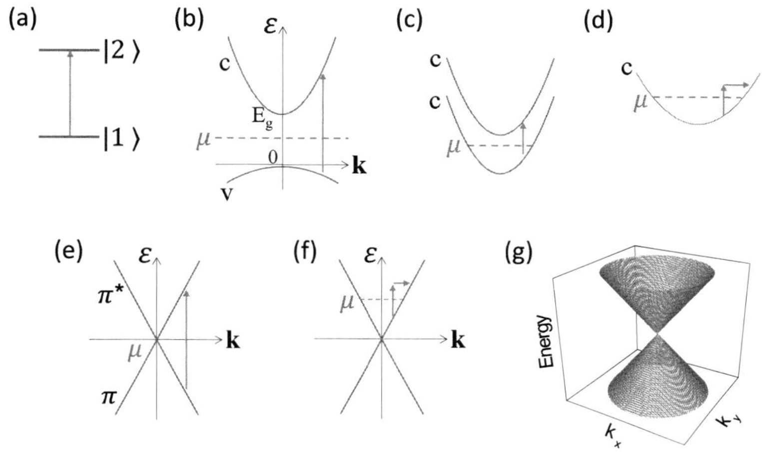

Fig. 1.1 (a)-(f) Sketch of optical transitions and (g) sketch of the Dirac cone.

In any quantum mechanical system, the optical properties are determined by transitions between two energy eigenstates. In the simplest case of a two-level atom the transitions are between the lower and upper level that correspond to different atomic orbitals (Fig. 1.1a). Typically, in thermal equilibrium the lower state is occupied and the upper is not. In crystals, we have many eigenstates, labeled by the wave vector $\mathbf{k}$ and the band index ν (which corresponds roughly to the atomic orbitals that give rise to the energy band). In conventional direct-gap semiconductors, such as GaAs, the valence (occupied) and conduction (not occupied) bands are separated by an energy gap E_g, the bands are approximately parabolic, the chemical potential μ is between the bands. Optical transitions, then, are transitions between the v- and c-band, which are vertical in $\mathbf{k}$-space if the small momentum transfer provided by the light field is neglected, Fig. 1.1b. Typical frequencies of those transitions in direct-gap semiconductors are in the near-infrared to blue, but they can also be in the far infrared. More common for far-infrared and tera-Hertz (THz) transitions are transitions between subbands in semiconductor quantum wells, Fig. 1.1c. If the wave functions of the two subbands are localized in the plane of the quantum well, they can be used, for example, for THz light emission in quantum cascade lasers. If the upper of the two subbands has a wave function that is not restricted to the plane of the quantum well, the transitions can be used in quantum-well infrared photodetectors (QWIPs).

Conventional metals are often well described by a single band, which in thermal equilibrium is only partially filled. If we restrict ourselves to that single band (of course, in reality there are a lot more bands, but often energetically far away), then optical transitions cannot be vertical in **k**-space. They could be created with the small momentum transfer from the light, or through excitations that involve, in addition to light, lattice vibrations that can give non-zero momentum transfer, Fig. 1.1d. They could also involve spatial oscillations of the entire electron system or plasmons that provide the necessary momentum transfer.

When we look at the band structure of graphene, or more precisely that part of the bandstructure that made graphene so famous, we see that it is not like any of the systems that we just looked at. It has no band gap and it does not have parabolic bands. But intrinsic graphene, i.e. graphene that is not doped with impurity atoms acting as donors or acceptors, is also not like a metal with a partially filled band. The Fermi energy (i.e. the chemical potential μ at zero temperature) coincides with the point of the conical intersection of the valence and conduction band (the two bands are often more properly referred to as π and π^* bands). It is clear from the figure that at the low end of the spectrum all frequencies contribute to the optical absorption, as there is no gap in the system. This has motivated a large body or research into the possibility of THz absorption and more generally THz spectroscopy. In practise, the chemical potential is usually not exactly at the Dirac point. In the absence of dopants but at non-zero temperature, it moves into the valence band, and in the presence of dopants like nitrogen atoms it can move into the conduction band, as shown in Fig. 1.1f.

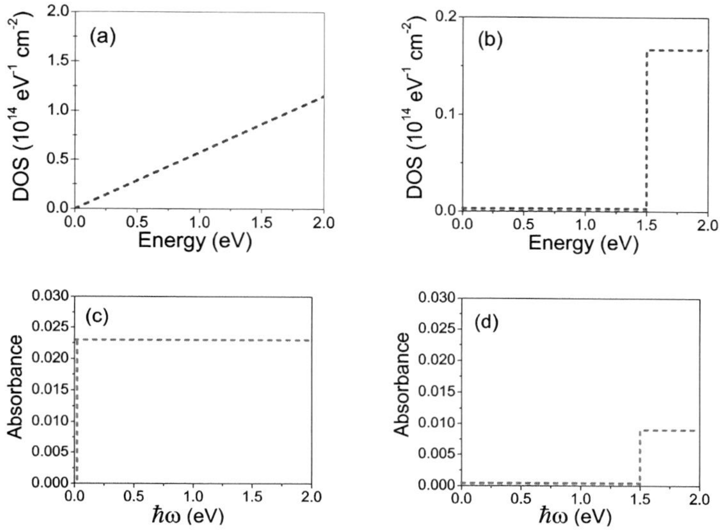

Fig. 1.2 (a) Joint density of states (DOS) and (c) absorbance of graphene. Corresponding results for a simplified quantum well model are shown in (b) and (d). Note that a more realistic model for the quantum well absorption would include Coulomb enhancement, which increases the band-to-band absorption approximately by a factor of 2, and exciton resonances below the bandgap.

Note that in Fig. 1.1e,f the wave vector $\mathbf{k}$ is two-dimensional, $\mathbf{k} = (k_x, k_y)$, and the bandstructure consists of two cones that touch each other at the so-called Dirac point, as shown in Fig. 1.1g. This is also known as the Dirac cone. For example, the conduction band energy is conventionally written as

$$\varepsilon_c(\mathbf{k}) = \hbar v_F k \tag{1.1}$$

where $k = \sqrt{k_x^2 + k_y^2}$ and v_F is the so-called Fermi velocity. The bottom of this band is here chosen to be the zero-point of the energy scale. The valence band energy is just the negative of the conduction band energy, $\varepsilon_v(\mathbf{k}) = -\varepsilon_c(\mathbf{k})$. Taking the bands to be symmetric in this sense is not strictly correct, but a good approximation.

The density of states (DOS), which is the number of states per unit energy per unit area, is proportional to the number of states in a ring with radius k, which in turn is proportional to $\varepsilon_c(\mathbf{k})$. In other words, the conduction band DOS, $g_c(\varepsilon)$, is proportional to $\varepsilon_c(\mathbf{k})$. Similarly, the joint

DOS, which counts the number of states for a vertical transition between v and c band, is proportional to $\varepsilon_c(\mathbf{k}) - \varepsilon_v(\mathbf{k})$. This is shown in Fig. 1.2a. As mentioned above, the mathematical details of the bandstructure and the DOS will be presented in a later section of this Introductory Tutorial, Sec. 1.2.

It is useful to compare graphene to semiconductor quantum wells. For that purpose we also show in Fig. 1.2 the joint DOS for a simplified quantum well model. Here, the energies are parabolic, and the joint DOS is a constant. In quantum wells, it is often a good assumption to assume that the optical dipole matrix elements that describe the light coupling between v and c bands are independent of the wave vector $\mathbf{k}$, and that the transition energies $\varepsilon_c(\mathbf{k}) - \varepsilon_v(\mathbf{k})$ are approximately given by the bandgap energy E_g, which is valid if we restrict ourselves to optical frequencies near the bandgap, $\hbar\omega \simeq E_g$. In this approximation, the quantum well absorbance becomes proportional to the joint DOS, and is therefore constant above the bandgap. Of course there is the Coulomb interaction between the electron in the c-band and the hole in the v-band, and that may lead to exciton resonances, but in the present discussion we want to omit those complications. Instead, we want to ask whether the absorbance of graphene is proportional to its joint DOS. The answer is no. The reason is based on the fact that the absorbance carries extra frequency factors compared to the joint DOS, and in contrast to a semiconductor quantum well with a non-zero energy gap E_g, we have to keep the frequency dependence due to those factors. It turns out that the optical susceptibility carries a factor of $1/\omega^2$, and calculating the absorbance yields another factor of ω. Hence, the combined frequency factors entering the absorbance are $1/\omega$. Furthermore, the squared momentum matrix element, which enters the susceptibility but not the joint DOS, becomes independent of the wavevector at small wavevectors and hence does not provide a qualitative change to the frequency dependence. That means that the absorbance is proportional to the product of $1/\omega$ and the joint DOS. Since the joint DOS is proportional to ω, the absorbance becomes a constant, independent of ω. Again, the somewhat hand-waving arguments made in this section will be put on a solid mathematical basis below, see Sec. 1.5. Regarding the comparison between graphene and semiconductor quantum wells, we should also mention that not only graphene has elastic properties that allows bending or other deviations from the ideal 2-dimensional planar structure. Such deformation can also be achieved in quantum wells, which are then often called nanomembranes, see for example [64–74].

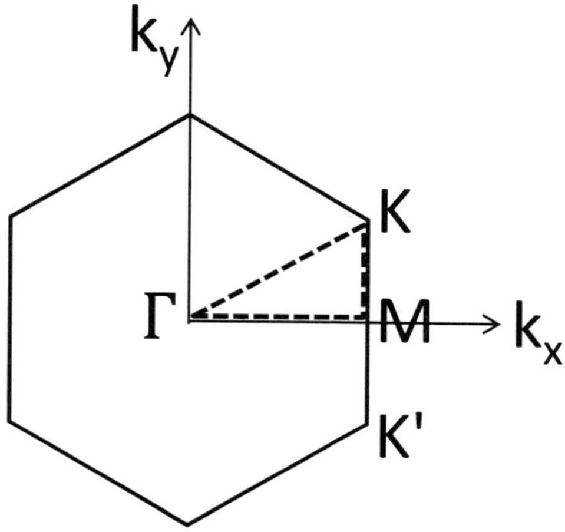

Fig. 1.3 The Brillouin zone of graphene with alternating Dirac points K and K′. Another K′ point is on the opposite side of Γ from K, therefore in the literature K′ is frequently denoted as −K. In Fig. 1.4 the simplified symmetric two-band bandstructure is shown for the entire BZ. The dashed line shows the path for which the bandstructure is shown in Fig. 1.5

Figure 1.2 reveals one other important aspect of graphene. The low-frequency absorbance of a single graphene layer is approximately 2.3%. In other words, if we shine light in normal incidence on graphene, the transmitted intensity is reduced by a factor T, and in addition we have a reflected beam of intensity R times the incident intensity. The absorbance is then defined as $A = 1 - R - T$, which is 2.3%. To very good accuracy we can neglect R and say that the transmitted intensity is 97.7% of the incident intensity. It is interesting that the low-frequency absorbance can be expressed by a very simple combination of natural constants, namely the product of π and the fine structure constant

$$A = \pi \frac{e^2}{\hbar c} \tag{1.2}$$

where e is the magnitude (absolute value) of the electron charge and c the speed of light in vacuum. The value of the fine structure constant is

$$\frac{e^2}{\hbar c} \simeq \frac{1}{137} \tag{1.3}$$

(Note that we are using Gauss units. In SI units the fine structure constant is written as $e^2/(4\pi\epsilon_0 \hbar c)$ where ϵ_0 is the permittivity of free space.) The question now is: is an absorbance of 2.3% large or small? The answer is: both! The fact that a single atomic layer can absorb 2.3% of light is actually rather remarkable, this is certainly a large number. Fig. 1.2 shows that a GaAs quantum well, which has typically, say, 20 atomic layers, can have a smaller absorbance. So, stacking graphene layers on top of each other can yield a very opaque medium (e.g. [75]) which can be beneficial for photodetector applications. But on the other hand, if we have only one layer, we can say that it is rather transparent. If the layer can be used, for example, to conduct electricity, we have a transparent electrode (e.g. [10]).

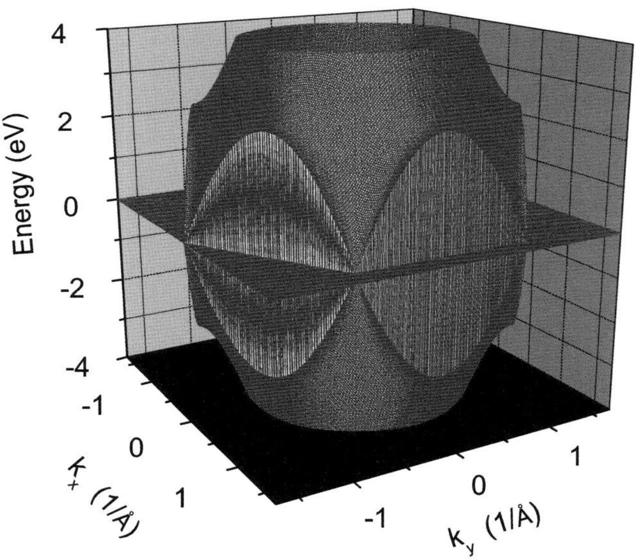

Fig. 1.4 Bandstructure of graphene within a simplified symmetric two-band model. The linear dispersion close to the Dirac point is clearly visible.

We want to mention some additional differences between graphene and semiconductor quantum wells. Many quantum wells are made out of binary compounds comprising materials from groups III and V of the periodic table, and many of those III-V semiconductors are direct-gap semiconductors with fundamental gaps at the Γ-point, i.e. at $\mathbf{k} = 0$. On the other hand, the Dirac point and Dirac cone in graphene are not at the Γ-point, but on the boundary of the Brillouin zone, at the so-called K (or K′) point, see

Fig. 1.3. A simplified 2-band model of the bandstructure, which we evaluate in detail in Sec. 1.2, yields the bandstructure shown in Figs. 1.4 and 1.5. It is clear from those figures that outside the immediate vicinity of the Dirac points, the bandstructure is not linear anymore, and hence we should expect a deviation from the simple behavior of the DOS and absorbance shown above in Fig. 1.2. A more realistic result is shown in Fig. 1.6. As we increase the light frequency ω, we approach transitions close to the M point. The band structure around this point has the shape of a saddle, and hence is called the saddle point. At the saddle, the DOS approaches infinity, and therefore we should expect infinite absorbance at the frequency that corresponds to vertical transitions at the saddle point. In practice, we will not see a singularity or infinite absorbance, but we do see a maximum. The singularity is 'smeared out' by several physical effects, which include strong dephasing due to electron-phonon coupling and excitonic effects between the electron in the c-band and the hole in the v-band. Figure 1.6 shows results where the dephasing has been included as a phenomenological parameter γ. In the case of $\gamma = 40$meV, the spectrum shows a pronounced peak at about 4.5eV. This peak would be even narrower and higher if we had chosen an even smaller value for the dephasing. A more realistic model is to assume the dephasing to be frequency dependent, $\gamma = \gamma(\omega)$, since at small frequencies the scattering processes are less and less efficient (due to the so-called reduction of phase space), whereas close to the saddle point the scattering processes are very efficient. The phenomenological frequency dependence used to compute the spectrum in Fig. 1.6a (solid line) is shown in Fig. 1.6b. In spite of the simplicity of the theory used to generate this spectrum, it is already quite realistic and similar to experimental data.

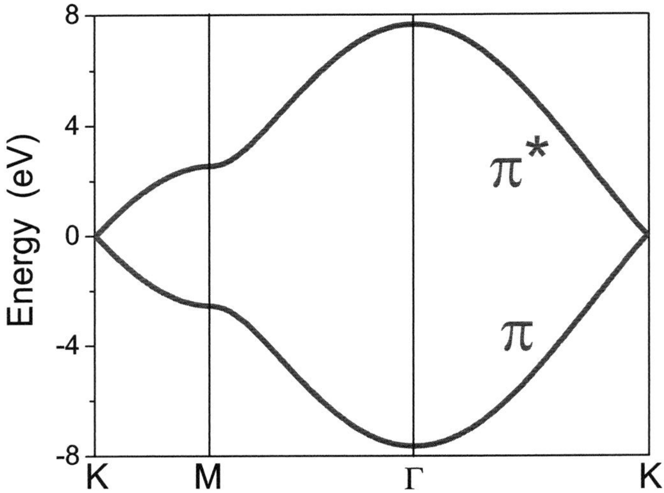

Fig. 1.5 Bandstructure (simplified symmetric two-band model) along selected lines in the Brillouin zone.

We should also mention another aspect that makes graphene such an intriguing system. Above, we said that the Dirac point is the K-point of the BZ, and without further explanation we distinguished between the K and K′ point, Fig. 1.3. The question then arises, what is different between K and K′. Regarding their contributions to linear optical spectra, there is actually no physical difference between those two points. However, the concepts involved in calculating optical spectra still allow us to comment on the difference between K and K′. We mentioned above that the theory of optical transitions involves the optical matrix elements, for example the momentum matrix element or, alternatively, a generalized dipole matrix element (remember that in 2-level atoms, it is most convenient to describe optical transitions within the so-called dipole approximation, in which the relevant optical matrix element is that of the position operator multiplied by the electron charge). It turns out that in graphene, the generalized dipole matrix element is given in terms of the gradient of a phase function $\phi(\mathbf{k})$, which is given as a simple function of the nearest-neighbor vectors of the honeycomb lattice. We show this phase function in Fig. 1.7, together with the bandstructure. The bandstructure is periodic in the reciprocal lattice as can be seen from Fig. 1.7a. In contrast, $\phi(\mathbf{k})$ is not periodic. Importantly, $\phi(\mathbf{k})$ is not even defined at the Dirac points, in other words the Dirac points are points of phase singularities. To be precise, what is shown in Fig. 1.7b is the principal value for $\mathrm{Im}[\ln e^{i\phi(\mathbf{k})}]$ which is between $-\pi$ and π. If we would 'walk' in a small circle around the Dirac point K in counter-clockwise direction, we would see the phase monotonically increasing (expect of course at the point where it jumps back from π to $-\pi$ in our specific choice of phase interval). Now, if we would do the same walk around K$'$, the phase would monotonically decrease. To make it really simple to see this, one could say that if we want to go from red to green (in the colors of Fig. 1.7b), we have to walk counter-clockwise around the K point, whereas around the K$'$ point we have to walk clockwise. This interesting geometrical or topological behavior of the phase, which is not only related to the dipole matrix element, but more generally to the Bloch wave functions in the honeycomb crystal, has motivated a large number of investigations of electronic and spin (or pseudo-spin) related phenomena. By now, many investigations have been devoted to effects associated with the Berry phase and Berry curvature, see for example Chapter 7 on page 213 of Ref. [18], that are based on the details of Bloch wave functions. The wave functions involve superpositions from two sub-lattices (since graphene has 2 atoms in the basis) and can thus be viewed in analogy to a 2-state spin system (spin up and spin down), except that here they are called pseudo-spin states.

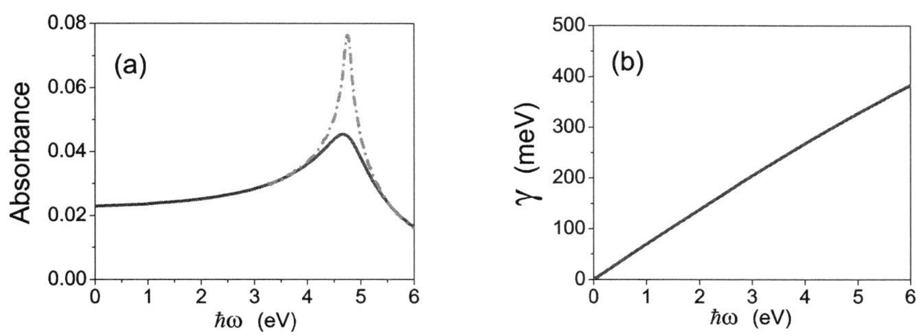

Fig. 1.6 (a) Theoretical linear absorption spectrum of graphene for fixed dephasing (dash-dotted line) with $\gamma = 40$meV, and with frequency dependent dephasing (solid) line as shown in (b).

It is interesting to note that the pseudo-spin analogy and importance of geometrical phase effects was discovered long before graphene became such a popular and widely-studied material, see Ref. [76–78]. Furthermore, the fact that there are two inequivalent valleys, K and K′, has motivated the concept of valley pseudo-spins, where now the two pseudo-spin states are the K and K′ valleys. This has led to the discovery of intriguing physical phenomena not only in graphene, but also in related material systems such as transition metal dichalcogenides, and has led to the term valley-spintronics. It is likely that in the future, the interplay between electronic, spin, mechanical and optical properties of quasi two-dimensional systems, as well as the combination of different kinds of two-dimensional materials in new so-called van der Waals heterostructures [79] yield many more discoveries, and it is also likely that the inherent optical properties, optical manipulation and control, and optical detection will play a crucial role in those developments.

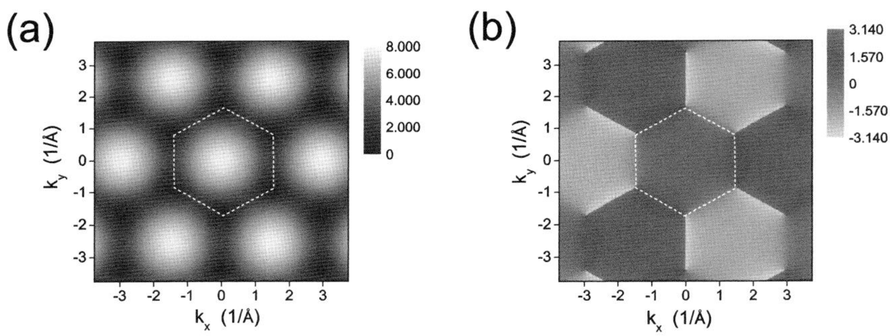

Fig. 1.7 (a) The conduction band energies (bandstucture) in units of eV across multiple Brillouin zones. The first BZ is indicated by the white dashed hexagon. (b) The phase $\phi(\mathbf{k})$ in units of radians across the same area of wave vector space. (For color reference, go to page 485.)

1.2 Bandstructure of Graphene

In this and the following sections, we provide an introductory text-book level treatment of the theory that is relevant to the understanding of the electronic bandstructure and linear optical properties of graphene.

Graphene is a 2-dimensional crystal with the carbon atoms forming a honeycomb structure, as shown in Fig. 1.8. The honeycomb structure consists of a hexagonal Bravais lattice with two basis atoms. This is shown in Fig. 1.9, where we indicate the conventional unit cell of the hexagonal Bravais lattice. The primitive translation vectors are $\mathbf{a}_1$ and $\mathbf{a}_2$, the lattice constant is a_L, and a is the carbon-carbon bondlength ($a \simeq 1.42\text{Å}$), which is related to the lattice constant a_L via $a_L = \sqrt{3}a$, see Fig. 1.10.

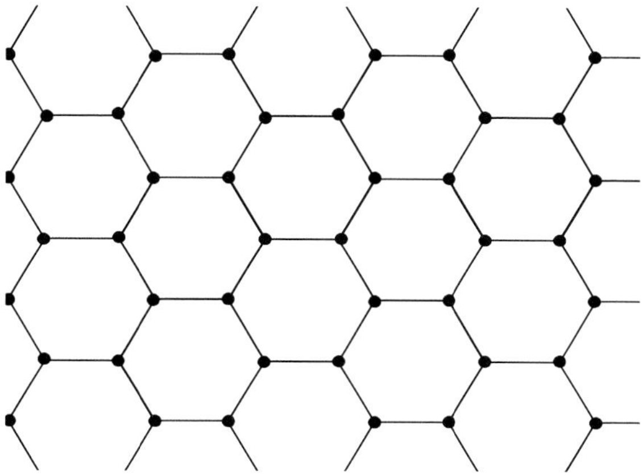

Fig. 1.8 Sketch of the hexagonal honeycomb crystal structure of graphene. The solid dots indicate the positions of the carbon atoms.

Let us derive the electronic bandstructure of graphene, or at least a simplified version thereof. A powerful yet simple approach is the tight-binding formalism. Three of the four valance electrons of carbon can occupy three s-p hybrid orbitals with orbital orientations in the two-dimensional plane forming covalent bonds with the three nearest neighbors. We are not interested in those. We are only interested in the fourth electron that can occupy an out-of-plane p_z-orbital $\Phi_p(\mathbf{r})$. In a single carbon atom this orbital fulfills the Schrödinger equation

$$\left\{-\frac{\hbar^2\nabla^2}{2m} + V(\mathbf{r})\right\}\Phi_p(\mathbf{r}) = \varepsilon_p^{(atom)}\Phi_p(\mathbf{r}) \tag{1.4}$$

where $V(\mathbf{r})$ can be thought of as an effective Coulomb potential, one that includes the effects of all but the outer p_z electrons. If the bond length a was very large, we wouldn't have any spatial overlap between adjacent p_z-orbitals, and each electron was stuck at its site without the ability to move. However, there is a small overlap (hence the term 'tight-binding formalism'), allowing electrons to move from site to site (this overlap is schematically indicated in Fig. 1.10b).

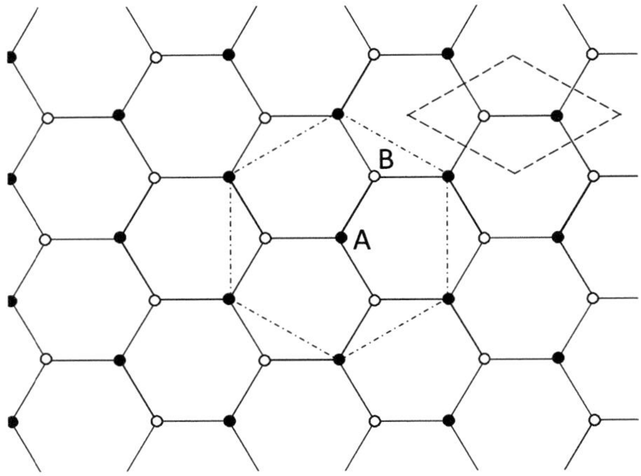

Fig. 1.9 In this sketch, the fact that graphene is a hexagonal lattice with two basis atoms, A and B, is shown. The conventional unit cell of the hexagonal lattice is indicated by the thin dashed-dotted line (note that in the hexagonal lattice, in addition to the points on the hexagon there is a lattice point at the center of the hexagon, which is different from the hexagons in the honeycomb structures). The thin dashed line shows a primitive cell spanned by the primitive translation vectors.

The energy eigenfunctions of the crystal (Bloch functions) are labeled by the wavevector $\mathbf{k}$, and can approximately be constructed as

$$\varphi_{n\mathbf{k}}^{(0)}(\mathbf{r}) = \frac{1}{\sqrt{N}} \sum_j e^{i\mathbf{k}\cdot\mathbf{R}_{j,n}} \Phi_p(\mathbf{r} - \mathbf{R}_{j,n}) \tag{1.5}$$

We have two such states, $n = 1, 2$, because we have two sublattices that consists only or either A or B atoms, and each sublattice gives rise to one set of Bloch functions $\varphi_{n\mathbf{k}}^{(0)}$. The site vectors of the ideal lattice positions are $\mathbf{R}_{j1} = \mathbf{R}_j$ and $\mathbf{R}_{j2} = \mathbf{R}_j + \vec{\delta}_1$, and $\mathbf{R}_j = j_1\mathbf{a}_1 + j_2\mathbf{a}_2$ are the positions of the unit cells, where the $\mathbf{a}_i$ are the primitive translation vectors in real space. There are various choices for the primitive translation vectors that can be found in the literature. We are using

$$\mathbf{a}_1 = \frac{a}{2}\left(3, \sqrt{3}\right), \mathbf{a}_2 = \frac{a}{2}\left(3, -\sqrt{3}\right)$$

implying the following choice for the three nearest-neighbor vectors (Fig. 1.10a): $\vec{\delta}_1 = \frac{a}{2}\left(1, \sqrt{3}\right)$, $\vec{\delta}_2 = \frac{a}{2}\left(1, -\sqrt{3}\right)$ and $\vec{\delta}_3 = a(-1, 0)$. In Eq. (1.5) j runs over all lattice sites, and we assume we have a total of $N = N_1 N_2$ sites (N_1 being the total number of unit cells in $\mathbf{a}_1$ direction, and similarly for N_2). A finite number of sites requires us to use periodic boundary conditions. The Bloch function has the periodicity dictated by Bloch's theorem,

$$\varphi_{n\mathbf{k}}^{(0)}(\mathbf{r} + \mathbf{R}_\ell) = e^{i\mathbf{k}\cdot\mathbf{R}_\ell} \varphi_{n\mathbf{k}}^{(0)}(\mathbf{r}) \tag{1.6}$$

and the periodicity of the crystal

$$\varphi_{n\mathbf{k}}^{(0)}(\mathbf{r}+N_1\mathbf{a}_1)=\varphi_{n\mathbf{k}}^{(0)}(\mathbf{r}+N_2\mathbf{a}_2)=\varphi_{n\mathbf{k}}^{(0)}(\mathbf{r}) \tag{1.7}$$

provided $N_1\mathbf{k}\cdot\mathbf{a}_1$and $N_2\mathbf{k}\cdot\mathbf{a}_2$ are multiple integers of 2π. The Bloch functions (1.5) suggest that the A and B sublattices are uncoupled, but we said already that there is a small wave function overlap between adjacent atoms. Hence, the true Bloch wave function must be a superposition of the form

$$\varphi_{\mathbf{k}}(\mathbf{r})=\sum_{n'}c_{n'}(\mathbf{k})\,\varphi_{n'\mathbf{k}}^{(0)}(\mathbf{r}) \tag{1.8}$$

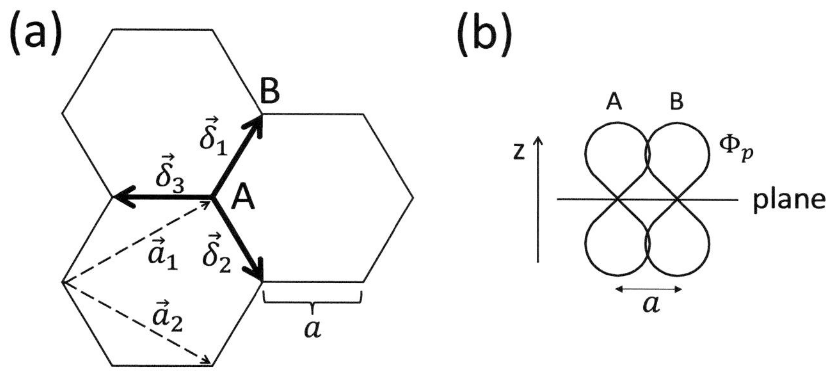

Fig. 1.10 (a) Defintitions of the two primitive translation vectors $\mathbf{a}_i$ and three nearest-neighbor vectors $\vec{\delta}_i$. (b) Sketch of the out-of-plane p_z orbitals of nearest neighbor atoms. A small wave function overlap, which motivates the use of the tight-binding theory, is indicated.

To determine the expansion coefficients $c_n(\mathbf{k})$, we insert (1.8) in the Schrödinger equation for the crystal,

$$\left\{-\frac{\hbar^2\nabla^2}{2m}+V_L(\mathbf{r})\right\}\varphi_{\mathbf{k}}(\mathbf{r})=\varepsilon_{\mathbf{k}}\varphi_{\mathbf{k}}(\mathbf{r}) \tag{1.9}$$

where $V_L(\mathbf{r})$ is the total lattice periodic potential felt by the outer valence electrons and $\varepsilon_{\mathbf{k}}$ denotes the band energies. We then multiply the equation by $\varphi_{n\mathbf{k}}^{(0)*}(\mathbf{r})$ and integrate over the crystal area. Writing the matrix elements in bra-ket notation, $\langle \mathbf{r}|n\mathbf{k}\rangle = \varphi_{n\mathbf{k}}^{(0)}(\mathbf{r})$, we have

$$\sum_{n'} \langle n\mathbf{k}|\widehat{h}_0|n'\mathbf{k}\rangle c_{n'}(\mathbf{k}) = \varepsilon_{\mathbf{k}} \sum_{n'} \langle n\mathbf{k}|n'\mathbf{k}\rangle c_{n'}(\mathbf{k}) \tag{1.10}$$

where we defined $\widehat{h}_0 = \frac{\widehat{\mathbf{p}}^2}{2m} + V_L(\mathbf{r})$, which includes the canonical momentum operator $\widehat{\mathbf{p}} = -i\hbar\vec{\nabla}$. We need to write the matrix elements entering (1.10) in terms of the original p_z-orbitals, so that we can make approximations based on the small overlap. For the Hamiltonian matrix elements, we have, using (1.5),

$$\langle n\mathbf{k}|\widehat{h}_0|n'\mathbf{k}\rangle = \frac{1}{N}\sum_{jj'} e^{i\mathbf{k}\cdot(\mathbf{R}_{jn}-\mathbf{R}_{j'n'})}\langle jn|\widehat{h}_0|j'n'\rangle \tag{1.11}$$

where

$$\langle jn|\widehat{h}_0|j'n'\rangle = \int d^2r \Phi_p^*(\mathbf{r}-\mathbf{R}_{jn})\,\widehat{h}_0\Phi_p(\mathbf{r}-\mathbf{R}_{j'n'}) \tag{1.12}$$

and similarly for the Bloch function overlap we have

$$\langle n\mathbf{k}|n'\mathbf{k}\rangle = \frac{1}{N}\sum_{jj'} e^{i\mathbf{k}\cdot(\mathbf{R}_{jn}-\mathbf{R}_{j'n'})}\langle jn|j'n'\rangle \tag{1.13}$$

The simplest non-trivial approximation we can make is to assume that there is only nearest-neighbor overlap in (1.11), meaning that only those terms are nonzero where $\mathbf{R}_{jn}-\mathbf{R}_{j'n'}$ is either $\vec{\delta}_1$ or $\vec{\delta}_2$ or $\vec{\delta}_3$, and in (1.13) we neglect any overlap and approximate $\langle n\mathbf{k}|n'\mathbf{k}\rangle = \delta_{n,n'}$. The Hamiltonian matrix elements (1.11) now contain two contributions. One is from the same site, $j' = j$ and $n' = n = 1$ (or $n' = n = 2$ which yields the same result)

$$\langle 1,\mathbf{k}|\widehat{h}_0|1,\mathbf{k}\rangle\Big|_{on-site} = \frac{1}{N}\sum_{j}\langle j,1|\widehat{h}_0|j,1\rangle = \varepsilon_p \tag{1.14}$$

where ε_p differs from $\varepsilon_p^{(atom)}$ because $\widehat{h}_0$ contains the lattice periodic potential, rather than the atomic potential. The second contribution comes from the nearest-neighbors, denoted by $<jj'>$ in the summation,

$$\begin{aligned}\langle 1,\mathbf{k}|\widehat{h}_0|2,\mathbf{k}\rangle\Big|_{nearest-neighbor} &= \frac{1}{N}\sum_{<jj'>} e^{i\mathbf{k}\cdot(\mathbf{R}_{j,1}-\mathbf{R}_{j',2})}\langle j,1|\widehat{h}_0|j',2\rangle \\ &= -t\sum_{i=1}^{3} e^{i\mathbf{k}\cdot\vec{\delta}_i}\end{aligned} \tag{1.15}$$

where we used the fact that, for nearest neighbors, the transfer or hopping matrix elements $\langle j,1|\widehat{h}_0|j',2\rangle$ are the same, independent of j and j', and denoted by $-t$. Defining $f(\mathbf{k}) = |f(\mathbf{k})|e^{i\phi(\mathbf{k})} = \sum_{i=1}^{3} e^{i\mathbf{k}\cdot\vec{\delta}_i}$, the Schrödinger equation (1.10) becomes

$$\begin{pmatrix} \varepsilon_p & -tf(\mathbf{k}) \\ -tf^*(\mathbf{k}) & \varepsilon_p \end{pmatrix} \begin{pmatrix} c_1^{(\nu)} \\ c_2^{(\nu)} \end{pmatrix} = \varepsilon_{\nu\mathbf{k}} \begin{pmatrix} c_1^{(\nu)} \\ c_2^{(\nu)} \end{pmatrix} \tag{1.16}$$

which, after diagonalization of the 2×2 matrix, finally yields the band-structure

$$\varepsilon_{\nu\mathbf{k}} = \varepsilon_{\pm,\mathbf{k}} = \varepsilon_p \pm t|f(\mathbf{k})| \tag{1.17}$$

and the coefficients determining the wave functions

$$\mathbf{c}^{(\nu)} = \frac{1}{\sqrt{2}} \begin{pmatrix} 1 \\ -\nu e^{-i\phi(\mathbf{k})} \end{pmatrix} \tag{1.18}$$

Since we started with two basis atoms and one orbital per atom, we end up having two bands, $\nu = \pm 1$. The valence band (or π band) corresponds to $\nu = -1$, and the conduction band (or π^* band) to $\nu = +1$.

We first look at the bandstructure. It is customary to set $\varepsilon_p = 0$, so that the band energies are $\varepsilon_{\nu,\mathbf{k}} = \nu t|f(\mathbf{k})|$. We can obtain an analytical expression for the band structure if we evaluate the sum over the three nearest-neigbor vectors in the expression for $f(\mathbf{k})$. We find

$$\begin{aligned} f(\mathbf{k}) &= 2\cos(k_x a/2)\cos(\sqrt{3}k_y a/2) + \cos(k_x a) \\ &\quad + i\left[2\sin(k_x a/2) + \cos(\sqrt{3}k_y a/2) - \sin(k_x a)\right] \end{aligned} \tag{1.19}$$

and, for later reference, we note the result for the phase that enters the wave functions,

$$\phi(\mathbf{k}) = \tan^{-1}\left(\frac{2\sin\left(\frac{k_x}{2}a\right)\cos\left(\frac{\sqrt{3}}{2}k_y a\right) - \sin(k_x a)}{2\cos\left(\frac{k_x}{2}a\right)\cos\left(\frac{\sqrt{3}}{2}k_y a\right) + \cos(k_x a)}\right) \tag{1.20}$$

The band structure is shown in Fig. 1.4 and 1.5. The reciprocal space consists of a hexagonal lattice with unit vectors

$$\mathbf{b}_1 = \frac{2\pi}{3a}\left(1, \sqrt{3}\right), \mathbf{b}_2 = \frac{2\pi}{3a}\left(1, -\sqrt{3}\right) \tag{1.21}$$

The reciprocal lattice vectors are superpositions of the form

$$\mathbf{G}_m = m_1\mathbf{b}_1 + m_2\mathbf{b}_2 \tag{1.22}$$

where m_1 and m_2 are integers. The first Brillouin zone (BZ) is the hexagon centered at $\mathbf{k} = 0$. As can be seen from the figure, close to the Dirac points,

$\mathbf{K} = (\frac{2\pi}{3a}, \frac{2\pi}{3\sqrt{3}a})$ or $\mathbf{K}' = (\frac{2\pi}{3a}, -\frac{2\pi}{3\sqrt{3}a})$, the bandstructure is linear. A Taylor expansion of $f(\mathbf{K}+\mathbf{q})$ close to the Dirac point $\mathbf{K}$, in other words $\mathbf{q} \to 0$ (we omit the details of that straightforward derivation) yields, to first order in q,

$$\begin{aligned} \varepsilon_{\nu,\mathbf{K}+\mathbf{q}} &= \nu\frac{3}{2}taq \\ &= \nu\hbar v_F q \end{aligned} \tag{1.23}$$

The factor $3ta/(2\hbar)$ is called the Fermi velocity v_F. The linear bandstructure (1.23) is of central importance, because, as mentioned above, it sets graphene apart from typical two-band semiconductors with parabolic valence and conduction bands that are separated by a bandgap. A similar Taylor expansion for the phase $\phi(\mathbf{k})$ yields

$$\phi(\mathbf{K}+\mathbf{q}) = \pi + \Theta_{\mathbf{q}} - \Theta_{\mathbf{K}} \tag{1.24}$$

Here, $\Theta_{\mathbf{q}}$ and $\Theta_{\mathbf{K}}$ are the angles of $\mathbf{q}$ and $\mathbf{K}$, respectively, relative to a fixed axis in the plane of the crystal, for example the laboratory-frame x-axis. In other words, we write $\mathbf{q} = q(\cos\Theta_{\mathbf{q}}, \sin\Theta_{\mathbf{q}})$ and similarly for $\mathbf{K}$. The expression (1.24) shows clearly the phase singularity of the wave function at the Dirac points, as the limit $\mathbf{q} \to 0$ yields $\phi(\mathbf{K})$ that depends on the direction of $\mathbf{q}$ at which we approach the Dirac point. Hence, exactly at the Dirac point $\phi(\mathbf{K})$ is ill-defined, but the phase is well-defined in the vicinity of the K-point. As mentioned above, we could imagine 'walking' on a small circle counter-clockwise around the K-point. According to (1.24) the phase would increase (because $\Theta_{\mathbf{q}}$ increases), and after one full circle we would pick up a phase increase of 2π. However, the situation is different if we would walk around the Dirac point K$'$. Here,

$$\phi(\mathbf{K}'+\mathbf{q}) = \pi - (\Theta_{\mathbf{q}} - \Theta_{\mathbf{K}'}) \tag{1.25}$$

and the phase decreases upon walking counter-clockwise around K$'$. Or, in other words, if we want to walk around K$'$ and want the phase to increase, we have to walk in the clockwise direction. This was shown above in Fig. 1.7. So, while the energies (or bandstructure) are the same at K and K$'$, the "helicity" related to the wave functions is different.

The fact that each Bloch state is a superposition of two kinds of p-orbitals, one from the A sublattice ($n = 1$) and one from the B sublattice ($n = 2$), which led us to the 2×2 Hamiltonian in (1.16), suggests the analogy with a 2-state spin model, which in this case is called a pseudo-spin model. A general spin wave function, for the real physical spin, can be written as a spinor of the form $a|\uparrow\rangle + b|\downarrow\rangle$, or simply $\binom{a}{b}$. In our present case, we write the Bloch wave function as a superposition of the two basis functions from the A and B atoms and the 2-dimensional vector (1.18) is the pseudo-spin analogue to $\binom{a}{b}$. The analogy can further be extended if we focus on the vicinity of the Dirac point, say K. Using again the Taylor expansion for $f(\mathbf{k})$,

$$f(\mathbf{K}+\mathbf{q}) \simeq -\frac{3}{2}aqe^{i(\Theta_{\mathbf{q}}-\Theta_{\mathbf{K}})} \tag{1.26}$$

and choosing a coordinate frame in which $\Theta_{\mathbf{K}} = 0$, we can write the 2×2 Hamiltonian in (1.16) as

$$H^{DiracCone} = \begin{pmatrix} \varepsilon_p & \frac{3}{2}ta(q_x + iq_y) \\ \frac{3}{2}ta(q_x - iq_y) & \varepsilon_p \end{pmatrix} \tag{1.27}$$

which we now call the 'Dirac cone' Hamiltonian. We have already seen that the bandstructure close to the Dirac points consists of cones, but the formal justification for the name Dirac point can be further clarified in the following way. We use Pauli spin matrices,

$$I = \begin{pmatrix} 1 & 0 \\ 0 & 1 \end{pmatrix}, \sigma_x = \begin{pmatrix} 0 & 1 \\ 1 & 0 \end{pmatrix}, \sigma_y = \begin{pmatrix} 0 & -i \\ i & 0 \end{pmatrix}, \sigma_z = \begin{pmatrix} 1 & 0 \\ 0 & -1 \end{pmatrix} \tag{1.28}$$

and with that re-write the Hamiltonian as

$$H^{DiracCone} = \varepsilon_p I + \hbar v_F(q_x\sigma_x + q_y\sigma_y^*) = \varepsilon_p I + \hbar v_F \mathbf{q} \cdot \vec{\sigma}^* \tag{1.29}$$

where we have used $\frac{3}{2}ta = \hbar v_F$. Similarly we obtain for the K′ point

$$H^{DiracCone} = \varepsilon_p I + \hbar v_F(q_x\sigma_x + q_y\sigma_y) = \varepsilon_p I + \hbar v_F \mathbf{q} \cdot \vec{\sigma} \tag{1.30}$$

If we would Fourier transform from wave vector to real space (configuration space), we would have to replace $\mathbf{q}$ by $-i\vec{\nabla}$ and obtain a Hamiltonian that is analogous to a 2-dimensional Dirac Hamiltonian for relativistic electrons, but of course instead of the speed of light we now have the Fermi velocity v_F entering the Hamiltonian [76].

Secondly, we take a closer look at the complete Bloch wave functions in real space. Writing the Bloch function as a product of a plane-wave factor and a lattice-periodic part $u_{\nu\mathbf{k}}(\mathbf{r})$,

$$\varphi_{\nu\mathbf{k}}(\mathbf{r}) = \frac{1}{\sqrt{A}}e^{i\mathbf{k}\cdot\mathbf{r}}u_{\nu\mathbf{k}}(\mathbf{r}) \tag{1.31}$$

we have

$$\begin{aligned} u_{\pm1,\mathbf{k}}(\mathbf{r}) = \sqrt{\frac{A_0}{2}} \sum_j \Big[& e^{-i\mathbf{k}\cdot(\mathbf{r}-\mathbf{R}_{j,1})}\Phi_p(\mathbf{r}-\mathbf{R}_{j,1}) \\ & \mp e^{-i(\mathbf{k}\cdot(\mathbf{r}-\mathbf{R}_{j,2})+\phi(\mathbf{k}))}\Phi_p(\mathbf{r}-\mathbf{R}_{j,2})\Big] \end{aligned} \tag{1.32}$$

where $A_0 = A/N$ is the unit cell area. Since the p-orbitals are strongly localized, which means that in the first (second) term only positions for which $\mathbf{r} - \mathbf{R}_{j,1} \simeq 0$ $(\mathbf{r} - \mathbf{R}_{j,2} \simeq 0)$ are relevant, we can further simplify the exponents in (1.32) and obtain

$$u_{\pm 1,\mathbf{k}}(\mathbf{r}) = \sqrt{\frac{A_0}{2}} \sum_j \left[\Phi_p(\mathbf{r} - \mathbf{R}_{j,1}) \mp e^{-i\phi(\mathbf{k})} \Phi_p(\mathbf{r} - \mathbf{R}_{j,2}) \right] \tag{1.33}$$

This form of writing the lattice-periodic part of the Bloch wave function might help to clarify the physical basis of the (relatively abstract) pseudo-spin model, as it clearly shows the linear superposition in terms of the two sublattice wave functions (replacing the two spin states in a real spin system).

We finally note that the bands discussed so far have a two-fold spin degeneracy. The electron spin (not the pseudo-spin!) is $s = \pm\frac{1}{2}\hbar$ (or $\uparrow$ and $\downarrow$, for short) , and there is one set of conduction and valence bands for electrons with spin up and another for electrons with spin down. These two sets are not coupled as long as we do not account for spin-orbit interaction in graphene, which has been studied for example in Ref. [80]. Furthermore, the optical excitation does not couple those two sets, and below we discuss matrix elements and interband polarizations without explicitly accounting for spin. Of course, the total optical response needs to include the effect of spin, but that will be simply a factor of 2 on the interband polarization, as the light field acts in the same manner on both sets of bands.

1.3 Optical Matrix Elements

Now that we have the bandstructure, the next step is to determine the matrix elements between conduction and valence bands. We restrict ourselves to vertical transitions in wave vector space, which means that the light field, which drives the transitions, does not change the momentum of the charge carriers. This is certainly true if the incident field is in normal incidence, meaning that it has zero in-plane momentum. But even for oblique incidence, the light wave vector is still very small in most practical applications, and the neglect of the in-plane wave vector transfer from the light to the carriers is a good approximation. So our task is to evaluate

$$\langle c, \mathbf{k} | \widehat{h}' | v, \mathbf{k} \rangle = \int d^2 r \varphi^*_{c\mathbf{k}}(\mathbf{r}) \widehat{h}' \varphi_{v\mathbf{k}}(\mathbf{r}) \tag{1.34}$$

where $\widehat{h}'$ is the light-matter coupling operator, and the integral extends over the finite area of the graphene crystal (remember we are using periodic boundary conditions!). The light-matter coupling $\widehat{h}'$ is obtained from the minimal substitution $\widehat{\mathbf{p}} \rightarrow \widehat{\mathbf{p}} + \frac{e}{c}\mathbf{A}$, where $\mathbf{A}$ is the vector potential, the electron charge is $-e$, and c is the speed of light (we are using cgs Gauss units in this Introductory Tutorial). That yields

$$\widehat{h} = \frac{\left(\widehat{\mathbf{p}} + \frac{e}{c}\mathbf{A}\right)^2}{2m} + V_L\left(\mathbf{r}\right) = \widehat{h}_0 + \widehat{h}' \tag{1.35}$$

with

$$\widehat{h}' = \frac{e}{mc}\mathbf{A}\cdot\widehat{\mathbf{p}} \tag{1.36}$$

where the $\mathbf{A}^2$ term has been neglected. In atomic, i.e. localised systems, it is possible to use $\widehat{h}' = e\mathbf{r}\cdot\mathbf{E}$, where $\mathbf{E}$ is the light field amplitude, but for the present case of an extended crystal we will use the $\mathbf{A}\cdot\widehat{\mathbf{p}}$ coupling. At this point, we could try to operate with the spatial gradient operator ($\widehat{\mathbf{p}} = -i\hbar\vec{\nabla}$) on the Bloch function (1.31) to evaluate (1.34), but it turns out that there is a rather elegant way to obtain an analytical expression that does not require knowledge of spatial gradients. The elegant way is based on the fact that the momentum matrix element can be expressed in terms of the k-gradient matrix elements as

$$\langle c, \mathbf{k}|\widehat{\mathbf{p}}|v, \mathbf{k}\rangle = \frac{i}{\hbar}m(\varepsilon_{c\mathbf{k}} - \varepsilon_{v\mathbf{k}})\langle u_{c,\mathbf{k}}|\ i\vec{\nabla}_{\mathbf{k}}|u_{v,\mathbf{k}}\rangle_{cell} \tag{1.37}$$

where, on the right-hand side, the matrix element is given as an integral over the unit cell,

$$\langle u_{c,\mathbf{k}}|\ i\vec{\nabla}_{\mathbf{k}}|u_{v,\mathbf{k}}\rangle_{cell} = \int_{cell} d^2r u^*_{c\mathbf{k}}(\mathbf{r}) i\vec{\nabla}_{\mathbf{k}} u_{v\mathbf{k}}(\mathbf{r}) \tag{1.38}$$

We are intentionally writing $i\vec{\nabla}_{\mathbf{k}}$ (with the factor i) inside the matrix element, because it is easy to remember that for plane waves, $e^{-i\mathbf{k}\cdot\mathbf{r}}$, the operator $i\vec{\nabla}_{\mathbf{k}}$ is identical to the spatial displacement operator $\mathbf{r}$. Of course, in our case we have Bloch wave functions, not plane waves, and hence in general $i\vec{\nabla}_{\mathbf{k}} \neq \mathbf{r}$. Moreover,

$$\langle c, \mathbf{k}|\widehat{\mathbf{p}}|v, \mathbf{k}\rangle \neq \frac{i}{\hbar}m(\varepsilon_{c\mathbf{k}} - \varepsilon_{v\mathbf{k}})\langle c, \mathbf{k}|\ \mathbf{r}|v, \mathbf{k}\rangle \tag{1.39}$$

Note the not-equal sign in (1.39)! This relation is actually valid (with an equal sign) in an atomic system with localised wave functions. It would also be valid in the crystal if we had chosen to work with an infinite crystal area (instead of using a finite area and periodic boundary conditions), but in that case our wave functions would be normalized to delta-functions and the matrix elements in (1.39), in which the difference in wave vectors between conduction and valence band states is zero, would be infinite (since a delta function at zero argument is infinite). That is not a real problem, as integrals over $\mathbf{k}$ of the matrix elements with different wave vector arguments are still well-defined, but from a practical point of view it is less convenient than dealing with finite normalization boxes. More details about the matrix elements can be found in Ref. [81]. Using the results for the lattice-periodic parts of the Bloch functions, Eq. (1.33), it is now relatively easy to evaluate the matrix element (1.38), since the only k-dependence in (1.33) is in the exponent. We neglect overlap integrals between $n = 1$ (A) and $n = 2$ (B) atoms, and note that

$$\sum_{jj'} \int_{cell-j_0} d^2r \Phi_p^*(\mathbf{r} - \mathbf{R}_{j,2}) \Phi_p(\mathbf{r} - \mathbf{R}_{j',2}) \simeq \sum_{jj'} \delta_{j,j'} \delta_{j,j_0} = 1 \qquad (1.40)$$

and then obtain

$$\langle u_{c,\mathbf{k}} | \, i\vec{\nabla}_{\mathbf{k}} | u_{v,\mathbf{k}} \rangle_{cell} = -\frac{1}{2} \vec{\nabla}_{\mathbf{k}} \phi(\mathbf{k}) \qquad (1.41)$$

Using (1.20) we have

$$\begin{aligned} \vec{\nabla}_{\mathbf{k}} \phi(\mathbf{k}) &= \frac{a}{|f(k)|^2} \\ &\times [(\cos(\sqrt{3} k_y a) - \cos(\frac{3}{2} k_x a) \cos(\frac{\sqrt{3}}{2} k_y a)) \widehat{\mathbf{x}} \\ &- \sqrt{3} \sin(\frac{3}{2} k_x a) \sin(\frac{\sqrt{3}}{2} k_y a) \widehat{\mathbf{y}}] \end{aligned} \qquad (1.42)$$

Looking again at the vicinity of the Dirac point, we can expand the numerator and denominator to first order in q (both are zero exactly at the Dirac point) and have

$$\vec{\nabla}_{\mathbf{k}} \phi(\mathbf{K} + \mathbf{q}) = \frac{1}{q} (-\sin \Theta_{\mathbf{q}}, \cos \Theta_{\mathbf{q}}) \qquad (1.43)$$

Just as the phase $\phi(\mathbf{K})$, we see that the k-gradient, and therefore the matrix element of the k-gradient operator, is ill-defined at the Dirac point. The analogous result for the K$'$ point is

$$\vec{\nabla}_{\mathbf{k}} \phi(\mathbf{K}' + \mathbf{q}) = -\frac{1}{q} (-\sin \Theta_{\mathbf{q}}, \cos \Theta_{\mathbf{q}}) \qquad (1.44)$$

which differs only by a minus sign from the result at the K point, and hence shows that the squared matrix element is the same for the K and K$'$ point.

1.4 Graphene as a many-electron system

Now that we have the bandstructure, single-electron wave functions and optical matrix elements, we can proceed to derive the optical susceptibility. To do this, we discuss graphene as a many-electron system in which the valence band is filled and the conduction band empty (this is true at zero temperature). Furthermore, we assume the external light field to be in normal incidence and to have an infinite cross section, so that the optical excitation is spatically homogeneous in the plane of the graphene layer. In second quantization, the Hamiltonian for the many-electron system is

$$\widehat{H} = \int d^2r \sum_{s} \psi^{\dagger}(\mathbf{r})\widehat{h}\psi(\mathbf{r}) \tag{1.45}$$

$$= \sum_{\nu\mathbf{k}} \varepsilon_{\nu,\mathbf{k}} a^{\dagger}_{\nu,\mathbf{k}} a_{\nu,\mathbf{k}} + \sum_{\nu\nu'\mathbf{k}} \langle \nu', \mathbf{k}|\widehat{h}'|\nu, \mathbf{k}\rangle a^{\dagger}_{\nu',\mathbf{k}} a_{\nu,\mathbf{k}} \tag{1.46}$$

where we have used the expansion of the field operators $\psi(\mathbf{r})$ and $\psi^{\dagger}(\mathbf{r})$, which annihilate and create an electron at position $\mathbf{r}$, respectively, in terms of the Bloch functions,

$$\psi(\mathbf{r}) = \sum_{\nu\mathbf{k}} \varphi_{\nu\mathbf{k}}(\mathbf{r}) a_{\nu,\mathbf{k}} \tag{1.47}$$

$$\psi^{\dagger}(\mathbf{r}) = \sum_{\nu\mathbf{k}} \varphi^{*}_{\nu\mathbf{k}}(\mathbf{r}) a^{\dagger}_{\nu,\mathbf{k}} \tag{1.48}$$

and the fact that the Bloch functions are eigenstates of $\widehat{h}_0$, see Eq. (1.9). The creation and annihilation field operators $\psi^{\dagger}(\mathbf{r})$ and $\psi(\mathbf{r})$, respectively, obey the anti-commutation relations $[\psi(\mathbf{r}), \psi^{\dagger}(\mathbf{r}')]_{+} = \delta(\mathbf{r} - \mathbf{r}')$, $[\psi(\mathbf{r}), \psi(\mathbf{r}')]_{+} = 0$ and $[\psi^{\dagger}(\mathbf{r}), \psi^{\dagger}(\mathbf{r}')]_{+} = 0$, where $[A, B]_{+} = AB + BA$. Correspondingly, the creation and annihilation operators for the Bloch states obey $[a_{\nu,\mathbf{k}}, a^{\dagger}_{\nu',\mathbf{k}'}]_{+} = \delta_{\mathbf{k},\mathbf{k}'}\delta_{\nu,\nu'}$, $[a_{\nu,\mathbf{k}}, a_{\nu',\mathbf{k}'}]_{+} = 0$ and $[a^{\dagger}_{\nu,\mathbf{k}}, a^{\dagger}_{\nu',\mathbf{k}'}]_{+} = 0$. We note again that we have two sets of conduction and valence bands, one for spin up and one for spin down, and the present theory is valid for each set separately. Using the equation of motion for Heisenberg operators,

$$i\hbar \frac{d}{dt} a_{\nu,\mathbf{k}} = \varepsilon_{\nu,\mathbf{k}} a_{\nu,\mathbf{k}} + \sum_{\nu\nu'} \langle \nu', \mathbf{k}|\widehat{h}'|\nu, \mathbf{k}\rangle a_{\nu',\mathbf{k}} \tag{1.49}$$

the equation of motion for the interband transition amplitude $P_{cv}(\mathbf{k}) = \langle a^{\dagger}_{v,\mathbf{k}} a_{c,\mathbf{k}}\rangle$ reads

$$i\hbar \frac{d}{dt} P_{cv}(\mathbf{k}, t) = \{\varepsilon_{c,\mathbf{k}} - \varepsilon_{v,\mathbf{k}} - i\gamma\} P_{cv}(\mathbf{k}, t) + \langle c, \mathbf{k}|\widehat{h}'|v, \mathbf{k}\rangle \tag{1.50}$$

In deriving (1.50) we have assumed that the graphene crystal is at zero temperature, and hence the conduction band occupation function is zero for all wave vectors, $\langle a^{\dagger}_{c,\mathbf{k}} a_{c,\mathbf{k}} \rangle = 0$, and the valence band is completely occupied, occupation function $\langle a^{\dagger}_{v,\mathbf{k}} a_{v,\mathbf{k}} \rangle = 1$. We have also included a phenomenological damping (or dephasing) term with a dephasing rate γ. We write the matrix element of $\widehat{h}'$ using the following steps. The vector potential is related to the electric field (light) amplitude via

$$\mathbf{E} = -\frac{1}{c}\frac{\partial}{\partial t}\mathbf{A} - \vec{\nabla}\Phi = -\frac{1}{c}\frac{\partial}{\partial t}\mathbf{A} \tag{1.51}$$

where the scalar potential Φ is set to zero in the so-called radiation gauge [82]. For monochromatic light fields, we write the vector potential (and analogously the E-field amplitude) as

$$\mathbf{A} = \mathbf{A}^{(-)} + \mathbf{A}^{(+)} = \frac{1}{2}\mathbf{A}_0 e^{-i\omega t} + \frac{1}{2}\mathbf{A}_0^{*} e^{+i\omega t} \tag{1.52}$$

which means that we have negative ($\mathbf{A}^{(-)}$) and positive ($\mathbf{A}^{(+)}$) frequency source terms in (1.50). Since the natural (undriven) oscillation of $P_{cv}(\mathbf{k}, t)$ is proportional to $e^{-i(\varepsilon_{c,\mathbf{k}} - \varepsilon_{v,\mathbf{k}})t}$ with $\varepsilon_{c,\mathbf{k}} - \varepsilon_{v,\mathbf{k}} > 0$, we keep only the the negative-frequency term in (1.52), which is called the rotating-wave approximation. Using $\mathbf{A}^{(-)} = \frac{c}{i\omega}\mathbf{E}^{(-)}$ and using the fact that the time-dependence of the steady-state solution of $P_{cv}(\mathbf{k}, t)$ driven by the external field is $P_{cv}(\mathbf{k}, t) \sim e^{-i\omega t}$, we have

$$P_{cv}(\mathbf{k}, t) = \frac{1}{\hbar\omega - [\varepsilon_{c,\mathbf{k}} - \varepsilon_{v,\mathbf{k}}] + i\gamma}\frac{e}{im\omega}\mathbf{E}^{(-)}(t) \cdot \langle c, \mathbf{k}|\widehat{\mathbf{p}}|v, \mathbf{k}\rangle \tag{1.53}$$

The macroscopic polarization $\mathbf{P}$ can be defined via the macroscopic electric current density $\mathbf{J}$ as [83]

$$\frac{\partial}{\partial t}\mathbf{P}(\mathbf{R}, t) = \mathbf{J}(\mathbf{R}, t) \tag{1.54}$$

The macroscpic electric current density is a spatial average of the expectation value of the current operator

$$\widehat{\mathbf{j}}(\mathbf{r}) = \frac{(-e)}{2m}\psi^{\dagger}(\mathbf{r})\widehat{\mathbf{p}}\psi(\mathbf{r}) + h.c. \tag{1.55}$$

where *h.c.* denotes the hermitian conjugate. Since we are presently dealing with the simple case of a spatially homogeneously excited system, the macroscopic current is independent of the position $\mathbf{R}$. Therefore, we can restrict ourselves to the crystal average

$$\mathbf{J}(t) = \frac{1}{A}\int_{crystal} d^2r \langle\widehat{\mathbf{j}}(\mathbf{r})\rangle \tag{1.56}$$

Inserting (1.47,1.48), and finally multiplying a factor of 2 to account for the two spin-degenerate sets of conduction and valence bands, we have

$$\mathbf{J}(t) = 2\frac{(-e)}{2m}\frac{1}{A}\sum_{\mathbf{k}\mathbf{k}'\nu\nu'} \langle \nu, \mathbf{k}|\widehat{\mathbf{p}}|\nu', \mathbf{k}'\rangle \langle a^{\dagger}_{\nu,\mathbf{k}} a_{\nu',\mathbf{k}'}\rangle + c.c. \tag{1.57}$$

where $c.c.$ denotes the complex conjugate. Since we are only interested in cases where interband transitions are created optically, we omit in the following the intraband contributions to (1.57), i.e. the terms where $\nu = \nu'$. The interband momentum matrix element, $\langle c, \mathbf{k}|\widehat{\mathbf{p}}|v, \mathbf{k}'\rangle$, is diagonal in $\mathbf{k}$, in other words proportional to $\delta_{\mathbf{k},\mathbf{k}'}$ [81]. We therefore have

$$\mathbf{J}_{inter}(t) = 2\frac{(-e)}{m}\frac{1}{A}\sum_{\mathbf{k}} \langle c, \mathbf{k}|\widehat{\mathbf{p}}|v, \mathbf{k}\rangle^{*} \langle a^{\dagger}_{v,\mathbf{k}} a_{c,\mathbf{k}}\rangle + c.c. \tag{1.58}$$

where the first term is the negative-frequency term and the second ($c.c.$) the corresponding positive-frequency term. The corresponding interband polarization is

$$\mathbf{P}^{(-)}_{inter}(t) = \frac{1}{-i\omega}\mathbf{J}^{(-)}_{inter}(t) \tag{1.59}$$

We write the light field amplitude in terms of its unit vector and magnitude, $\mathbf{E}(t) = \widehat{\mathbf{e}}_0 E(t)$ and take the projection of $\mathbf{P}^{(-)}$ onto $\widehat{\mathbf{e}}_0$, i.e. $P^{(-)} = \widehat{\mathbf{e}}_0^{*} \cdot \mathbf{P}^{(-)}(t)$. Using the steady-state solution (1.53), we have

$$P^{(-)}_{inter}(t) = -\frac{1}{\omega^2}\frac{2}{A}\sum_{\mathbf{k}} \frac{e^2}{m^2}\frac{|\widehat{\mathbf{e}}_0 \cdot \langle c, \mathbf{k}|\widehat{\mathbf{p}}|v, \mathbf{k}\rangle|^2}{\hbar\omega - [\varepsilon_{c,\mathbf{k}} - \varepsilon_{v,\mathbf{k}}] + i\gamma} E^{(-)}(t) \tag{1.60}$$

$$= \chi(\omega) E^{(-)}(t) \tag{1.61}$$

where in the second line we define the single-layer optical interband susceptibility $\chi(\omega)$. Using (1.37), this can be written as

$$\chi(\omega) = -\frac{2}{A}\sum_{\mathbf{k}} \left(\frac{\varepsilon_{c,\mathbf{k}} - \varepsilon_{v,\mathbf{k}}}{\hbar\omega}\right)^2 \frac{|\widehat{\mathbf{e}}_0 \cdot \mathbf{d}_{cv}(k)|^2}{\hbar\omega - [\varepsilon_{c,\mathbf{k}} - \varepsilon_{v,\mathbf{k}}] + i\gamma} \tag{1.62}$$

with the effective interband dipole matrix element (more strictly k-gradient interband matrix element)

$$\mathbf{d}_{cv}(k) = (-e)\langle u_{c,\mathbf{k}}|\, i\vec{\nabla}_{\mathbf{k}}|u_{v,\mathbf{k}}\rangle_{cell} \tag{1.63}$$

$$= e\frac{1}{2}\vec{\nabla}_{\mathbf{k}}\phi(\mathbf{k}) \tag{1.64}$$

The general form of the susceptibility (1.62) is valid for any two-band model, but the specific form of the bandstructure ($\varepsilon_{c,\mathbf{k}}$ and $\varepsilon_{v,\mathbf{k}}$) and the effective interband dipole matrix element in graphene yields the specific characteristics that distinguish graphene from conventional (parabolic) semiconductors. This will become clear in the next section, where we derive the absorbance of a single graphene layer.

We end this section by noting that the many-particle Hamiltonian is often written in a local basis instead of the Bloch basis that we used in (1.46). In the local basis we have

$$H = \sum_{jn} \varepsilon_p c_{jn}^{\dagger} c_{jn} - \sum_{<jn,j'n'>} t c_{jn}^{+} c_{j'n'} \tag{1.65}$$

This is the conventional form of a tight-binding Hamiltonian. The operators $c_{jn}^{\dagger}$ and c_{jn} create and annihilate, respectively, electrons at site jn, i.e. at the n-atom in unit cell j. The first term is the on-site energy, and the second stems from the nearest-neighbor hopping (again, the notation $< jn, j'n' >$ means that the summation is restricted to nearest neighbors). In the basis of the local orbitals, the field operators are expanded as

$$\psi(\mathbf{r}) = \sum_{jn} \Phi_p \left(\mathbf{r} - \mathbf{R}_{jn} \right) c_{jn} \tag{1.66}$$

and the transformation between the annihilation operators from the Bloch basis to the local basis is

$$c_{jn} = \frac{1}{\sqrt{N}} \sum_{\nu \mathbf{k}} e^{i\mathbf{k} \cdot \mathbf{R}_{j,n}} c_n^{(\nu)}(\mathbf{k}) a_{\nu,\mathbf{k}} \tag{1.67}$$

1.5 Absorbance of graphene

As mentioned above, we wish to model the absorption of a single graphene layer excited by a light field of frequency ω under normal incidence. We use a relatively simple transfer matrix formulation that solves Maxwell's propagation equation for monochromatic fields. The formalism yields the reflectance and transmittance of a field incident on the graphene layer (here in normal incidence). The deviation of the transmittance from unity then yields the layer's absorbance. One advantage of this procedure is that it can be easily generalized, for example to oblique incidence or multiple layers or both. Furthermore, the approach is valid for arbitrary vectorial polarization of the light field (i.e. linearly polarized light, circularly polarized light etc.).

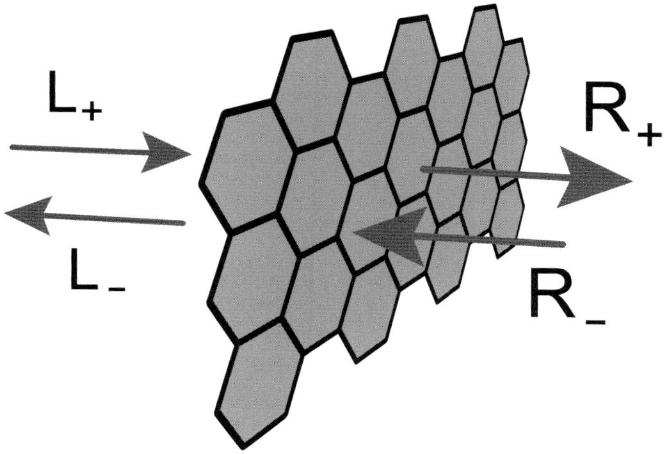

Fig. 1.11 Sketch of the transfer matrix configuration.

We start with the macroscopic Maxell equations

$$\vec{\nabla} \times \mathbf{E} = -\frac{1}{c}\frac{\partial}{\partial t}\mathbf{B} \tag{1.68}$$

and

$$\vec{\nabla} \times \mathbf{H} = \frac{1}{c}\frac{\partial}{\partial t}\mathbf{E} + \frac{4\pi}{c}\frac{\partial}{\partial t}\mathbf{P}_{3D} \tag{1.69}$$

where the magnetic field and magnetic induction are the same $\mathbf{H} = \mathbf{B}$ (since graphene is not magnetic), and $\mathbf{P}_{3D}$ denotes the polarization of the layer. Even though it is a single atomic layer, the atomic orbitals form a 3-dimensional (3D) system and give rise to the response $\mathbf{P}_{3D}$, which in general is a function of x, y and z. We assume the fields to be homogeneous in the x-y plane and to vary only in the z-direction. Eliminating the magnetic field we have the propagation equation

$$\left\{\frac{1}{c^2}\frac{\partial^2}{\partial t^2} - \frac{\partial^2}{\partial z^2}\right\}\mathbf{E} = -\frac{4\pi}{c^2}\frac{\partial^2}{\partial t^2}\mathbf{P}_{3D} \tag{1.70}$$

The graphene layer is taken to be at the position $z = z_{layer}$, and the optical response is limited to the position of the layer. A simple mathematical model for that is to use a delta-function response of the form [84]

$$\mathbf{P}_{3D}(z,t) = \delta(z - z_{layer})\mathbf{P}(t) \tag{1.71}$$

Since we are interested in weak monochromatic fields, the propagation equation is valid separately for the negative and positive frequency contributions (here we restrict ourselves to the negative frequency contributions). We also take again the projection onto $\widehat{\mathbf{e}}_0$, and have

$$\left\{-\frac{\omega^2}{c^2} - \frac{\partial^2}{\partial z^2}\right\} E^{(-)} = -4\pi\frac{\omega^2}{c^2}\delta(z - z_{layer})P^{(-)} \tag{1.72}$$

Integrating twice across the layers (i.e. from $z_{layer} - \eta$ to $z_{layer} + \eta$ where $\eta \to 0$), we have

$$\frac{\partial}{\partial z}E^{(-)}\bigg|_R - \frac{\partial}{\partial z}E^{(-)}\bigg|_L = -4\pi\frac{\omega^2}{c^2}P^{(-)} \tag{1.73}$$

and

$$E^{(-)}\Big|_R - E^{(-)}\Big|_L = 0 \tag{1.74}$$

Here, R and L denote the left ($z_{layer} - \eta$) and right ($z_{layer} + \eta$) side of the layer. On both sides of the layer, we choose plane wave solutions to Maxwell's equations,

$$E^{(-)}(z,t) = E_+^{(-)}(z,t) + E_-^{(-)}(z,t) \tag{1.75}$$

where $E_+^{(-)}(z,t)$ is the wave travelling to the right,

$$E_+^{(-)}(z,t) \sim e^{i(qz-\omega t)} \tag{1.76}$$

and $E_-^{(-)}(z,t)$ is the one travelling to the left

$$E_-^{(-)}(z,t) \sim e^{i(-qz-\omega t)} \tag{1.77}$$

Here, $q = \frac{\omega}{c}$ is the wave vector of the light in vacuum. The z-derivative is

$$\frac{\partial}{\partial z}E^{(-)}(z,t) = iqE_+^{(-)}(z,t) - iqE_-^{(-)}(z,t) \tag{1.78}$$

Denoting, for simplicity, the field adjacent to the layer as $E_+^{(-)}\Big|_R = R_+$, $E_+^{(-)}\Big|_L = L_+$, $E_-^{(-)}\Big|_R = R_-$, and $E_-^{(-)}\Big|_L = L_-$, we have the waves as indicated in Fig. 1.11. Solving for the waves on the right side of the layer in terms of those on the left side of the layer and the layer polarization, we have

$$\begin{pmatrix} R_+ \\ R_- \end{pmatrix} = \begin{pmatrix} 1 & 0 \\ 0 & 1 \end{pmatrix} \begin{pmatrix} L_+ \\ L_- \end{pmatrix} + \begin{pmatrix} 1 \\ -1 \end{pmatrix} 2\pi i\frac{\omega}{c}P^{(-)} \tag{1.79}$$

The polarization is driven by the total field at the position of the layer (which is continuous across the layer according to (1.74)),

$$P^{(-)} = \chi[L_+ + L_-] = \chi[R_+ + R_-] \tag{1.80}$$

In a transmission experiment, there is no field incident from the right, so $R_- = 0$. This yields the total transfer matrix across the layer as

$$\begin{pmatrix} R_+ \\ 0 \end{pmatrix} = \begin{pmatrix} 1 - i\beta & -i\beta \\ i\beta & 1 + i\beta \end{pmatrix} \begin{pmatrix} L_+ \\ L_- \end{pmatrix} \tag{1.81}$$

with $\beta = -2\pi \frac{\omega}{c} \chi$. Denoting the amplitude reflection coefficient

$$r = \frac{L_-}{L_+} \tag{1.82}$$

and the amplitude transmission coefficient, not to be confused with the hopping integral in Sec. (1.2),

$$t = \frac{R_+}{L_+} \tag{1.83}$$

we have

$$\begin{pmatrix} t \\ 0 \end{pmatrix} = \begin{pmatrix} 1 - i\beta & -i\beta \\ i\beta & 1 + i\beta \end{pmatrix} \begin{pmatrix} 1 \\ r \end{pmatrix} \tag{1.84}$$

The transmittance of the layer is defined for the intensity, not the amplitude, and hence $T = |t|^2$. Similarly, the reflectance is $R = |r|^2$. If the layer would not absorb any light, $T + R = 1$. However, if the layer does absorb light, the absorbance is defined as

$$A = 1 - T - R \tag{1.85}$$

Solving (1.84) yields

$$T = |t|^2 = \left| \frac{1}{1 + i\beta} \right|^2 \simeq 1 + 2\mathrm{Im}\beta \tag{1.86}$$

and

$$R = |r|^2 = \left| -\frac{i\beta}{1 + i\beta} \right|^2 \simeq |\beta|^2 \tag{1.87}$$

where we used the fact that the response of a single layer, of order β, is much less than 1, and, similarly, quadratic terms of order β^2 can be neglected against linear terms of order β. We then have

$$A \simeq -2\mathrm{Im}\beta = 4\pi \frac{\omega}{c} \mathrm{Im}\chi \tag{1.88}$$

Using a numerical evaluation of the k-integral in (1.62), we obtain the absorbance over a large range of frequencies shown in Fig. 1.6. At small frequencies, optical transitions between bands that are almost linear in k dominate. At higher frequencies (about 4eV) transitions close to the saddle point dominate. As one can see from the figure, the absorbance does not tend to zero in the limit of zero frequency, but to a finite value of about 2.3%. This is one of the most remarkable features of graphene, because this number is solely given by natural constants,

$$\lim_{\omega \to 0} A(\omega) = \frac{\pi e^2}{\hbar c} \simeq \frac{\pi}{137} \simeq 0.023 \tag{1.89}$$

where we have used the fact that the fine structure constant, $\frac{e^2}{\hbar c}$, is approximately 1/137. To see how (1.89) comes about, we first note that the phenomenological damping that we included in (1.50) is actually a function of frequency. It tends to zero as ω approaches zero. Hencs, for small ω we can use the fact that

$$\lim_{\gamma \to 0} \mathrm{Im} \frac{1}{x + i\gamma} = -\pi\delta(x)$$

In the small dephasing limit, the absorption with (1.62) becomes

$$\lim_{\gamma \to 0} A(\omega) = 4\pi \frac{\omega}{c} \frac{2}{A} \sum_{\mathbf{k}} \left(\frac{\varepsilon_{c,\mathbf{k}} - \varepsilon_{v,\mathbf{k}}}{\hbar\omega} \right)^2 |\widehat{\mathbf{e}} \cdot \mathbf{d}_{cv}(\mathbf{k})|^2 \pi\delta(\hbar\omega - [\varepsilon_{c,\mathbf{k}} - \varepsilon \text{[illegible]}$$

([illegible])

The sum over $\mathbf{k}$ is a short-hand notation for

$$\frac{1}{A} \sum_{\mathbf{k}} = \int_{BZ} \frac{d^2k}{(2\pi)^2} \quad \text{([illegible])}$$

where the integral extends of the Brillouin zone. In the small-frequency limit, only the Dirac points contribute to the integral. However, the conventional choice of the BZ is not very convenient for the evaluation of the integral, as it touches 6 different Dirac points. It is more convenient in the present situation to shift the integration region such that exactly two Dirac points are included. Such a shift is possible because the integrand has the periodicity of the reciprocal lattice. For such an integrand, performing an integral over one unit cell in reciprocal space is independent of the choice of the unit cell. Moreover, integrating over the vicinity of one K and one K′ point yields the same contribution. Hence, we can limit the integral to one Dirac point and multiply the result by a factor of 2 in order to account for the effect of the other Dirac point. As before, we limit ourselves to the vicinity of $\mathbf{K}$ by writing $\mathbf{k} = \mathbf{K} + \mathbf{q}$ and considering only small $\mathbf{q}$. Then, using (1.23), we have

$$\varepsilon_{c,\mathbf{K}+\mathbf{q}} - \varepsilon_{v,\mathbf{K}+\mathbf{q}} = 2\hbar v_F q \tag{1.93}$$

and, for linearly polarized light, choosing $\widehat{\mathbf{e}} = \widehat{\mathbf{x}}$ and using (1.43) an (1.64),

$$|\widehat{\mathbf{x}} \cdot \mathbf{d}_{cv}(\mathbf{K} + \mathbf{q})|^2 = e^2 \frac{\sin^2 \Theta_{\mathbf{q}}}{4q^2} \tag{1.94}$$

we have

$$\lim_{\omega,\gamma\to 0} A(\omega) = \frac{4e^2 v_F^2}{c\omega} \int_0^\infty dqq\delta(\hbar\omega - 2\hbar v_F q) \int_0^{2\pi} d\Theta_{\mathbf{q}} \sin^2\Theta_{\mathbf{q}} \qquad (1.95)$$

which yields (1.89).

As mentioned in Sec. 1.1, the expression for the absorbance is similar to that of the joint density of states (DOS). The latter is given by

$$g_J(\varepsilon) = 2\frac{1}{A}\sum_{\mathbf{k}} \delta(\hbar\omega - [\varepsilon_{c,\mathbf{k}} - \varepsilon_{v,\mathbf{k}}]) \qquad (1.96)$$

where the factor of 2 accounts for the spin degeneracy. Near one of the Dirac points, we readily obtain

$$g_J(\varepsilon) = \frac{1}{\pi(2\hbar v_F)^2}\varepsilon\Theta(\varepsilon) \qquad (1.97)$$

In Fig. 1.2 we use an approximate value for the Fermi velocity, $v_F = 10^6$m/s.

To see the origin for the qualitative difference between the absorbance and the joint DOS, note that the absorbance is proportional to $\omega \mathrm{Im}\chi(\omega)$, and $\chi(\omega)$ carries one factor of $1/\omega$ due the transformation from the vector potential to the E-field amplitude and another factor of $1/\omega$ due to the transformation from the interband current to the interband polarization, see Eq. (1.60). Furthermore, the dipole matrix element squared yields a factor of $1/q^2$, which is cancelled by the factor of q^2 from the interband transition energy squared, see Eqs. (1.93) and (1.94), which implies that the momentum matrix element is independent of q in the limit of small q. Hence, the absorbance differs from the joint DOS by a factor of $1/\omega$. Since in the small frequency limit the joint DOS is proportional to ω, the absorbance is independent of ω.

For the comparison with a GaAs quantum well in Fig. (1.2), we evaluate the equations for the absorbance and joint DOS, (1.91) and (1.96), respectively, with

$$\frac{\varepsilon_{c,\mathbf{k}} - \varepsilon_{v,\mathbf{k}}}{\hbar\omega} \simeq \frac{E_g^{QW}}{E_g^{QW}} = 1 \qquad (1.98)$$

and

$$\mathbf{d}_{cv}(\mathbf{k}) \simeq \mathbf{d}_{cv}(\mathbf{0}) \qquad (1.99)$$

We then obtain the simple result that the absorbance of the quantum well is, within the range of validity of the approximations made, proporptional to the DOS:

$$A(\omega) = 4\pi^2\frac{e^2}{\hbar c}E_g^{QW}|\widehat{\mathbf{e}}\cdot\mathbf{r}_{cv}(\mathbf{0})|^2 g_J(\hbar\omega) \qquad (1.100)$$

Here, we have defined $\mathbf{d}_{cv}(\mathbf{0}) = e\mathbf{r}_{cv}(\mathbf{0})$, but we remember that the matrix element $\mathbf{r}_{cv}$ is that of the k-gradient operator rather than the displacement operator, see Eq. (1.63). Again, we see that the fine structure constant plays a central role in the relation between absorbance and DOS, but there are also material dependent parameters, namely the energy gap and dipole matrix element, that enter the relation. For linearly polarized light and GaAs, we use $|\widehat{\mathbf{x}} \cdot \mathbf{r}_{cv}(\mathbf{0})| \simeq \frac{1}{\sqrt{2}} 5\ \mathring{A}$, and we set the quantum well bandgap energy to $E_g^{QW} = 1.5\text{eV}$. Evaluating the k-integral in (1.96) with parabolic energies,

$$\varepsilon_{c,\mathbf{k}} - \varepsilon_{v,\mathbf{k}} = \frac{\hbar^2 k^2}{2m_e} + E_g^{QW} + \frac{\hbar^2 k^2}{2m_h} = \frac{\hbar^2 k^2}{2m_r} + E_g^{QW} \tag{1.101}$$

where m_r is the reduced mass, we obtain

$$g_J(\varepsilon) = \frac{m_r}{\pi\hbar^2}\Theta(\varepsilon - E_g^{QW}) \tag{1.102}$$

This function is shown in Fig. 1.2.

For a direct comparison of the absorbance of a graphene layer with that of a single semiconductor quantum well, we can define

$$\varepsilon_d = \frac{\hbar^2}{4m_r r_{cv}(\mathbf{0})^2} \tag{1.103}$$

and write the quantum well absorbance in the vicinity of the bandgap ($\hbar\omega \simeq E_g^{QW}$) as

$$A(\omega) \simeq \frac{\pi e^2}{\hbar c}\frac{E_g^{QW}}{\varepsilon_d}\Theta(\hbar\omega - E_g^{QW}) \tag{1.104}$$

Here, the first factor is the graphene absorbance, followed by a quantum-well-specific factor and the step function indicating that the quantum well absorbance is zero for frequencies below the bandgap. In the numerical example shown in Fig. 1.2, we use $m_e = 0.067m_0$ and $m_h = 0.1m_0$, and obtain $\varepsilon_d = 3.8\text{eV}$, which is about 2.5 times larger than the bandgap of 1.5eV. Hence, the single-layer graphene absorbance is larger than that of the quantum well, even though the quantum well is not a single atomic layer, but typically consists of about 20 atomic layers.

Acknowledgement

We gratefully acknowledge partial financial support from the US NSF under grant ECCS-1406673.

Bibliography

[1] K. S. Novoselov, V. I. Fal'ko, L. Colombo, P. R. Gellert, M. G. Schwab and K. Kim, A roadmap for graphene, *Nature* **490**, 7419, pp. 192–200 (2012).

[2] F. Bonaccorso, Z. Sun, T. Hasan and A. C. Ferrari, Graphene photonics and optoelectronics, *Nature Photonics* **4**, 9, pp. 611–622 (2010).

[3] P. Avouris, Graphene: Electronic and photonic properties and devices, *Nano Lett.* **10**, 11, pp. 4285–4294 (2010).

[4] E. O. Polat and C. Kocabas, Broadband optical modulators based on graphene supercapacitors, *Nano Letters* **13**, pp. 5851 – 5857 (2013).

[5] S. H. Mousavi, I. Kholmanov, K. B. Alici, D. Purtseladze, N. Arju, K. Tatar, D. Y. Fozdar, J. W. Suk, Y. F. Hao, A. B. Khanikaev, R. S. Ruoff and G. Shvets, Inductive tuning of fano-resonant metasurfaces using plasmonic response of graphene in the mid-infrared, *Nano Letters* **13**, 3, pp. 1111–1117 (2013).

[6] I. T. Kim, A. Magasinski, K. Jacob, G. Yushin and R. Tannenbaum, Synthesis and electrochemical performance of reduced graphene oxide/maghemite composite anode for lithium ion batteries, *Carbon* **52**, pp. 56–64 (2013).

[7] S. H. Lee, J. Choi, H. D. Kim, H. Choi and B. Min, Ultrafast refractive index control of a terahertz graphene metamaterial, *Scientific Reports* **3**, p. 2135 (2013).

[8] A. Saynatjoki, L. Karvonen, J. Riikonen, W. Kim, S. Mehravar, R. A. Norwood, N. Peyghambarian, H. Lipsanen and K. Kieu, Rapid large-area multiphoton microscopy for characterization of graphene, *ACS Nano* **7**, 10, pp. 8441–8446 (2013).

[9] I. Baylam, M. N. Cizmeciyan, S. Ozharar, E. O. Polat, C. Kocabas and A. Sennaroglu, Femtosecond pulse generation with voltage-controlled graphene saturable absorber, *Optics Letters* **39**, 17, pp. 5180–5183 (2014).

[10] W. S. Koh, C. H. Gan, W. K. Phua, Y. A. Akinov and P. Bai, The potential of graphene as an ito replacement in organic solar cells: An optical perspective, *IEEE J. Quant. Electr.* **20**, p. 4000107 (2014).

[11] T. Chu and Z. H. Chen, Understanding the electrical impact of edge contacts in few-layer graphene, *ACS Nano* **8**, 4, pp. 3584–3589 (2014).

[12] A. Y. Nikitin, P. Alonso-Gonzalez and R. Hillenbrand, Efficient coupling of

light to graphene plasmons by compressing surface polaritons with tapered bulk materials, *Nano Letters* **14**, 5, pp. 2896–2901 (2014).

[13] N. Dabidian, I. Kholmanov, A. B. Khanikaev, K. Tatar, S. Trendafilov, S. H. Mousavi, C. Magnuson, R. S. Ruoff and G. Shvets, Electrical switching of infrared light using graphene integration with plasmonic fano resonant metasurfaces, *ACS Photonics* **2**, pp. 216–227 (2015).

[14] A. Bostwick, T. Ohta, T. Seyller, K. Horn and E. Rotenberg, Quasiparticle dynamics in graphene, *Nature Physics* **3**, 1, pp. 36–40 (2007).

[15] A. H. Castro-Neto, F. Guinea, N. Peres, K. Novoselov and A. Geim, The electronic properties of graphene, *Rev. Mod. Phys.* **81**, pp. 110 – 155 (2009).

[16] J. Warner, F. Schäffel, A. Bachmatiuk and M. Rümmeli, *Graphene - Fundamentals and emergent applications.* Elsevier, Waltham (2013).

[17] A. Knorr, *Graphene and Carbon Nanotubes: Ultrafast Optics and Relaxation Dynamics*, 1st edn. Wiley-VCH, Heidelberg (2013).

[18] H. Aoki and M. Dresselhaus (eds.), *Physics of Graphene.* Springer-Verlag, Berlin (2014).

[19] A. C. Ferrari, Raman spectroscopy of graphene and graphite: Disorder, electron-phonon coupling, doping and nonadiabatic effects, *Solid State Communications* **143**, 1-2, pp. 47–57 (2007).

[20] D. Graf, F. Molitor, K. Ensslin, C. Stampfer, A. Jungen, C. Hierold and L. Wirtz, Spatially resolved Raman spectroscopy of single- and few-layer graphene, *Nano Letters* **7**, 2, pp. 238–242 (2007).

[21] L. M. Malard, J. Nilsson, D. C. Elias, J. C. Brant, F. Plentz, E. S. Alves, A. H. Castro and M. A. Pimenta, Probing the electronic structure of bilayer graphene by raman scattering, *Phys. Rev. B* **76**, 20, p. 201401 (2007).

[22] L. Malard, M. Pimenta and G. Dreselhaus, Raman spectroscopy in graphene, *Phys. Rep.* **473**, pp. 51 – 87 (2009).

[23] J. C. Chacon-Torres, L. Wirtz and T. Pichler, Manifestation of charged and strained graphene laters in the raman response of graphite intercalation compounds, *ACS Nano* **7**, pp. 9249 – 9259 (2013).

[24] J. M. Dawlaty, S. Shivaraman, M. Chandrashekhar, F. Rana and M. G. Spencer, Measurement of ultrafast carrier dynamics in epitaxial graphene, *Appl. Phys. Lett.* **92**, 4, p. 042116 (2008).

[25] D. Sun, Z. K. Wu, C. Divin, X. B. Li, C. Berger, W. A. de Heer, P. N. First and T. B. Norris, Ultrafast Relaxation of Excited Dirac Fermions in Epitaxial Graphene Using Optical Differential Transmission Spectroscopy, *Phys. Rev. Lett.* **101**, p. 157402 (2008).

[26] R. W. Newson, J. Dean, B. Schmidt and H. M. van Driel, Ultrafast carrier kinetics in exfoliated graphene and thin graphite films, *Opt. Exp.* **17**, pp. 2326–2333 (2009).

[27] Y. Zhang, T.-T. Tang, C. Girit, Z. Hao, M. C. Martin, A. Zettl, M. F. Crommie, Y. R. Shen and F. Wang, Direct observation of a widely tunable bandgap in bilayer graphene, *Nature* **459**, pp. 820 – 823 (2009).

[28] B. A. Ruzicka, S. Wang, L. K. Werake, B. Weintrub, K. P. Loh and H. Zhao, Hot carrier diffusion in graphene, *Phys. Rev. B* **82**, 19, p. 195414 (2010).

[29] M. Breusing, S. Kuehn, T. Winzer, E. Malic, F. Milde, N. Severin, J. P.

Rabe, C. Ropers, A. Knorr and T. Elsaesser, Ultrafast nonequilibrium carrier dynamics in a single graphene layer, *Phys. Rev. B* **83**, 15, p. 153410 (2011).
[30] D. Sun, C. Divin, C. Berger, W. A. de Heer, P. N. First and T. B. Norris, Hot carrier cooling by acoustic phonons in epitaxial graphene by ultrafast pump-probe spectroscopy, *Phys. Stat. Sol. C* **8**, pp. 1194 – 1197 (2011).
[31] B. Y. Sun, Y. Zhou and M. W. Wu, Dynamics of photoexcited carriers in graphene, *Phys. Rev. B* **85**, 12, p. 125413 (2012).
[32] K. M. Dani, J. Lee, R. Sharma, A. D. Mohite, C. M. Galande, P. M. Ajayan, A. M. Dattelbaum, H. Htoon, A. J. Taylor and R. P. Prasankumar, Intraband conductivity response in graphene observed using ultrafast infrared-pump visible-probe spectroscopy, *Phys. Rev. B* **86**, 12, p. 125403 (2012).
[33] T. Stroucken, J. H. Gronqvist and S. W. Koch, Excitonic resonances as fingerprint of strong coulomb coupling in graphene, *Journal of the Optical Society of America B-Optical Physics* **29**, 2, pp. A86–A94 (2012).
[34] D. Sun, C. Divin, M. Mihnev, T. Winzer, E. Malic, A. Knorr, J. E. Sipe, C. Berger, W. A. de Heer, P. N. First and T. B. Norris, Current relaxation due to hot carrier scattering in graphene, *New Journal of Physics* **14**, p. 105012 (2012).
[35] B. Sun and M. Wu, Negative differential transmission in graphene, *Phys. Rev. B* **88**, pp. 235422 –(10) (2013a).
[36] B. Sun and M. Wu, Microscopic theory of ultrafast dynamics of carriers photoexcited by THz and near-infrared linearly polarized laser pulses in graphene, *New Journal of Physics* **15**, pp. 083038 –(30) (2013b).
[37] L. Malard, K.F.Mal, A. C. Neto, N. Peres and T. Heinz, Observation of intra- and inter-band transitions in the transient optical response of graphene, *New Journal of Physics* **15**, pp. 015009 –(14) (2013).
[38] J. L. Cheng, N. Vermeulen and J. E. Sipe, Third order optical nonlinearity of graphene, *New Journal of Physics* **16**, p. 053014 (2014).
[39] I. Al-Naib, J. E. Sipe and M. M. Dignam, High harmonic generation in undoped graphene: Interplay of inter- and intraband dynamics, *Phys. Rev. B* **90**, 24, p. 245423 (2014).
[40] K. F. Mak, F. H. da Jornada, K. L. He, J. Deslippe, N. Petrone, J. Hone, J. Shan, S. G. Louie and T. F. Heinz, Tuning many-body interactions in graphene: The effects of doping on excitons and carrier lifetimes, *Phys. Rev. Lett.* **112**, 20, p. 207401 (2014).
[41] F. Kadi, T. Winzer, E. Malix, A. Knorr, F. Gottfert, M. Mittendorff, S. Winnerl and M. Helm, Microscopic description of intraband absorption in graphene: The occurrence of transient negative differential transmission, *Phys. Rev. Lett.* **113**, pp. 035502 –(5) (2014).
[42] P. Bowlan, E. Martinez-Moreno, K. Reimann, T. Elsaesser and M. Woerner, Ultrafast terahertz response of multilayer graphene in the nonperturbative regime, *Phys. Rev. B* **89**, 4, p. 041408 (2014).
[43] A. Majumdar, J. Kim, J. Vuckovic and F. Wang, Graphene for tunable nanophotonic resonators, *IEEE J. Select. Top. Quant. Electr.* **20**, p. 4600204 (2014).
[44] V. N. Kotov, B. Uchoa, V. M. Pereira, F. Guinea and A. H. Castro-Neto,

Electron-electron interactions in graphene: Current status and perspectives, *Reviews of Modern Physics* **84**, 3, pp. 1067–1125 (2012).

[45] E. G. Mishchenko, Effect of electron-electron interactions on the conductivity of clean graphene, *Phys. Rev. Lett.* **98**, 21, p. 216801 (2007).

[46] R. A. Jishi, L. Venkataraman, M. S. Dresselhaus and G. Dresselhaus, Phonon modes in carbon nanotubules, *Chemical Physics Letters* **209**, 1-2, pp. 77–82 (1993).

[47] M. Scheuch, T. Kampfrath, M. Wolf, K. von Volkmann and C. Frischkorn, Temperature dependence of ultrafast phonon dynamics in graphite, *Applied Physics Letters* **99**, pp. 211908 –(4) (2011).

[48] P. Araugo, D. Mafra, K. Saito, J. Kong and M. Dresselhaus, Phonon self-energy corections to nonzero wave-vector phonon modes in single-layer graphene, *Phys. Rev. Lett.* **109**, pp. 046801 –(5) (2012).

[49] G. D. Sanders, A. R. T. Nugraha, K. Sato, J. H. Kim, J. Kono, R. Saito and C. J. Stanton, Theory of coherent phonons in carbon nanotubes and graphene nanoribbons, *Journal of Physics-Condensed Matter* **25**, 14, p. 144201 (2013).

[50] B. V. Cunning, K. Ishioka, C. L. Brown and D. Kielpinski, Size-dependent hot-phonon dynamics in graphene flakes, *Applied Physics Letters* **104**, 18, p. 181907 (2014).

[51] C.-H. Park, F. Giustino, M. L. Cohen and S. G. Louie, Velocity renormalization and carrier lifetime in graphene from the electron-phonon interaction, *Phys. Rev. Lett.* **99**, pp. 086804 –(4) (2007).

[52] M. Calandra and F. Mauri, Electron-phonon coupling and electron self-energy in electron-doped graphene: Calculation of angular-resolved photoemission spectra, *Phys. Rev. B* **76**, 20, pp. 205411 – (9) (2007).

[53] C. H. Park, F. Giustino, M. L. Cohen and S. G. Louie, Electron-phonon interactions in graphene, bilayer graphene, and graphite, *Nano Letters* **8**, 12, pp. 4229–4233 (2008).

[54] K. Kaasbjerg, K. Thygesen and K. Jacobsen, Unraveling the acoustic electron-phonon interaction in graphene, *Phys. Rev. B* **85**, pp. 165440 – (8) (2012).

[55] D. A. Siegel, C. Y. Hwang, A. V. Fedorov and A. Lanzara, Electron-phonon coupling and intrinsic bandgap in highly-screened graphene, *New Journal of Physics* **14**, pp. 095006 – (8) (2012).

[56] C. H. Park, N. Bonini, T. Sohier, G. Samsonidze, B. Kozinsky, M. Calandra, F. Mauri and N. Marzari, Electron-phonon interactions and the intrinsic electrical resistivity of graphene, *Nano Letters* **14**, 3, pp. 1113–1119 (2014).

[57] T. Sohier, M. Calandra, C. H. Park, N. Bonini, N. Marzari and F. Mauri, Phonon-limited resistivity of graphene by first-principles calculations: Electron-phonon interactions, strain-induced gauge field, and Boltzmann equation, *Phys. Rev. B* **90**, 12, p. 125414 (2014).

[58] D. H. Chae, T. Utikal, S. Weisenburger, H. Giessen, K. von Klitzing, M. Lippitz and J. Smet, Excitonic Fano resonance in free-standing graphene, *Nano Letters* **11**, 3, pp. 1379–1382 (2011).

[59] K. Mak, J. Shan and T. Heinz, Seeing many-body effects in single- and few-

layer graphene: Observation of two-dimensional saddle-point excitons, *Phys. Rev. Lett.* **106**, pp. 046401 – (4) (2011).
[60] L. Yang, J. Deslippe, C. H. Park, M. L. Cohen and S. G. Louie, Excitonic effects on the optical response of graphene and bilayer graphene, *Phys. Rev. Lett.* **103**, 18, p. 186802 (2009).
[61] K. Oum, T. Lenzer, M. Scholz, D. Y. Jung, O. Sul, B. J. Cho, J. Lange and A. Muller, Observation of ultrafast carrier dynamics and phonon relaxation of graphene from the deep-ultraviolet to the visible region, *Journal of Physical Chemistry C* **118**, 12, pp. 6454–6461 (2014).
[62] A. T. Roberts, R. Binder, N. H. Kwong, D. Golla, D. Cormode, B. J. LeRoy, H. O. Everitt and A. Sandhu, Optical characterization of electron-phonon interactions at the saddle point in graphene, *Phys. Rev. Lett.* **112**, 18, p. 187401 (2014).
[63] R. Saito, G. Dresselhaus and M. Dresselhaus, *Physical Properties of Carbon Nanotubes*, 1st edn. World Scientific Publishing Co., London (1998).
[64] N. Ohtani, K. Kishimoto, K. Kubota, S. Saravanan, Y. Sato, S. Nashima, P. Vaccaro, T. Aida and M. Hosoda, Uniaxial-strain-induced transition from type-ii to type-i band configuration of quantum well microtubes, *Physica E* **21**, pp. 732 – 736 (2004).
[65] Y. Sun, E. Menard and J. A. Rogers, Gigahertz operation in flexible transistors on plastic substrates, *Appl. Phys. Lett.* **88**, p. 183509 (2006).
[66] L. Zhang, L. Dong and B. J. Nelson, Bending and buckling of rolled-up SiGe/Si microtubes using nanorobotic manipulation, *Appl. Phys. Lett.* **92**, p. 243102 (2008).
[67] H. C. Ko, M. P. Stoykovich, J. Song, V. Malyarchuk, W. M. Choi, C.-J. Yu, J. B. Geddes, J. Xiao, S. Wang, Y. Huang and J. A. Rogers, A hemispherical electronic eye camera based on compressible silicon optoelectronics, *Nature* **454**, pp. 748 – 753 (2008).
[68] Y. Mei, D. J. Thurmer, C. Deneke, S. Kiravittaya, Y.-F. Chen, A. Dadgar, F. Bertram, B. Bastek, A. Krost, J. Christen, T. Reindl, M. Stoffel, E. Coric and O. G. Schmidt, Fabrication, self-assembly, and properties of ultrathin AlN/GaN porous crystalline nanomembranes: Tubes, spirals, and curved sheets, *ACS Nano* **3**, pp. 1663 – 1668 (2009).
[69] P. Cendula, S. Kiravittaya, Y. Mei, C. Deneke and O. Schmidt, Bending and wrinkling as competing relaxation pathways for strained free-hanging films, *Phys. Rev. B* **79**, pp. 085429–1 – 085429–4 (2009).
[70] C. Deneke, A. Malachias, S. Kiravittaya, M. Benyoucef, T. Metzger and O. Schmidt, Strain states in a quantum well embedded into a rolled-up microtube: X-ray and photoluminescence studies, *Appl. Phys. Lett.* **96**, pp. 143101–1 – 143101–3 (2010).
[71] M. Hossein-Zadeh and K. J. Vahala, An optomechanical oscillator on a silicon chip, *IEEE J. Select. Topics in Quant. Electr.* **16**, pp. 276–287 (2010).
[72] K. Usami, A. Naesby, T. Bagci, B. M. Nielsen, J. Liu, S. Stobbe, P. Lodahl and E. S. Polzik, Optical cavity cooling of mechanical modes of a semiconductor nanomembrane, *Nature Physics* **8**, pp. 168 – 172 (2012).
[73] J. Snyder, J. Getpreecharsawas, D. Fang, T. Gaborski, C. Striemer,

P. Fauchet, D. Borkholder and J. McGrath, High-performance, low-voltage electroosmotic pumps with molecularly thin silicon nanomembranes, *Proc. Natl. Acad. Sci. USA* **110**, pp. 18425 – 18430 (2013).

[74] A. Salmon, M. Capener, J. Baumberg and S. Elliott, Rapid microcantilever-thickness determination by optical interferometry, *Meas. Sci. Technol.* **25**, p. 015202 (2014).

[75] S. Chugh, M. Man, Z. Chen and K. Webb, Ultra-dark graphene stack metamaterials, *Appl. Phys. Lett.* **106**, p. 061102 (2015).

[76] D. DiVincenzo and E. Mele, Self-consistent effective-mass theory for intralayer screening in graphite intercalation compounds, *Phys. Rev. B* **29**, pp. 1686 – 1693 (1984).

[77] T. Ando, T. Nakanishi and R. Saito, Berry's phase and absence of back scattering in carbon nanotubes, *J. Phys. Soc. Japan* **67**, pp. 2857 – 2862 (1998).

[78] P. McEuen, M. Bockrath, D. Cobden, Y. Yoon and S. Louie, Disorder, pseudospin, and backscattering in carbon nanotubes, *Phys. Rev. Lett.* **83**, pp. 5098 – 5101 (1999).

[79] A. Geim and I. Grigorieva, Van der waals heterostructures, *Nature* **499**, pp. 419 – 425 (2013).

[80] H. Min, J. Hill, N. Sinitsyn, B. Sahu, L. Kleinman and A. MacDonald, Intrinsic and Rashba spin-orbit interaction in graphene sheets, *Phys. Rev. B* **74**, p. 165310 (2006).

[81] B. Gu, N. Kwong and R. Binder, Relation between the interband dipole and momentum matrix elements in semiconductors, *Phys. Rev. B* **87**, pp. 125301 – (17) (2013).

[82] M. O. Scully and M. S. Zubairy, *Quantum Optics*. Cambridge University Press, Cambridge (1997).

[83] J. Sipe, Bulk-selvedge coupling theory for the optical properties of surfaces, *Phys. Rev. B* **10**, pp. 1589 – 1599 (1980).

[84] G. Khitrova, H. M. Gibbs, F. Jahnke, M. Kira and S. W. Koch, Nonlinear optics of normal-mode coupling semiconductor microcavities, *Reviews of Modern Physics* **71**, pp. 1591–1639 (1999).

Part 1

Linear Optical Response and Raman Spectroscopy

Chapter 2

Microscopic Theory for the Groundstate and Linear Optical Response of Novel Two-Dimensional Materials with Hexagonal Symmetry

Tineke Stroucken* and Stephan W. Koch
Department of Physics and Scientific Center for Material Sciences, Philipps University Marburg, Renthof 5, 35032 Marburg/Germany
**tineke.stroucken@physik.uni-marburg de*

This chapter summarizes recent theoretical work to determine the groundstate and the linear optical properties of quasi-two-dimensional materials with hexagonal lattice symmetry. The main ingredients of the fully relativistic tight-binding model analysis for graphene and transition metal dichalcogenites are summarized. Dirac-Bloch equations are derived and solved in the regimes of weak and strong Coulomb coupling. Whereas single-layer graphene always seems to be in the weak Coulomb-coupling regime, realistic parameter analysis of WSe_2 and WS_2 identifies these materials to be in the regime of strong Coulomb coupling. Consequently, they are predicted to exhibit an excitonic insulator groundstate with optically active p-excitons. On this basis, a unified understanding and excellent agreement of the theoretical predictions with a large variety of systematic experiments is obtained.

2.1 Introduction

Novel 2D materials have emerged as a new class of material systems whose thickness is just a single unit cell. The firstly discovered –and most famous– representative of this material class is graphene whose remarkable properties – it is flexible, transparent, stronger than steel, more conductive than copper– have been celebrated since is its first isolation by Novosolov and Geim 2004. [Novoselov *et al.* (2004)] Soon after, other novel 2D materials could be fabricated, among them transition metal dichalcogenides (TMDCs). Similar to graphene, TMDCs are layered materials with strong in-plane bonding and weak out-of-plane interactions enabling exfoliation into two-dimensional layers. Within the layers, the atoms are arranged in a honeycomb lattice with hexagonal symmetry.

The reduction of the thickness down to a monolayer can alter the physical properties drastically, making these material systems interesting both for fundamental material science and technological applications. Because of its unique properties, graphene has been proposed as a candidate for many applications such as transparent conducting electrodes, touchscreens, liquid crystal displays, solar cells, saturable absorbers, just to name a few. However, the lack of a gap in graphene's bandstructure makes it a poor material for digital electronics and optoelectronic devices.

In contrast to graphene, TMDCs monolayers such as MoS_2, $MoSe_2$, WS_2, and WSe_2 have sizable bandgaps, allowing for applications such as transistors. Moreover, since TMDCs exhibit pronounced light-matter coupling, they could become attractive for optoelectronic applications, e.g. in photodetectors and electroluminescent devices.

One of the important requirements for the basic understanding of the fundamental material properties is the knowledge of the electronic bandstructure. Whereas in a mesoscopic system, such as a semiconductor quantum well (QW), the quantum confinement effects can be treated within the envelope function approximation, the bandstructure of a single monolayer differs not only quantitatively but also qualitatively from that of the corresponding bulk material.

In graphene, the single-particle bandstructure exhibits two distinct crossing points at the Fermi level where electrons and holes are degenerate. In the vicinity of these Dirac points, the dispersion is a cone similar to the light cone in relativistic mechanics with the Fermi velocity v_F replacing the speed of light. The occurrence of the cones results from the symmetry between the two equivalent sublattices constituting the overall honeycomb lattice. The sublattice wavefunctions can be combined into a pseudospinor whose properties are determined by the ultra-relativistic Dirac equation. As a consequence, the quasiparticles in graphene can be considered as massless ultra-relativistic Dirac-Fermions, characterized by the pseudospin σ, the electric charge e, and the Fermi velocity which is typically in the range $v_F \propto c/300$. The nonequivalent Dirac points are related by the parity or time-reversal transformation with opposite pseudospins in the different valleys. Similar to the real electron spin, the pseudospin is associated with an angular momentum and couples to the optical field. As a result, different valleys are addressed by oppositely circular polarized electromagnetic fields.

In TMDCs, the bandstructure changes from indirect to direct as one reduces the thickness from bulk to a monolayer. This transition has been predicted by bandstructure calculations [Cheiwchanchamnangij and Lambrecht (2012); Cappelluti *et al.* (2013)] and was confirmed experimentally by photoluminescence (PL) experiments that show an intensity increase of several orders of magnitude when the material thickness approaches a monolayer. [Mak *et al.* (2010); Zeng *et al.* (2013)] As in graphene, the direct gap is located at the two Dirac points of the Brillioun-zone, and the near bandgap quasiparticles can be described as relativistic Dirac Fermions with a pseudospin that couples to the light field. However, unlike in graphene where the different sublattices are occupied with the same species of atoms, the nonequivalence of the two sublattices in TMDCs leads to the opening of a gap that can be interpreted as rest energy of a massive electron-hole pair. Moreover, a strong spin-orbit interaction leads to a valley dependent splitting of the valence-band states with different (real) spin. The valley dependent spin splitting allows one to address distinct spin and valley states selectively by resonant excitation with circularly polarized light and suggests the possibility of so called valleytronics. [Yao *et al.* (2008); Cao *et al.* (2012); Mak *et al.* (2012a); Zeng *et al.* (2013); Shan *et al.* (2015)]

While it is generally accepted that the electronic properties of graphene and TMDCs can be assigned to the existence of relativistic, chiral quasiparticles, the role of the electron-electron Coulomb interaction and its influence on the optical properties of graphene and graphene-like materials is still not well understood. Within the noninteracting picture, the optical spectrum of graphene is characterized by a constant absorption in the visible and a pronounced absorption peak in the ultraviolet spectral range which is attributed to the van Hove singularity at the M point of the Brillouin zone. [Kravets *et al.* (2010); Yang *et al.* (2009); Yang (2011); Malic *et al.* (2011)]

Microscopic calculations including electron-electron interactions [Yang *et al.* (2009); Yang (2011); Malic *et al.* (2011)] predict the formation of so-called saddle-point excitons, manifesting themselves in an electron-hole-correlation induced redshift of the van Hove singularities that has been observed experimentally. [Mak *et al.* (2011); Chae *et al.* (2011)] At the same time, the constant absorption in the visible range is amazingly robust against Coulomb interaction effects.

In TMDCs, a number of recent linear und nonlinear optical experiments observed excitons with unusually large binding energies and strong deviations from the Rydberg series known from conventional semiconductors [Chernikov *et al.* (2014); Ye *et al.* (2014); Zhu *et al.* (2014); Wang *et al.* (2014)]. Interestingly, the reported experimental values for the binding energy of the strongest exciton may differ by as much as a factor of two for nominally identical samples.

In the framework of a relativistic description, a measure for the relative strength of the Coulomb interaction is the effective fine-structure constant $\alpha =$

$e^2/\kappa\hbar v_F$, where κ combines the dielectric constant of the environment with intrinsic screening effects. In quantum electrodynamics (QED), $\alpha = 1/137$ is a small parameter that is often used in perturbation theory. In graphene, $\alpha_G \approx 2.41/\kappa$, and in a typical TMDC, α lies somewhere between $3/\kappa$ and $5/\kappa$, indicating prominent Coulomb interaction effects.

Generally, one can distinguish between Coulomb modifications of the electronic groundstate and Coulombic signatures in the excitation properties. From strongly correlated systems it is known that the Coulomb interaction can induce a transition from a semimetal to a Mott insulator where, unlike in conventional semiconductors, the gap results from the electron-electron rather than the electron-ion interactions. In particular, the electron exchange interaction has been identified as the dominant mechanism responsible for the opening of a gap. [Engel and Schmid (2009)]

In a conventional semiconductor, the bandstructure is fixed by the crystalline composition and lattice arrangement. The electronic groundstate is characterized by full valence and empty conduction bands. Optical excitation leads to electron-hole pair creation. At low excitation levels, the attractive electron-hole Coulomb interaction leads to the formation of bound pairs, i.e. excitons, which, in the limit of not too strong Coulomb attraction (Wannier limit), are mathematically described by a hydrogen-like equation, which is known as the Wannier equation. [Haug and Koch (2009)]

In the linear absorption spectrum of such GaAs-like systems, the excitonic resonances form a Rydberg series energetically below the fundamental absorption edge. Due to the different effective masses and the effective screening of the Coulomb interaction in a semiconductor, the binding energy is much smaller than in hydrogen, typically a few meV. Correspondingly, the Bohr radius is much larger than in hydrogen and extends over a large number of unit cells. If the Bohr radius exceeds the sample dimensions, quantum confinement leads to an enhancement of the excitonic binding [Haug and Koch (2009)] and in an isolated truly 2D material, the reduction of the effective screening causes an additional binding enhancement. While the confinement induced modifications are well known from quantum wells and other mesoscopic structures, the modified screening is intrinsic to truly 2D materials and is a subject of ongoing research [Keldysh (1978); Cudazzo *et al.* (2011); Wehling *et al.* (2011)].

Considering a semimetal as a semiconductor with vanishing band gap, the existence of bound electron-hole-pair states and a possible transition into a gapped, strongly coupled phase become intimately related. Clearly, the existence of bound exciton states in a gap-less system allows for the possibility to reduce the system's energy by the creation of excitons, contradicting the assumption that the gap-less dispersion corresponds to the systems groundstate. Indeed, quite generally, if the exciton binding energy exceeds the noninteracting gap energy, excitons may form spontaneously and the system is expected to undergo a phase transition into an excitonic state. [Keldysh and Kopaev (1965); Jerome *et al.* (1967); Halperin and Rice (1968)] As the quasiparticle spectrum exhibits a gap exceeding that of the noninteracting groundstate, this state is known as excitonic insulator.

As a consequence of its zero-gap single-particle bandstructure, single-layer graphene (SLG) seems a promising candidate for an excitonic insulator. Indeed, the possible existence of a strongly coupled phase in graphene has been discussed already in 1960 [Pauling (1960); CastroNeto *et al.* (2009b)]. In the recent literature, this discussion has resurfaced with partly contradicting results. While perturbative studies based on a renormalization group analysis yield a stable semimetallic groundstate [Gonzalez *et al.* (1999); Sheehy and Schmalian (2007); Fritz *et al.* (2008); Sinner and Ziegler (2010)], nonperturbative methods predict a semimetal-to-insulator transition at sufficiently high coupling strengths [Khveshchenko (2001); Khveshchenko and Shively (2006); Juricic (2009); Drut and Lähde (2009a,b); Sabio *et al.* (2010); Stroucken *et al.* (2011, 2012)] with critical values ranging from 0.5 to 1.62.

For Schrödinger-like quasiparticles, bound exciton solutions exist for arbitrarily weak Coulomb interaction. However, the analysis of the Wannier equation with a linear dispersion shows that the existence of ultra-relativistic bound excitons and hence a phase transition into an excitonic phase requires a minimal effective Coulomb coupling strength, in general agreement with other nonperturbative theoretical investigations. [Grönqvist *et al.* (2012)] Combination of the static limit of the standard Lindhard formula with experimentally measured values for the Fermi velocity predicts values for the Coulombic coupling that are slightly larger than the critical value, thus allowing for a small but finite gap. However, despite intense investigations, so far no experimental evidence has been presented for the occurrence of a gap in the spectrum of SLG.

Though the electronic properties of TMDCs are very similar to those of graphene, the optical properties are widely believed to be comparable to those of conventional semiconductors despite the observed deviations from the excitonic Rydberg series. In general, the existence of such a Rydberg series requires three features: (i) Schrödinger-like quasiparticles with quadratic dispersion, (ii) local screening preserving the $1/r$ dependence of the Coulomb interaction, and (iii) the occurrence of s-type bright states. Since the large gap in TMDCs – typically ranging between 1.5 and 3 eV – seems to allow for a Taylor expansion of the relativistic single-particle dispersion, and the apparent observation of s-type bright states, the occurring deviations from the standard Rydberg series have been mainly assigned to nonlocal screening effects. [Berkelbach *et al.* (2013); Chernikov *et al.* (2014); He *et al.* (2014); Latini *et al.* (2015)]

Treating a monolayer as a thin dielectric sheet with constant polarizability, the reduction of the dimensionality leads to a dielectric function that increases linearly with momentum q, thus reducing the Coulomb interaction strength at small electron-hole distances r. [Keldysh (1978); Cudazzo *et al.* (2011); Berkelbach *et al.* (2013)] However, this anti-screening model is based on the ad-hoc assumption of a constant sheet polarizibility which lacks a rigorous microscopic foundation. Recent *ab initio* calculations confirm the assumed linear dependence of the dielectric function on the momentum q for small values of q, but at the same time show that it drastically overestimates the effects of screening for the strongly bound excitons with an extension in reciprocal space beyond the linear regime of the dielectric function. [Latini *et al.* (2015)] In the strict 2D limit, simple scaling arguments can be used to show that the anti-screening model is valid for a quadratic single-particle dispersion only, i.e. in the nonrelativistic limit, whereas in the ultrarelativistic limit all screening effects are again purely local.

As the fine-structure constant relates the exciton Bohr radius a_B to the Compton wavelength λ_C via $a_B = \lambda_C/\alpha$, it is also a measure for the importance of relativistic effects. Within a relativistic treatment, the optical field couples to the total angular momentum j that combines the orbital angular momentum with the pseudospin of the quasiparticles and lifts the degeneracy between states with different angular momentum. Most interestingly, $\alpha = 1$ marks a critical value where the binding of the lowest exciton state with $j = \pm 1/2$ exactly equals the gap. This leads to the surprising conclusion that the condition for the excitonic phase transition depends solely on the relative coupling strength α. The size of the gap enters only indirectly via screening of the Coulomb interaction, which is in general less effective in wide gap semiconductors.

The transition into an excitonic insulator state has direct consequences for the optical selection rules. The usual s-type optical selection rules are based on the implicit assumption that the system is excited from the non-interacting GaAs-like groundstate and excitons are created via one-photon absorption processes. However, if the groundstate itself is excitonic, the dipole allowed transitions correspond to intra-excitonic transitions, leading to a Lymann series instead of a Rydberg series. [Stroucken and Koch (2015)] Hence, the validity of all three assumptions mentioned above depend on the importance of relativistic effects, i.e. on the value of the effective fine-structure constant α.

In this chapter, we investigate the groundstate and the linear optical properties of graphene and TMDC monolayers as a function of the effective coupling strength. To this end, we extend existing methods to describe the excitation properties of conventional semiconductors to allow for a simultaneous and self-consistent treatment of a correlated groundstate and the excitation dynamics. In order to derive and understand the mechanisms of exciton condensation and its effects on the optical properties, we use the simplified model Hamiltonian known as massive Dirac-Fermion model and often consider the fine-structure constant as a free parameter. Of course, to apply the model results to real materials, a proper understanding of the interplay between screening and excitonic effects is crucial. Though some considerations concerning screening in real 2D materials are included in this chapter, a thorough analysis of screening can only be provided by first-principle calculations, which are beyond the scope of this chapter. Despite this incompleteness of the current state of our analysis, we can perform some comparisons of the theoretical results with available experimental data. The agreement we obtain is remarkable and indicates that despite some obvious simplifications, our model captures the relevant physical processes in these novel material systems.

2.2 Hamiltonian

2.2.1 *The massive Dirac Hamiltonian*

The hexagonal lattice structure is common to graphene and the semiconducting TMDCs considered in this article. In graphene, the honeycomb lattice results from a sp^2 hybridization of the 2s, $2p_x$, and $2p_y$ orbitals of the constituting carbon atoms. The $2p_z$ orbitals perpendicular to the plane are responsible for the electronic and optical properties. Similarly, in a group VI TMDC MX_2, the M and X atoms form a 2D hexagonal lattice where the M atoms are s-p bonded to six adjacent X atoms, and the partly filled d-bands of the M atoms build the conduction and valence bands.

The atoms in the hexagonal lattice are localized at the corners of a honeycomb and each corner is related to its next three neighbor via a translation $\boldsymbol{\delta}_i$. The honeycomb lattice in itself is not a Bravais lattice but consists of two sublattices which are usually referred to as A and B lattice. The sublattices are connected by the translation vectors $\boldsymbol{\delta}_i$, that can be used to characterize the A and B lattices. Defining the A-lattice by $\boldsymbol{\delta}_1 = \frac{a}{2\sqrt{3}}(1, \sqrt{3})$, $\boldsymbol{\delta}_2 = \frac{a}{2\sqrt{3}}(1, -\sqrt{3})$ and $\boldsymbol{\delta}_3 = \frac{a}{\sqrt{3}}(-1, 0)$, the B lattice is then defined by $\boldsymbol{\delta}'_i = -\boldsymbol{\delta}_i$. Thus, the two sublattices are related by the parity transformation. In reciprocal space, the first Brillouin zone has the same honeycomb geometry as in real space. Similar to the real-space configuration, the corners of the Brillouin zone, i.e. the $\boldsymbol{K}^{\pm}$ Dirac points, are related by the parity transformation.

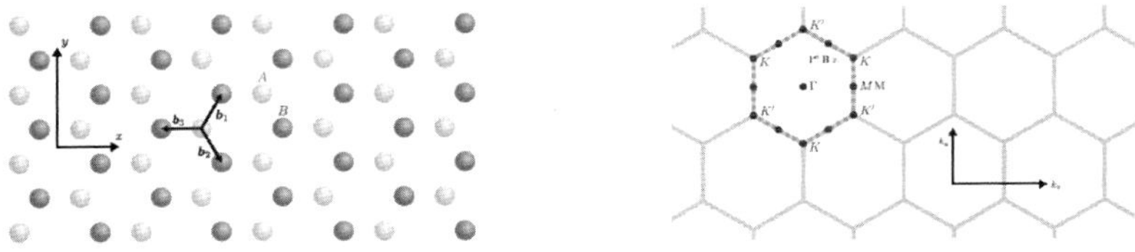

Fig. 2.1 Lattice structure of a graphene sheet. The left part shows the real space lattice consisting of the equivalent sublattices A and B. The sublattices are connected by the translation $\boldsymbol{\delta}_i$. The right part shows the reciprocal lattice with the Dirac points $\boldsymbol{K}^{\pm}$, the Γ, and the M point. In our representation, the first Brillouin zone is centered at the Γ point.

Based on general symmetry considerations, it has been shown in Ref. [Xiao *et al.* (2012)], that the minimal two-band Hamiltonian in lowest-order $\mathbf{k}\cdot\mathbf{p}$ theory is given by

$$H_0 = \sum_{\tau,\mathbf{k}} \hat{\Psi}^{\dagger}_{\tau\mathbf{k}} \left(at\mathbf{k}\cdot\hat{\sigma}_{\tau} + \frac{\Delta}{2}\hat{\sigma}_z - \tau\lambda\frac{\hat{\sigma}_z - 1}{2}\hat{s}_z \right) \hat{\Psi}_{\tau\mathbf{k}}. \tag{2.1}$$

Here, $\tau = \pm 1$ is the so-called valley index identifying the two non-equivalent Dirac points and the field operator $\hat{\Psi}^{\dagger}_{\tau\mathbf{k}}$ is the tensor product of the electron spin state and a two-component quasi-spinor. The Pauli matrices $\hat{\sigma}_{\tau} = (\tau\hat{\sigma}_x, \hat{\sigma}_y)$ and $\hat{\sigma}_z$ act in the pseudospin space and $\hat{s}_z$ in the real spin space, respectively. The parameters Δ and 2λ correspond to the gap and an effective spin splitting of the valence bands, respectively. The parameters t and a denote the effective hopping matrix element and the lattice constant, respectively. These can be taken either from first-principle calculations or from experiment.

For graphene, the Hamiltonian (2.5) can be derived straightforwardly from a tight-binding (TB) approach. Using the carbon 2p$_z$ orbitals

$$\phi(\mathbf{r}) = \frac{1}{d^{5/2}}\sqrt{\frac{1}{32\pi}}\cos\vartheta\, r\, e^{-r/2d}$$

centered around the A and B atoms as basis functions and taking into account nearest-neighbor hopping only, the free-particle part of the Hamiltonian can be expressed in terms of the hopping matrix element $\gamma = \int d^3r\phi^*(\mathbf{r} - \boldsymbol{\delta}_i)\,\hat{H}_0\,\phi(\mathbf{r})$ and the function $f(\mathbf{k}) = \sum_{i=1}^{3} e^{i\mathbf{k}\cdot\boldsymbol{\delta}_i}$ that reflects the symmetry properties of the honeycomb lattice. Diagonalization of the resulting single-particle Hamiltonian [Wallace (1947)] yields the well known TB band structure

$$E^{c/v}_{\mathbf{k}} = \pm\gamma|f(\mathbf{k})| \tag{2.2}$$

consisting of symmetric electron-hole bands. The electron-hole bands touch at the roots of $|f(\mathbf{k})|$, defining the two nonequivalent Dirac points

$\boldsymbol{K}^{\pm} = (2\pi/\sqrt{3}a, \pm 2\pi/3a)$. In the vicinity of these nodes, a first-order Taylor expansion gives

$$f(\boldsymbol{K}^{\pm} + \mathbf{k}) = -\frac{\sqrt{3}a}{2} e^{-i\pi/6}(k_x \pm ik_y).$$

We combine the operators $\hat{a}^{\dagger}_{\mathbf{k}}$ and $\hat{b}^{\dagger}_{\mathbf{k}}$ that create a particle in the state $\mathbf{k}$ on the respective sublattice $\{A\}$ and $\{B\}$ into a pseudospinor $\hat{\Psi}^{\dagger}_{\mathbf{k}} = (\hat{a}^{\dagger}_{\mathbf{k}}, \hat{b}^{\dagger}_{\mathbf{k}})$. This way, we can express the single-particle part of the Hamiltonian in the vicinity of the Dirac point as

$$H_0 = v_F \sum_{\tau\mathbf{k}} \hat{\Psi}^{\dagger}_{\tau\mathbf{k}} \hbar\mathbf{k} \cdot \boldsymbol{\sigma}_{\tau} \hat{\Psi}_{\tau\mathbf{k}}, \tag{2.3}$$

which corresponds to Eq. (2.5) with vanishing gap and spin-orbit splitting.

For TMDCs, the derivation of Eq. (2.5) from a tight-binding model is not as straightforward as for graphene but requires additional projection techniques. A detailed derivation can be found e.g. in Ref. [Liu (2013)]. The resulting basis functions for the spinors are given by linear combinations of the M d-orbitals $|\psi_c\rangle = |d_{z^2}\rangle$ and $|\psi_v^{\tau}\rangle = \frac{1}{\sqrt{2}}\left(|d_{x^2-y^2}\rangle + i\tau\,|d_{xy}\rangle\right)$, where the hopping integrals not only contain the direct d-d interactions, but also indirect interactions mediated by the X-p orbitals. In analogy to graphene, we refer to the operators that create a particle with pseudospin up and down as $a^{\dagger}$ and $b^{\dagger}$ respectively.

Equation (2.5) is the starting point for our analysis and formally equivalent to the 2D Dirac Hamiltonian for massive Fermions with the Fermi velocity by $v_F = at/\hbar$ replacing the speed of light, and a valley and spin dependent effective mass $m_{s\tau} = (\Delta - s\tau\lambda)/2v_F^2 \equiv \Delta_{s\tau}/2v_F^2$. The eigenstates have the relativistic dispersion

$$\epsilon_{s\tau k} = \pm\sqrt{(\Delta_{s\tau}/2)^2 + (\hbar v_F k)^2}\,, \tag{2.4}$$

which reduces to the linear dispersion of graphene for a vanishing gap and can be approximated reasonably well by a parabolic dispersion close to the Dirac points of a wide gap TMDC.

2.2.2 *Light-matter interaction*

According to the minimal substitution, the light-matter (LM) Hamiltionian is given by

$$H_I = -e\frac{v_F}{c} \sum_{\tau\mathbf{k}} \hat{\Psi}^{\dagger}_{\tau\mathbf{k}} \mathbf{A} \cdot \hat{\sigma}_{\tau} \hat{\Psi}_{\tau\mathbf{k}}. \tag{2.5}$$

Thus, excitation with left(right)-handed circular polarized light induces a spin-flip in the pseudospin space, i.e. a jump between the basis functions with $\Delta m = 0, \pm 2$

for graphene and TMDCs, respectively. The missing angular momentum quantum required to satisfy angular momentum conservation is supplied by the change of angular momentum associated with the flip of the pseudospin.

The eigenstates of H_0 can be choosen as eigenstates of the total angular momentum operator $\hat{J}_z = \hat{L}_z + \frac{\tau}{2}\hat{\sigma}_z$:

$$\Psi_{s\tau cjk} = \begin{pmatrix} e^{i(j-\tau/2)\theta_{\mathbf{k}}} u_k \\ e^{i(j+\tau/2)\theta_{\mathbf{k}}} v_k \end{pmatrix}, \quad \Psi_{s\tau\nu jk} = \begin{pmatrix} e^{i(j-\tau/2)\theta_{\mathbf{k}}} v_k \\ -e^{i(j+\tau/2)\theta_{\mathbf{k}}} u_k \end{pmatrix}, \tag{2.6}$$

with $u_k = \sqrt{(\epsilon_k + \Delta/2)/2\epsilon_k}$, $v_k = \sqrt{(\epsilon_k - \Delta/2)/2\epsilon_k}$, and $\theta_{\mathbf{k}}$ is the k-space angle defined by $\tan\theta_{\mathbf{k}} = \frac{k_y}{k_x}$. Excitation with left(right)-handed circular polarized light increases (decreases) the z- component of $\hat{J}$ in the K^+-valley by one, and vice versa at the K^- valley. Together with the change of orbital angular momentum of the basis functions, the total change in angular momentum is ± 1 for excitation with right (left) handed circular polarized light, in agreement with the general optical selection rules. The optically induced current can be obtained from the LM Hamiltonian via $\mathbf{j} = -c\langle\frac{\delta H_I}{\delta \mathbf{A}}\rangle$. Using the spinor basis functions, one finds the simple expression $j_\tau = \sum_{\mathbf{k}} \langle a^\dagger_{\tau\mathbf{k}} b_{\tau\mathbf{k}}\rangle + \langle b^\dagger_{-\tau\mathbf{k}} a_{-\tau\mathbf{k}}\rangle$, which is valid for all values for the gap and the Fermi velocity.

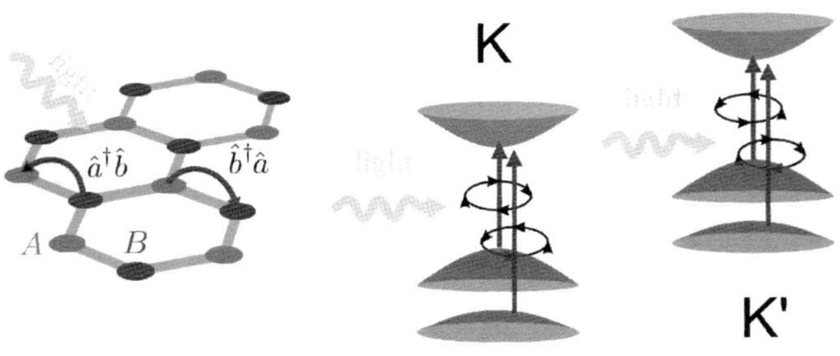

Fig. 2.2 Schematic representation of the light–matter interaction. Photon absorption or emission is associated with with a pseudo-spin flip, i.e. an electron is transferred between the two basis functions spanning the pseudo-spin space. In graphene, this transition corresponds to a hopping between the sublattices and is best represented in real space. In TMDCs, the basis functions correspond approximately to the valence and conduction band states and a representation in the band picture is most appropiate.

Using the eigenstates of H_0, the LM Hamiltonian can be written as

$$H_I = -e\frac{v_F}{c}\sqrt{2}\sum_{s\tau\mathbf{k}} \tau u_k v_k \left(e^{-i\tau\theta_{\mathbf{k}}} A^\tau + e^{i\tau\theta_{\mathbf{k}}} A^{-\tau}\right)\left(c^\dagger_{s\tau\mathbf{k}} c_{s\tau\mathbf{k}} - \nu^\dagger_{s\tau\mathbf{k}} \nu_{s,\tau,\mathbf{k}}\right)$$
$$- e\frac{v_F}{c}\sqrt{2}\sum_{s,\tau,\mathbf{k}} \tau \left\{\left(v_k^2 e^{-i\tau\theta_{\mathbf{k}}} A^\tau - u_k^2 e^{i\tau\theta_{\mathbf{k}}} A^{-\tau}\right) c^\dagger_{s\tau\mathbf{k}} \nu_{s\tau\mathbf{k}} + h.c.\right\}. \tag{2.7}$$

Here, $c^\dagger_{s\tau\mathbf{k}}$ ($\nu^\dagger_{s\tau\mathbf{k}}$) creates an electron with spin s and valley index τ in the conduction (valence) band and $A^\tau = (A_x + i\tau A_y)/\sqrt{2}$. In Eq. (2.7), the first line describes the intraband and the second line the interband transitions, respectively. Contributions proportional to A^τ correspond to the absorption of a photon with circular polarization σ^τ. In general, the interaction Hamiltonian contains both intra- and interband contributions and reduces to a simple dipole interaction only if $u_k^2 = 1 - v_k^2 = 1$. In a wide-gap system, the leading orders of the transition matrix elements are given by $u_k^2 \approx 1$, $u_k v_k \approx \frac{\hbar v_F k}{\Delta_{s\tau}}$ and $v_k^2 \approx (\hbar v_F k)^2/\Delta_{s\tau}^2$ and the dipole approximation is valid in the close vincinity of the Dirac points. In contrast, in zero-gap materials like graphene or silicene, the transition matrix elements for both intra- and interband absorption are equal in magnitude. Differences in the transition probabilities exclusively result from the different occupation numbers in the initial state.

2.2.3 *Coulomb interaction*

Probably one of the most interesting aspects of the mathematical equivalence between the quasiparticle and the Dirac Hamiltonian is the possibility to study an ultrarelativistic interacting system in solid state physics. For the effectively 2D system, the Coulomb interaction between the quasiparticles is described by

$$H_C = \frac{1}{2} \sum_{\tau\mathbf{k}\mathbf{k}'\mathbf{q}} : \hat{\Psi}^\dagger_{\tau\mathbf{k}+\mathbf{q}} \hat{\Psi}_{\tau\mathbf{k}} V(\mathbf{q}) \hat{\Psi}^\dagger_{\tau\mathbf{k}'-\mathbf{q}} \hat{\Psi}_{\tau\mathbf{k}'} : \tag{2.8}$$

where $: \cdot :$ denotes normal ordering. In terms of the spinor wavefunctions, the Coulomb Hamiltonian describes scattering between particles with a given pseudospin component. Expanding the density operator in terms of the eigenfunctions of the single-particle part of the Hamiltonian yields

$$\sum_{\mathbf{k}} \hat{\Psi}^\dagger_{\tau\mathbf{k}-\mathbf{q}} \hat{\Psi}_{\tau\mathbf{k}} = \sum_{\mathbf{k}} \sum_{\alpha\alpha'} \langle \alpha\mathbf{k} - \mathbf{q} | \alpha'\mathbf{k} \rangle c^\dagger_{\tau,\alpha,\mathbf{k}-\mathbf{q}} c_{\tau,\alpha',\mathbf{k}}. \tag{2.9}$$

Within this basis, the overlap matrix elements include a valley dependent phase factor and describe 2^4 processes, not all of which are fundamentally different. The 10 nonequivalent processes include the repulsive electron-electron and hole-hole interactions as well as the attractive electron-hole interactions. Here, we have the cases where (i) both electrons remain in their initial bands as well as (ii) processes where both particles change bands and (iii) Auger-processes where either the electron or hole is scattered within a band under simultaneous creation or annihilation of an electron-hole pair. In a wide gap semiconductor, the processes where at least one quasiparticle is scattered across the gap are very unlikely due to energy conservation, whereas in a gapless system like graphene, all matrix elements are equal in magnitude. A schematic representation of the various contributions is presented in Fig. 2.3. As usual, processes with zero momentum transfer describe a divergent self-interaction that cancels with the electron-ion and ion-ion interaction in the Jellium limit and has to be excluded from the Coulomb Hamiltonian.

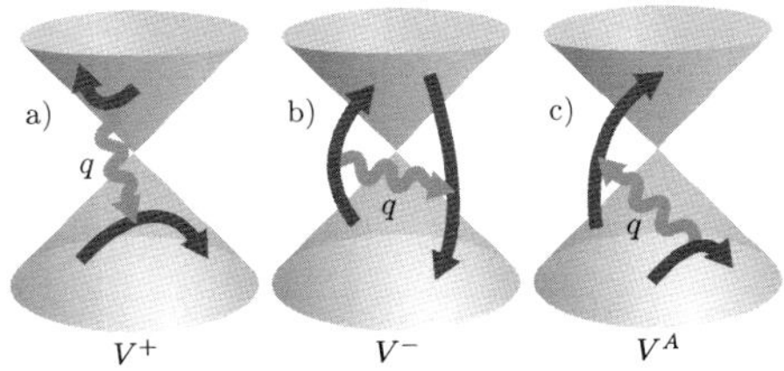

Fig. 2.3 Schematic representation of the Coulomb matrix elements. Reprinted (figure 3) with permission from [T. Stroucken, J. H. Grönqvist, and S. W. Koch, *Phys. Rev. B* **84**, 205445 (2011).] Copyright (2011) by the American Physical Society.

The effective 2D Coulomb matrix element

$$V(\mathbf{q}) = \frac{2\pi e^2}{\mathcal{A}\kappa q} F(qd) \tag{2.10}$$

depends on the momentum transfer q. It is given by the bare 2D Coulomb potential modified by the effective background dielectric constant κ and multiplied by a form factor which accounts for the finite extension of the wavefunctions in the z-direction perpendicular to the layer. Within the two-band model, the effective dielectric constant contains contributions from the system's environment, such as substrate screening etc., and possible nonresonant intrinsic contributions arising from all bands other than those explicitly treated in the model Hamiltonian. Screening from the explicitly included bands has to be computed from the microscopic polarization function and should not be lumped into the background dielectric constant.

Considering a thin layer on a substrate with dielectric constant ϵ_S and covered by air, the strict 2D limit of the effective dielectric constant is given by the half-space value $\kappa = (\epsilon_S + 1)/2$, regardless of the intrinsic background of the layer. Inclusion of a finite thickness gives a q-dependent value of the effective background screening that smoothly interpolates between the half-space value at $q = 0$ and the intrinsic value ϵ_M at $qd >> 1$. Additionally, the finite extension of the atomic orbitals leads to a further modification of the 2D Coulomb potential that is independent of the dielectric screening and enters into the form factor $F(qd)$ which is a monotonically decreasing function with $F(0) = 1$.

Using scaling arguments, one can show that the main effect of $F(qd)$ is to provide an intrinsic length scale. Introducing the scaled quantities $\mathbf{k} = \bar{\mathbf{k}}/\lambda$,

$\bar{c}_{s\tau\alpha\bar{\mathbf{k}}} = c_{s\tau\alpha\mathbf{k}}$ yields the dimensionless many-body Hamiltonian

$$\frac{\lambda}{\hbar v_F}(H_0 + H_C) = \frac{\lambda}{\lambda_C}\left(\sum_{\tau\bar{\mathbf{k}}}\sqrt{1+(\lambda_C\bar{\mathbf{k}}/\lambda)^2}\left(\bar{c}^\dagger_{s\tau\bar{\mathbf{k}}}\bar{c}_{s\tau\bar{\mathbf{k}}} - \bar{\nu}^\dagger_{s\tau\bar{\mathbf{k}}}\bar{\nu}_{s\tau\bar{\mathbf{k}}}\right)\right.$$
$$\left. + \frac{\alpha}{2}\frac{\lambda_C}{\lambda}\frac{2\pi}{\bar{A}}\sum_{\bar{\mathbf{q}}} : \bar{\rho}_{\tau\bar{\mathbf{q}}}\frac{F(\bar{q}d/\lambda)}{\bar{q}}\bar{\rho}_{\tau-\bar{\mathbf{q}}} :\right). \tag{2.11}$$

where $\alpha = e^2/\kappa\hbar v_F$ is the effective fine-structure constant and $\lambda_C = 2\hbar v_F/\Delta$ is the Compton wavelength. In a typical TMDC, the Compton wavelength is on the order of a lattice constant. We can see that the scaled Hamiltonian is characterized by α as a measure for the relative importance of Coulombic effects and the Compton wavelength is the intrinsic length scale indicating the importance of relativistic effects. Choosing the scaling length as $\lambda = \lambda_C/\alpha = a_B$, which is the Bohr radius of a particle with mass $m = \Delta/2v_F^2$, the fine-structure constant is also a measure for the importance of relativistic and finite size effects. If $\alpha << 1$, a second-order Taylor expansion of the kinetic energy gives a constant rest energy and a kinetic and a Coulombic part that are both proportional to α^2, yielding the familiar nonrelativistic result. In the ultra-relativistic limit $\alpha >> 1$ or $\Delta = 0$, the RHS of Eq. (2.11) depends on the scaling length only via the form factor and the only intrinsic length scale is the effective sheet thickness d. In general, finite-size effects become important if $a_B/d \lesssim 1$.

2.3 Weak Coulomb coupling

A convenient method to describe the optical properties of semiconductors and semiconductor heterostructures is given by the semiconductor Bloch equations (SBE) [Haug and Koch (2009)] for the microscopic polarizations and the conduction- and valence-band occupation probabilities. As is well known, the two-particle Coulomb interaction couples the dynamics for the N-particle expectation values to those of the $N+1$-particle expectation values, which is known as the hierarchy problem. In order to obtain a closed set of equations, this hierarchy must be truncated. A systematic truncation scheme is provided by the cluster expansion, which has been proven to work well under many different excitation conditions. [Kira and Koch (2006)] The coupled set of equations for the relevant correlation functions allows us to compute the system dynamics after an external excitation. However, before we can do this, we have to chose a proper basis for the expansion of the field operators and we have to know the initial state of the system before any external perturbation is applied.

In a typical experiment, the system is excited from its groundstate and the response to an externally applied field is measured. Thus, to analyze such experiments theoretically, it is crucial to have an adequate approximation for the groundstate. Usually, bandstructure calculations provide a suitable basis to describe both the groundstate and the optical excitations. Using the single-particle expectation values

$$\begin{aligned} g_{s\tau\mathbf{k}} &= \langle \nu^{\dagger}_{s\tau\mathbf{k}} \nu_{s\tau\mathbf{k}} \rangle - \langle c^{\dagger}_{s\tau\mathbf{k}} c_{s\tau\mathbf{k}} \rangle \\ P_{s\tau\mathbf{k}} &= \langle c^{\dagger}_{s\tau\mathbf{k}} \nu_{s\tau\mathbf{k}} \rangle, \end{aligned} \tag{2.12}$$

as dynamical variables, the dipole interaction couples the dynamics for the vector potential to the expectation value of the interband transitions $P_{s\tau\mathbf{k}}$ which are in turn coupled to the population inversion $g_{s\tau\mathbf{k}}$. Here, the noninteracting groundstate is characterized by the expectation values $g_{s\tau\mathbf{k}} = 1$ and $P_{s\tau\mathbf{k}} = 0$.

In a weakly interacting system, the Coulomb interaction leads to excitation-dependent renormalizations of the single-particle energy dispersion but it does not change the groundstate expectation values of the interband polarization and the occupation numbers. In contrast, in a strongly correlated system the eigenstates of the noninteracting Hamiltonian are no longer eigenstates of the Coulomb-interacting system and an expansion of the interacting Hamiltonian in terms of the noninteracting basis functions yields nondiagonal contributions that lead to a finite expectation value of $P_{s\tau\mathbf{k}}$ in the interacting groundstate. As a result, the groundstate expectation values of the dynamical variables are no longer determined by the simple relations given above and have to be computed consistently, i.e., at the same level of approximation as the changes induced by an external optical field.

To determine the groundstate in the interacting system, we apply the variational principle

$$\delta\langle \hat{H} \rangle \Big|_{A=0} = 0, \tag{2.13}$$

where $\delta\langle \hat{H} \rangle|_{A=0}$ is the expectation value of the total energy in the absence of external fields. Due to the Coulomb interaction, the expectation value of the Hamiltonian contains the same many-body correlations as the Heisenberg equations of motion and cannot be solved exactly. As the groundstate is a stationary state of the isolated system, the solution of the variational equation is equivalent to determining a stationary solution for the equations of motion without optical field.

2.3.1 *The Dirac-Bloch equations*

In this section, we summarize the Dirac-Bloch equations (DBE) for the massive Dirac Hamiltonian with Coulomb interaction. Using the eigenstates of the single-particle part of the Hamiltonian as a basis, the Heisenberg equations of motion for the dynamical variables $g_{s\tau\mathbf{k}}$ and $P_{s\tau\mathbf{k}}$ are given by

$$i\hbar\frac{\mathrm{d}}{\mathrm{d}t}P_{s\tau\mathbf{k}} = 2\left(\Sigma_{s\tau\mathbf{k}} - \frac{1}{c}\mathbf{A}\cdot\mathbf{j}_{s\tau\mathbf{k}}\right)P_{s\tau\mathbf{k}} - g_{s\tau\mathbf{k}}\Omega_{s\tau\mathbf{k}} - i\hbar\frac{\mathrm{d}}{\mathrm{d}t}P_{s\tau\mathbf{k}}\bigg|_{\mathrm{coll}} \tag{2.14}$$

$$\hbar\frac{\mathrm{d}}{\mathrm{d}t}g_{s\tau\mathbf{k}} = 4\Im\left[P^*_{s\tau\mathbf{k}}\Omega_{s\tau\mathbf{k}}\right] - \hbar\frac{\mathrm{d}}{\mathrm{d}t}g_{s\tau\mathbf{k}}\bigg|_{\mathrm{coll}}. \tag{2.15}$$

As usual, the Coulomb interaction leads to a renormalization of the single particle energy and the Rabi frequency,

$$\begin{aligned}
\Sigma_{s\tau\mathbf{k}} &= \epsilon_{s\tau\mathbf{k}} + \frac{1}{2}\sum_{\mathbf{k}'} V(|\mathbf{k}-\mathbf{k}'|)\Big[W_{cccc}(\mathbf{k},\mathbf{k}') - W_{c\nu\nu c}(\mathbf{k},\mathbf{k}')\Big]g_{s\tau\mathbf{k}'} \\
&\quad + \sum_{\mathbf{k}'} V(\mathbf{k}-\mathbf{k}')\Big[W_{cc\nu c}(\mathbf{k},\mathbf{k}')P_{s\tau\mathbf{k}'} + c.c.\Big], \qquad (2.16)\\
\Omega_{s\tau\mathbf{k}} &= \sum_{\mathbf{k}'} V(|\mathbf{k}-\mathbf{k}'|) \\
&\quad \times \left[W_{cc\nu\nu}(\mathbf{k},\mathbf{k}')P_{s\tau\mathbf{k}'} + W_{c\nu c\nu}(\mathbf{k},\mathbf{k}')P^*_{s\tau\mathbf{k}'} + W_{c\nu\nu\nu}(\mathbf{k},\mathbf{k}')g_{s\tau\mathbf{k}'}\right] \\
&\quad + \tau\sqrt{2}\frac{ev_F}{c}\left(v_k^2 \mathrm{e}^{-i\tau\theta_{\mathbf{k}}}A^{\tau} - u_k^2\mathrm{e}^{i\tau\theta_{\mathbf{k}}}A^{-\tau}\right). \qquad (2.17)
\end{aligned}$$

Here

$$W_{\alpha\alpha'\beta\beta'}(\mathbf{k},\mathbf{k}') = \langle\alpha\mathbf{k}|\alpha'\mathbf{k}'\rangle\langle\beta\mathbf{k}'|\beta'\mathbf{k}\rangle$$

contains the overlap matrix elements between the conduction and valence bands. In Eqs. (2.14) and (2.15), the terms $\mathrm{d}/\mathrm{d}t|_{\mathrm{coll}}$ refer to incoherent scattering contributions beyond the Hartree-Fock approximation and $\mathbf{j}_{s\tau\mathbf{k}} = -\tau\frac{e}{\hbar}\nabla_{\mathbf{k}}\epsilon_{s\tau\mathbf{k}}$ is the intraband current matrix element.

The renormalized single-particle energy and Rabi frequency differ from the standard expressions in the SBE by the Coulomb matrix elements for scattering processes across the bands. Some typical examples are

$$\begin{aligned}
W_{cccc}(\mathbf{k},\mathbf{k}') &= u_k^2u_{k'}^2 + v_k^2v_{k'}^2 + 2u_kv_ku_{k'}v_{k'}\cos(\theta_{\mathbf{k}}-\theta_{\mathbf{k}'}) \\
W_{cc\nu\nu}(\mathbf{k},\mathbf{k}') &= 2u_kv_ku_{k'}v_{k'} + u_k^2u_{k'}^2\mathrm{e}^{i(\theta_{\mathbf{k}}-\theta_{\mathbf{k}'})} + v_k^2v_{k'}^2\mathrm{e}^{-i(\theta_{\mathbf{k}}-\theta_{\mathbf{k}'})} \\
W_{ccc\nu}(\mathbf{k},\mathbf{k}') &= v_ku_{k'}\left(u_ku_{k'}\left(1-\mathrm{e}^{-i(\theta_{\mathbf{k}}-\theta_{\mathbf{k}'})}\right) - v_kv_{k'}\left(1-\mathrm{e}^{i(\theta_{\mathbf{k}}-\theta_{\mathbf{k}'})}\right)\right) \\
W_{c\nu c\nu}(\mathbf{k},\mathbf{k}') &= 2u_kv_ku_{k'}v_{k'} - v_k^2u_{k'}^2\mathrm{e}^{i(\theta_{\mathbf{k}}-\theta_{\mathbf{k}'})} - u_k^2v_{k'}^2\mathrm{e}^{-i(\theta_{\mathbf{k}}-\theta_{\mathbf{k}'})} \qquad (2.18)
\end{aligned}$$

where the first line describes electron-electron scattering, the second line electron-hole interaction, the third line Auger processes, and the last line electron-hole-pair creation and annihilation, respectively. These processes induce coupling between the interband polarization and the population inversion. As a result, the noninteracting groundstate expectation values in general do not solve the equations of motion for the isolated system and thus do not correspond to the true groundstate of the Coulomb interacting system. This can be seen by inserting $g_{s\tau\mathbf{k}} = 1$, $P_{s\tau\mathbf{k}} = 0$ and $\mathbf{A} = 0$ as initial condition into the equation for the interband polarization Eq. (2.14), yielding

$$
\begin{aligned}
i\hbar\frac{\mathrm{d}}{\mathrm{d}t}P_{s\tau\mathbf{k}} &= \sum_{\mathbf{k}'} V(|\mathbf{k}-\mathbf{k}'|)W_{cvvv}(\mathbf{k},\mathbf{k}') \\
&= \sum_{\mathbf{k}'} V(|\mathbf{k}-\mathbf{k}'|)v_k u_{k'}\left(u_k u_{k'} - v_k v_{k'}\right)\left(1-\cos(\theta_{\mathbf{k}}-\theta_{\mathbf{k}'})\right) .
\end{aligned}
$$

Only if $u_k = v_k$, the RHS vanishes and the noninteracting groundstate is a stationary state of the equations without optical field. In a gapped system, the magnitude of the matrix elements decreases with the number of particles n changing bands, and the leading contribution is proportional to v_k^n. Only if $v_k \ll u_k$, i.e. in a wide-gap system, both Auger processes and pair creation and annihilation can be neglected and, apart from a phase factor that can be be removed by a change of variables, the conventional SBE are recovered.

An obvious way to circumvent the difficulties in specifying the initial conditions for the DBE is to use the experimentally observed values for the gap and Fermi-velocity instead of those resulting from bandstructure calculations. While doing so, renormalizations of the single-particle energy and Rabi frequency by groundstate contributions are already contained in the 'bare' single-particle quantities and should be subtracted. Thus, the population inversion $g_{s\tau\mathbf{k}}$ is replaced by $1-\Delta g_{s\tau\mathbf{k}}$ eliminating the Auger contributions resulting from the groundstate. This way, the renormalized quantities contain optically induced interband transition amplitudes only.

2.3.2 *Mass and Fermi-velocity renormalization*

In this section, we want to show how the interacting groundstate can be determined from the equations of motion for the isolated system, i.e., the Coulomb interacting system without optical field, and how this approach is related to the variational principle. Defining the variables

$$\Gamma_{s\tau\mathbf{k}} = f^b_{s\tau\mathbf{k}} - f^a_{s\tau\mathbf{k}} = \langle b^\dagger_{s\tau\mathbf{k}} b_{s\tau\mathbf{k}}\rangle - \langle a^\dagger_{s\tau\mathbf{k}} a_{s\tau\mathbf{k}}\rangle, \tag{2.19}$$

$$\Pi_{s\tau\mathbf{k}} = \langle b^\dagger_{s\tau\mathbf{k}} a_{s\tau\mathbf{k}}\rangle, \tag{2.20}$$

the mean-field expectation value of the Hamiltonian for the isolated system is given by

$$\begin{aligned}\langle H_0 + H_C \rangle = &\\ &\sum_{s\tau\mathbf{k}} \left\{ \frac{\Delta_{s\tau}}{2} \left(f^a_{s\tau\mathbf{k}} - f^b_{s\tau\mathbf{k}} \right) + \tau\hbar v_F k \mathrm{e}^{-i\tau\theta_{\mathbf{k}}} \Pi^*_{s\tau\mathbf{k}} + \tau\hbar v_F k \mathrm{e}^{i\tau\theta_{\mathbf{k}}} \Pi_{s\tau\mathbf{k}} \right\} \\ &- \frac{1}{2} \sum_{s\tau\mathbf{k}\mathbf{k}'} V(\mathbf{k}-\mathbf{k}') \left\{ f^a_{s\tau\mathbf{k}} f^a_{s\tau\mathbf{k}'} + f^b_{s\tau\mathbf{k}} f^b_{s\tau\mathbf{k}'} + 2\Pi^*_{s\tau\mathbf{k}} \Pi_{s\tau\mathbf{k}'} \right\} . \end{aligned} \tag{2.21}$$

From the variational principle, we obtain the equation

$$\begin{aligned} 0 &= \delta \langle H_0 + H_C \rangle \\ &= \sum_{s\tau\mathbf{k}} \left\{ \left(\frac{\Delta_{s\tau}}{2} - \hat{V}[f^a] \right) \delta f^a_{s\tau\mathbf{k}} - \left(\frac{\Delta_{s\tau}}{2} + \hat{V}[f^b] \right) \delta f^b_{s\tau\mathbf{k}} \right\} \\ &+ \sum_{s\tau\mathbf{k}} \left\{ \left(\tau\hbar v_F k \mathrm{e}^{-i\tau\theta_{\mathbf{k}}} - \hat{V}[\Pi] \right) \delta\Pi^*_{s\tau\mathbf{k}} + \left(\tau\hbar v_F k \mathrm{e}^{i\tau\theta_{\mathbf{k}}} - \hat{V}[\Pi^*] \right) \delta\Pi_{s\tau\mathbf{k}} \right\} \end{aligned} \tag{2.22}$$

where we defined the functional relation

$$\hat{V}[f] \equiv \sum_{\mathbf{k}'} V(\mathbf{k}-\mathbf{k}') f(\mathbf{k}').$$

Clearly, f^a, f^b, Π and Π^* cannot be varied independently of each other and we have to supply some additional conditions. In order to be radiatively stable, the macroscopic optical polarization $\sum_{s\tau\mathbf{k}} \left(\Pi_{s\tau\mathbf{k}} + \Pi^*_{s-\tau\mathbf{k}} \right) \boldsymbol{e}_\tau$ associated with the groundstate must vanish. Furthermore, the groundstate clearly has to be a stationary state. Therefore, the expectation value of any operator in this state has to be static such that the time derivative of the basic variables vanishes. Thus, we impose the stationary solution of the equations of motion as constraints to assure a steady state.

To this end, we derive the equations of motion for the variational variables in the absence of an external optical field in the form

$$\begin{aligned} i\hbar\partial_t \Pi_{s\tau\mathbf{k}} &= \left(\Delta + \hat{V}[\Gamma] \right) \Pi_{s\tau\mathbf{k}} + \left(\tau\hbar v_F k \mathrm{e}^{-i\tau\theta_{\mathbf{k}}} - \hat{V}[\Pi] \right) \Gamma_{s\tau\mathbf{k}} \\ i\hbar\partial_t f^a_{s\tau\mathbf{k}} &= -i\hbar\partial_t f^b_{s\tau\mathbf{k}} = -\Pi_{s\tau\mathbf{k}} \left(\tau\hbar v_F k \mathrm{e}^{i\tau\theta_{\mathbf{k}}} - \hat{V}[\Pi^*] \right) - c.c. \end{aligned}$$

Here, we again made the dynamical Hartree-Fock approximation to be consistent with the equations derived in the previous section. The groundstate distributions are then obtained from the steady-state condition

$$0 = \tilde{\Delta}_{s\tau\mathbf{k}} \Pi_{s\tau\mathbf{k}} + \tau\hbar\tilde{v}_F k \mathrm{e}^{-i\tau\theta_{\mathbf{k}}} \Gamma_{s\tau\mathbf{k}} \tag{2.23}$$

$$0 = 2\Im\left[\Pi_{s\tau\mathbf{k}}\tau\hbar\tilde{v}_F e^{i\tau\theta_\mathbf{k}}\right], \tag{2.24}$$

where we defined the renormalized gap and Fermi-velocity by

$$\tilde{\Delta}_{s\tau\mathbf{k}} = \Delta_{s\tau} + \hat{V}[\Gamma] \tag{2.25}$$

and

$$\tau\hbar\tilde{v}_{s\tau\mathbf{k}}k e^{-i\tau\theta_\mathbf{k}} = \tau\hbar v_F k e^{-i\tau\theta_\mathbf{k}} - \hat{V}[\Pi]. \tag{2.26}$$

Together with the relation $\Gamma^2_{s\tau\mathbf{k}} = 1-4\left|\Pi_{s\tau\mathbf{k}}\right|^2$, which follows from the equations of motion and is valid for any coherent state, this yields the algebraic relations

$$\Pi_{s\tau\mathbf{k}} = -\frac{\tau\hbar\tilde{v}_{s\tau\mathbf{k}}k}{2\varepsilon_{s\tau\mathbf{k}}}e^{-i\tau\theta_\mathbf{k}}, \tag{2.27}$$

$$\Gamma_{s\tau\mathbf{k}} = \frac{\tilde{\Delta}_{s\tau\mathbf{k}}}{2\varepsilon_{s\tau\mathbf{k}}}, \tag{2.28}$$

with

$$\varepsilon_{s\tau\mathbf{k}} = \sqrt{\left(\frac{\tilde{\Delta}_{s\tau\mathbf{k}}}{2}\right)^2 + \left(\hbar\tilde{v}_{s\tau\mathbf{k}}k\right)^2}. \tag{2.29}$$

Inserting these relations into the definitions of the renormalized gap and Fermi-velocity yields the integral equations

$$\tau\hbar\tilde{v}_{s\tau\mathbf{k}}k = \tau\hbar v_F k + \frac{1}{2}\sum_{\mathbf{k}'}V(\mathbf{k}-\mathbf{k}')\frac{\tau\hbar\tilde{v}_{s\tau\mathbf{k}'}k'}{\varepsilon_{s\tau\mathbf{k}'}}e^{i\tau(\phi_\mathbf{k}-\phi'_\mathbf{k})}, \tag{2.30}$$

$$\tilde{\Delta}_{s\tau\mathbf{k}} = \Delta_{s\tau} + \frac{1}{2}\sum_{\mathbf{k}'}V(\mathbf{k}-\mathbf{k}')\frac{\tilde{\Delta}_{s\tau\mathbf{k}'}}{\varepsilon_{s\tau\mathbf{k}'}} \tag{2.31}$$

that determine the renormalized Fermi-velocity and the gap within the mean-field approximation. [Stroucken *et al.* (2011); Sabio *et al.* (2010)]

Clearly, the renormalized Fermi-velocity is real and the macroscopic optical polarization vanishes due to the angle dependence of Eq. (2.27). As can be recognized, the renormalizations always occur simultaneously except for the special case where the noninteracting gap vanishes. In this case, Eq. (2.31) always has the trivial solution $\tilde{\Delta}_{s\tau\mathbf{k}} = 0$ and the Coulomb interaction only leads to a renormalization of the Fermi-velocity. In the strict 2D limit, this renormalization is logarithmically divergent for small values of k in agreement with predictions based on the renormalization group theory. [Gonzalez *et al.* (1999); Sheehy and Schmalian (2007); Fritz *et al.* (2008); Sinner and Ziegler (2010)] In general, the integrals on the RHS of Eqs. (2.30) and (2.31) are divergent in the strict 2D limit and the inclusion of the form factor $F(qd)$ is crucial to obtain finite results. [Stroucken *et al.* (2011); Grönqvist *et al.* (2012)]

A particularly interesting situation arises if we have no noninteracting gap and the gap equation (2.31) has a nontrivial solution. In this case Eq. (2.31) predicts the opening of a gap with a finite expectation value for $\Gamma_{s\tau\mathbf{k}} = f^b_{s\tau\mathbf{k}} - f^a_{s\tau\mathbf{k}}$. As the gapless single-particle dispersion corresponds to equivalent sublattices A and B, a finite expectation value for $\Gamma_{s\tau\mathbf{k}}$ in the groundstate corresponds to a spontaneous symmetry breaking where the electron density has a lower symmetry than the lattice. However, this will happen for each K-point separately. Since the contributions from the different K-points are opposite in sign, the combined electron density of both K-points still has the full lattice symmetry. In section 2.4.1, we will investigate the conditions for this phase transition in detail.

The stationary solutions of Eqs. (2.30) and (2.31) correspond to eigenstates of the mean-field Hamitonian

$$H_0 = \sum_{s\tau,\mathbf{k}} \hat{\Psi}^\dagger_{s\tau\mathbf{k}} \left(a\tilde{t\mathbf{k}} \cdot \hat{\sigma}_\tau + \frac{\tilde{\Delta}_{s\tau\mathbf{k}}}{2} \hat{\sigma}_z \right) \hat{\Psi}_{s\tau\mathbf{k}} \tag{2.32}$$

with the eigenvalues $\pm\varepsilon_{s\tau\mathbf{k}}$. The corresponding eigenstates are given by Eq. (2.6), where the noninteracting gap and Fermi-velocity are replaced by the renormalized values, and the groundstate is characterized by a completely filled valence and empty conduction band, respectively. As $\varepsilon_{s\tau\mathbf{k}} > \epsilon_{s\tau\mathbf{k}}$, the total energy lies below the energy of the noninteracting groundstate. Similar to the case where we replace the single-particle energies by the experimental values, the SBE in the renormalized bands correspond to those derived in the previous section where the single-particle energy and the Fermi-velocity have been replaced by the respective renormalized values. The additional renormalizations result from the optically induced changes to the groundstate.

2.3.3 *Linear optical response: Dirac excitons*

Taking the weakly interacting groundstate as initial state, the equation for the linear interband transition amplitude is given by

$$\begin{aligned} i\hbar\frac{\mathrm{d}}{\mathrm{d}t}P_{s\tau\mathbf{k}} &= 2\varepsilon_{s\tau\mathbf{k}}P_{s\tau\mathbf{k}} \\ &- \sum_{\mathbf{k}'} V(|\mathbf{k}-\mathbf{k}'|)\left[W_{cc\nu\nu}(\mathbf{k},\mathbf{k}')P_{s\tau\mathbf{k}'} + W_{c\nu c\nu}(\mathbf{k},\mathbf{k}')P^*_{s\tau\mathbf{k}'}\right] \\ &- \sqrt{2}\tau\frac{ev_F}{c}\left(v_k^2 \mathrm{e}^{-i\tau\theta_\mathbf{k}}A^\tau - u_k^2 \mathrm{e}^{i\tau\theta_\mathbf{k}}A^{-\tau}\right) - i\hbar\gamma P_{s\tau\mathbf{k}} \end{aligned} \tag{2.33}$$

whereas the populations are at least second order in the exciting field. For simplicity, we treat here the effects of the collision terms as a phenomenological dephasing. The macroscopic optical polarization is obtained from the microscopic transition amplitudes via

$$j^\tau = \tau\sqrt{2}ev_F \sum_{s\mathbf{k}} \mathrm{e}^{i\tau\theta_\mathbf{k}} \left\{u_k^2\left(P_{s-\tau\mathbf{k}} - P^*_{s\tau\mathbf{k}}\right) + v_k^2\left(P_{s\tau\mathbf{k}} - P^*_{s-\tau\mathbf{k}}\right)\right\} \quad (2.34)$$

and contains a phase factor resulting from the flip of the pseudospin. In the frequency domain, the linear optical polarization is related to the vector potential via $j^\tau = \frac{\omega^2}{c}\chi^\tau(\omega)A^\tau(\omega)$ and the absorption is given by [Stroucken *et al.* (2012)]

$$\alpha(\omega) = \frac{\frac{4\pi\omega}{c}\Im[\chi(\omega)]}{\left|1 - i\frac{2\pi\omega}{c}\chi(\omega)\right|^2}. \quad (2.35)$$

As can be recognized from Eq. (2.34), in a gapped system with $u_k > v_k$, the resonant contributions to the optical polarization predominantly stem from the $K^{-\tau}$ valley whereas in graphene both valleys contribute equally to the linear absorption.

To obtain the linear response, we expand the optical polarization in terms of the eigenfunctions of the Dirac-Wannier equation

$$2\varepsilon_{s\tau\mathbf{k}}\phi_{s\tau\lambda}(\mathbf{k}) - \sum_{\mathbf{k}'} V(|\mathbf{k}-\mathbf{k}')\left[W_{cc\nu\nu}(\mathbf{k},\mathbf{k}')\phi_{s\tau\lambda}(\mathbf{k}') + W_{c\nu c\nu}(\mathbf{k},\mathbf{q})\phi^*_{s\tau\lambda}(\mathbf{k}')\right]$$
$$= E_{s\tau\lambda}\phi_{s\tau\lambda}(\mathbf{k})\,. \quad (2.36)$$

Here, the optical field couples to the p-type parts of the eigenstates of the Dirac-Wannier equation. Equation (2.36) is the relativistic analogue of the Wannier equation known from semiconductors. The coupling between resonant and non-resonant states corresponds to the coupling between the positive and negative energy solutions in relativistic quantum mechanics.

Usually, the solutions of the Wannier equation with a sperically symmetric potential can be chosen as eigenstates of the z-component of the angular momentum operator $\hat{L}_z$. However, for the chiral particles considered here, the spherically symmetric Hamiltonian only commutes with the z-component of the total angular momentum $\hat{J}_z$ which includes the pseudospin of the quasiparticles. Nevertheless, one can define the projection onto the p-type contributions

$$\phi^\pm_{s\tau\lambda}(\mathbf{k}) = \int \frac{\mathrm{d}\theta_\mathbf{k}}{2\pi}\mathrm{e}^{\mp i\tau\theta_\mathbf{k}}\phi_{s\tau\lambda}(k). \quad (2.37)$$

With the aid of the projected wave functions, we derive the generalized Elliot formula for the the linear susceptibility as [Stroucken and Koch (2015)]

$$\chi^\tau(\omega) = -\frac{2e^2v_F^2}{\omega^2}\sum_{s,\lambda}\left(\frac{\left|\sum_\mathbf{k} u_\mathbf{k}^2\phi^-_{s-\tau\lambda}(k)\right|^2}{\hbar(\omega+i\gamma) - E_{s-\tau\lambda}} + \frac{\left|\sum_\mathbf{k} v_\mathbf{k}^2\phi^-_{s\tau\lambda}(k)\right|^2}{\hbar(\omega+i\gamma) - E_{s\tau\lambda}}\right)$$
$$+\frac{2e^2v_F^2}{\omega^2}\sum_{s,\lambda}\left(\frac{\left|\sum_\mathbf{k} v_\mathbf{k}^2\phi^+_{s-\tau\lambda}(k)\right|^2}{\hbar(\omega+i\gamma) + E_{s-\tau\lambda}} + \frac{\left|\sum_\mathbf{k} u_\mathbf{k}^2\phi^+_{s\tau\lambda}(k)\right|^2}{\hbar(\omega+i\gamma) + E_{s\tau\lambda}}\right). \quad (2.38)$$

2.3.3.1 *Graphene*

For graphene, the overlap matrix elements are given by $W_{cc\nu\nu}(\mathbf{k}, \mathbf{k}') = (1 + \cos(\theta_\mathbf{k} - \theta'_\mathbf{k}))/2$ and $W_{c\nu\nu c} = (1 - \cos(\theta_\mathbf{k} - \theta'_\mathbf{k}))/2$, respectively. Thus, the Coulomb interaction couples the eigenstates of $\hat{L}_z$ with $m = \pm 1$ and we make the ansatz

$$\phi_{s\tau\lambda}(\mathbf{k}) = \phi^+_{s\tau\lambda}(\mathbf{k})\mathrm{e}^{i\tau\theta_\mathbf{k}} + \phi^-_{s\tau\lambda}(\mathbf{k})\mathrm{e}^{-i\tau\theta_\mathbf{k}}.$$

As can be shown by scaling arguments, in the strict 2D limit the Wannier equation has no intrinsic length scale and the functions $\phi^\pm$ can only be normalization constants. Inserting the ansatz into the Wannier equation for any fixed value of $\mathbf{k}$, we find even and odd solutions with $\phi^+ = \pm\phi^- = \frac{1}{\sqrt{2\mathcal{N}}}$ with $\mathcal{N} = \sum_\mathbf{k}$ and the eigenvalues

$$E_k^{even} = 2\hbar\tilde{v}_{s\tau\mathbf{k}}k - \sum_{\mathbf{k}'} V(\mathbf{k} - \mathbf{k}')\cos(\theta_\mathbf{k} - \theta_{\mathbf{k}'}) = 2\hbar v_F k, \tag{2.39}$$

$$\begin{aligned} E_k^{odd} &= 2\hbar\tilde{v}_{s\tau\mathbf{k}}k - \sum_{\mathbf{k}'} V(\mathbf{k} - \mathbf{k}')\cos^2(\theta_\mathbf{k} - \theta_{\mathbf{k}'}) \\ &= 2\hbar v_F k + \sum_{\mathbf{k}'} V(\mathbf{k} - \mathbf{k}')\cos(\theta_\mathbf{k} - \theta_{\mathbf{k}'})\,(1 - \cos(\theta_\mathbf{k} - \theta_{\mathbf{k}'}))\,. \end{aligned} \tag{2.40}$$

Interestingly, for the even solution, the Coulomb modifications of the energy eigenvalues cancel exactly with the renormalization of the Fermi velocity. The resulting free-carrier dispersion leads to a constant and universal contribution to the optical absorption which is independent of any material parameter. In contrast, the eigenvalue of the odd solution contains not only the renormalized Fermi velocity but also an additional renormalization. While the renormalized Fermi velocity is logarithmically divergent in the strict 2D limit, the additional renormalization diverges linearly. Using the inverse effective sheet thickness $1/d$ as cut-off parameter, the Coulomb integral can be evaluated, yielding

$$\begin{aligned} E_k^{odd} &= \frac{\hbar v_F}{d}\left\{2kd\left[1 + \alpha\left(\frac{C}{2} - \frac{1}{4}\ln kd\right)\right] - \frac{\alpha}{2}\right\} \\ &= 2\hbar v_F^*(\alpha, k)k - \frac{e^2}{2\kappa d} \end{aligned} \tag{2.41}$$

with

$$C = \frac{1}{2\pi}\left[\frac{1}{3}I_1 + \sum_{n=2}^{\infty}\frac{2n+1}{(n+2)(n-1)}I_n\right] \approx -0.1408, \tag{2.42}$$

$$I_n = \int d\theta P_n(\cos\theta)\cos\theta\,(1 - \cos\theta)\,. \tag{2.43}$$

The first line of Eq. (2.41) shows that the energy dispersion scales strictly with the graphene energy unit $E_G = \hbar v_F/d$ and can be considered a function of the dimensionless variable kd and the fine structure constant α. The second line of Eq. (2.41) allows for a simple interpretation in terms of a Fermi-velocity renormalization and a constant energy shift that can be understood as the electrostatic energy of two interacting electrons separated by $d/2$. It is notable, that the magnitude of the energy shift is half of the mass or band-edge renormalization in a wide-gap system but opposite in sign, thus suggesting the interpretation as a 'negative' bandgap renormalization.

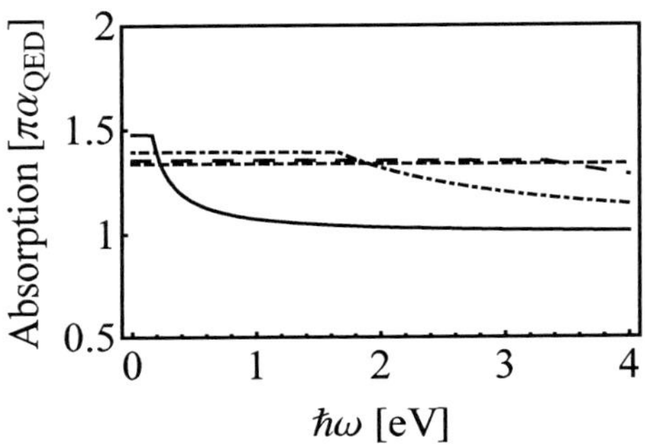

Fig. 2.4 Linear absorption spectrum of graphene in the low energy range for $\alpha = 0.01$ (solid line), $\alpha = 0.1$ (dot-dashed line), $\alpha = 0.2$ (long-dashed line), and $\alpha = 0.5$ (short-dashed line) respectively. The horizontal axis presents the excitation energy in eV, using a sheet thickness $d = 1.8$ Å, see text for explanation.

Inserting the solutions and eigenvalues into the Elliot formula (2.38), we find the linear absorption coefficient

$$\alpha(\omega) = \frac{\pi\alpha_{QED}}{2}\left[1 + \left(\frac{v_F}{v_F^*}\right)^2 \frac{\hbar\omega + e^2/2\kappa d}{\hbar\omega} + \left(\frac{v_F}{v_F^*}\right)^2 \frac{\hbar\omega - e^2/2\kappa d}{\hbar\omega}\theta(e^2/2\kappa d - \hbar\omega)\right], \tag{2.44}$$

where the renormalized Fermi velocity has to be evaluated at the poles of the susceptibility. In Eq. (2.44), the second contribution stems from the resonant part of the odd solution whereas the last contribution corresponds to the nonresonant part, respectively. If the excitation energy is below the negative bandgap, i.e. $\hbar\omega < \alpha E_G/2$, the combined resonant and nonresonant contributions give an enhanced absorption with slight deviations from a constant value due to the renormalization of the Fermi velocity. For higher energies, the absorption decreases towards the free-particle absorption.

2.3.3.2 *Gapped transition metal dichalcogenides*

If the gap is large enough to neglect pair contributions $\propto v_k^2$, the coupling between the different angular momentum states becomes negligible and $\phi^+_{s\tau\lambda}(\mathbf{k})$ and $\phi^-_{s\tau\lambda}(\mathbf{k})$ are independent though nondegenerate eigenstates of the Wannier equation. The relevant eigenfunctions for the optically active states obey the equation

$$2\varepsilon_{s\tau\mathbf{k}}\phi^+_{s\tau\lambda}(\mathbf{k}) - \sum_{\mathbf{k}'} V(|\mathbf{k}-\mathbf{k}')\phi^+_{s\tau\lambda}(\mathbf{k}-\mathbf{q}) = E^+_{s\tau\lambda}\phi^+_{s\tau\lambda}(\mathbf{k}), \tag{2.45}$$

and equivalently for $\phi^-_{s-\tau\lambda}(\mathbf{k})$. Equation (2.45) is the hydrogen equation with relativistic dispersion. If the exciton wavefunction is strongly localized in k space, a Taylor expansion of the single-particle dispersion gives the nonrelativistic hydrogen equation and the corresponding Rydberg series of the optically active states. However, it should be noted that the exciton binding energy is proportional to the renormalized reduced mass, i.e., the renormalized gap, rather than to the bare mass. This is qualitatively different to what is known from GaAs-like semiconductors where the renormalization of the single-particle energy leads to a shift but does not affect the exciton binding.

The fully relativistic Wannier equation corresponds to the k-space representation of the Dirac Coulomb problem. In real space, the wavefunctions are two-component spinors [Pereira *et al.* (2008); Rodin and Castro Neto (2013)]

$$\hat{\Psi}^{s\tau}_{jn}(\mathbf{r}) = \begin{pmatrix} e^{i(j-\tau/2)\varphi}A^{s\tau}_{jn}(r) \\ e^{i(j+\tau/2)\varphi}B^{s\tau}_{jn}(r) \end{pmatrix}$$

obeying the wave equation

$$\left(2v_F\hat{\sigma}\cdot\mathbf{p} + \Delta_{s\tau}\hat{\sigma}_z - \frac{e^2}{r}\right)\Psi^{s\tau}_{jn}(\mathbf{r}) = E^{s\tau}_{jn}\Psi^{s\tau}_{jn}(\mathbf{r}). \tag{2.46}$$

Here j is the eigenvalue of the total angular momentum $\hat{L}_z + \frac{\tau}{2}\hat{\sigma}_z$, and n is the principle quantum number. Thus, we can identify the set of quantum numbers $\{\lambda\} = (j,n)$ such that, e.g., an s -type wave function has $j = \pm 1/2$. The eigenfunctions $\phi^+_{s\tau jn}$ of Eq. (2.45) correspond to the solutions of Eq. (2.46) with positive energy and are given by

$$\begin{aligned}\phi^+_{s\tau jn}(\mathbf{k}) &= u^2_{\mathbf{k}}e^{i(j-\tau/2)\theta_{\mathbf{k}}}A^{s\tau}_{jn}(k) + v^2_{\mathbf{k}}e^{i(j+\tau/2)\theta_{\mathbf{k}}}B^{s\tau}_{jn}(k)\\ &\approx e^{i(j-\tau/2)\theta_{\mathbf{k}}}A^{s\tau}_{jn}(k),\end{aligned} \tag{2.47}$$

while the solutions corresponding to negative energies are of no relevance here.

For the 2D Coulomb interaction potential, the eigenvalues of the relativistic hydrogen problem are given by [Pereira *et al.* (2008); Rodin and Castro Neto

(2013)].

$$E_{nj}^{s\tau} = \tilde{\Delta}_{s\tau} \left\{ \frac{n + \sqrt{j^2 - \left(\frac{\alpha}{2}\right)^2}}{\sqrt{\left(\frac{\alpha}{2}\right)^2 + \left(n + \sqrt{j^2 - \left(\frac{\alpha}{2}\right)^2}\right)^2}} \right\} \tag{2.48}$$

with $n = 0, 1, \ldots$ and $j = \pm 1/2, \pm 3/2, ...$ For small values of α, the eigenvalue spectrum reduces to that of the nonrelativistic Rydberg series

$$\begin{aligned} E_{nj}^{s\tau} &= \Delta_{s\tau} - \alpha^2 \Delta_{s\tau}/8(n + |j|)^2 \\ &= \Delta_{s\tau} - \alpha^2 \Delta_{s\tau}/8(N - 1/2)^2 \end{aligned} \tag{2.49}$$

with $N = n + |j| + 1/2 = 1, 2, \ldots.$

Interestingly, if $\alpha > 1$, the eigenvalues with $j = \pm 1/2$ are complex and therefore do not correspond to stable resonances in the optical spectrum. At the same time, the real part of the lowest eigenvalue vanishes , i.e., it is degenerate with the noninteracting groundstate. Thus, the value $\alpha = 1$ marks a critical value, separating the weakly and strongly interacting regimes.

2.4 Regime of Strong Coulomb Coupling

2.4.1 *Excitonic instability of the weakly interacing groundstate*

As pointed out before, a nontrivial solution of the gap equation for graphene leads to the opening of a gap in the single-particle dispersion. The opening of a gap is associated with a phase transition into a strongly correlated state, which is known as excitonic insulator. This excitonic insulator state resembles in many ways a Bose-Einstein condensate (BEC) of excitons. Such an excitonic BEC has been predicted already in the 1960's as it was realized that an exciton binding energy exceeding the value of the noninteracting gap signals the possibility for the system to lower its energy by the spontaneous formation of excitons [Keldysh and Kopaev (1965); Keldysh and Kozlov (1968); Jerome *et al.* (1967); Halperin and Rice (1968)]. In this section, we show that a convenient criterion for the phase transition into an excitonic insulator state is the existence of negative or zero energy solutions for the relativistic Wannier equation and not only occurs in a gapless system but also in a gapped TMDC.

To this end, we consider the infinitesimal transformation $\hat{U}(\beta) = \exp(\beta \hat{B}_{s\tau\lambda}^{\dagger})$ generated by the exciton operator $\hat{B}_{s\tau\lambda}^{\dagger} = \sum_{\mathbf{k}} \phi_{s\tau\lambda}(\mathbf{k}) c_{s\tau\mathbf{k}}^{\dagger} \nu_{s\tau\mathbf{k}}$. The operator $\hat{B}_{s\tau\lambda}^{\dagger}$ creates an electron–hole pair with zero center-of-mass momentum in the state described by the normalized wavefunction $\phi_{s\tau\lambda}(\mathbf{k})$. Applying $\hat{U}$ to the weakly interacting groundstate, the infinitesimal transformation yields an energy shift

$$\delta E = \beta^2 \Bigg(\sum_{\mathbf{k}} 2\varepsilon_{s\tau\mathbf{k}} \, |\phi_{s\tau\lambda}(\mathbf{k})|^2 - \sum_{\mathbf{k}\mathbf{k}'} \phi^*_{s\tau\lambda}(\mathbf{k}) V(\mathbf{k}-\mathbf{k}') W_{cc\nu\nu}(\mathbf{k},\mathbf{k}') \, \phi_{s\tau\lambda}(\mathbf{k}') \\ - \frac{1}{2} \sum_{\mathbf{k}\mathbf{k}'} V(\mathbf{k}-\mathbf{k}') \Big[W_{c\nu c\nu}(\mathbf{k},\mathbf{k}') \phi^*_{s\tau\lambda}(\mathbf{k}) \phi^*_{s\tau\lambda}(\mathbf{k}') + c.c. \Big] \Bigg) \\ + \mathcal{O}(\beta^4), \tag{2.50}$$

where the renormalized single-particle energy corresponds to that of the weakly interacting system. To lowest order, the energy shift is proportional to the created exciton density β^2. Subsequent variation of $E/\beta^2 - E_{s\tau\lambda} \sum_{\mathbf{k}} \phi_{s\tau\lambda}(\mathbf{k})^2$ yields the Dirac-Wannier equation (2.36). In the limit of vanishing density, the system energy is $E = \beta^2 E_{s\tau\lambda} = N E_{s\tau\lambda}$ where $N = \beta^2$ is the total number of pairs in the low density limit. Hence, if the Wannier equation allows for negative energy solutions with $E_{s\tau\lambda} < 0$, the system can gain energy by forming bound electron-hole-pair states.

While we searched for p-type solutions of the Dirac-Wannier equation for the optically active states in the weakly coupled regime, for the spontaneously formed groundstate excitons we are not restricted by such symmetry considerations and we may look for the energetically lowest exciton. As the coupling of the bands increases with decreasing gap, we may expect a qualitative dependence of the critical value of α on the size of the gap. [Pereira *et al.* (2008)] In the following, we will focus on the two limiting cases (i) where the gap is wide enough to neglect the band-overlap which applies for TMDCs and (ii) the graphene case where the noninteracting gap vanishes.

For the gapped TMDCs, the optically active s-type eigenstates of the relativistic hydrogen equation indeed correspond to the lowest lying eigenstates of the Dirac Wannier equation. Thus, for TMDCs, we obtain a critical value $\alpha_c = 1$, where the lowest lying bright excitons are degenerate with the noninteracting groundstate.

For graphene, the lowest bound exciton corresponds to a real isotropic configuration for which the Wannier equation reduces to

$$2\varepsilon_{s\tau\mathbf{k}} \, \phi_{s\tau\lambda}(\mathbf{k}) - \sum_{\mathbf{k}'} V(\mathbf{k}-\mathbf{k}') \phi_{s\tau\lambda}(\mathbf{k}') = E_\lambda \, \phi_{s\tau\lambda}(\mathbf{k}). \tag{2.51}$$

In principle, this equation is equivalent to the relativistic hydrogen equation for the optically active states in TMDCs with $\Delta = 0$ and naively, one might expect a critical value $\alpha_c = 1$ as well. However, as the energy levels of all eigenstates of the gapless hydrogen equation are degenerate with the noninteracting groundstate, the collapse of the lowest 1s state is not a necessary condition for the opening of a gap. Therefore, we determined the critical value of α for graphene by numerically

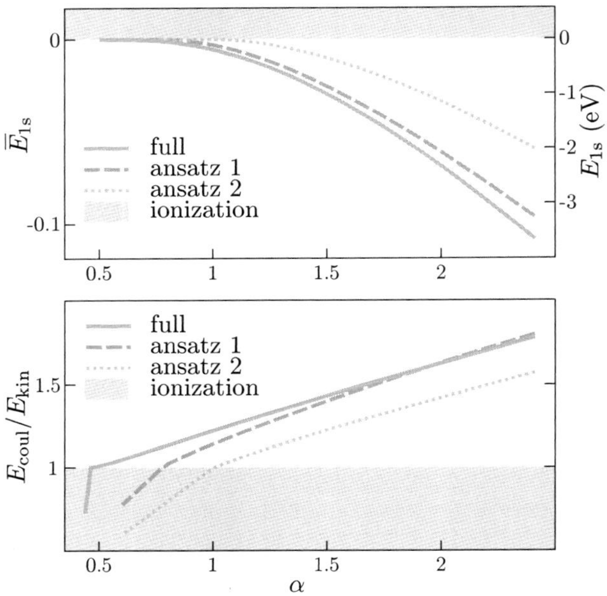

Fig. 2.5 (top) Electron–hole pair binding energy vs. coupling strength α. The left axis presents the scaled energies and the right axis gives the energy in eV, calculated using $d = 0.18\,\text{Å}$. The full numerical solution (solid line) is compared with results obtained using the ansatz wave functions $f(y) \propto (1+y^2)^{-3/2}$ (dashed line) and $f(y) \propto e^{-y^2}$ (dotted line). (bottom) Ratio of the Coulomb and kinetic energy vs. α. The shaded areas indicate the region of ionized solutions. Reprinted (figure 2) from [J.H. Grönqvist, T.Stroucken, M. Lindberg, and S.W. Koch, "Wannier excitons signalling strong Coulomb coupling in graphene", *Eur. Phys. J. B* **85**, 395 (2012).] Copyright (2012) by The European Physical Journal. With kind permission of The European Physical Journal (EPJ). (For color reference, go to page 485.)

solving the Wannier equation with linear dispersion. [Grönqvist *et al.* (2012)] In order to obtain finite results, we included a form factor accounting for the finite extension of the carbon wave function in the z-direction. As a result, all energies scale strictly with the graphene energy unit E_G. In Fig. 2.5, the resulting scaled lowest eigenvalue E_{1s} is shown as a function of the coupling strength α. Here, $\alpha = 2.4$ corresponds to the nominal value of the fine structure constant calculated with the experimentally observed Fermi velocity $v_F = 9.07 \cdot 10^5$ m/s without screening ($\kappa = 1$). The numerical results are compared to analytical estimates using a variational ansatz with a single parameter trial wavefunction $\bar{\phi}(\bar{\mathbf{k}}) \to f(y)$. Fig. 2.5 shows the results for $f(y) = \sqrt{8\pi}(1+y^2)^{-3/2}$ giving an ionization threshold

$\alpha_c = \pi/4$ (dashed line) and $f(y) = \sqrt{8\pi}e^{-y^2}$ with $\alpha_c = 1$ (dotted line). For the full numerical solution, a critical value $\alpha_c \approx 0.46$ is found, in good agreement with the analytical values in Ref. [Pereira *et al.* (2008)].

2.4.2 *Determination of the strongly interacting groundstate: Gap equations*

As we have seen in the previous section, the existence of a critical α_c is an inherent property of the relativistic exciton problem and we expect a phase transition into an excitonic groundstate above this critical value. To determine the groundstate properties, we proceed as in the weakly interacting case and solve the steady state DBE without optical field. The derivations follow exactly the steps of Section 2.3.2, yielding

$$0 = 2\bar{\Sigma}_{s\tau\mathbf{k}}\bar{P}_{s\tau\mathbf{k}} - \bar{g}_{s\tau\mathbf{k}}\bar{\Omega}_{s\tau\mathbf{k}} \tag{2.52}$$

$$0 = 4\Im\left[\bar{P}^*_{s\tau\mathbf{k}}\bar{\Omega}_{s\tau\mathbf{k}}\right]. \tag{2.53}$$

which we combine with $\bar{g}^2_{s\tau\mathbf{k}} = 1 - 4\left|\bar{P}_{s\tau\mathbf{k}}\right|^2$, to obtain the algebraic relations

$$\bar{P}_{s\tau\mathbf{k}} = \frac{\bar{\Omega}_{s\tau\mathbf{k}}}{2\mathcal{E}_{s\tau\mathbf{k}}}, \tag{2.54}$$

$$\bar{g}_{s\tau\mathbf{k}} = \frac{\bar{\Sigma}_{s\tau\mathbf{k}}}{\mathcal{E}_{s\tau\mathbf{k}}}, \tag{2.55}$$

with

$$\mathcal{E}_{s\tau\mathbf{k}} = \sqrt{\bar{\Sigma}^2_{s\tau\mathbf{k}} + \bar{\Omega}^2_{s\tau\mathbf{k}}}. \tag{2.56}$$

Inserting these into the definitions of the renormalized energy and Rabi frequency yields the gap equations

$$\begin{aligned}\bar{\Sigma}_{s\tau\mathbf{k}} = \varepsilon_{s\tau\mathbf{k}} + \frac{1}{2}\sum_{\mathbf{k}'} V(|\mathbf{k}-\mathbf{k}'|)\left[W_{cccc}(\mathbf{k},\mathbf{k}') - W_{c\nu\nu c}(\mathbf{k},\mathbf{k}')\right]\\ \times\left(1 - \frac{\bar{\Sigma}_{s\tau\mathbf{k}'}}{\mathcal{E}_{s\tau\mathbf{k}'}}\right)\end{aligned} \tag{2.57}$$

$$\bar{\Omega}_{s\tau\mathbf{k}} = \frac{1}{2}\sum_{\mathbf{k}'} V(|\mathbf{k}-\mathbf{k}'|)\left[W_{cc\nu\nu}(\mathbf{k},\mathbf{k}')\frac{\bar{\Omega}_{s\tau\mathbf{k}'}}{\mathcal{E}_{s\tau\mathbf{k}'}} + W_{c\nu c\nu}(\mathbf{k},\mathbf{k}')\frac{\bar{\Omega}^*_{s\tau\mathbf{k}'}}{\mathcal{E}_{s\tau\mathbf{k}'}}\right] \tag{2.58}$$

where we omitted the Auger contribtions. In graphene, the Auger contributions vanish exactly for the assumed real, isotropic groundstate, whereas for TMDCs Auger contributions are generally small.

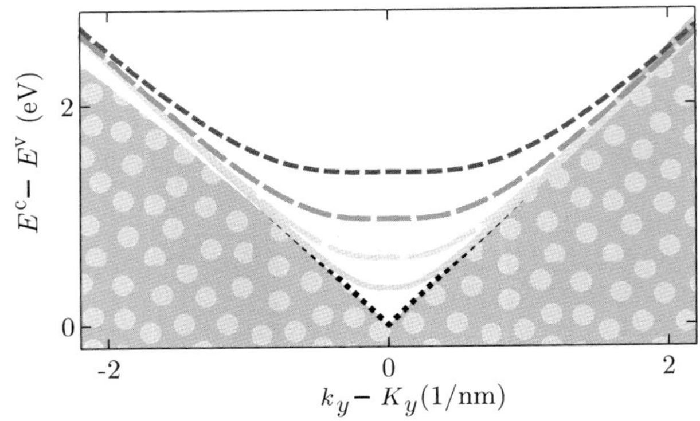

Fig. 2.6 Dependence of the quasiparticle bandstructure $E^{\mathrm{c}}_{\mathbf{k}} - E^{\mathrm{v}}_{\mathbf{k}}$ on the effective fine-structure constant α_G. The curves shown are for $\alpha_G = 0.9$ (solid yellow line), 1.0 (long-dashed, green line), 1.1 (medium-dashed, blue line), and 1.2 (short-dashed, purple line), together with the linear tight-binding dispersion (dotted black line). The shaded, dotted area shows the full tight-binding dispersion. Reprinted (figure 9) with permission from [T. Stroucken, J. H. Grönqvist, and S. W. Koch, *Phys. Rev. B* **84**, 205445 (2011).] Copyright (2011) by the American Physical Society. (For color reference, go to page 485.)

The solution of the gap equation defines the mean-field Hamiltonian

$$\hat{H}^{\mathrm{MF}} = \sum_{s\tau\mathbf{k}} \bar{\Sigma}_{s\tau\mathbf{k}} \left(c^{\dagger}_{s\tau\mathbf{k}} c_{s\tau\mathbf{k}} - \nu^{\dagger}_{s\tau\mathbf{k}} \nu_{s\tau\mathbf{k}} \right) + \bar{\Omega}_{s\tau\mathbf{k}} c^{\dagger}_{s\tau\mathbf{k}} \nu_{s\tau\mathbf{k}} + \bar{\Omega}^{*}_{s\tau\mathbf{k}} \nu^{\dagger}_{s\tau\mathbf{k}} c_{s\tau\mathbf{k}}.$$

The groundstate of the interacting mean-field Hamiltonian can be constructed from the weakly interacting groundstate $|\Psi\rangle_0$ by the Bogoliubov transformation

$$|\Psi\rangle_{\mathrm{BCS}} = \prod_{\mathbf{k}} \left(\tilde{u}_k + \tilde{v}_k c^{\dagger}_{s\tau\mathbf{k}} \nu_{s\tau\mathbf{k}} \right) |\Psi\rangle_0, \tag{2.59}$$

with $\tilde{u}_k = \sqrt{(\mathcal{E}_{s\tau\mathbf{k}} + \bar{\Sigma}_{s\tau\mathbf{k}})/\mathcal{E}_{s\tau\mathbf{k}}}$ and $\tilde{v}_k = \sqrt{(\mathcal{E}_{s\tau\mathbf{k}} - \bar{\Sigma}_{s\tau\mathbf{k}})/\mathcal{E}_{s\tau\mathbf{k}}}$, and corresponds to a BCS-like condensate of excitons. [Stroucken *et al.* (2011)] Diagonalization of the interacting mean-field Hamiltonian yields the Bogoliubov single-particle dispersion $E^{\mathrm{c/v}}_{s\tau\mathbf{k}} = \pm\mathcal{E}_{s\tau\mathbf{k}}$. As one can recognize, a nontrivial solution for the Rabi frequency in the absence of an external field increases the gap. Indeed, for graphene the gap equations are exactly equivalent to the equations for the mass and Fermi-velocity renormalization derived in 2.3.2. This equivalence shows that a gap opening in graphene can be interpreted as the transition into an excitonic groundstate and the binding energy of the condensed excitons corresponds to the gap. Similarly, any nontrivial solution in a gapped system not only increases the gap but also leads to a renormalization of the effective mass.

In Fig. 2.6, we show the dispersion of the Bogoliubov band in graphene for different values of the coupling strength. As can be recognized, the dispersion exhibits a slight Mexican hat shape which is typical for an excitonic insulator. Apart from this shape, the dispersion is similar to the relativistic dispersion of a massive quasiparticle with mass $\tilde{\Delta}_{s\tau\mathbf{k}}/2v_F^2$.

2.4.3 *Optical excitations of the condensate*

To obtain the optical response in the regime of strong Coulomb coupling, we make the ansatz

$$P_{s\tau\mathbf{k}} = \bar{P}_{s\tau\mathbf{k}} + \Delta P_{s\tau\mathbf{k}}, \tag{2.60}$$

$$g_{s\tau\mathbf{k}} = \bar{g}_{s\tau\mathbf{k}} + \Delta g_{s\tau\mathbf{k}} \tag{2.61}$$

dividing the distributions into a static part arising from the excitonic groundstate and a dynamical part due to the optical excitations. As there is no experimental evidence for the observation of a gap in graphene, we restrict the following analysis to TMDCs in the wide gap approximation. For the graphene case, we refer to Refs. [Stroucken *et al.* (2011)] and [Stroucken *et al.* (2012)].

For the TMDCs, the dominant part of the Coulomb interaction contains the phase factor resulting from the pseudo-spin. Applying a transformation of both the dynamical and groundstate interband polarization $P_{s\tau\mathbf{k}} \to e^{-i\tau\theta_{\mathbf{k}}} P_{s\tau\mathbf{k}}$, the resulting generalization of the linearized DBE for the strong Coulomb-coupling regime is obtained as

$$\begin{aligned}
i\hbar\partial_t P^{[1]}_{s\tau\mathbf{k}} &= 2\bar{\Sigma}_{s\tau\mathbf{k}} P^{[1]}_{s\tau\mathbf{k}} - \bar{g}_{s\tau\mathbf{k}} \sum_{\mathbf{k}'} V_{\mathbf{k},\mathbf{k}'} P^{[1]}_{s\tau\mathbf{k}'} \\
&\quad + 2\bar{P}_{s\tau\mathbf{k}} \sum_{\mathbf{k}'} V_{\mathbf{k},\mathbf{k}'} g^{[1]}_{s\tau\mathbf{k}'} - \bar{\Omega}_{s\tau\mathbf{k}'} g^{[1]}_{s\tau\mathbf{k}} \\
&\quad - e\frac{v_F}{c} \bar{g}_{s\tau\mathbf{k}} A^{-\tau} - \frac{2}{c} \mathbf{A} \cdot \mathbf{j}_{s\tau\mathbf{k}} \bar{P}_{s\tau\mathbf{k}},
\end{aligned} \tag{2.62}$$

$$i\hbar\partial_t g^{[1]}_{s\tau\mathbf{k}} = 2\bar{\Omega}_{s\tau\mathbf{k}'} P^{[1]*}_{s\tau\mathbf{k}} + 2\bar{P}^{*}_{s\tau\mathbf{k}} \sum_{\mathbf{k}'} V_{\mathbf{k},\mathbf{k}'} P^{[1]}_{s\tau\mathbf{k}'} - c.c. \tag{2.63}$$

The static groundstate polarizations and populations induce a coupling between the linearized equations for the optical polarization and populations and add an additional source term to the equation of motion for the optical interband transition amplitude. As the additional source term is proportional to the intraband current matrix element with odd parity, it mixes states with different parity. Though the odd terms of the linear interband polarization do not couple to the optical field directly, they serve as source terms for the optically induced intraband current.

The total optically induced macroscopic optical polarization in terms of the transformed variables is given by

$$j^{\tau} = \tau\sqrt{2}ev_F \sum_{s\mathbf{k}} \left\{ (P_{s-\tau\mathbf{k}} - P^*_{s\tau\mathbf{k}}) - \mathrm{e}^{i\tau\theta_{\mathbf{k}}} j^{\tau}_{s\tau\mathbf{k}} g_{s\tau\mathbf{k}} \right\} \tag{2.64}$$

where $j^{\tau}_{s\tau\mathbf{k}} = \boldsymbol{e}^*_{\tau} \cdot \boldsymbol{j}_{s\tau\mathbf{k}}$ is the projection of the intraband current matrix element of the unit vector corresponding to a circularly polarized state. Thus, the optical field couples to the s-type contributions of the transformed optical interband polarization and the p-type optically induced intraband current.

Equations (2.62) and (2.63) can be decoupled by the transformation

$$\Pi^{[1]}_{s\tau\mathbf{k}} = \frac{\mathcal{E}_{s\tau\mathbf{k}} + \bar{\Sigma}_{s\tau\mathbf{k}}}{\mathcal{E}_{s\tau\mathbf{k}}} P^{[1]}_{s\tau\mathbf{k}} - \frac{\mathcal{E}_{s\tau\mathbf{k}} - \bar{\Sigma}_{s\tau\mathbf{k}}}{\mathcal{E}_{s\tau\mathbf{k}}} P^{[1]*}_{s\tau\mathbf{k}} - \frac{\bar{\Omega}_{s\tau\mathbf{k}}}{\mathcal{E}_{s\tau\mathbf{k}}} g^{[1]}_{s\tau\mathbf{k}},$$
$$\Gamma^{[1]}_{s\tau\mathbf{k}} = \frac{\bar{\Omega}_{s\tau\mathbf{k}}}{\mathcal{E}_{s\tau\mathbf{k}}} \left(P^{[1]}_{s\tau\mathbf{k}} + P^{[1]*}_{s\tau\mathbf{k}} \right) + \frac{\bar{\Sigma}_{s\tau\mathbf{k}}}{\mathcal{E}_{s\tau\mathbf{k}}} g^{[1]}_{s\tau\mathbf{k}},$$

which is equivalent to a transformation into the Bogoliubov conduction and valence bands.

As result we obtain

$$i\hbar\partial_t \Pi^{[1]}_{s\tau\mathbf{k}} = 2\mathcal{E}_{s\tau\mathbf{k}} - \sum_{\mathbf{k}'} V_{\mathbf{k},\mathbf{k}'} \Pi^{[1]}_{s\tau\mathbf{k}'} - 2e\frac{v_F}{c} \bar{g}_{s\tau\mathbf{k}} A^{-\tau} - \frac{4}{c} \mathbf{A} \cdot \mathbf{j}_{s\tau\mathbf{k}} \bar{P}_{s\tau\mathbf{k}}, \tag{2.65}$$

$$i\hbar\partial_t \Gamma^{[1]}_{s\tau\mathbf{k}} = 0. \tag{2.66}$$

Similary as in the weak coupling regime, the solutions of Eq. (2.65) can be expanded in terms of the eigenfunctions of the Bogoliubov-Wannier equation

$$2\mathcal{E}_{s\tau\mathbf{k}} \psi^{s\tau}_{\mu}(\mathbf{k}) - \sum_{\mathbf{k}'} V_{\mathbf{k},\mathbf{k}'} \psi^{s\tau}_{\mu}(\mathbf{k}') = E^{s\tau}_{\mu} \psi^{s\tau}_{\mu}(\mathbf{k}'), \tag{2.67}$$

where the dispersion of the noninteracting bands has been replaced by the Bogoliubov quasiparticle dispersion.

The groundstate polarization obeys $2\mathcal{E}_{s\tau\mathbf{k}} \bar{P}_{s\tau\mathbf{k}} - \sum_{\mathbf{k}'} V_{\mathbf{k},\mathbf{k}'} \bar{P}_{s\tau\mathbf{k}'} = 0$ such that it can be expanded into the eigenfunctions of the Bogoliubov-Wannier equation with the eigenvalue $E^{s\tau}_{\mu} = 0$. Hence, we make the ansatz

$$\bar{P}_{s\tau\mathbf{k}} = \sum_{\mu, E_{\mu}=0} \bar{P}_{s\tau\mu} \psi^{s\tau}_{\mu}(\mathbf{k}), \tag{2.68}$$

$$\Pi^{[1]}_{s\tau\mathbf{k}} = \sum_{\mu} \Pi^{[1]}_{s\tau\mu} \psi^{s\tau}_{\mu}(\mathbf{k}), \tag{2.69}$$

Inserting the ansatz into Eq. (2.65) yields

$$\Pi^{[1]}_{s\tau\mu}(\omega) = -2e\frac{v_F}{c}A^{-\tau}\frac{(\gamma_\mu^{s\tau})^*}{\hbar(\omega+i\gamma)-E_\mu} - \frac{4}{c}\frac{\mathbf{A}\cdot\mathbf{j}_\mu^{s\tau}}{\hbar(\omega+i\gamma)-E_\mu}. \tag{2.70}$$

Here, $\gamma_\mu^{s\tau} = \sum_{\mathbf{k}} \bar{g}_{s\tau\mathbf{k}}\psi_\mu(\mathbf{k})$ is the coupling strength of the macroscopic interband polarization to the optical field. For an s-type groundstate ($1 < \alpha < 3$), the populations also have spherical symmetry and the coupling strength vanishes for all non s-type excition states, i.e. for all real exciton resonances. Generally, if $\alpha < (2n+1)/2$, the groundstate populations are composed of exciton states with $|j| \leq (2n+1)/2$, while real exciton resonances require $|j| \geq (2n+1)/2$. Thus, in the strong Coulomb coupling regime, the first term on the RHS of Eq. (2.70) vanishes.

Reversing the transformation (2.65), we can compute the optical susceptibility for the excitonic insulator state

$$\chi^\sigma(\omega)_{EI} = -\frac{1}{\omega^2}\sum_{s,\mu}\frac{|\mathbf{j}_\mu^{s-\sigma}|^2}{\hbar(\omega+i\gamma)-E_\mu^{s,-\sigma}} + \frac{1}{\omega^2}\sum_{s,\mu} -\frac{|\mathbf{j}_\mu^{s\sigma}|^2}{\hbar(\omega+i\gamma)+E_\mu^{s,\sigma}} \quad . \tag{2.71}$$

Here, all contributions have intraband character.

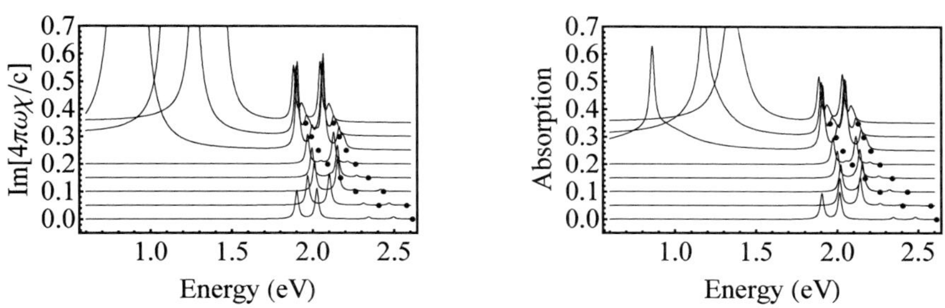

Fig. 2.7 Linear optical spectra for different values of α. The graphs have been obtained using the material parameters for MoS_2 given in Ref. [Xiao *et al.* (2012)], a constant dephasing rate $\gamma = 0.0125eV/\hbar$ and a momentum cut-off of $(7\text{Å})^{-1}$. The coupling constant is given as $\alpha = 4.11/\kappa$ and κ increases from $\kappa = 2$ (lowest line), until to $\kappa = 5.5$ (upper line) in steps of $\Delta\kappa = 0.5$. For a better visibility, the curves have been shifted along the vertical axis. In the strong coupling regime $\kappa \leq 4.11$, the exciton envelope function of the optically bright states have p-type character, whereas in the weak coupling regime $\kappa > 4.11$ optically bright states have s-type exciton envelopes. The dots denote the values for the renormalized gaps of the A and B excitons respectively. Left: Imaginary part of the susceptibility. Right: True absorption.

As can be recognized from Eq. (2.71), the poles of the linear susceptibility for the excitonic insulator state occur at the eigenvalues of the exciton Hamiltonian with Bogoliubov dispersion. The dominant effect of the exciton condensation on the single quasiparticle dispersion is a renormalization of the gap. Hence, the spectrum can be obtained in very good approximation from Eq. (2.46), provided one uses the appropriate renormalized gap. The major difference as compared to the optical susceptibility, Eq.(2.38), for the uncorrelated groundstate is the oscillator strength, which is now given by the intraband current matrix element between the excited exciton state and the excitonic groundstate

$$\mathbf{j}_{\mu}^{s\tau} = \sum_{\mathbf{k}} \psi_{\mu}^{s\tau\,*}(\mathbf{k})\mathbf{j}_{s\tau\mathbf{k}} P_{s\tau\mathbf{k}}^{[0]} \equiv \langle \mu s\tau|\mathbf{j}|0\rangle .$$

This matrix element connects the part of the groundstate with largest $|j| \leq \alpha/2$ to the excited states with $|j'| = |j| + 1 > \alpha/2$ such that dipole allowed transitions change the total angular momentum by one. Explicitly, in the regime with $1 < \alpha < 3$, optical transitions connect the s-type excitonic groundstate $j = \tau/2, l = 0$ with the excited excitonic states $j' = 3\tau/2, l = \tau$, corresponding to p-like states. Thus, combining the results of the weakly and strongly interacting regimes, the modified optical selection rules can be summarized by the condition

$$(2|l| - 1) < \alpha < (2|l| + 1) \tag{2.72}$$

for the optically active states.

In Fig. 2.7 we show the linear absorption spectra for different values of the coupling strength using the material parameters for MoS_2 [Xiao *et al.* (2012)] with a nominal value of $\alpha = 4.11/\kappa$. We increased the (constant) background screening from $\kappa = 2$ until $\kappa = 5.5$ in steps of $\Delta\kappa = 0.5$. To evaluate the renormalized gap, we used a momentum cut-off of $1/d = 1/7$ Å^{-1}. For $\kappa \leq 4.11$, the system is in the strong Coulomb coupling regime and the dominant peaks occur around 2 eV and correspond to the two spin-split p-type eigenstates of the relativistic hydrogen equation with $(n, j) = (0, \pm 3/2)$. If $\kappa > 4.11$, the system is in the weak coupling regime. The peaks around 2 eV are correspond to the s-type excited states of the spin-split excitons and the lowest lying $1s$-excitons with $(n, j) = (0, \pm 1/2)$ appear energetically below, where the spectra in the strong coupling regime are feature-less. Due to the large coupling strength, the absorption spectrum in this spectral region differs significantly from the imaginary part of the linear susceptibility. In particular, as a result of the modification of the refractive index by the real part of the susceptibility, the absorption lines of the spin-split $1s$-exciton resonance appear as a single broad spectral feature.

2.5 Screening

As we have seen in the previous section, within the massive Dirac Fermion model, the importance of excitonic effects and their influence on the optical properties of

2D materials with hexagonal symmetry depend parametrically on the single parameter α. To determine the actual value of α in a given material system, it is not sufficient to know the screening of the dielectric environment, but intrinsic screening effects should be included as well. In general, screening effects arise (i) from filled valence bands that are not considered explicitly in the model Hamiltonian and (ii) from contributions of the conduction and valence band under consideration. The Coulomb matrix elements used to set up the interacting model Hamiltonian should contain only the first type (i) of contributions which can be considered to lead to a constant background dielectric constant. However, for a film of vanishing thickness d, this part becomes negligible. [Wehling *et al.* (2011)] However, the contributions (ii) should then be computed selfconsistently, using the Coulomb matrix elements screened by the background dielectric constant.

In the strict 2D limit, the screened Coulomb potential can be obtained from the Poisson equation

$$\epsilon(z)\left(\mathbf{q}_{\|}^2 - \partial_z^2\right)\phi(\mathbf{q}_{\|}, z, \omega) = 4\pi i \mathbf{q}_{\|} \cdot \mathbf{P}_{\|}^L(\mathbf{q}_{\|}, \omega)\delta(z) \tag{2.73}$$

where $\epsilon(z)$ is the background dielectric constant of the environment and

$$\mathbf{P}_{\|}^L(\mathbf{q}_{\|}, \omega) = -i\mathbf{q}_{\|}\chi^L(\mathbf{q}_{\|}, \omega)\phi(\mathbf{q}_{\|}, z = 0, \omega)$$

contains the resonant contributions to the longitudinal polarization of the 2D layer. Here, $\chi^L(\mathbf{q}_{\|}, \omega)$ is the longitudinal dielectric function which should be computed microscopically from the bands under consideration. The 2D Coulomb potential is then obtained as

$$V_s^{2D}(\mathbf{q}_{\|}, \omega) = \phi(\mathbf{q}_{\|}, z = 0, \omega) = \frac{2\pi e^2}{\epsilon_B q_{\|}} \frac{1}{1 + 2\pi q_{\|}\chi^L(\mathbf{q}_{\|}, \omega)/\epsilon_B} \tag{2.74}$$

where $\epsilon_B = (\epsilon_1 + \epsilon_2)/2$ is the effective dielectric constant for a sheet embedded between two different dielectric layers of permittivity ϵ_1 and ϵ_2, respectively.

The longitudinal susceptibility is related to the polarization function $P(\mathbf{q}_{\|}, \omega)$ of the 2D layer via $\chi_L(\mathbf{q}_{\|}, \omega) = -e^2 P(\mathbf{q}_{\|}, \omega)/q_{\|}^2$, which is convienently treated within the random phase approximation (RPA).

For graphene, the static dielectric function has been derived in Ref. [Hwang and Das Sarma (1997)], giving $\epsilon = \epsilon_B(1 + \pi\alpha/2)$. The constant dielectric function in graphene reflects the lack of an intrinsic legth scale and gives a maximum value of the effective coupling constant

$$\alpha_{\text{eff}} = \frac{\alpha}{1 + \pi\alpha/2} \leq \frac{2}{\pi} \approx 0.64,$$

which is slightly larger than the critical value. Inserting the nominal value of $\alpha = 2.4$ for graphene gives an effective value of the coupling strength $\alpha_{\text{eff}} = 0.5$

which is exactly the critical value, thus explaining why a gapped dispersion in graphene has not been observed.

For the massive Dirac-Hamiltonian (2.5), the static limit of the RPA dielectric function of the TMD monolayer is given by [Rodin and Castro Neto (2013)],

$$\epsilon(q) = \epsilon_B \left(1 + \frac{\alpha}{2\pi} \sum_{s\tau} \frac{P(\lambda_C q)}{\lambda_C q}\right) \tag{2.75}$$

where $\lambda_C = 2\hbar v_F / \Delta_{s\tau}$ is the spin and valley dependent Compton wavelength. An analytical expression for $P(Q)$ is given in Ref. [Rodin and Castro Neto (2013)].

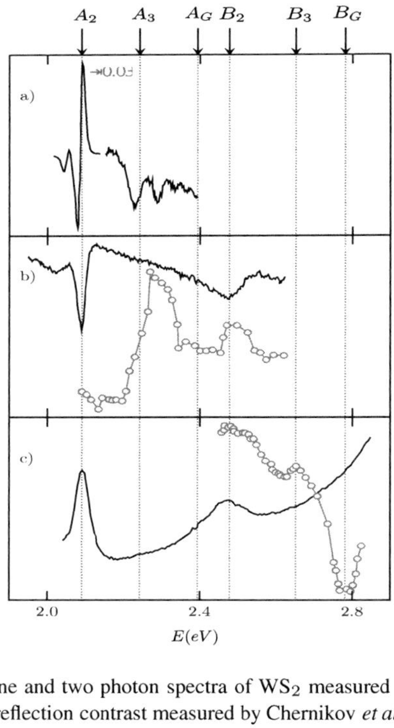

Fig. 2.8 Comparsion of one and two photon spectra of WS_2 measured by different groups. Part a) shows the derivative of the reflection contrast measured by Chernikov *et al.* [Chernikov *et al.* (2014)], part b) displays the linear reflection contrast (black line) together with the TP-PLE spectrum (red dots) measured by Ye *et al.* [Ye *et al.* (2014)], and part c) presents the linear absorption (black line) and TP-PLE (red dots) measured by Zhu et al. [Zhu *et al.* (2014)]. To compensate for temperature dependent shifts of the band gaps, we shifted the data such that the lowest resonance of the A excitons are energetically aligned. On the upper axis, we assigned the quantum numbers assuming bright p states. The respective bandgaps are denoted by A_G and B_G. The dotted line are guides to the eye and show the energetic position of some selected states. Reprinted from [T. Stroucken and S.W. Koch, *J. Phys.: Condens. Matter* **27**, 345003 (2015)]. Copyright (2015) by The Institute of Pysics.

In the ultrarelativistic regime, $\lambda_C q \gg 1$, $P(Q) \approx \pi^2 Q/4$ and the dielectric function $\epsilon(q) = \epsilon_B(1 + \pi\alpha/2)$ reduces to the constant of graphene. In the nonrelativistic limit, $\lambda_C q \ll 1$, $P(Q) \approx \frac{\pi Q^2}{3}$, yielding $\epsilon(q) = \epsilon_B(1 + \frac{1}{3}\alpha^2 a_B q)$ where a_B is the Bohr radius of an electron-hole pair with reduced mass $m_r = 2\lambda_C/\alpha$. In this regime, the intrinsic screening corresponds to the anti-screening proposed by [Keldysh (1978); Cudazzo *et al.* (2011)] with an anti-screening length $r_0 = \frac{1}{3}\alpha^2 a_B$, Using the nominal values of α in suspended TMDCs, the anti-screening length is typically $\propto 1 - 1.5\, a \approx 3 - 5$ Å, which corresponds roughly to the physical thickness of the monolayer. For supported samples, the anti-screening length is reduced correspondingly. The anti-screening weakens the strength of the Coulomb interaction on a length scale small compared to the anti-screening length. Thus, for the weakly bound excitons with a spatial extension large compared to the 2D-exciton Bohr radius, we expect the anti-screening effect to be of minor importance.

2.6 Comparison with Experiment

To test the applicability of our model, we compare its predictions with some recent experimental observations. As we have seen in the previous section – and as seems to be the case in all reported experiments– graphene is always in the regime of weak Coulomb coupling. Therefore, we focus here on the available experimental data for TMDCs. For these, we adopt the commonly used notation of A and B excitons for the lower and higher direct exciton transitions with a spin and valley index combination $s\tau = 1, -1$ respectively.

To determine the effective fine-structure constants, we use the material parameters given in Ref. [Xiao *et al.* (2012)] and neglect all screening effects apart from the substrate screening. Furthermore, we need the additional knowledge of the Coulombic bandgap renormalization to compute the energetic positions of the optically bright states. Whereas the bandgap renormalization diverges for an ideal, strictly 2D Coulomb potential, it is finite for any real system where the Coulomb matrix elements must be evaluated from the gap equations with the form factor calculated with the atomic orbitals used to represent the pseudo spinor space in Eq. (2.5). Since its explicit expression depends on the specific atomic orbitals contributing to the valence and conduction band, we do not calculate the bandgap renormalization but rather use it as a single fit parameter.

While the small spin-orbit splitting in MoX_2 complicates the experimental determination of the fundamental gap, the large spin splitting in WX_2 allows the experimentalists to observe not only the dominant A and B exciton peaks, but also the observation of an entire excitonic series from which the gap can be deduced. Therefore, we compare the predictions of our theoretical model with experimental data on WS_2 and WSe_2. With the material parameters for WS_2 $t = 1.37$eV, $2\lambda = 0.43$ eV, $a = 3.197$Å and $t = 1.19$eV, $2\lambda = 0.46$ eV, $a = 3.310$Å for WS_2 respectively, [Xiao *et al.* (2012)] and an effective dielectric constant $\kappa = (1+3.9)/2$ for the SiO_2 substrate, we find an effective coupling constant $\alpha = 1.34$ for WS_2 and $\alpha = 1.49$ for WSe_2, respectively. Hence, we conclude that these two systems are both in the strong Coulomb coupling regime and we expect the observation of bright excitonic states that correspond to the p-type excited states of the relativistic hydrogen model.

We start with the analysis of the bright excitonic states in WS_2 on SiO_2 as observed in several recent experiments. [Chernikov *et al.* (2014); Zhu *et al.* (2014); Ye *et al.* (2014)] The experiments detect the linear reflection contrast [Chernikov *et al.* (2014); Ye *et al.* (2014)], the linear absorption [Zhu *et al.* (2014)], and the two-photon photoluminescence (TP-PLE) at different temperatures. All experiments observe two clearly recognizable peaks that can be attributed to the lowest bright transitions of the A and B exciton, respectively. Though the absolute spectral positions in the different experimental setups differ slightly, the energetic separation of the two dominant peaks is in very good mutual agreement. The differences in the absolute positions are most likely due to different bandgap renormalizations under different experimental conditions. In Fig. 2.8, we show the combined experimental data where we shifted the energy scales such that the position of the dominant peak agrees.

In the linear reflection spectra of Ref. [Chernikov *et al.* (2014)], five bright resonances are observed that can be assigned to the A exciton, all of them energetically below the lowest bright B exciton transition. From these measurements, we estimate the value $E_{gap} = 2.36$ eV for the A exciton, which we use as a single fit parameter to produce the energetic position of all A and B exciton resonances. With these parameters, we find a binding energy of $0.25eV$ for the lowest bright exciton transition. Adopting the nomenclature commonly used for the nonrelativistic states, this state corresponds to the $2p$ exciton. The computed energetic positions of the dominant peaks are shown as vertical lines in Fig. 2.8, with a labeling resulting from our analysis in terms of the p-type bright states. Please note that these assignements do not correspond to the labels in the original publications.

Additionally, we present in Fig. 2.9 the computed result for the spectral position of the lowest five bright A-exciton transitions and compare them with the experimentally observed resonance positions. The diamonds show the experimental data points taken from Ref. [Chernikov *et al.* (2014)]. For comparison, we also show the predicted values assuming nonrelativistic bright s-states. Note that

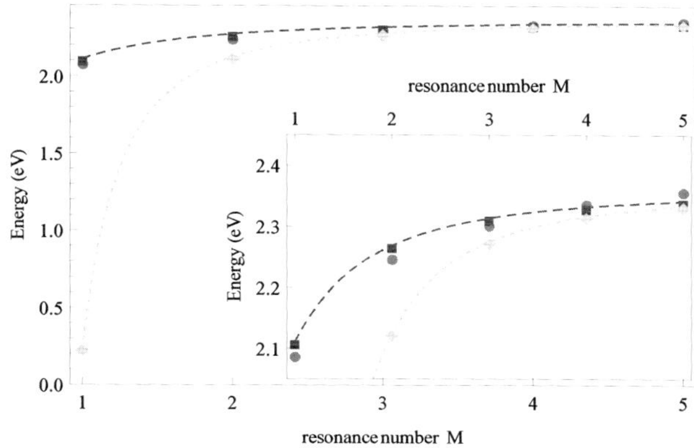

Fig. 2.9 Spectral position of the energetically lowest bright excitonic transitions with $j = \pm 3/2$ in WS_2. The squares are the computed results for p-type excitons, the dots show the experimental data taken from [Chernikov *et al.* (2014)], and the diamonds indicate the theoretical predictions if one assumes s- type bright states, respectively. The dashed and dotted lines are guides to the eye. The x-axis label denotes the number M of the experimentally observed bright exciton resonances and is related to the priniple quantum number via $N = M - 1$. The inset to the figure shows the results for the energetically higher states with a finer energy resolution. Reprinted from [T. Stroucken and S.W. Koch, *J. Phys.: Condens. Matter* **27**, 345003 (2015)]. Copyright (2015) by The Institute of Pysics.

the $j = \pm 1/2$ series gives complex eigenvalues of the relativistic wave equation and thus cannot be used for comparison.

Our computed results show remarkably good agreement with the experimental observations given the fact that only the effective bandgap was used as a single fit parameter. In contrast, the results based on the assumption that the series has s-type character are completely off the scale, in particular for the energetically lowest states.

In Ref. [Wang *et al.* (2014)], TP-PLE measurements have been performed in conjunction with SHG experiments on WSe_2. This material system has very similar properties as WS_2, in particular it has the same symmetries and similar spin-splitting of the valence bands. Using PLE, the resonance position of the A exciton is determined as 1.75 eV, whereas in TP-PLE, the dominant peak occurs at approximately 1.9 eV. In agreement with the observation in Ref. [Ye *et al.* (2014)], the TP-PLE spectrum shows no significant signal at the transition energy of the lowest bright A exciton. In contradiction to the results of Ref. [Ye *et al.* (2014)], the TP-PLE spectrum does not show a significant signal at the transition energy of the lowest bright B exciton. In contrast, the SHG spectrum exhibits two dominant peaks at 1.75 eV, i.e, at the lowest bright resonance, and at 2.16 eV, and less prominent peaks at 1.9, 2.02 and 2.3 eV.

In Ref. [He *et al.* (2014)], linear absorption and TP-PLE measurements have been performed at room temperature on WSe_2. The dominant peaks corresponding to the lowest bright A and B exciton are observed at 1.65 eV and 2.08 eV respectively, and excited states of the A exciton at 1.82, 1.88, 1.91 and 1.94 e showing an overall red-shift of ca. 0.1 eV as compared to the data of [Wang *et* (2014)]. As reported in the supplemental material, at low temperatures the re nances are blue shifted and are in almost perfect agreement with the data of [W *et al.* (2014)].

To model the experimental data, we choose the gap to match the theoret value of the lowest bright A exciton transition with the experimentally obser value of 1.75 eV. This gives a gap of $2.01eV$ for the A exciton, which allows to compute the whole exciton series of both spin and valley components. The sults are summarized in Table 2.1 together with the experimentally available da Resonances that could not be resolved by the specific experiment remain as emp spaces in Table 2.1 , while those resonances that could in principle occur in the experiment but are not observed are indicated by an 'x'. As can be recognized from the table, not only the agreement between theory and experiment is remarkable, but also the data obtained by the different experimental techniques on different samples is brought into almost perfect mutual agreement as soon as one adopts our interpretation in terms of optically active p-states.

2.7 Summary

In summary, the presented general theory allows for a unified analysis of the groundstate and the near bandgap optical properties of quasi two-dimensional systems with hexagonal lattice symmetry. Using a fully relativistic analysis of a simple model Hamiltonian for the Coulomb interacting electron-hole excitations, we identify the relevant parameter combinations needed to characterize the coupling regime for the respective material systems. Whereas the currently realized single-layer graphene configurations seem to belong into the regime of weak Coulomb interaction, several of the novel transition metal dichalcogenides are in the strong Coulomb coupling regime. For these systems, we identify the excitonic insulator groundstate and attribute the optically active one-photon transitions to intra-excitonic excitations into states with a p-like symmetry of the excitonic wavefunctions. With this assignment, we not only obtain a remarkably good agreement with recent experiments but we can also resolve apparent contradictions between different experimental observations.

Despite the obvious limitations of the applicability of our simplified model to real material systems, the results should be useful to gain a basic understanding

Table 2.1 Excitonic resonance positions extracted from experimental data together with the theoretical predictions where the gap of the A-exciton has been used as single fit parameter to match the experimental data. For WS_2, the bandgap has been fitted to match the data of Ref. [Chernikov *et al.* (2014)] (see Fig. 1 and the discussion in the text). For WSe_2, the lowest bright resonance position has been fitted to the data of Ref. [Wang *et al.* (2014)]. The data of Ref. [Zhu *et al.* (2014)] have been obtained at room temperature and show a red shift of approximately 0.1 eV as compared to the linear reflection spectra at 10K. The labeling corresponds to the main quantum number N of the Nth excitonic level, which is usually used in semiconductor optics. Reprinted from [T. Stroucken and S.W. Koch, *J. Phys.: Condens. Matter* **27**, 345003 (2015)]. Copyright (2015) by The Institute of Pysics.

	E_2^A	E_3^A	E_4^A	E_G^A	E_2^B	E_3^B	E_4^B	E_G^B	method
WS_2	2.11	2.27	2.31	2.36	2.49	2.68	2.74	2.79	theory[a]
	2.09	2.25	2.3		2.5				linear reflection[b]
	2.02				2.4				linear absorption[c]
					2.4	2.6		2.73	TP-PLE[c]
	2.04				2.45				linear reflection[d]
	x	2.25	2.29	2.35	2.46				TP-PLE[d]
WSe_2	1.75	1.91	1.96	2.01	2.15	2.35	2.41	2.47	theory[a]
	1.75								PLE[e]
	x	1.9			x				TP-PLE[e]
	1.75	1.91	1.98	2.02	2.17	2.3			SHG[e]
	1.65	1.82	1.88		2.08				linear reflection[f]

[a] [Stroucken and Koch (2015)]. [b] [Chernikov *et al.* (2014)] [c] [Zhu *et al.* (2014)] [d] [Ye *et al.* (2014)] [e] [Wang *et al.* (2014)] [f] [He *et al.* (2014)]

of the underlying physical processes and the nature of the excitonic states. At this stage, there are still a number of open questions needing further research. In particular, it would be interesting to go beyond the simple massive Dirac-Fermion model employed in our analysis and to perform quantitative first-principle studies that fully include all of the relevant interaction and screening effects. On this basis, it should then be possible to develop truly quantitative models that allow us to compute both the linear and the excitation dependent nonlinear optical properties reliably. In close collaboration with experimental investigations, this should enable quantitative theory-experiment comparisons for this fascinating class of material systems.

Acknowledgement

This work is supported by the Sonderforschungsbereich 1083 funded by the Deutsche Forschungsgemeinschaft.

Bibliography

[1] Timothy C. Berkelbach, Mark S. Hybertsen, and David R. Reichman *Phys. Rev. B* **88**, 045318 (2013).

[2] Timothy C. Berkelbach, Mark S. Hybertsen, and David R. Reichman *Phys. Rev. B* **92**, 085413 (2015).

[3] T. Cao, G. Wang, W. Han, H. Ye, C. Zhu, J. Shi, Q. Niu, P. Tan, E. Wang, B. Liu, and J. Feng, *Nature Communications* **3**, 887 (2012).

[4] E. Cappelluti, R. Roldan, J. A. Silva-Guillen, P. Ordejon, and F. Guinea, *Phys. Rev. B* **88**, 075409 (2013).

[5] A.H. Castro Neto, F. Guinea, N.M.R. Peres, K.S. Novoselov, and A.K. Geim, *Rev. Mod. Phys.* **81**, 109 (2009).

[6] Antonio H. Castro Neto, *Physics* **2**, 30 (2009).

[7] Dong-Hun Chae, Tobias Utikal, Siegfried Weisenburger, Harald Giessen, Klaus v. Klitzing, Markus Lippitz, and Jurgen Smet, *Nano Lett.* **11**, 13791382 (2011).

[8] Tawinan Cheiwchanchamnangij and Walter R. L. Lambrecht, *Phys. Rev. B* **85**, 205302 (2012).

[9] A. Chernikov, T. C. Berkelbach, H. M. Hill, A. Rigosi, Y. Li, O. B. Aslan, D. R. Reichman, M. S. Hybertsen, and T. F. Heinz, *Phys. Rev. Lett.* **113**, 076802 (2014).

[10] P. Cudazzo, I. V. Tokatly, and A. Rubio, *Phys. Rev. B* **84**, 085406 (2011).

[11] J.E. Drut and T.A. Lähde, *Phys. Rev. Lett.* **102**, 026802 (2009).

[12] J.E. Drut and T.A. Lähde, *Phys. Rev. B* **79**, 165425 (2009).

[13] E. Engel and R. N. Schmid *Phys. Rev. Lett.* **103**, 036404 (2009).

[14] L. Fritz, J. Schmalian, M. Müller, and S. Sachdev, *Phys. Rev. B* **78**, 085416 (2008).

[15] J. González, F. Guinea, M.A.H. Vozmediano, *Phys. Rev. B* **59**, R2474 (1999).

[16] J. H. Grönqvist, T. Stroucken, M. Lindberg, and S. W. Koch, *Eur. Pphs. J. B* **85**, 395 (2012).

[17] B. I. Halperin and T. M. Rice, *Rev. Mod. Phys.* **40**, 755 (1968).

[18] see e.g. Chap. 10 in: H. Haug and S.W. Koch, *Quantum Theory of the Optical and Electronic Properties of Semiconductors*, 5th edition, World Scientific, Singapore (2009).

[19] Keliang He, Nardeep Kumar, Liang Zhao, Zefang Wang, Kin Fai Mak, Hui Zhao, and Jie Shan *Phys. Rev. Lett.* **113**, 026803 (2014).

[20] E. H. Hwang and S. Das Sarma *Phys. Rev. B* **75**, 205418 (2007).

[21] D. Jerome, T. M. Rice, and W. Kohn, *Phys. Rev.* **158**, 462 (1967).

[22] V. Juricic, I.F. Herbut, and G.W. Semenoff, *Phys. Rev. B* **80**, R081405 (2009).

[23] Keldysh, L. V. and Kopaev, Yu. V. (1964). *Fiz. Tv. Tela* **6**, 279 [*Sov. Phys. Solid State* **6**, 2219 (1965)].

[24] Keldysh, L. V. and Kozlov, A. N., *Zh. Eksp. Teor. Fiz.* **54**, 978 [Sov. Phys. JETP 27, 521] (1968).

[25] L. V. Keldysh, *JETP Lett.* **29**, 658 (1978).

[26] D.V. Khveshchenko, *Phys. Rev. Lett.* **87**, 246802 (2001).

[27] D.V. Khveshchenko and W.F. Shively, *Phys. Rev. B* **73**, 115104 (2006).

[28] M. Kira and S.W. Koch, *Progress in Quantum Electronics* **30**, 155 (2006).

[29] Andor Kormanyos, Viktor Zolyomi, Neil D. Drummond, Peter Rakyta, Guido Burkard, and Vladimir I. Falko, *Phys. Rev. B* **88**, 045416 (2013).

[30] Andor Kormanyos, Guido Burkard, Martin Gmitra, Jaroslav Fabian, Viktor Zolyomi, Neil D Drummond and Vladimir Falko, *2D Materials* **2**, 022001 (2015).

[31] V. G. Kravets, A. N. Grigorenko, R. R. Nair, P. Blake, S. Anissimova, K. S. Novoselov, and A. K. Geim *Phys. Rev. B* **81**, 155413 (2010).

[32] S. Latini, T. Olsen, and K. S. Thygesen, *Phys. Rev. B* **92**, 245123 (2015).

[33] Li and Heinz, *Phys. Rev. Lett.* **113**, 266804 (2014).

[34] Gui-Bin-Liu, wen-Yu Shan, Yugui Yao, Wang Yao, and Di Xiao, *Phys. Rev. B* **88**, 085433 (2013).

[35] K. F. Mak, C. Lee, J. Hone, J. Shan, and T. F. Heinz, *Phys. Rev. Lett.* **105**, 136805 (2010).

[36] Kin Fai Mak, Jie Shan, and Tony F. Heinz *Phys. Rev. Lett.* **106**, 046401 (2011).

[37] K.F. Mak, K. He, J. Shan, and T. F. Heinz, *Nature Nanotechnology* **7**, 494498 (2012)

[38] K. F. Mak, K. He, C. Lee, G. H. Lee, J. Hone, T. F. Heinz, and J. Shan, *Nature Materials* **12**, 207211 (2013).

[39] E. Malic, J. Maultzsch, S. Reich, and A. Knorr, *Phys. Rev. B* **82**, 035433 (2010).

[40] Ermin Malic, Torben Winzer, Evgeny Bobkin, and Andreas Knorr *Phys. Rev. B* **84**, 205406 (2011).

[41] K. S. Novoselov, A. K. Geim, S. V. Morozov, D. Jiang, Y. Zhang, S. V. Dubonos, I. V. Grigorieva, and A. A. Firsov, Science 306, 666 (2004).

[42] L. Pauling *The Nature of the Chemical Bond and the Structure of Molecules and Crystals: An Introduction to Modern Structural Chemistry*, Cornell University Press (1960).

[43] V. M. Pereira, V. N. Kotov, and A. H. Castro Neto, *Phys. Rev. B* **78**, 085101 (2008).

[44] A. Ramasubramaniam, *Phys. Rev. B* **86**, 115409 (2012).

[45] A. S. Rodin, A. H. Castro Neto, *Phys. Rev. B* **88**, 195437 (2013).

[46] Habib Rostami, Ali G. Moghaddam, and Reza Asgari *Phys. Rev. B* **88**, 085440 (2013).

[47] J. Sabio, F. Sols, and F. Guinea *Phys. Rev. B* **82**, 121413(R) (2010).

[48] G.W. Semenoff, *Phys. Rev. Lett.* **53**, 2449 (1984).

[49] Wen-Yu Shan, Jianhui Zhou, and Di Xiao *Phys. Rev. B* **91**, 035402 (2015).

[50] D.E. Sheehy and J. Schmalian, *Phys. Rev. Lett.* **99**, 226803 (2007).

[51] A. Sinner and K. Ziegler, *Phys. Rev. B* **82**, 165453 (2010).

[52] T. Stroucken, J. H. Grönqvist, and S. W. Koch, *Phys. Rev. B* **84**, 205445 (2011).

[53] T. Stroucken, J. H. Grönqvist, and S. W. Koch, *JOSA B*, **29**, A86-A94 (2012).

[54] T. Stroucken, J. H. Grönqvist, and S. W. Koch *Phys. Rev. B* **87**, 245428 (2013).

[55] T. Stroucken and S.W. Koch, *J. Phys.: Condens. Matter* **27**, 345003 (2015).

[56] See e.g. Q. Tang, Z. Zhou, *Progress in Materials Science* **58**, 12441315 (2013), and refrences therein.
[57] P.R. Wallace, *Phys. Rev.* **71**, 622, (1947).
[58] G. Wang, X. Marie, I. Gerber, T. Amand, D. Lagarde, L. Bouet, M. Vidal, A. Balocchi, B. Urbaszek, *Phys. Rev. Lett.* **114**, 097403 (2015).
[59] G. Wang, L. Bouet, M. M. Glazov, T. Amand, E. L. Ivchenko, E. Palleau, X. Marie, and B. Urbaszek, *2D Materials* **2**, 034002 (2015)
[60] T. O. Wehling, E. Sasioglu, C. Friedrich, A. I. Lichtenstein, M. I. Katsnelson, and S. Blügel *Phys. Rev. Lett.* **106**, 236805 (2011).
[61] D. Xiao, G.-B. Liu, W. Feng, X. Xu, and W. Yao, *Phys. Rev. Lett.* **108**, 196802 (2012).
[62] L. Yang, J. Deslippe, C.-H. Park, M.L. Cohen, and S.G. Louie, *Phys. Rev. Lett.* **103**, 186802 (2009).
[63] Li Yang *Phys. Rev. B* **83**, 085405 (2011).
[64] Wang Yao, Di Xiao, and Qian Niu *Phys. Rev. B* **77**, 235406 (2008).
[65] Z. Ye, T. Cao, K. O'Brien, H. Zhu, X. Yin, Y. Wang, S. G. Louie, X. Zhang, *Nature* **513**, 214218 (2014).
[66] H. Zeng, J. Dai, W. Yao, D. Xiao, and X. Cui, *Nature Nano*, **7**, pp 490-493,(2012).
[67] H. Zeng, G.-B. Liu, J. Dai, Y. Yan, B. Zhu, R. He, L. Xie, S. Xu, X. Chen, W. Yao, and X. Cui, *Scientific Reports* **3**, 1608 (2013).
[68] Bairen Zhu, Xi Chen, Xiaodong Cui, *Scientific Reports* **5**, 9218 (2015).

Chapter 3

Raman Spectroscopy of Graphene

Sven Reichardt[1,2] and Ludger Wirtz[1]

[1] *Physics and Materials Science Research Unit, Université du Luxembourg, 1511 Luxembourg, Luxembourg*

[2] *JARA-FIT and 2nd Institute of Physics, RWTH Aachen University, 52074 Aachen, Germany*

Raman spectroscopy of graphene is reviewed from a theoretical perspective. After an introduction of the building blocks (electronic band structure, phonon dispersion, electron-phonon interaction, electron-light coupling), Raman intensities are calculated using time-dependent perturbation theory. The analysis of the contributing terms allows for an intuitive understanding of the Raman peak positions and intensities. The Raman spectrum of pure graphene only displays two principle peaks. Yet, their variation as a function of internal and external parameters and the occurrence of secondary, defect-related peaks, conveys a lot of information about the system. Thus, Raman spectroscopy is used routinely to analyze layer number, defects, doping and strain of graphene samples. At the same time, it is an intriguing playground to study the optical properties of graphene.

3.1 Introduction

Raman scattering is the inelastic scattering of light. The frequency of a photon can change by transferring energy to and/or receiving energy from the lattice vibrations of the material. In quantum mechanics, the "allowed" energies of an oscillation are quantized and - in the case of a harmonic oscillation - equidistant. The vibrations of the lattice can thus be described in the language of quasi-particles and the term *phonon* is usually used. The incoming photon changes its energy by effectively exciting or absorbing one or several phonons. If a phonon is excited, the photon looses energy and the process is called *Stokes* scattering. In the case of the absorption of a phonon, the photon gains energy and the process is referred to as *Anti-Stokes* scattering. The spectrum of the inelastically scattered light therefore features discrete peaks, whose positions can be directly associated with certain vibrational modes of the crystal.

In first order Raman scattering, only optical phonons at the Brillouin zone center can be excited. Furthermore, selection rules only allow for vibrations with special symmetry properties. For graphene, only one peak, the so-called G peak, is due to first-order Raman scattering. An additional line, the so-called $2D$ line, is due to second order Raman scattering, where two phonons are excited at the same time. The overall Raman spectrum of graphene thus consists mainly of two peaks (see Fig. 3.4) and is quite similar to the spectrum of graphite and of carbon nanotubes. Nevertheless, via the exact peak positions, their width, their dispersion as a function of the excitation laser energy, and the occurrence of side peaks, Raman spectroscopy yields a surprising amount of information about graphene samples: For example, it gives information on the number of layers, their stacking order, the underlying substrate, defects, impurities, doping, and strain. Since Raman scattering is a fast and non-invasive method, it has evolved into one of the principal characterization tools for graphene and related graphitic materials.

Raman spectroscopy is often categorized as "vibrational spectroscopy" because it probes primarily the phonon frequencies. However, a quantitative description of the Raman intensities is deeply connected to the optical (absorption and emission) properties of the concerned material. This is particularly the case for resonant Raman spectroscopy, where the emitted/absorbed light is in resonance with electronic transitions. In graphene, due to the linear crossing of the optically active π and π^* bands, the Raman effect is resonant for all laser frequencies in the infrared, visible, and near UV ranges. Different "quantum pathways" [Chen *et al.* (2011)] (corresponding to contributions of different microscopic processes in perturbation theory) contribute to the intensity of a given peak. This gives rise to intriguing quantum interference effects with sometimes surprising consequences for the Raman intensities (e.g., as a function of laser energy and of sample doping). Furthermore, the intensities and the widths of the Raman peaks are strongly influenced by the lifetimes of intermediate electronic excitations. Raman scattering thus also yields information on the complex dynamics of electron-hole pairs in graphene. All this makes a chapter on

Raman spectroscopy an integral part of a book about the optical properties of graphene.

A number of books/book chapters [Saito *et al.* (1998); Reich *et al.* (2004)] and review articles [Ferrari (2007); Malard *et al.* (2009); Ferrari and Basko (2013)] have already been devoted to the topic of Raman scattering in graphitic materials. This book chapter puts particular emphasis on the theoretical foundations of resonant Raman scattering in graphene. It explains the basic theory for Raman scattering in graphene using perturbation theory. Together with the study of the groundbreaking theoretical research articles (in particular [Basko (2008, 2009); Venezuela *et al.* (2011); Herziger *et al.* (2014)]), it should enable the newcomer to the field to learn how an exact quantitative calculation can be achieved in principle. For the computationally less ambitious reader, the chapter gives detailed qualitative understanding how the different terms in perturbation theory depend on external stimuli (such as doping, defects, or strain) and what conclusions one can draw from the Raman spectra.

The chapter is structured as follows: In Section 3.2, we first introduce the basic building blocks for the understanding of resonant Raman spectra in graphene: The electronic band structure, the phonon dispersion, the coupling between electrons and phonons, and the coupling of the electronic states to light. Afterwards, the formalism of time-dependent perturbation theory is applied to give the reader an intuitive understanding and enable her/him to study the relevant research literature. In Section 3.3, the theory is then applied to understand qualitatively and quantitatively how Raman spectroscopy can be used to study the quality of graphene flakes as well as to understand the effects of environment, doping and strain.

3.2 Theoretical Description of the Raman Spectrum of Graphene

In this first section, we will have a look at the microscopic origin of the Raman spectrum of graphene, in particular the behavior, shape, and size of its various peaks. To this end, we will use time-dependent perturbation theory to calculate expressions for the Raman scattering amplitudes. This approach has the advantage that the individual terms of the perturbation series can be represented by so-called *Feynman diagrams*. The graphical representation of the calculation allows for easier bookkeeping of the terms in the perturbation series and at the same time gives a simple physical picture for the microscopic processes governing Raman scattering.

In the first part of this section, we will introduce the basic building blocks needed for the perturbation theory calculation. We briefly discuss the electronic band structure of graphene before having a more detailed look at its phonon dispersion. Thereafter, we will discuss the coupling between electrons and phonons and their treatment in perturbation theory and show how to include the effects of an external electromagnetic field, i.e. of incoming and outgoing photons. This first part will be concluded by briefly touching upon electron-defect scattering and a summary in which we give a graphical representation of the mathematical expressions for the various building blocks in terms of Feynman diagrams.

The second part contains an overview over the kinematics of Raman scattering. We will derive what kind of peaks are expected in the Raman spectrum of graphene based on kinematical considerations alone.

In the third subsection, we will focus on the actual calculation of the Raman spectrum. Using time-dependent perturbation theory and the language of Feynman diagrams, we will demonstrate how the Raman spectrum of graphene can be calculated. In this section we will mainly focus on the two most prominent Raman peaks and study their behavior from the scattering amplitude obtained in perturbation theory.

3.2.1 *Basic building blocks*

We start by introducing the basic ingredients needed for a theoretical description of the Raman spectrum of graphene. These ingredients are the electronic band structure, the phonon dispersion, the electron-phonon coupling, the electron-light coupling, and (if the sample is not perfect) electron-defect scattering. With the exception of the latter, we will have a look at each of these building blocks before we conclude this sub-section with a summary.

3.2.1.1 *Electronic band structure*

Graphene is a monolayer of carbon atoms arranged in a two-dimensional hexagonal lattice. Each carbon atom possesses four valence electrons, three of which occupy hybridized sp^2 orbitals, which form strong, in-plane covalent $\sigma-$bonds. The forth valence electron occupies a p_z-orbital, which form π-bonds, in which the electrons are delocalized. The electronic band structure, obtained from an *ab initio* calculation using density functional theory (DFT) is shown in Fig. 3.1.

Since we will be interested in describing and calculating the Raman spectrum for excitation energies in or near the visible spectral range

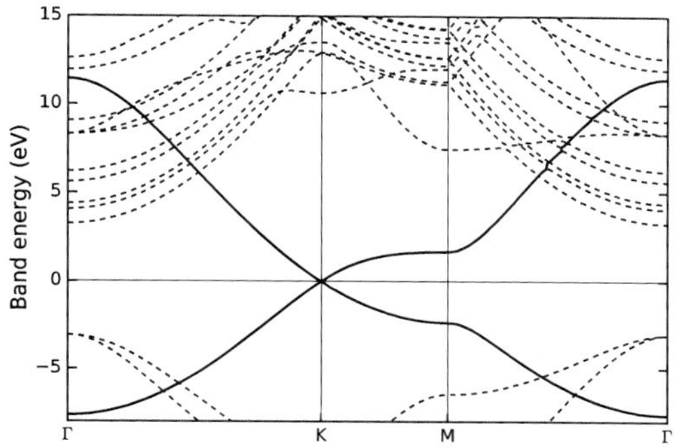

Fig. 3.1: **Electronic band structure of monolayer graphene.** *Ab initio* electronic band structure of monolayer graphene along the high symmetry line $\Gamma - K - M - \Gamma$ obtained from density functional theory (DFT) in the generalized gradient approximation (GGA). The full lines are the π and π^* bands that are by far the most important bands for the Raman scattering process.

(~1.5-4 eV), it is sufficient to focus on the bands near the Fermi energy, which are the π and π^* bands (full lines in Fig. 3.1). Note that for a quantitatively accurate calculation of the Raman spectrum, the DFT band structure needs to be corrected to properly include electron-electron-interaction effects. These corrections can be computed, for example, within the *GW* approximation [Hedin (1965); Onida *et al.* (2002)] and their main effect is a steepening of the conic π bands around the K point with their slope increasing by roughly 18% [Grüneis *et al.* (2008); Venezuela *et al.* (2011)].

3.2.1.2 *Phonon dispersion*

Besides the electronic band structure, the phonon dispersion of graphene also plays a very important role for the understanding of the Raman spectrum. Since the lattice structure of graphene can be described by a hexagonal lattice with two atoms per unit cell, the phonon band structure consists of 2×3=6 branches: Three acoustic ones, which can be further divided into an in-plane longitudinal (LA), an in-plane transverse (TA) and an out-of-plane transverse (ZA) branch, and three optical branches, which can also be separated into in-plane longitudinal (LO), in-plane transverse (TO), and out-of-plane transverse (ZO). The terms "longitudinal" and "transverse" refer to displacements in directions parallel and perpendicular to the quasi-momentum $\mathbf{q}$ of the phonon, respectively. Note that these notions only make sense for small phonon wave vectors, i.e for $\mathbf{q}$ near Γ. Away from Γ, the character of the modes is a mixture of acoustic (in-phase oscillations of neighboring atoms) and optical (opposite phase oscillations of neighboring atoms) behavior. However, for simplicity the labels LA etc. are usually also used to designate the phonon branches away from Γ by following the specific branch along the high-symmetry lines. The phonon dispersion of graphene as obtained from a density functional perturbation theory (DFPT) calculation is shown in Fig. 3.2a.

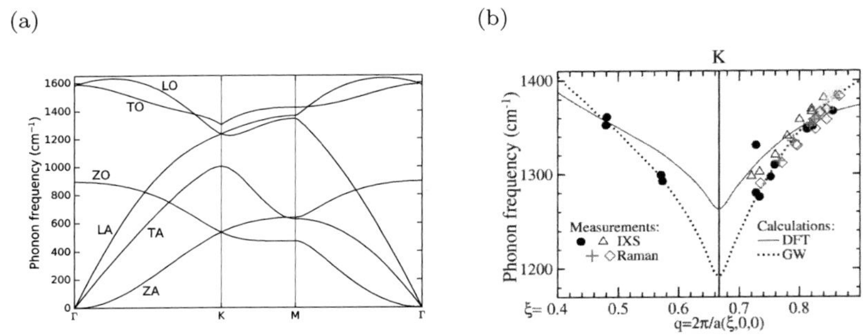

Fig. 3.2: **Phonon dispersion of monolayer graphene.** (a) *Ab initio* phonon dispersion of monolayer graphene along the high symmetry line $\Gamma - K - M - \Gamma$ obtained from density functional perturbation theory (DFPT) in the local density approximation (LDA). The phonon frequencies are given in units of cm^{-1} to allow for an easier comparison to Raman shifts, which are traditionally reported as (shifts of) the inverse wave length (see the next subsection for a conversion formula between the two). (b) Zoom-in into the phonon dispersion for phonon wave vectors around K. Symbols denote experimental data obtained from inelastic X-ray scattering (IXS) and Raman scattering. The full lines represent the TO branch obtained from density functional theory within the generalized gradient approximation (GGA). The dashed lines represents the TO branch dispersion corrected within perturbation theory, where the electron-phonon coupling was obtained on the level of the GW approximation. (Figure taken from [Lazzeri *et al.* (2008)].)

The most prominent features of the phonon dispersion are the kink in the TO branch at K and the positive slope of the LO branch at Γ. As we will see in a later section, these are precisely the parts of the phonon dispersion that play the most important role in the description of the Raman spectrum, as the phonons that contribute the most to it are the two degenerate in-plane optical phonons at Γ and the phonons of the TO branch around the K point, whose lattice vibration patterns are shown in Fig. 3.3. For a correct description of the Raman spectrum it is therefore of utmost importance to accurately describe these parts of the phonon dispersion.

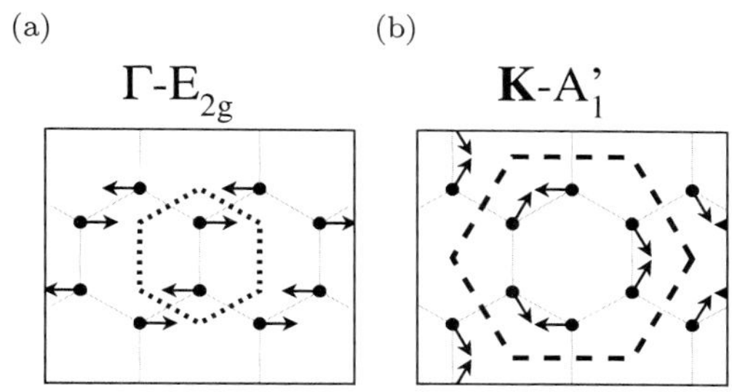

Fig. 3.3: **Lattice vibration patterns.** (a) Vibration pattern corresponding to one of the two degenerate, optical in-plane phonons at Γ. In the commonly used *Mulliken notation*, this mode is called an E_{2g} mode, as it transforms in the E_{2g} representation of the point group of Γ. The dotted line marks the boundary of the Wigner-Seitz cell of the graphene lattice. (b) Vibration pattern corresponding to the TO branch at K. In the Mulliken notation, this mode is denoted as A_1', as it transforms in the A_1' representation of the point group of K. The dashed lines denote the boundary of a supercell of six atoms. (Figure taken from [Lazzeri *et al.* (2008)].)

The pronounced kinks arise due to so-called *Kohn anomalies* [Kohn (1959)]. When the lattice ions vibrate according to a phonon mode with wave vector $\mathbf{q}$, they induce an electronic charge density of the same periodicity, whose magnitude becomes very small when $\mathbf{q}$ connects two points on the Fermi surface. Since the Fermi surface of graphene consists of the two points K and K' at the corners of the Brillouin zone (which are separated by a vector $\mathbf{K}$), phonons with $\mathbf{q} = \mathbf{0}$ or $\mathbf{q} = \mathbf{K}$ lead to these small, periodic electronic charge densities. This decrease of the magnitude of the periodic electronic charge densities in turn lead to descreased screening of the ionic charges, leading to a great enhancement of electron-phonon interaction. This large electron-phonon coupling for $\mathbf{q}$ near Γ or K is responsible for the comparatively steep slope of the phonon dispersion (i.e., the kinks) in these parts of the Brillouin zone, since it can be shown that at Γ and K, the slope of the phonon dispersion is directly proportional to the squared electron-phonon coupling between the corresponding phonon and electrons in the π and π^* bands averaged over a small circle of electronic $\mathbf{k}$ vectors around the Fermi points K and K' [Piscanec *et al.* (2004)].

The increased electron-phonon coupling at Γ and especially at K is further strongly enhanced by electronic correlation effects [Lazzeri *et al.* (2008)]. This manifests itself in the inaccurate quantitative description of the phonon dispersion near K in density functional theory, where correlation effects due to electron-electron interaction are not well-described by exchange-correlation functionals in the commonly used local density or generalized gradient approximations. As will be commented upon in the next subsection, it is possible to obtain a more accurate description of the electron-phonon coupling for phonons at Γ and K on the level of the GW approximation. These corrected values for the electron-phonon coupling lead to a correction of the phonon frequencies within perturbation theory. The most important result is a steepening of the kink near K and a decrease of the phonon frequency of the TO branch at K by more than 100 cm^{-1} [Lazzeri *et al.* (2008)], which matches the experimental data well (compare Fig. 3.2b).

3.2.1.3 *Electron-phonon coupling*

Besides a description of the electron and phonon band structures, one needs a description of the coupling between the two. For this, we consider the electronic Hamiltonian (discarding electron-electron interaction for the moment), which has the form

$$H_0 = \frac{\mathbf{p}^2}{2m_e} + V(\mathbf{r}; \{\mathbf{R}_{n\alpha,0}\}), \tag{3.1}$$

where $\mathbf{r}$ and $\mathbf{p}$ denote the position and momentum operators, respectively, and V is the lattice potential which parametrically depends on the position of the nuclei, which we collectively denote with $\{\mathbf{R}_{n\alpha}\}$, where n labels the unit cell and α specifies the atom within the unit cell. The index "0" refers to the equilibrium positions of the nuclei.

When the lattice atoms are oscillating, the position of the nucleus at position $\mathbf{R}_{n\alpha,0}$ changes by a small displacement: $\mathbf{R}_{n\alpha,0} \to \mathbf{R}_{n\alpha} = \mathbf{R}_{n\alpha,0} + \mathbf{u}(\mathbf{R}_{n\alpha,0})$, where $\mathbf{u}(\mathbf{R}_{n\alpha,0})$ denotes the displacement of the nucleus at equilibrium position $\mathbf{R}_{n\alpha,0}$. As a consequence, the lattice potential changes as well and for small displacements $\mathbf{u}(\mathbf{R}_{n\alpha,0})$ the potential can be expanded into a Taylor series:

$$\begin{aligned} V(\mathbf{r};\{\mathbf{R}_{n\alpha}\}) &= V(\mathbf{r};\{\mathbf{R}_{n\alpha,0}\}) + \sum_{n,\alpha} \frac{\partial V}{\partial \mathbf{R}_{n\alpha}}(\mathbf{r};\{\mathbf{R}_{n\alpha,0}\}) \cdot \mathbf{u}(\mathbf{R}_{n\alpha,0}) \\ &+ \frac{1}{2} \sum_{\substack{n,m\\ \alpha,\beta}} \sum_{i,j} \frac{\partial^2 V}{\partial R_{n\alpha,i} \partial R_{m\beta,j}}(\mathbf{r};\{\mathbf{R}_{n\alpha,0}\}) u_i(\mathbf{R}_{n\alpha,0}) u_j(\mathbf{R}_{m\beta,0}) \\ &+ \mathcal{O}(u^3). \end{aligned} \tag{3.2}$$

Here, the sums run over all unit cells n, all atoms α within the unit cell, and all cartesian coordinates i. The electronic Hamiltonian thus acquires two more terms (up to quadratic order in the displacement), describing the coupling of electrons to one and two phonons, respectively:

$$H_0 \to H_0 + H^{(1)}_{\mathrm{el-ph}} + H^{(2)}_{\mathrm{el-ph}}, \tag{3.3}$$

where $H^{(i)}_{\mathrm{el-ph}}$ is given by the ith order term in the expansion of the potential given above.

After quantizing the phonon field, the operator corresponding to a displacement of the atom at equilibrium position $\mathbf{R}_{n\alpha,0}$ can be written as a superposition of the vibrational eigenmodes:

$$\mathbf{u}(\mathbf{R}_{n\alpha,0}) = \sum_{\mathbf{q},\lambda} \sqrt{\frac{\hbar}{2M_C \omega_{\mathbf{q},\lambda}}} \left[b_{\lambda,\mathbf{q}} \mathbf{v}^{\alpha}_{\lambda,\mathbf{q}} + b^{\dagger}_{\lambda,-\mathbf{q}} \left(\mathbf{v}^{\alpha}_{\lambda,-\mathbf{q}}\right)^* \right] \mathrm{e}^{i\mathbf{q}\cdot\mathbf{R}_n}, \tag{3.4}$$

where $\mathbf{q}$ and λ denote the quasimomentum and phonon branch (e.g. λ=TO for the transverse optical branch) of the eigenmode, respectively, $\mathbf{R}_n$ is the position, i.e., lattice site, of the nth unit cell, and $b^{(\dagger)}_{\lambda,\mathbf{q}}$ is an operator that destroys (creates) one phonon of branch λ with quasimomentum $\mathbf{q}$. The vector $\mathbf{v}^{\alpha}_{\lambda,\mathbf{q}}$ denotes one of the two three-dimensional parts of the six-dimensional eigenvector of the dynamical matrix describing the displacement of atom α of the unit cell.

To describe the Raman scattering amplitude within perturbation theory, one needs the matrix elements of the electron-phonon Hamiltonian for the π and π^* states and for one or two specific phonons. For example, the matrix elements of the electron-one-phonon Hamiltonian for a phonon of branch λ and with quasimomentum $\mathbf{q}$ in the final state are given by

$$\begin{aligned} g^{\lambda}_{s,\mathbf{k}-\mathbf{q};s',\mathbf{k}} &= \langle s,\mathbf{k}-\mathbf{q};\lambda,\mathbf{q}|H^{(1)}_{\mathrm{el-ph}}|s',\mathbf{k}\rangle \\ &= \sqrt{\frac{\hbar}{2M_C\omega_{\mathbf{q},\lambda}}} \sum_{n,\alpha} \mathrm{e}^{i\mathbf{q}\cdot\mathbf{R}_n} \left(\mathbf{v}^{\alpha}_{\lambda,\mathbf{q}}\right)^* \\ &\quad \cdot \left\langle s,\mathbf{k}-\mathbf{q} \middle| \frac{\partial V}{\partial \mathbf{R}_{n\alpha}}(\mathbf{r};\{\mathbf{R}_{n\alpha,0}\}) \middle| s',\mathbf{k} \right\rangle, \end{aligned} \tag{3.5}$$

where s and s' specify the electronic band and $|s, \mathbf{k} - \mathbf{q}; \lambda, \mathbf{q}\rangle$ is a state consisting of an unperturbed electronic state specified by the band index s and Bloch wave vector $\mathbf{k} - \mathbf{q}$ and a phonon from branch λ with quasi-momentum $\mathbf{q}$. For the Raman spectrum of graphene, only the π and π^* bands are important, i.e. $s, s' = \pi, \pi^*$ and for every $\mathbf{k}$, $\mathbf{q}$, and λ the electron-phonon coupling matrix elements can be written as a 2×2-matrix $\mathbf{g}^{\lambda}_{\mathbf{k}-\mathbf{q},\mathbf{q}} = \left(g^{\lambda}_{s,\mathbf{k}-\mathbf{q};s',\mathbf{k}}\right)_{s,s'=\pi,\pi^*}$ in the space of the π and π^* bands.

By the same procedure, one can obtain a 2×2-matrix containing the matrix elements of the electron-two-phonon interaction Hamiltonian $H^{(2)}_{\text{el-ph}}$. For the purpose of this chapter, only those matrix elements that contain the two phonons in the final state are important. Again, these can conveniently be written as a 2×2-matrix in the space of π and π^* bands, which we will denote by $\tilde{\mathbf{g}}^{\lambda_1,\lambda_2}_{\mathbf{k}-\mathbf{q}_1-\mathbf{q}_2,\mathbf{k}}$, where $\lambda_{1(2)}$ and $\mathbf{q}_{1(2)}$ are the quantum numbers of phonon 1(2).

Numerical values for the electron-phonon matrix elements can be obtained from density functional perturbation theory. It should be noted, however, that the underestimation of electron-electron interaction effects in DFT leads to an underestimation of the electron-phonon matrix elements as well, especially for $\mathbf{k}$ and $\mathbf{q}$ near K, where long-range electronic correlation effects play an important role. For specific $\mathbf{k}$ and $\mathbf{q}$, such as $\mathbf{k} = \mathbf{q} = K$, it is possible to incorporate these effects by relating the electron-phonon matrix elements at these high-symmetry points to changes of the band energies when the lattice ions are displaced statically according to the phonon displacement pattern shown in Fig. 3.3b. For instance, for a phonon from the TO branch with quasimomentum $\mathbf{q} = \mathbf{K}$, the square of the electron-phonon coupling for a π and π^* state with $\mathbf{k} = 2\mathbf{K} = \mathbf{K}'$ obeys [Piscanec *et al.* (2004); Lazzeri *et al.* (2008)]:

$$\left|g^{\text{TO}}_{\pi,\mathbf{K};\pi^*,2\mathbf{K}}\right|^2 \propto \lim_{d\to 0}\left(\frac{\left[\varepsilon^{\pi^*}_{\mathbf{K}} - \varepsilon^{\pi}_{\mathbf{K}}\right](d)}{d}\right)^2, \tag{3.6}$$

where d is the amplitude of the displacement and $\varepsilon^{\pi(\pi^*)}_{\mathbf{K}}(d)$ is the energy of the π (π^*) band at K when the atoms are displaced.

When incorporating electron-electron interaction effects on the level of the GW approximation into the band energies, this procedure allows one to calculate the electron-phonon matrix elements at these high-symmetry points while taking into account correlation effects more accurately. This procedure, however, is limited to high-symmetry points for $\mathbf{k}$ and $\mathbf{q}$ as only there the electron-phonon matrix elements can directly be related to changes of the band energies.

3.2.1.4 *Electron-light coupling*

The final ingredient needed for a theoretical description of Raman spectra is the coupling of electrons to light, i.e. an external electromagnetic field. An external electromagnetic field can be described in terms of a vector potential $\mathbf{A}(\mathbf{r})$ and a scalar potential $\phi(\mathbf{r})$. In Coulomb gauge $\nabla \cdot \mathbf{A}(\mathbf{r}) = 0$, the scalar potential must be zero for fields vanishing at infinity. The external electromagnetic field can therefore be described entirely in terms of a vector potential $\mathbf{A}(\mathbf{r})$. The Coulomb gauge imposes one constraint on $\mathbf{A}(\mathbf{r})$, leaving two degrees of freedom for the external electromagnetic field, which can be identified with the two possible linearly independent polarization directions of light.

The coupling of electrons to an external vector potential $\mathbf{A}(\mathbf{r})$ can be introduced via the *minimal coupling* prescription [1] $\mathbf{p} \to \mathbf{p} + \frac{e}{c}\mathbf{A}(\mathbf{r})$. The electronic Hamiltonian then becomes:

$$\begin{aligned} H &= \frac{\left[\mathbf{p} + \frac{e}{c}\mathbf{A}(\mathbf{r})\right]^2}{2m_e} + V(\mathbf{r}; \{\mathbf{R}_0\}) \\ &= H_0 + \frac{e}{cm_e}\mathbf{A}(\mathbf{r}) \cdot \mathbf{p} + \frac{e^2}{2c^2 m_e}[\mathbf{A}(\mathbf{r})]^2 \\ &= H_0 + H^{(1)}_{\text{el-R}} + H^{(2)}_{\text{el-R}}. \end{aligned} \tag{3.7}$$

The coupling to the external electromagnetic field generates two more terms in the Hamiltonian that can be treated in perturbation theory.

To obtain the matrix elements for the coupling of electrons to individual photons, the electromagnetic field has to be quantized. For this, one expands the vector potential in the eigenmodes of a free electromagnetic field, i.e. into plane waves, and promotes it to an operator:

$$\mathbf{A}(\mathbf{r}) = \int \frac{\mathrm{d}^3 k}{(2\pi)^3} \sum_{\mu} \sqrt{\frac{4\pi c^2 \hbar}{2\omega_k}} \left[\boldsymbol{\epsilon}_{\mu,\mathbf{k}} a_{\mu,\mathbf{k}} + (\boldsymbol{\epsilon}_{\mu,-\mathbf{k}})^* a^\dagger_{\mu,-\mathbf{k}}\right] \mathrm{e}^{i\mathbf{k}\cdot\mathbf{r}}, \tag{3.8}$$

where μ labels the two possible polarizations of a photon, $\omega_k = c|\mathbf{k}|$ and $\boldsymbol{\epsilon}_{\mu,\mathbf{k}}$ are the frequency and the polarization vector of a photon with polarization μ and wave vector $\mathbf{k}$, respectively, and $a^{(\dagger)}_{\mu,\mathbf{k}}$ is an operator that destroys (creates) a photon with polarization μ and wave vector $\mathbf{k}$.

[1] We work with *Gaussian* (or *cgs*) units. In SI units, the factor e/c is to be replaced by e.

The matrix elements of the electron-one-photon Hamiltonian between two unperturbed electronic states and with one photon with polarization μ and wave vector $\mathbf{k}'$ in the initial state then reads:

$$\begin{aligned}\gamma^{\mu}_{s,\mathbf{k}+\mathbf{k}';s',\mathbf{k}} &= \langle s,\mathbf{k}-\mathbf{k}';\mu,\mathbf{k}'|H^{(1)}_{\mathrm{el-R}}|s',\mathbf{k}\rangle \\ &= \frac{e}{cm_e}\sqrt{\frac{4\pi c^2\hbar}{2\omega_{k'}}}\boldsymbol{\epsilon}^*_{\mu,\mathbf{k}'}\cdot\left\langle s,\mathbf{k}-\mathbf{k}'\middle|\mathrm{e}^{i\mathbf{k}'\cdot\mathbf{r}}\mathbf{p}\middle|s',\mathbf{k}\right\rangle,\end{aligned} \tag{3.9}$$

where $|s,\mathbf{k}-\mathbf{k}';\mu,\mathbf{k}'\rangle$ denotes a state consisting of an unperturbed electronic state specified by the band index s and Bloch wave vector $\mathbf{k}-\mathbf{k}'$ and a photon with polarization μ with wave vector $\mathbf{k}'$. This expression is often simplified further by the *dipole approximation* $\exp(i\mathbf{k}'\cdot\mathbf{r})\approx 1$, which is justified by the fact that the wave length of the light (~300-800 nm for light in the visible spectrum) is much larger than the typical crystal length scale given by the lattice constant (~2.46 Å for graphene), i.e. $\mathbf{k}'\cdot\mathbf{r}\approx 0$ if $\mathbf{r}$ varies on the crystal scale only as it does in the matrix elements between Bloch states. In other words, for the description of Raman spectra, we can treat the external photons as having zero wave vector (i.e. we set $\mathbf{k}'=0$), yet finite frequency. Again, we will collect the matrix elements between the possible combinations of π and π^* states for fixed $\mathbf{k}$ and light polarization μ in a 2×2-matrix $\gamma^{\mu}_{\mathbf{k}}=\left(\gamma^{\mu}_{s,\mathbf{k};s',\mathbf{k}}\right)_{s,s'=\pi,\pi^*}$. The matrix elements for the *absorption* of a photon can be obtained by hermitian conjugation of the matrix.

Finally, one can follow a similar procedure to obtain the matrix elements between π and π^* states for the electron-two-photon Hamiltonian $H^{(2)}_{\mathrm{el-R}}$. For a calculation of the Raman spectrum only those matrix elements are important which involve one photon with polarization μ in the initial state and one photon with polarization ν in the final state. We will denote the corresponding 2×2-matrix of matrix elements in the basis of π and π^* states by $\tilde{\gamma}^{\mu\nu}_{\mathbf{k}}$.

3.2.1.5 *Electron-defect scattering*

The final building block needed for a complete description of the Raman spectrum within the framework of perturbation theory is a description of electron-defect scattering. If the graphene sheet contains defects, electrons may scatter from them elastically, leading to additional observable peaks in the Raman spectrum. Since a first, qualitative understanding of these defect-assisted peaks does not require a detailed model for electron-defect scattering, we will not have a closer look at this issue but merely refer the reader to the relevant literature [Hwang and Das Sarma (2008); Venezuela *et al.* (2011)].

3.2.1.6 *Summary and Feynman rules*

To conclude this section, we summarize the elemental building blocks that are needed for the calculation of the Raman spectrum of graphene. Each of these elementary processes can be represented graphically by a so-called *Feynman diagram*. In these diagrams each involved (quasi-)particle is represented by a specific line. We represent electrons (and holes) by full lines with arrows, phonons by dashed lines, and photons by wavy lines. The elemental Feynman diagrams for the various building blocks are shown in Table 3.1.

In addition to the matrix elements for scattering processes between particles, there also appears a diagram for the propagation of an unperturbed electron state between two scattering events. This amplitude is given by the *retarded Green's function*, defined as:

$$G^R_{s,\mathbf{k}}(t) = \begin{cases} -ie^{-i\varepsilon^s_{\mathbf{k}}t/\hbar}\left(1 - n_{\mathrm{F}}(\varepsilon^s_{\mathbf{k}})\right), & t > 0 \\ +ie^{-i\varepsilon^s_{\mathbf{k}}t/\hbar}n_{\mathrm{F}}(\varepsilon^s_{\mathbf{k}}), & t < 0, \end{cases} \tag{3.10}$$

where $\varepsilon^s_{\mathbf{k}}$ is the energy of the state $|s = \pi, \pi^*, \mathbf{k}\rangle$ and $n_{\mathrm{F}}(\varepsilon)$ denotes the Fermi-Dirac distribution. The retarded Green's function describes the propagation of an unperturbed electron state for positive times ($t > 0$) and of an unperturbed hole state for negative times $t < 0$. Since the perturbation theory calculation will be done in frequency space, we also give the Fourier transform of the retarded Green's function:

$$G_{s,\mathbf{k}}(\omega) = \frac{n_{\mathrm{F}}(\varepsilon^s_{\mathbf{k}})}{\hbar\omega - \varepsilon^s_{\mathbf{k}} - \frac{i}{2}\gamma^s_{\mathbf{k}}} + \frac{1 - n_{\mathrm{F}}(\varepsilon^s_{\mathbf{k}})}{\hbar\omega - \varepsilon^s_{\mathbf{k}} + \frac{i}{2}\gamma^s_{\mathbf{k}}}. \tag{3.11}$$

Here, we accounted for the fact that an electronic state $|s, \mathbf{k}\rangle$ has a finite lifetime $\gamma^s_{\mathbf{k}}$, due to electron-phonon, electron-defect, and electron-electron scattering. We also included an additional factor of $1/\hbar$ into the Fourier transform, so that the Green's function in Fourier space has the dimensions of inverse energy. As in the sections before, it is convenient to write the Fourier-transformed Green's function as a 2×2-matrix in the space of π and π^* states at fixed $\mathbf{k}$: $\mathbf{G}_{\mathbf{k}}(\omega) = \left(G_{s,\mathbf{k}}(\omega)\delta_{s,s'}\right)_{s,s'=\pi,\pi^*}$, where $\delta_{s,s'}$ is the Kronecker delta.

	Diagram	Feynman rule
Propagation of electron	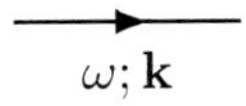	$i\mathbf{G}_{\mathbf{k}}(\omega)$
Electron-light interaction	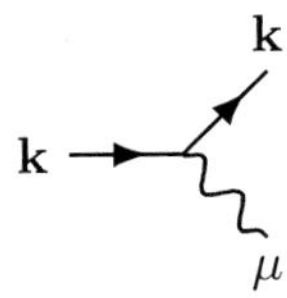	$(-i)\gamma^{\mu}_{\mathbf{k}}$
	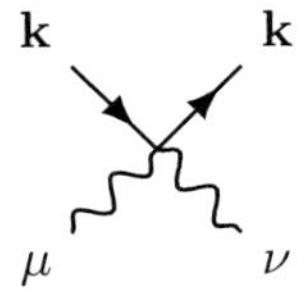	$(-i)\tilde{\gamma}^{\mu\nu}_{\mathbf{k}}$
Electron-phonon interaction	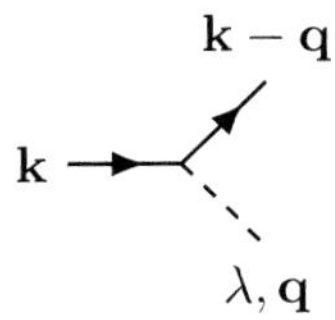	$(-i)\mathbf{g}^{\lambda}_{\mathbf{k}-\mathbf{q},\mathbf{k}}$
	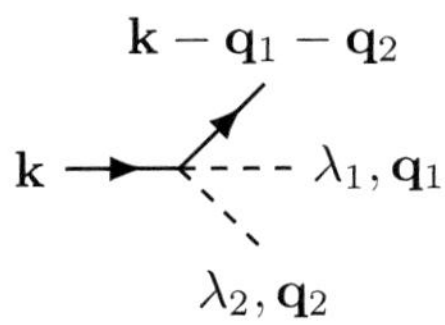	$(-i)\tilde{\mathbf{g}}^{\lambda_1\lambda_2}_{\mathbf{k}-\mathbf{q}_1-\mathbf{q}_2,\mathbf{k}}$

Table 3.1: **Feynman rules for the basic building blocks for Raman scattering.** Feynman diagrams and associated factors appearing in the perturbation series for the Raman scattering amplitudes. Each factor is a 2×2-matrix in the basis of π and π^* states, i.e., each element of the 2×-matrix is a matrix element between π and π^* states.

3.2.2 *General kinematic considerations and possible Raman processes*

Before we use the elementary processes and their associated matrix elements to compute the quantum mechanical amplitude for the various possible Raman processes, we will first discuss some kinematic aspects of Raman scattering in general and discuss which Raman processes are expected on kinematic grounds. In a general Raman process, the initial state consists of one photon of energy $\varepsilon_{\text{in}} = \hbar\omega_{\text{in}}$ and polarization μ and the final state contains one photon of energy $\varepsilon_{\text{out}} = \hbar\omega_{\text{out}}$ and polarization ν plus n phonons with wave vectors $\mathbf{q}_1, \ldots, \mathbf{q}_n$ from branches $\lambda_1, \ldots, \lambda_n$. The energy conservation law determines the observed Raman shift

$$\omega_{\text{in}} - \omega_{\text{out}} = \sum_{i=1}^{n} \omega_{\mathbf{q}_i,\lambda_i} \tag{3.12}$$

as the sum of the frequencies of the involved phonons. Traditionally, Raman shifts (and phonon frequencies) are converted to shifts of the inverse wavelength, $\Delta\lambda^{-1} = \lambda_{\text{in}}^{-1} - \lambda_{\text{out}}^{-1}$, and given in units of cm^{-1}, where the inverse wavelength can be expressed in terms of the frequency ω via the relation $\lambda^{-1} = \omega/(2\pi c)$. For reference, an energy of $\hbar\omega = 1$ meV corresponds to a Raman shift of approximately 8 cm^{-1}. The wave vectors of the phonons in the final state are further restricted by momentum conservation. In case there are no defects in the sample, the sum of the wave vectors of the phonons must be zero, since the wave vectors of the photons are negligible and hence do not need to be considered in the momentum conservation law:

$$\sum_{i=1}^{n} \mathbf{q}_i = 0 \qquad \text{(no defects).} \tag{3.13}$$

In the presence of defects, the sum of the momenta of the phonons need not be equal to zero anymore since electronic scattering at defects can make up for it. If we limit ourselves to a maximum of two phonons, these kinematic considerations lead to the following categorization of Raman processes and their corresponding designations:

a) One-phonon processes without defects (G peak)
Momentum conservation forces the phonon momentum to be equal to zero, i.e. the phonon must be a phonon from the Brillouin zone center, i.e. the Γ point. Non-zero Raman shifts further require the phonon to be an

optical one since the frequencies of acoustic phonons vanish at Γ. Furthermore, horizontal mirror symmetry forbids processes with one out-of-plane phonon. Thus the phonons that contribute to the one-phonon processes in the absence of defects are the two degenerate, optical in-plane phonons at Γ. These are doubly degenerate due to rotation symmetry, i.e., they possess the same energy leading to the same Raman shift for both. The corresponding peak in the Raman spectrum, located at around 1580 cm^{-1} in pristine graphene, is universally called the G peak as it was first observed in graphite. Note that, as will be discussed in a later section, a breaking of rotation symmetry, for example by strain, leads to the two in-plane optical phonons having different frequencies and induces a splitting of the G peak.

b) Two-phonon processes without defects (most prominently the $2D$, $2D'$, and $D + D''$ peaks)
Here momentum conservation requires the wave vectors of the involved phonons to be opposite, i.e. $\mathbf{q}_1 = \mathbf{q} = -\mathbf{q}_2$. The vector $\mathbf{q}$, however, is not constrained by any further condition and can in principle correspond to any point in the Brillouin zone. This explains why the corresponding Raman peaks are much broader than the G peak, as a continuum of phonon frequencies contributes to them. The fact that there are still clear, observable peaks is related to the fact that certain phonon wave vectors and frequencies lead to a resonance in the Raman amplitude as will be discussed in the next section. The peaks that belong to this category are the $2D$ peak, whose dominant contribution comes from two phonons of the TO branch with $\mathbf{q}$ near K is and located at around 2700 cm^{-1}, the $2D'$ peak at around 3200 cm^{-1}, whose main contribution stems from two phonons of the LO branch near Γ, and the $D + D''$ peak, which is dominated by one phonon of the TO branch with $\mathbf{q}$ near K and one phonon from the LA branch with opposite wave vector and appears at around 2450 cm^{-1}.

c) One-phonon processes with defects (most prominently the D and D' peaks)
In contrast to the G peak, the presence of defects allows phonons away from the Γ point to participate in a one-phonon Raman process as the momentum of the phonon can be compensated by electron-defect scattering (see next section for an example). Thus, in principle, any phonon can contribute to the Raman spectrum. However, just as for the peaks in *b)*, the spectrum is dominated by a few selected phonons that lead to a resonance in the Raman amplitude. The two most prominent peaks that fall into this category are the D peak, which is mostly due to a TO phonon near K and

is located at around 1350 cm^{-1}, and the D' peak, which is mostly caused by a phonon from the LO branch near Γ and occurs at around 1600 cm^{-1}. As their names suggest, these phonons are the same as the ones observed in the two-phonon overtones $2D$ and $2D'$ from the previous paragraph.

d) Two-phonon processes with defects (most prominently the $D+D'$ peak) Finally, there are additional two-phonon Raman processes that are made possible by the presence of defects. In theses processes the two phonons do not need to have opposite wave vectors and can thus stem from different points in the Brillouin zone. The most prominent peak in this category is the $D + D'$ peak at around 2950 cm^{-1}, in which one phonon from the TO branch with a wave vector from the vicinity of K and an LA phonon with a wave vector close to Γ make up the resonant contribution.

A typical, experimentally obtained Raman spectrum for a sample with defects is shown in Fig. 3.4, where the most prominent peaks are labeled and the most dominant phonons are identified.

Having given an overview over the kinematic constraints that govern Raman spectrum, we will now turn to the dynamical aspects in the next section and show how to obtain expressions for the Raman amplitudes and how these can be used to determine which phonons play the most important role in the various processes.

3.2.3 *Diagrammatic calculation of the Raman amplitudes*

In this section we will illustrate the technique of diagrammatic, time-dependent perturbation, theory to obtain mathematical expressions for the quantum mechanical amplitudes for the various Raman processes. The reader can find a more detailed account of this technique in [Fetter and Walecka (1971); Peskin and Schroeder (1995); Mahan (2000)]. We will mainly focus on two examples: One-phonon and two-phonon processes without defects. Afterward, we will briefly comment on the defect-assisted peaks.

3.2.3.1 *One-phonon, defect-free processes (G peak)*

For the first example, consider a defect-free process with one photon with polarization μ and energy $\hbar\omega_{\mathrm{in}}$ in the initial state and one photon with polarization ν and energy $\hbar\omega_{\mathrm{out}}$ plus a phonon of branch λ and with momentum $\mathbf{q}$ in the final state. This is the process that gives rise to the

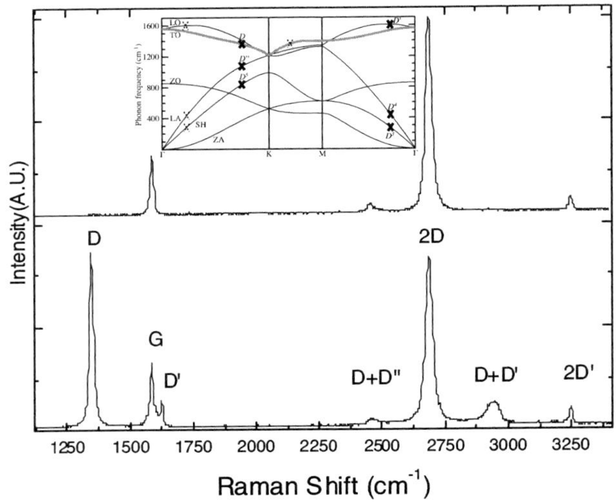

Fig. 3.4: **Example of a Raman spectrum and contributing phonons.** Two examples of a Raman spectrum for a sample without (upper spectrum) and with defects (lower spectrum). The most prominent peaks as described in the text have been labeled. Inset: Phonon dispersion of graphene, including the corrected TO branch. The phonons which most prominently contribute to the various Raman peaks are marked by crosses. (Figure assembled from figures taken from Refs. [Ferrari and Basko (2013)] and [Venezuela *et al.* (2011)].)

G peak. The three leading-order, topologically inequivalent Feynman diagrams that can be assembled from the elementary diagrams of Table 3.1 are shown in Fig. 3.5.

Note that momentum conservation dictates $\mathbf{q} = 0$ since the momentum of photons is negligible as discussed earlier. The wave vectors of the involved virtual electronic states are therefore all the same. The condition $\mathbf{q} = 0$ also fixes the type of the phonon to be optical as acoustic phonons with zero momentum do not couple to π and π^* states [Mañes (2007)]. The phonon branch index λ is thus restricted to the two branches TO and LO which are degenerate at the Γ point. Energy conservation fixes the frequency of the outgoing photon to $\omega_{\text{out}} = \omega_{\text{in}} - \omega_{\mathbf{q}=0,\lambda}$. The finite lifetime, i.e., decay width, of the phonon can be included via the substitution $\omega_{\mathbf{q}=0,\lambda} \to \omega_{\mathbf{q}=0,\lambda} - i/(2\hbar)\gamma_{\mathbf{q}=0,\lambda}$, where the relation between lifetime and decay width is given by $\gamma = \hbar/\tau$. With these kinematic constraints, application of

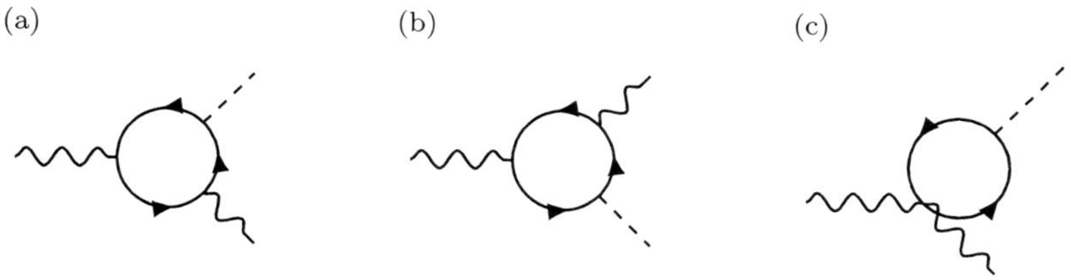

Fig. 3.5: **Leading-order Feynman diagrams for the G peak.** The three leading-order Feynman diagrams that can be constructed from the basic building blocks of the theory.

the Feynman rules yields the following mathematical expressions for the amplitudes corresponding to the three diagrams:

$$\mathcal{M}_1^G = \sum_{\text{spin}} \sum_{\mathbf{k}} \int \frac{\mathrm{d}\omega}{2\pi} \, \mathrm{e}^{+i\omega 0^+} (-1) \mathrm{tr} \Big[\mathbf{G}_{\mathbf{k}}(\omega) \, (\gamma_{\mathbf{k}}^{\mu})^{\dagger} \, \mathbf{G}_{\mathbf{k}}(\omega - \omega_{\text{in}}) \times \mathbf{g}_{\mathbf{k},\mathbf{k}}^{\lambda} \mathbf{G}_{\mathbf{k}}(\omega - \omega_{\text{out}}) \gamma_{\mathbf{k}}^{\nu} \Big] \tag{3.14}$$

$$\mathcal{M}_2^G = \sum_{\text{spin}} \sum_{\mathbf{k}} \int \frac{\mathrm{d}\omega}{2\pi} \, \mathrm{e}^{+i\omega 0^+} (-1) \mathrm{tr} \Big[\mathbf{G}_{\mathbf{k}}(\omega) \gamma_{\mathbf{k}}^{\nu} \mathbf{G}_{\mathbf{k}}(\omega + \omega_{\text{out}}) \times \mathbf{g}_{\mathbf{k},\mathbf{k}}^{\lambda} \mathbf{G}_{\mathbf{k}}(\omega + \omega_{\text{in}}) \, (\gamma_{\mathbf{k}}^{\mu})^{\dagger} \Big] \tag{3.15}$$

$$\mathcal{M}_3^G = \sum_{\text{spin}} \sum_{\mathbf{k}} \int \frac{\mathrm{d}\omega}{2\pi} \, \mathrm{e}^{+i\omega 0^+} (-1) \mathrm{tr} \Big[\mathbf{G}_{\mathbf{k}}(\omega) \mathbf{g}_{\mathbf{k},\mathbf{k}}^{\lambda} \mathbf{G}_{\mathbf{k}}(\omega - \omega_{\mathbf{q}=0,\lambda}) \tilde{\gamma}_{\mathbf{k}}^{\mu\nu} \Big], \tag{3.16}$$

where it is understood that ω_{out} is expressed through ω_{in} and $\omega_{\mathbf{q}=0,\lambda}$ and the latter is a complex number that includes the finite lifetime of the phonon, as discussed before. Notice that, in our convention, where the dimensions of the electron-phonon and electron-light matrix elements are those of an energy and the dimension of the Green's function in Fourier space is inverse energy, the dimension of each matrix element is that of a frequency, i.e. a rate.

To obtain the amplitude for each diagram from the elementary Feynman rules, one goes against the fermion flow (i.e. against the direction of the arrows) and writes down, from left to right the factors for each piece one encounters, imposing momentum and energy (frequency) conservation at each vertex. Since all elementary pieces are 2×2 matrices, the final result will be a matrix. After adding a factor of (-1) that accompanies a fermion loop, one sums over all internal degrees of freedom, which includes a sum over spin, taking the trace of the matrix product, and summing over any undetermined internal momenta and frequencies.

The integral over the loop frequency ω can be evaluated with the residue theorem of complex analysis. Note that in the leading order diagrams for Raman scattering, only the electronic Green's functions have poles that contribute to the contour integral. The factor $\exp(i\omega 0^+)$ prescribes that the contour be closed in the upper half of the complex plane and therefore only those electronic states contribute that feature a positive imaginary part, which are those states that are occupied. Thus the integration over ω picks out the occupied states. After the integral has been carried out, the expression for each Feynman diagram becomes a sum of terms. For example, the amplitude $\mathcal{M}_2^G$ becomes a sum of three terms:

$$\mathcal{M}_2^G = \sum_{\text{spin}} \sum_{\mathbf{k}} \left(-\frac{i}{\hbar}\right) \sum_{s=\pi,\pi^*} n_{\mathrm{F}}(\varepsilon_{\mathbf{k}}^s)$$
$$\times \left\{ \left[\gamma_{\mathbf{k}}^{\nu} \mathbf{G}_{\mathbf{k}}(\omega + \omega_{\mathrm{out}}) \mathbf{g}_{\mathbf{k},\mathbf{k}}^{\lambda} \mathbf{G}_{\mathbf{k}}(\omega + \omega_{\mathrm{in}}) \left(\gamma_{\mathbf{k}}^{\mu}\right)^{\dagger}\right]_{ss} \Big|_{\hbar\omega = \varepsilon_{\mathbf{k}}^s + \frac{i}{2}\gamma_{\mathbf{k}}^s} \right.$$
$$+ \left[\mathbf{g}_{\mathbf{k},\mathbf{k}}^{\lambda} \mathbf{G}_{\mathbf{k}}(\omega + \omega_{\mathrm{in}}) \left(\gamma_{\mathbf{k}}^{\mu}\right)^{\dagger} \mathbf{G}_{\mathbf{k}}(\omega) \gamma_{\mathbf{k}}^{\nu}\right]_{ss} \Big|_{\hbar\omega = -\hbar\omega_{\mathrm{out}} + \varepsilon_{\mathbf{k}}^s + \frac{i}{2}\gamma_{\mathbf{k}}^s}$$
$$\left. + \left[\left(\gamma_{\mathbf{k}}^{\mu}\right)^{\dagger} \mathbf{G}_{\mathbf{k}}(\omega) \gamma_{\mathbf{k}}^{\nu} \mathbf{G}_{\mathbf{k}}(\omega + \omega_{\mathrm{out}}) \mathbf{g}_{\mathbf{k},\mathbf{k}}^{\lambda}\right]_{ss} \Big|_{\hbar\omega = -\hbar\omega_{\mathrm{in}} + \varepsilon_{\mathbf{k}}^s + \frac{i}{2}\gamma_{\mathbf{k}}^s} \right\} \tag{3.17}$$

The number of terms generated in this way is equal to the number of internal propagators (i.e. the lines that make up the loop). For a simple loop, as it is encountered in the diagrams for the leading order terms for Raman scattering, the number of internal lines is equal to the number of vertices. This means that a diagram with three external lines that do not meet in a common vertex will lead to a sum of three terms after the loop frequency has been integrated over. Some books and articles, e.g. the commonly cited [Yu and Cardona (2010)], prefer to work directly with these terms and write down a diagram for each one of them. These kinds of diagrams are known as Goldstone diagrams and should not be confused with the Feynman diagrams used in this text. In fact, every Feynman diagram can be viewed as a short-hand notation for the sum of several Goldstone diagrams with the sum being written as an integral over the loop frequency. Unfortunately, the distinction between Goldstone and Feynman diagrams is not always made in the literature and one sometimes encounters Goldstone diagrams being labeled as Feynman diagrams and *vice versa*. As pointed out, both are essentially equivalent as long as one uses the corresponding set of rules to translate a diagram into its corresponding amplitude and as long as one makes sure to write down all possible topologically non-equivalent diagrams of the chosen approach. In the Goldstone diagram formulation this entails several different time-ordering of the external lines, while in the Feynman diagram formalism the loop integral takes care of this technicality. We thus prefer to work with the Feynman diagram formalism as it greatly reduces the number of diagrams that need to be drawn. In our case, the three Feynman diagrams for the process giving rise to the G peak correspond to eight Goldstone diagrams.

Once one has obtained the quantum mechanical amplitudes for the individual diagrams, the total amplitude for the process can be obtained by adding the individual amplitudes:

$$\mathcal{M}^{\mu\nu\lambda}_{G\text{ peak}}(\omega_{\text{in}};\gamma_{\text{el}};\omega_{\text{ph}},\gamma_{\text{ph}}) = \sum_{d=1}^{3} \mathcal{M}_d, \tag{3.18}$$

where we explicitly noted all quantities which the total matrix element depends on. These are the frequency of the incoming photon, the lifetimes of the involved electronic states (summarily denoted as γ_{el}), and the frequency and decay width of the phonon, abbreviated as ω_{ph} and γ_{ph}, respectively. Note that the phonon frequency of an optical phonon at Γ is independent of the branch λ, as noted before. Finally, Fermi's golden rule yields the total rate of the Raman process, i.e., the probability per unit time for the process to happen, which is directly proportional to the recorded Raman intensity:

$$\begin{aligned} I_G(\omega_{\text{out}}) &\propto \frac{\mathrm{d}P_{G\text{ peak}}}{\mathrm{d}t} \\ &= 2\pi \sum_{\lambda=\text{LO,TO}} \left|\mathcal{M}^{\mu\nu\lambda}_{G\text{ peak}}(\omega_{\text{in}};\gamma_{\text{el}};\omega_{\text{ph}},\gamma_{\text{ph}})\right|^2 \times \delta(\omega_{\text{in}}-\omega_{\text{out}}-\omega_{\text{ph}}) \end{aligned} \tag{3.19}$$

For a phonon with finite lifetime, the δ-function has to be replaced by a Lorentzian:

$$\delta(\omega_{\text{in}}-\omega_{\text{out}}-\omega_{\text{ph}}) \to \frac{1}{\pi}\frac{\gamma_{\text{ph}}/(2\hbar)}{(\omega_{\text{in}}-\omega_{\text{out}}-\omega_{\text{ph}})^2+(\gamma_{\text{ph}}/(2\hbar))^2} \tag{3.20}$$

From this expression we can immediately see that the position and width of the peak are entirely determined by the frequency (approx. 1580 cm^{-1} in pristine graphene) and width (5-15 cm^{-1} depending on the sample quality) of the phonon. The intensity of the G peak, however, depends predominantly on the excitation frequency $\omega_{\rm in}$ as well as on the lifetime of the involved electronic states, while the frequency and lifetime of the phonon only have some minor influence on the intensity of the peak. Regarding the latter, it should be noted that due to the absence of a band gap in graphene, there are always electronic transitions that are in resonance with the energy of the incoming light. The condition for resonance is that the energy of the in- or outgoing light matches the energy of an electronic transition. To give an example of a resonant term appearing in the Raman amplitude, we take a look at one of the terms contained in the matrix product of the first term of $\mathcal{M}_2^G$:

$$\mathcal{M}_2^G \supset \sum_{\rm spin}\sum_{\mathbf{k}} \left(-\frac{i}{\hbar}\right) n_{\rm F}(\varepsilon_{\mathbf{k}}^{\pi}) \left(1 - n_{\rm F}(\varepsilon_{\mathbf{k}}^{\pi^*})\right) \left(1 - n_{\rm F}(\varepsilon_{\mathbf{k}}^{\pi^*})\right)$$
$$\times \frac{\langle \pi_{\mathbf{k}}; \omega_{\rm out}^{\nu} | H_{\rm el-R}^{(1)} | \pi_{\mathbf{k}}^* \rangle \langle \pi_{\mathbf{k}}^*; \omega_{\rm ph}^{\lambda} | H_{\rm el-ph}^{(1)} | \pi_{\mathbf{k}}^* \rangle}{\left[\hbar\omega_{\rm in} - (\varepsilon_{\mathbf{k}}^{\pi^*} - \varepsilon_{\mathbf{k}}^{\pi}) + \frac{i}{2}(\gamma_{\mathbf{k}}^{\pi} + \gamma_{\mathbf{k}}^{\pi^*})\right]} \tag{3.21}$$
$$\times \frac{\langle \pi_{\mathbf{k}}^* | H_{\rm el-R}^{(1)} | \pi_{\mathbf{k}}; \omega_{\rm in}^{\mu} \rangle}{\left[\hbar(\omega_{\rm in} - \omega_{\rm ph}) - (\varepsilon_{\mathbf{k}}^{\pi^*} - \varepsilon_{\mathbf{k}}^{\pi}) + \frac{i}{2}(\gamma_{\mathbf{k}}^{\pi} + \gamma_{\mathbf{k}}^{\pi^*} + \gamma_{\rm ph})\right]},$$

where we replaced the frequency of the outgoing light by the difference of the frequency of the incoming light and the phonon frequency and accounted for the finite lifetime of the phonon by adding a negative imaginary part to the phonon frequency. This term corresponds to a process where an electron in the π band is resonantly excited to the π^* band by the incoming photon from which it drops to a virtual state in the π^* band by emitting a phonon and then recombines with the hole by emitting the photon found in the final state. When the resonance condition $\Delta\varepsilon_{\mathbf{k}} = \hbar\omega_{\rm in}$ or $\hbar\omega_{\rm out}$ is met, the denominator becomes minimal, leading to the resonant behavior. Note that, since the band structure is monotonically increasing away from the Fermi surface (i.e. the K point), the density of states increase monotonically with energy as well and hence the number of states that are in resonance with the energy of the incoming light increases with $\omega_{\rm in}$, leading to the observed strong increase of I_G as a function of $\omega_{\rm in}$ [Cançado *et al.* (2007); Basko (2009)]. By contrast, Raman scattering in semi-conductors with sizable band gaps is mostly non-resonant, unless the excitation energy matches the energy of the band gap or an excitonic state within the gap, whereupon a strong resonance is recorded.

In the limit of small $\omega_{\rm in}$, the matrix element vanishes, which can easily be explained by the following argument: For small $\omega_{\rm in}$, the electronic states that are in resonance with the excitation energy are located very close to the K point. However, very close to the K point, the band structure of graphene can be considered to be conic (the conic band structure is often called *Dirac cone* due to the Hamiltonian reducing to the Dirac Hamiltonian in this limit). A conic band structure has a continuous in-plane rotation symmetry, which implies angular momentum conservation. If we consider circularly polarized [2] light and phonons at Γ, each of them has an angular momentum of $l_z = \pm\hbar$. The initial state thus has angular momentum $\pm\hbar$, while the final state can have an angular momentum of $+2\hbar$, $0\hbar$, or $-2\hbar$, which would violate angular momentum conservation. Therefore, in the case of a conic band structure, the G peak process would not be possible in leading order in perturbation theory as it would violate angular momentum conservation. If one goes one order beyond the linear, conic approximation for the band structure, however, the terms that break the continuous rotation symmetry down to the $120°$ rotation symmetry of the lattice cause a *trigonal warping* of the band structure. The reduced $120°$ symmetry means that angular momentum is only conserved modulo $\pm 3\hbar$, in other words, states with angular momenta that differ by an integer multiple of $3\hbar$ get mixed. This in turn leads to a non-vanishing Raman amplitude at the leading order in perturbation theory, since now the total angular momenta of the initial and final states can differ by an integer multiple of $3\hbar$ and hence an incoming photon with angular momentum $+\hbar$ can generate a phonon and a photon both with angular momentum $-\hbar$. This argument also illustrates that it is important to go beyond the Dirac cone approximation of the band structure when considering the G peak and that it is necessary to include the trigonal warping terms, which make the G peak process possible in the first place.

3.2.3.2 *Two-phonon, defect-free processes (e.g., $2D$, $2D'$, and $D+D''$ peaks)*

As a second example for the diagrammatic calculation of the Raman spectrum, consider again a defect-free process, this time with two phonons in the final state. This kind of process gives rise to the $2D$, $2D'$, and $D+D''$ peaks, among others. With the available building blocks, the amplitude in lowest order in perturbation theory can be constructed from eleven Feynman diagrams (corresponding to 38 Goldstone diagrams), four of which are shown in Fig. 3.6.

[2]The case of linearly polarized photons and phonons can be constructed by taking suitable linear combinations of the circular polarized modes.

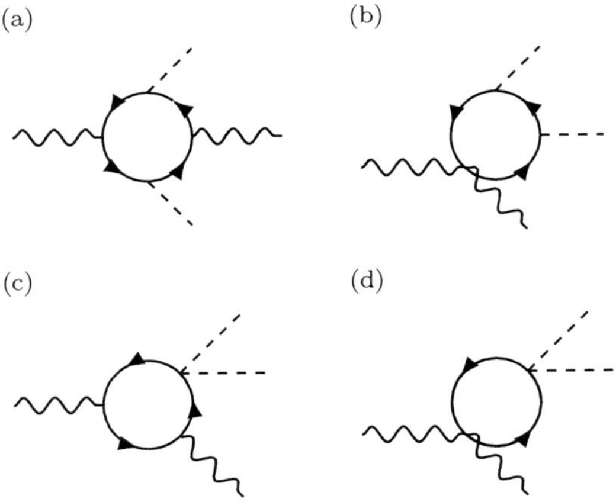

Fig. 3.6: **Four examples of leading-order Feynman diagrams for the two-phonon, defect-free peaks.** Four of the eleven Feynman diagrams that appear in leading order of perturbation theory. Of special importance is the type of diagram shown in (a) as it can become *double resonant* (see text).

Again, application of the Feynman rules leads to the following expressions for the four example diagrams:

$$\mathcal{M}_1^{2\text{ ph.}} = \sum_{\text{spin}} \sum_{\mathbf{k}} \int \frac{\mathrm{d}\omega}{2\pi}\, \mathrm{e}^{+i\omega 0^+} (-1)\mathrm{tr}\Bigg[\mathbf{G}_{\mathbf{k}}(\omega) \mathbf{g}^{\lambda_1}_{\mathbf{k},\mathbf{k}+\mathbf{q}} \mathbf{G}_{\mathbf{k}+\mathbf{q}}(\omega + \omega_{\mathbf{q},\lambda_1}) \gamma^{\nu}_{\mathbf{k}+\mathbf{q}} \times \mathbf{G}_{\mathbf{k}+\mathbf{q}}(\omega + \omega_{\text{out}} + \omega_{\mathbf{q},\lambda_1}) \mathbf{g}^{\lambda_2}_{\mathbf{k}+\mathbf{q},\mathbf{k}} \times \mathbf{G}_{\mathbf{k}}(\omega + \omega_{\text{in}}) \left(\gamma^{\mu}_{\mathbf{k}}\right)^{\dagger} \Bigg] \tag{3.22}$$

$$\mathcal{M}_2^{2\text{ ph.}} = \sum_{\text{spin}} \sum_{\mathbf{k}} \int \frac{\mathrm{d}\omega}{2\pi}\, \mathrm{e}^{+i\omega 0^+} (-1)\mathrm{tr}\Bigg[\mathbf{G}_{\mathbf{k}}(\omega) \mathbf{g}^{\lambda_1}_{\mathbf{k},\mathbf{k}+\mathbf{q}} \mathbf{G}_{\mathbf{k}+\mathbf{q}}(\omega + \omega_{\mathbf{q},\lambda_1}) \mathbf{g}^{\lambda_2}_{\mathbf{k}+\mathbf{q},\mathbf{k}} \times \mathbf{G}_{\mathbf{k}}(\omega + \omega_{\mathbf{q},\lambda_1} + \omega_{-\mathbf{q},\lambda_2}) \tilde{\gamma}^{\mu\nu}_{\mathbf{k}} \Bigg] \tag{3.23}$$

$$\mathcal{M}_3^{2\text{ ph.}} = \sum_{\text{spin}} \sum_{\mathbf{k}} \int \frac{\mathrm{d}\omega}{2\pi}\, \mathrm{e}^{+i\omega 0^+} (-1)\mathrm{tr}\Bigg[\mathbf{G}_{\mathbf{k}}(\omega) \tilde{\mathbf{g}}^{\lambda_1\lambda_2}_{\mathbf{k},\mathbf{k}} \mathbf{G}_{\mathbf{k}}(\omega + \omega_{\mathbf{q},\lambda_1} + \omega_{-\mathbf{q},\lambda_2}) \gamma^{\nu}_{\mathbf{k}} \times \mathbf{G}_{\mathbf{k}}(\omega + \omega_{\text{in}}) \left(\gamma^{\mu}_{\mathbf{k}}\right)^{\dagger} \Bigg] \tag{3.24}$$

$$\mathcal{M}_4^{2\text{ ph.}} = \sum_{\text{spin}} \sum_{\mathbf{k}} \int \frac{d\omega}{2\pi} e^{+i\omega 0^+} (-1)\text{tr}\Big[\mathbf{G}_{\mathbf{k}}(\omega)\tilde{\mathbf{g}}^{\lambda_1\lambda_2}_{\mathbf{k},\mathbf{k}} \mathbf{G}_{\mathbf{k}}(\omega + \omega_{\mathbf{q},\lambda_1} + \omega_{-\mathbf{q},\lambda_2})\tilde{\gamma}^{\mu\nu}_{\mathbf{k}}\Big], \tag{3.25}$$

The total amplitude for this process can again be constructed by summing all (in the leading order eleven) amplitudes:

$$\mathcal{M}^{\mu\nu\lambda_1\lambda_2}_{2\text{ ph.}}(\mathbf{q};\omega_{\text{in}};\gamma_{\text{el}};\omega_{\text{ph}},\gamma_{\text{ph}}) = \sum_{d=1}^{11} \mathcal{M}_d, \tag{3.26}$$

The intensity is then calculated form Fermi's golden rule after summing over all possible phonon branches λ_1, λ_2 and wave vectors $\mathbf{q}$:

$$\begin{aligned} I_{2\text{ ph.}}(\omega_{\text{out}}) &\propto \frac{dP_{2\text{ ph.}}}{dt} \\ &= 2\pi \sum_{\lambda_1,\lambda_2} \sum_{\mathbf{q}} \left|\mathcal{M}^{\mu\nu\lambda_1\lambda_2}_{2\text{ ph.}}(\mathbf{q};\omega_{\text{in}};\gamma_{\text{el}};\omega_{\text{ph}},\gamma_{\text{ph}})\right|^2 \\ &\quad \times \delta(\omega_{\text{in}} - \omega_{\text{out}} - \omega_{-\mathbf{q},\lambda_1} - \omega_{\mathbf{q},\lambda_2}). \end{aligned} \tag{3.27}$$

Again, for phonons with finite lifetimes, the δ-function is to be replaced by a Lorentzian. From the above expression, we can see that the two-phonon, defect-free process will result in a Raman spectrum that consists of a sum of Lorentzians, one for each combination of phonon branches λ_1, λ_2 and for each phonon wave vector $\mathbf{q}$. The position of each of these Lorentzians in the Raman spectrum (i.e., the recorded Raman shift) is given by the sum of the frequencies of the two involved phonons. Since the phonon frequencies for the different wave vectors $\mathbf{q}$ form a continuum, one does not record individual peaks, however, but rather a continuous distribution in the Raman spectrum. Note that the weight of each Lorentzian, i.e. the weight of the contribution of each wave vector $\mathbf{q}$ and of each of the branches λ_1 and λ_2, is given by the corresponding Raman matrix element, which depends again on the excitation frequency ω_{in}, the electronic lifetimes γ_{el}, and the phonon frequencies and lifetimes ω_{ph} and γ_{ph}, respectively.

As will be discussed in detail in the next paragraph, for each pair (λ_1, λ_2) of phonon branches considered, the Raman amplitude will have a dominating maximum due to a double resonance for specific phonon wave vectors $\mathbf{q}^{\lambda_1,\lambda_2}_{\text{res}}(\omega_{\text{in}})$ that depend on the excitation frequency. These will be the phonons that contribute the most to the Raman amplitude and hence the corresponding Lorerentzians will have a much bigger weight than those of the other phonons. This in turn leads to the observation of a clear peak for the two-phonon processes. This peak will have its maximum at the center of the Lorentzian corresponding to the $\mathbf{q}_{\text{res}}^{\lambda_1,\lambda_2}$, i.e. at $\omega_{\mathbf{q}_{\text{res}},\lambda_1} + \omega_{\mathbf{q}-\mathbf{q}_{\text{res}},\lambda_2}$. While the Raman matrix element for these specific wave vectors might be maximal, this does not mean, however, that other "nearby" phonons do not sizably contribute to the observed Raman amplitude. In fact, the condition that fixes $\mathbf{q}_{\text{res}}^{\lambda_1,\lambda_2}$ depends on the excitation laser energy, which in case of a resonance in the Raman amplitude is equal to some specific electronic excitation energy. Since the electronic levels are smeared out by a finite decay width γ_{el}, so is the resonance condition for $\mathbf{q}_{\text{res}}^{\lambda_1,\lambda_2}$. Thus, the range of the phonon wave vectors that give a sizable contribution to the Raman spectrum is determined by the electronic lifetime γ_{el} and since the range of $\mathbf{q}$ determines the range of phonon frequencies that contribute to the two-phonon peaks, the electronic lifetime has a major impact on the width of these Raman peaks. Note that $\gamma_{\text{el}} = \mathcal{O}(100 \text{ meV}) \gg \gamma_{\text{ph}} = \mathcal{O}(1 \text{ meV})$ and hence the phononic lifetimes hardly play any role for the width of the two-phonon peaks. This should be contrasted with the width of the G peak which is entirely determined by the width of the degenerate phonon involved and thus is much narrower than the two-phonon-induced peaks. For example, typical widths of the $2D$ peak range from 20-40 cm^{-1} compared to the typical widths of 5-15 cm^{-1} for the G peak. [Neumann *et al.* (2015)]

To understand the mentioned double-resonant behavior of the Raman amplitude for the two-phonon processes, we take a look at one of the amplitudes corresponding to one of the Feynman diagrams. In particular, we will look at one of the terms generated by the contour integration over ω that are contained in the matrix product in $\mathcal{M}_1^{2\text{ ph.}}$:

$$\begin{aligned}
\mathcal{M}_1^{2\text{ ph.}} \supset & \sum_{\text{spin}} \sum_{\mathbf{k}} \left(-\frac{i}{\hbar}\right) n_{\text{F}}(\varepsilon_{\mathbf{k}}^{\pi}) n_{\text{F}}(\varepsilon_{\mathbf{k}+\mathbf{q}}^{\pi}) \left(1 - n_{\text{F}}(\varepsilon_{\mathbf{k}}^{\pi^*})\right) \left(1 - n_{\text{F}}(\varepsilon_{\mathbf{k}+\mathbf{q}}^{\pi^*})\right) \\
& \times \frac{\langle \pi_{\mathbf{k}+\mathbf{q}}; \omega_{\text{out}}^{\nu} | H_{\text{el-R}}^{(1)} | \pi_{\mathbf{k}+\mathbf{q}}^* \rangle}{\left[\hbar\omega_1 - (\varepsilon_{\mathbf{k}+\mathbf{q}}^{\pi} - \varepsilon_{\mathbf{k}}^{\pi}) + \frac{i}{2}(-\gamma_{\mathbf{k}+\mathbf{q}}^{\pi} + \gamma_{\mathbf{k}}^{\pi} - \gamma_1)\right]} \\
& \times \frac{\langle \pi_{\mathbf{k}}; \omega_{\mathbf{q}}^{\lambda_1} | H_{\text{el-ph}}^{(1)} | \pi_{\mathbf{k}+\mathbf{q}} \rangle \langle \pi_{\mathbf{k}+\mathbf{q}}^*; \omega_{-\mathbf{q}}^{\lambda_2} | H_{\text{el-ph}}^{(1)} | \pi_{\mathbf{k}}^* \rangle}{\left[\hbar(\omega_{\text{out}} + \omega_1) - (\varepsilon_{\mathbf{k}+\mathbf{q}}^{\pi^*} - \varepsilon_{\mathbf{k}}^{\pi}) + \frac{i}{2}(\gamma_{\mathbf{k}}^{\pi} + \gamma_{\mathbf{k}+\mathbf{q}}^{\pi^*} - \gamma_1)\right]} \\
& \times \frac{\langle \pi_{\mathbf{k}}^* | H_{\text{el-R}}^{(1)} | \pi_{\mathbf{k}}; \omega_{\text{in}}^{\mu} \rangle}{\left[\hbar\omega_{\text{in}} - (\varepsilon_{\mathbf{k}}^{\pi^*} - \varepsilon_{\mathbf{k}}^{\pi}) + \frac{i}{2}(\gamma_{\mathbf{k}}^{\pi} + \gamma_{\mathbf{k}}^{\pi^*})\right]} .
\end{aligned} \tag{3.28}$$

It can be interpreted as describing the excitation of an electron with wave vector **k** from the π to the π^* band due to the absorption of a photon with frequency ω_{in} and polarization μ, followed by scattering of the excited electron to a state with wave vector $\mathbf{k}+\mathbf{q}$ in the π^* band by emitting a phonon with frequency $\omega_2 \equiv \omega_{-\mathbf{q},\lambda_2}$. In the meantime the empty state at wave vector **k** in the π band has been filled by an electron that was scattered from a state with wave vector $\mathbf{k}+\mathbf{q}$ in the π band by emitting a phonon with frequency $\omega_1 \equiv \omega_{\mathbf{q},\lambda_1}$, leaving behind a hole. Finally, this hole recombines with the scattered excited electron in the π^* band by via emission of a photon with frequency ω_{out} and polarization ν. This process is illustrated in Fig. 3.7a for two phonons of the same branch (TO) with wave vectors $\pm\mathbf{q}$ near K and electronic states with wave vectors **k** near K and $\mathbf{k}+\mathbf{q}$ near K', respectively.

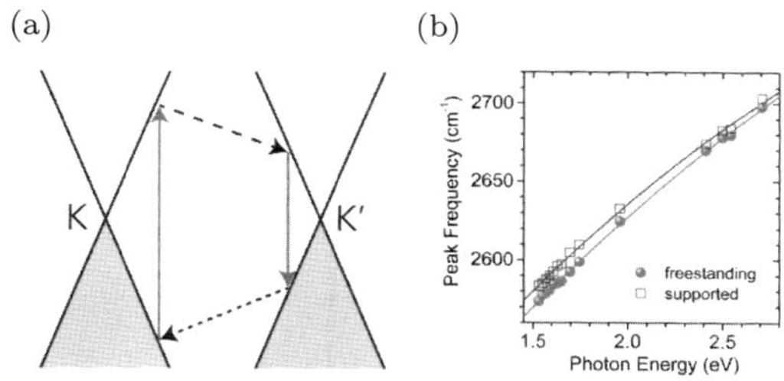

Fig. 3.7: **Illustration of one possible double-resonant Raman process and dispersion of the $2D$ line.** (a) Illustration of the exemplary microscopic Raman process described in the text, which involves two phonons with wave vectors $\pm\mathbf{q}$ near K and electronic states with wave vectors **k** near K and $\mathbf{k}+\mathbf{q}$ near K', respectively. Full lines represent the conic electronic band structure around K and K' and the shaded regions represent the filled π band. The vertical, full arrows represent the excitation and subsequent recombination of an electron-hole pair, while the diagonal, dashed and dotted arrows, represent the scattering of electron and hole, respectively, from a state with wave vector near K to a state with wave vector near K'. (Figure taken from [Ferrari and Basko (2013)].) (b) Observed dependence of the position of the $2D$ peak on the energy of the incoming photon. Dots and squares refer to data taken on freestanding and supported graphene, respectively. Full lines represent fits to the experimental data. (Figure taken from [Berciaud *et al.* (2013)].)

The amplitude for this process becomes maximal when one or several of the denominators become minimal, i.e. when their real part vanishes. While the amplitude for the G peak could only become single-resonant since both denominators could not vanish simultaneously, here the presence of electronic states with different wave vectors allows for two of the denominators to vanish simultaneously if the phonon wave vector obeys some resonance condition. Note that irrespective of the phonon wave vector $\mathbf{q}$, single resonance is always possible, as already discussed for the G peak. To derive the condition for double resonance for the phonon wave vector $\mathbf{q}$, we look at the amplitude for fixed $\mathbf{q}$ and fixed phonon branches λ_1 and λ_2 and require two of the three factors in the denominator of Equation 3.28 to vanish for some electronic wave vector $\mathbf{k}$. We will see that this is only possible for a specific choice of $\mathbf{q}$, which we already denoted as $\mathbf{q}_{\mathrm{res}}^{\lambda_1,\lambda_2}$ in the previous paragraph. In principle, any two combinations of the three denominators can vanish at once. We restrict ourselves to one of the three possible pairs of vanishing terms to illustrate how to find the resonance condition for the phonon wave vector $\mathbf{q}$. The two terms we seek to minimize are the first and third of the factors in the denominator. By setting the real part of the latter to zero, we immediately find a condition for the wave vector of the electronic states that are in resonance with the light:

$$\hbar\omega_{\mathrm{in}} = \varepsilon_{\mathbf{k}}^{\pi^*} - \varepsilon_{\mathbf{k}}^{\pi} \quad \Rightarrow \quad \mathbf{k} = \mathbf{k}_{\mathrm{res}}(\omega_{\mathrm{in}}). \tag{3.29}$$

Note that since $\mathbf{k}$ is a two-dimensional vector, this condition describes an entire resonance surface (a closed line in two dimensions) of $\mathbf{k}$-vectors in the Brillouin zone. For example, in the limit of a conic band structure, the set of $\mathbf{k}_{\mathrm{res}}$ describes a circle in the Brillouin zone. By also requiring that the first factor in the denominator is minimal, we finally arrive at a condition for $\mathbf{q}$:

$$\hbar\omega_{\mathbf{q},\lambda_1} = \varepsilon_{\mathbf{k}_{\mathrm{res}}(\omega_{\mathrm{in}})+\mathbf{q}}^{\pi} - \varepsilon_{\mathbf{k}_{\mathrm{res}}(\omega_{\mathrm{in}})}^{\pi} \quad \Rightarrow \quad \mathbf{q} = \mathbf{q}_{\mathrm{res}}(\omega_{\mathrm{in}}). \tag{3.30}$$

Again, $\mathbf{q}$ is a two-dimensional vector and thus this equation defines a resonance surface for $\mathbf{q}_{\mathrm{res}}$, which, for example, is a circle for a conic band structure and a conic phonon dispersion as found for $\mathbf{q}$ near K or Γ. Due to the shape of the graphene bands and the shape of the optical phonon dispersion near K, solutions for these equations always exist. By plugging the resonant wave vectors $\mathbf{k}_{\mathrm{res}}$ and $\mathbf{q}_{\mathrm{res}}$ and the constraints into the remaining factor that we did not explicitly minimize, we find by using $\omega_{\mathrm{in}} = \omega_{\mathrm{out}} + \omega_1 + \omega_2$:

$$\begin{aligned} &\hbar(\omega_{\mathrm{out}} + \omega_{\mathbf{q}_{\mathrm{res}},\lambda_1}) - (\varepsilon_{\mathbf{k}_{\mathrm{res}}+\mathbf{q}_{\mathrm{res}}}^{\pi^*} - \varepsilon_{\mathbf{k}_{\mathrm{res}}}^{\pi}) \\ &= -\hbar\omega_{\mathbf{q}_{\mathrm{res}},\lambda_2} + \varepsilon_{\mathbf{k}_{\mathrm{res}}}^{\pi^*} - \varepsilon_{\mathbf{k}_{\mathrm{res}}+\mathbf{q}_{\mathrm{res}}}^{\pi^*}. \end{aligned} \tag{3.31}$$

In case the two phonons stem from the same branch, i.e., $\lambda_1 = \lambda_2$, this equation can be further simplified by plugging in the constraint for $\omega_{\mathbf{q},\lambda_1}$:

$$\begin{aligned} & -\hbar\omega_{\mathbf{q}_{\text{res}},\lambda_2} + \varepsilon^{\pi^*}_{\mathbf{k}_{\text{res}}} - \varepsilon^{\pi^*}_{\mathbf{k}_{\text{res}}+\mathbf{q}_{\text{res}}} \\ =& (\varepsilon^{\pi}_{\mathbf{k}_{\text{res}}} + \varepsilon^{\pi^*}_{\mathbf{k}_{\text{res}}}) - (\varepsilon^{\pi}_{\mathbf{k}_{\text{res}}+\mathbf{q}_{\text{res}}} + \varepsilon^{\pi^*}_{\mathbf{k}_{\text{res}}+\mathbf{q}_{\text{res}}}). \end{aligned} \tag{3.32}$$

Notice that the terms in parentheses are a measure for the electron-hole asymmetry at $\mathbf{k}_{\text{res}}$ and $\mathbf{k}_{\text{res}} + \mathbf{q}_{\text{res}}$. Since $\mathbf{k}_{\text{res}}$ and $\mathbf{k}_{\text{res}} + \mathbf{q}_{\text{res}}$ have to be near K for the resonance conditions to be fulfilled [Maultzsch *et al.* (2004); Venezuela *et al.* (2011)], the third factor will be small as well since the electron-hole asymmetry is small near K.

In contrast, for two phonons of different branches, i.e. $\lambda_1 \neq \lambda_2$, the third factor will in general not be small and this explains why the Raman peaks due to phonons of the same branch are much more visible than the Raman bands with two phonons of different branches.

It should be noted that one can construct similar conditions for $\mathbf{q}^{\lambda_1,\lambda_2}_{\text{res}}$ by demanding that the two other possible pairs of terms be minimal. In general, this will result, however, in slightly different wave vectors $\mathbf{q}^{\lambda_1,\lambda_2}_{\text{res}}$. Also note that the resonant conditions resulting from the other diagrams have $\mathbf{q}$ depend on λ_2 or even on both λ_1 and λ_2, which justifies the notation $\mathbf{q}^{\lambda_1,\lambda_2}_{\text{res}}$. By explicitly evaluating the double-resonance conditions, one finds [Venezuela *et al.* (2011)] that the phonons that give the biggest contribution to the various Raman peaks are a phonon from the TO branch with $\mathbf{q}$ near K for the D peak, a phonon from the LO branch with $\mathbf{q}$ near Γ for the D' peak, and a phonon from the LA branch with $\mathbf{q}$ near K for the D'' peak.

Finally, we want to point out that $\mathbf{q}^{\lambda_1,\lambda_2}_{\text{res}}$ depends on the excitation frequency ω_{in} as mentioned before. Since the phonon wave vector determines the resonant contribution and hence the position of the maximum of the peak, the two-phonon peaks will shift when the excitation frequency ω_{in} is varied, which leads to the these peaks being called *dispersive*. In the approximation of a linear band structure (i.e. a Dirac cone) and a linear phonon dispersion near K and Γ and for the resonant electron and phonon wave vectors being parallel along high-symmetry lines, the two conditions for double resonance become linear equations for $|\mathbf{k}|$ and $|\mathbf{q}|$, which are also linear in ω_{in}. Hence the solutions to these equations, $|\mathbf{k}_{\text{res}}|$ and $|\mathbf{q}_{\text{res}}|$, are also linear in ω_{in} and because of the approximately linear dispersion of the phonons near K and Γ, the resonant phonon frequency and with it the peak position are linear in ω_{in} as well. This linear dispersion of the two-phonon frequencies has been observed experimentally, e.g., in [Thomsen and Reich (2000); Maultzsch *et al.* (2004); Berciaud *et al.* (2013)] and is shown in Fig. 3.7b. Note that contrary to the process for the G peak, the Raman amplitude does not vanish for a conic band structure, since the final state contains three particles (one photon, two phonons), so that angular momentum conservation as implied by a conic band structure can always be obeyed.

So far we have focused on the double-resonant contribution to the Raman amplitude and determined which phonon wave vectors $\mathbf{q}$ maximize it, which in turn determines the Raman shift at which the two-phonon Raman peaks are observed. However, we want to stress that while the double-resonant terms are undoubtedly of high importance for a correct description of the Raman peaks, quantum interference effects also play a crucial role in the correct description of the Raman spectrum [Venezuela *et al.* (2011)]. While the contributions of different phonons are added after squaring the matrix element (different phonons mean different final states), the sum over electronic wave vectors $\mathbf{k}$ has to be done before squaring the matrix element. As pointed out in the literature [Venezuela *et al.* (2011)], the double-resonant electronic states $\mathbf{k}_{\text{res}}$ can constructively or destructively interfere with one another, which can significantly influence the peak shape and height.

3.2.3.3 *Defect-assisted peaks (e.g., the D, D', and $D + D'$ peaks)*

After the detailed discussion of defect-free processes, we conclude this subsection by briefly addressing the defect-assisted processes. In the presence of defects, electron-phonon scattering events can be complemented or replaced by elastic electron-defect scattering events. For example, if one replaces one of the two electron-phonon scattering events in the defect-free two-phonon process by an electron-defect scattering event, one obtains a process that has one instead of two phonons in the final state. This phonon can in principle have an arbitrary wave vector $\mathbf{q}$, as already discussed in the section on kinematic constraints. A simple way to obtain the amplitude for such a process is to replace the electron-phonon matrix element that involves the second phonon by one of the defect-scattering amplitudes and set the corresponding phonon frequency in the denominator to zero.

While the double resonance conditions that involve the second phonon can thus no longer be fulfilled, the double resonance conditions that involve the first phonon, however, are unaffected by this. Therefore, the corresponding Raman process will still be doubly resonant for the same phonon wave vectors $\mathbf{q}_{\text{res}}$ we found for the two-phonon, defect-free case above. Accordingly, the maximum of the corresponding Raman peak will be located at the position of the frequency of the resonant wave vectors. Compared to the defect-free two-phonon process involving two phonons of the same branch, where the Raman shift of the peak maximum is given by the sum of the frequencies of the two phonons, the Raman shift for the corresponding one-phonon, defect-assisted peak will thus be halved. This explains the designation $2D$ and $2D'$ for two of the two-(same-)phonon peaks, as their corresponding defect-assisted peaks are denoted D (for resonant phonons of the TO branch with $\mathbf{q}$ near K) and D' (for resonant phonons of the LA branch near Γ), respectively. Since the double resonance conditions are the same for the two-phonon process and one-phonon process with defects, the position of the Raman peaks of the latter also depend on the excitation frequency ω_{in}.

In addition to the matrix element squared, the intensity of the defect-assisted peaks will also contain a factor of the defect concentration [Venezuela *et al.* (2011)]. This direct linear dependence of the height of the peak on the defect concentration is easily understandable since the height of the peak is proportional to the probability per unit time of the corresponding Raman process. Since the latter requires a defect in order to be non-zero, the probability per unit time for the process to happen is proportional to the defect concentration and hence the peak height is as well. A more detailed discussion of the dependence of the intensity of theses peaks can be found in the final section of this chapter.

Finally, there are also Raman processes which involve two phonons and additional electron-defect scattering events. This allows for more and different conditions for double resonance, leading to several Raman sidebands and peaks, the most prominent of which is the $D + D'$ peak.

3.3 Influence of Internal and External Parameters on the Raman Spectrum

In the final section of this chapter, we discuss how various internal properties (e.g. defects or layer number) of a graphene sample or externally controllable parameters (e.g. doping or strain) influence the Raman spectrum. We will restrict ourselves to the discussion of the arguably most common and arguably best understood external influences, namely the number of layers, the defect concentration, doping, and strain. Other influences not covered here include disorder and edge effects [Ferrari (2007)], substrate influences [Berciaud *et al.* (2013); Forster *et al.* (2013); Neumann *et al.* (2015)], and temperature [Calizo *et al.* (2007)], for which the interested reader is referred to the literature.

3.3.1 *Layer number*

In the previous parts of this chapter, we were only concerned with monolayer graphene. While it is still by far the most often studied graphene system in contemporary research and allows one to illustrate the general theoretical approach to the calculation of Raman spectra, multi-layer graphene also has received some attention in the literature. Indeed, one important application of Raman spectroscopy is the determination of the number of layers after sample exfoliation or growth. We therefore briefly summarize the most important changes to the Raman spectrum if the sample consists not of one but of several layers of graphene.

By increasing the number of layers, the number of atoms in the unit cell changes by 2 for each additional layer. The number of atoms in the unit cell determines the number of (optical) phonon branches. For N atoms in the unit cell the number of optical branches is given by $3N - 3$, two thirds of which are in-plane phonons. However, almost across the whole Brillouin zone, the new phonon branches that are relevant for Raman scattering are degenerate with the ones already existing in monolayer graphene. Only near the K point do the elsewhere degenerate TO branches split slightly [Herziger *et al.* (2014)].

By contrast, the electronic band structure undergoes a more severe change. Both the π and π^* bands split into N bands each. As the different layers of multilayer graphene are bound mostly by van der Waals forces, which are much weaker than the intralayer covalent forces, the splitting of the additional bands is small compared to the energy scale of the bands. Thus all of these bands now play a role in the Raman processes.

By taking both the changes to the phonon dispersion and the electronic band structure into account, we can summarize the changes to the various Raman processes and peaks as follows:

For the one-phonon, defect-free process, i.e. the G peak, very little change is observed as the position and decay width of the degenerate phonon at Γ are as good as unchanged. The separation in energy of the additional bands from the monolayer ones is so large that the additional possible electronic transitions have transition energies out of range of the phonon. Hence they couple only weakly to it, which means that the frequency and decay width of the phonon at Γ are not measurably different from the monolayer ones.

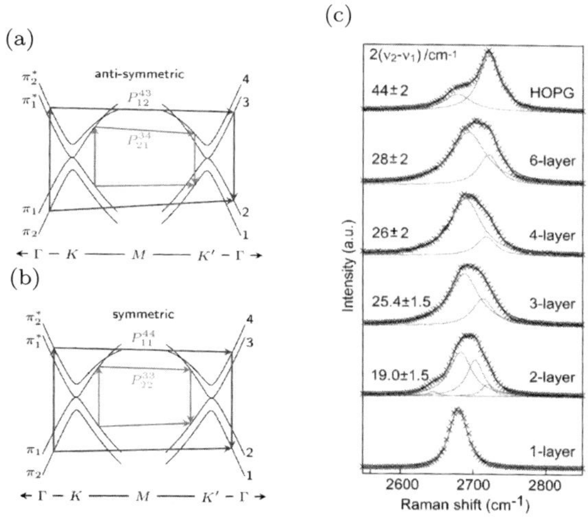

Fig. 3.8: **Change of the $2D$ peak width with layer number.** (a,b) Two examples of double-resonant processes for bilayer graphene. Shown are the two π and π^* bands of bilayer graphene and examples of electronic transitions that can become double-resonant. (Figure taken from [Herziger *et al.* (2014)].) (c) Evolution of the FWHM of the $2D$ peak with the number of layers, from monolayer graphene to highly oriented pyrolithic graphite (HOPG). (Figure taken from [Graf *et al.* (2007)].)

For the two-phonon and/or defect-assisted processes, however, the situation is different. Recall that the existence of a clear peak in monolayer graphene is attributed to the fact that for certain electron wave vectors $\mathbf{k}_{\text{res}}$ and phonon wave vectors $\mathbf{q}_{\text{res}}$ the Raman amplitude becomes double resonant. The additional electronic bands in multilayer graphene allow for additional double-resonance conditions with slightly different solutions $\mathbf{k}_{\text{res}}$ and $\mathbf{q}_{\text{res}}$ (compare Fig. 3.8a for an example of a double-resonant process in the case of bilayer graphene). In addition, the slight splitting of the TO branch of the phonon dispersion for $\mathbf{q}$ near K further allows for different $\mathbf{q}_{\text{res}}$. Altogether, this leads to more phonon wave vectors $\mathbf{q}$ that satisfy a double resonance condition and hence several close-by phonon frequencies now contribute significantly to the observed Raman shift. This leads to a measurable broadening of the two-phonon and defect-assisted peaks, in particular to a splitting/broadening of the most prominent peak, the $2D$ peak. As expected from the above considerations, the broadening of these peaks increases with every additional number of layers [Ferrari *et al.* (2006); Graf *et al.* (2007)] (see Fig. 3.8b) and converges towards the peak structure of bulk graphite.

The broadening of the $2D$ peak in particular is nowadays routinely used to reliably estimate the number of layers after sample exfoliation or growth.

3.3.2 *Defect concentration*

As already mentioned before, the presence of defects in a graphene sample changes the Raman spectrum in two ways: On the one hand, it leads to more peaks in the spectrum as the defect-assisted Raman processes become possible. On the other hand, it modifies the height and widths of the non-defect-assisted peaks. For both groups of Raman peaks, we will briefly discuss the dependence of their height and width on the defect concentration in this subsection.

For the *defect-assisted* peaks, e.g. the D, D', and $D + D'$ peaks, the peak height and width depends non-monotonically on the defect concentration n_D.

Firstly, there is the direct dependence of the intensity on n_D, which appears as a scaling factor in the intensity. This factor represents the fact that the probability for a defect-assisted Raman process is proportional to the number of defects.

Secondly, there is the indirect dependence of the quantum mechanical amplitude on n_D through the decay widths, i.e., the lifetimes, of the intermediate electronic states, $\gamma_{\text{el}} = \gamma_{\text{el}}(n_D)$, which appear in the denominator of the matrix element of the Raman process and thus higher widths lead to smaller quantum mechanical amplitudes. In general, the decay widths of electronic states contain several contributions related to the different ways in which an electron can scatter. The three most prominent ones are electron-electron interaction, i.e., Coulomb scattering with another electron, electron-phonon interaction, i.e., activation of a lattice vibration, and scattering from a defect. The latter contribution to the total decay width of an electronic state will be proportional to n_D, as the presence of more defects leads to a higher probability for an electron to scatter from a defect.

Depending on the defect concentration, the first or the second aspect will dominate [Venezuela *et al.* (2011)].

For low defect concentrations, the electronic decay widths will be dominated by the electron-electron and electron-phonon contributions and the dependence of γ_{el} on n_D will be negligible. The only dependence of the intensity of the defect-assisted peaks will then be due to the overall linear factor of n_D and the intensity of the defect-assisted peaks will increase linearly with n_D. Indeed, this behavior has been observed experimentally [Ferreira *et al.* (2010)].

For high defect concentrations, the electronic decay widths will be dominated by electron-defect scattering and hence will be proportional to n_D. Since the electronic decay widths appear in each of the multiple factors in the denominator of the Raman matrix element, this effect will eventually outweigh the overall linear factor of n_D appearing in the expression for the Raman intensity. Therefore, for large enough densities, the heights of the defect-assisted peaks will decrease.

While the peak heights thus behave non-monotonically as a function of n_D, the peak widths monotonically increase with n_D. Here, the only relevant effect is the increase of the electronic decay widths with n_D. As discussed before, the broadening of the electronic states leads to a smearing of the resonance condition and hence a wider range of phonon frequencies will non-negligibly contribute to the Raman peaks. Thus, the widths of the peaks increase monotonically with defect concentration.

The properties of the *non-defect-assisted* peaks, on the other hand, depend on the defect concentration in a purely monotonic fashion.

Their intensities monotonically decrease as a function of the defect concentration, since they only depend on n_D through the decay widths of the electronic states, which enter the quantum mechanical matrix element through factors in the denominator. Since for low defect concentrations, the dependence of the electronic decay widths on n_D is negligible, the height of the non-defect-assisted peaks, is independent of the defect concentration as well. The different behavior of the intensity of the defect-assisted and non-defect-assisted peaks as a function of n_D has been suggested as a measure for the amount of defects and disorder in a sample by measuring the ratio of the intensity of the D and G peaks [Cançado *et al.* (2011)] (see Fig. 3.9).

The behavior of the widths of the defect-free peaks under changes of n_D depends on the specific peak. The width of the G peak, as shown in the previous section, does not depend on the electronic decay widths as it

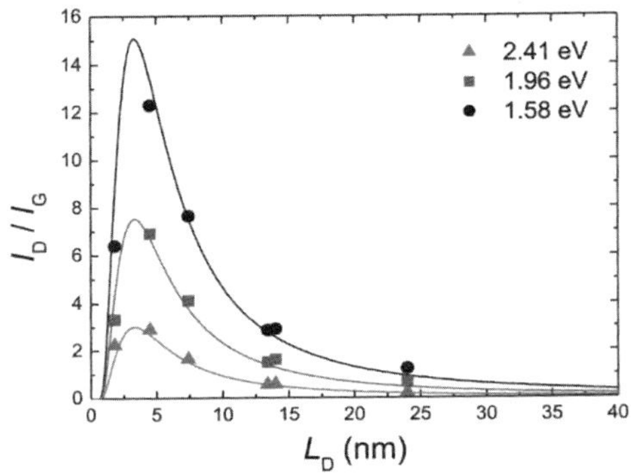

Fig. 3.9: **Measured dependence of the intensity of the Raman peaks on the defect concentration.** Ratio of the intensity of the D and G peaks as a function of the average distance between defects L_D, which is a measure for the defect concentration n_D. (Figure taken from [Cançado *et al.* (2011)].)

is entirely determined by the lifetime of the created phonon. The widths of the multi-phonon peaks, such as the $2D$, $2D'$, or $D+D''$ peaks, by contrast, increase with n_D, for the same reason as for the defect-assisted peaks.

3.3.3 *Doping*

Doping, i.e., changes of the charge carrier density, leads to a shift of the Fermi level. Hole (p-type) doping leads to states in the π band not being occupied, while electron (n-type) doping causes states in the π^* band to be occupied. This will influence the Raman spectrum in three ways:

Firstly, the electronic band structure is slightly affected as the screening of the electron-electron interaction is increased by the presence of additional charge carriers. The most important effect of this is decrease of the slope of the π and π^* bands near the K point [Elias *et al.* (2011)]. This effects the two-phonon and/or defect-assisted Raman processes by changing the condition for double resonance.

An additional effect is the increase of the decay width of excited electronic states (or equivalently a decrease of their lifetime) γ_{el}, as the scattering rate of an electron due to electron-electron interaction increases with additional charge carriers in the system. This will affect the height of the Raman peaks and correspondingly their area, since they are proportional

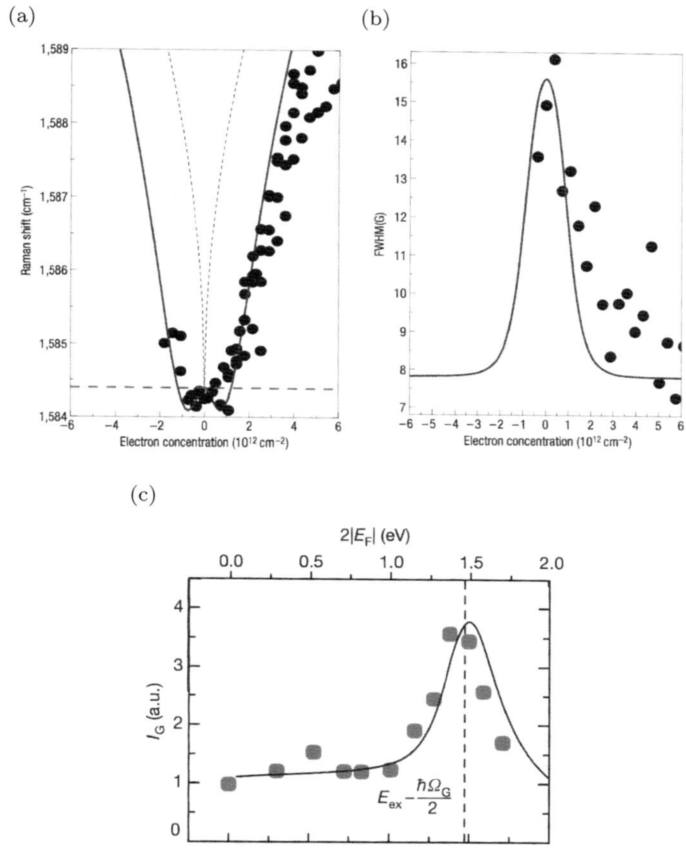

Fig. 3.10: **Effects of doping on the Raman *G* peak.** (a,b) Change of the frequency (a) and the decay width (b) of the in-plane optical phonons at Γ with doping, which determine the frequency (position) and width of the Raman G peak. The dots represent experimental data while the full lines represent the results of a theoretical calculation. (Figures taken from [Pisana *et al.* (2007)].) (c) Intensity of the Raman G peak as a function of Fermi energy, i.e., doping. The dots represent experimental data, while the full line represent the result of a theoretical calculation using a simplified version of the theory presented in the previous section. (Figure taken from [Chen *et al.* (2011)].) (For color reference, go to page 486.)

to the matrix element squared of the corresponding Raman process, which in turn decreases for higher decay widths of the electronic states as they appear in the denominator of the amplitude. Since the amplitude for the two-phonon processes (e.g., the $2D$ peak process) contains one more electronic Green's functions than the amplitude for the one-phonon, defect-free process (leading to the G peak), it contains one more factor of γ_{el}, and hence the height and area of the two-phonon peaks reduces much more with doping than the height and area of the G peak [Basko *et al.* (2009)]. In addition to the height, the width of the two-phonon and/or defect-assisted peaks changes as well, as experimentally observed for the $2D$ peak [Berciaud *et al.* (2013)], since it is also related to the electronic decay width, as mentioned in the last section.

Secondly, the phonon frequencies and widths change. This is most noticeable for those phonons which are affected by Kohn anomalies: the degenerate optical in-plane phonons at Γ and the TO branch at and around K. The phonon at Γ is affected by doping through a change of its *self-energy*, i.e., its change of frequency due to electron-phonon interaction. The commonly associated picture behind this is the temporary decay of the phonon into an electron-hole pair which recombines to produce the same phonon again. This process is only relevant if the electron or the hole has not been scattered in the meantime, i.e., if they are long-lived enough or, equivalently, if their decay width is low enough. Otherwise this virtual process does not play a role as recombination to the original phonon is not possible. The decay width is "low enough" if it is much lower than the transition energy of the electron-hole pair which is equal to the energy of the phonon. In other words, for this process to play a role, the lifetime of the excited electron or the hole must be much longer than the period of the phonon oscillation. If this is the case, the nuclei are oscillating while the electron stays in its excited state without 'instantly' dropping back to the ground state and without adiabatically following the nuclei, as assumed in the adiabatic Born-Oppenheimer approximation. In graphene, the decay widths of the exited electron and the hole are indeed much lower compared to the phonon energy, leading to the statement that the adiabatic Born-Oppenheimer approximation breaks down in graphene [Pisana *et al.* (2007)]. With doping, the occupation of the excited states and holes is changed and this in turn changes the self-energy of the phonon, resulting in a change of the phonon frequency of the degenerate optical branches at Γ [Ando (2006); Pisana *et al.* (2007)], as shown in Fig. 3.10a. Note that in other materials, the ratio of electronic decay widths and phonon frequencies is usually much larger than in graphene and hence the exited electrons and holes have time to relax to the ground state and adiabatically follow the oscillating nuclei and hence the self-energy correction in other materials is usually negligible.

At the same time, doping also blocks a possible decay channel of the phonons by preventing the decay into an electron-hole pair due to the Pauli principle. If the Fermi level exceeds half of the phonon energy, then the decay into an electron and a hole with each carrying away half of the phonon energy becomes impossible due to the Pauli exclusion principle since the electronic state that would be occupied after the decay is already occupied due to doping. Thus, if the Fermi level exceeds half of the phonon energy, the phonon decay widths drops noticeably (compare Fig. 3.10b), as the decay channel into electron-hole pairs makes up a sizable part of the total decay width of the phonons. In the case of finite temperature, this sharp drop of the decay width is smeared out by the Fermi-Dirac distribution. The strong decrease of the decay width of the phonon at Γ with doping directly translates into a strong decrease of the width of the Raman G peak, since the latter is equal to the decay width of the phonon, as explained in the theoretical section. By monitoring the G peak width, one thus has a qualitative measure to check if the Fermi level exceeds half of the phonon energy.

For the two-phonon and/or defect-assisted peaks, the physical picture is roughly the same as for the G peak. As mentioned earlier, the phonons involved in the processes leading to the D, D'', $D + D''$ and, most importantly, the $2D$ peak have wave vectors near K. In the vicinity of K, effects of the Kohn anomaly play an important role since it leads to a steepening of the slope of the phonon dispersion, which plays an important role in determining which phonons obey the double-resonance conditions and hence where the center of the Raman peaks is situated. The Kohn anomaly is very sensitive to the Fermi surface and thus a change of the Fermi level through doping means that it has a large influence on the phonon dispersion around K and consequently on the position of the Raman peaks. Indeed, it is found experimentally that the two-phonon and/or defect-assisted peaks [Das *et al.* (2008)] shift with doping, albeit not as strongly as the G peak.

The decay width of the phonons that fulfill the double-resonance condition, are not affected for low doping levels. The finite momentum of these phonons does not allow the phonon to decay into an electron-hole pair due to energy and momentum conservation, as the necessary hole states would have to have an energy of around half of the laser energy for which the phonon fulfills the double-resonance condition. For typical experimental laser energies of 2-3 eV, it would require a Fermi level shift of 1-1.5 eV to block the decay width of the phonons involved in the double-resonant process. The influence of doping on the decay widths of the phonons can thus be neglected for typical doping levels achieved in experiment.

Finally, doping influences the Raman amplitudes directly. As seen in the previous section, the Raman amplitudes contain factors of the Fermi-Dirac distribution of the involved electronic states. By changing the Fermi level and thus varying the occupation of the electronic states, some of the Raman amplitudes are set to zero. For the G peak amplitude, which does not rely primarily on the resonant terms to exist, doping in general has only minor effects. It has been demonstrated experimentally however [Chen *et al.* (2011)], that if the Fermi level comes close to half of the frequency of the outgoing light, the total amplitude for the G peak process increases dramatically. The underlying reason for this increase is the blocking of terms in the amplitude that destructively interfere with the resonant terms, leading to a large *increase* of the amplitude, as depicted in Fig. 3.10c.

For the two-phonon and/or defect-assisted processes, the situation is different. Here, the interference between the different terms in the amplitude is believed to be constructive [Chen *et al.* (2011)] and therefore raising the Fermi level to half of the frequency of the light only leads to a sharp *decrease* of the peak height and ultimately to its disappearance, as has been demonstrated for the $2D$ peak.

We can summarize the effects of doping on the Raman spectrum as follows: The G peak is expected to shift slightly, first to lower frequencies and than to higher ones. In addition, it will narrow once the Fermi level is bigger than $\hbar\omega_{\text{ph}}/2$ or lower than $-\hbar\omega_{\text{ph}}/2$, with the width of the peak decreasing to as low as 4-5 cm^{-1} [Neumann *et al.* (2015)]. The two-phonon and/or defect-assisted peaks, in particular the $2D$ peak, are expected to shift as well, albeit not as strong as the G peak. Furthermore the ratio of the areas of the $2D$ and G peaks decreases as well with doping.

3.3.4 *Strain*

Mechanical strain has a large influence on the Raman spectrum. This is hardly surprising as it changes the bond lengths of the atoms, which leads to both a change in the phonon frequencies as the effective 'springs' between the atoms are softened or hardened and to a change in the electronic band structure since the unit cell becomes distorted and hence the Brillouin zone does as well.

The G peak is less affected by these changes than the two-phonon and/or defect-assisted peaks as the electronic structure only directly influences the intensity, i.e., the peak height. The peak position and width, by contrast, are only determined by the frequency and width of the degenerate optical phonons at Γ. The width of these phonons remains almost completely unaffected by strain since it is mostly determined by the decay of the phonon into electron-hole pairs. The position, however, is rather sensitive to strain since it is equal to the frequency of the phonon. As mentioned before, strain leads to a change of the bond length between the atoms, which in turn modifies the effective spring constant between them and hence the phonon frequency. In general, compressive strain leads to a hardening of the bonds as the charge density becomes higher between the atoms thus binding the atoms stronger together, which results in a higher phonon frequency and therefore a larger Raman shift. Tensile strain, on the other hand, leads to the opposite effect and reduces the phonon frequency and the Raman shift. For experimentally accessible strain values, the relation between Raman shift of the G peak and strain is a linear one, with the constant of proportionality given by the *Grüneisen* parameter for the in-plane optical phonon at Γ. For the typical case of biaxial strain (see below), the Grüneisen parameter is on the order of 70 $\mathrm{cm}^{-1}/\%$.

The two-phonon and/or defect-assisted peaks such as the $2D$ peak, however, undergo a larger change with strain. Here, both the change in the phonon frequencies and the distortion of the electronic band structure play a role. When the electronic band structure is distorted, the conditions for double resonance are changed. When combined with the change of the phonon frequencies this leads to a much larger shift of the two-phonon and/or defect-assisted peak positions with strain compared to that of the G peak.

Additionally, one has to distinguish between *uniaxial* strain and *biaxial* strain:

Uniaxial strain means that strain is applied along one specific direction. The fact that a specific direction is now singled out has a big influence on the various Raman peaks. For the G peak, the most important effect is the lifting of the degeneracy of the in-plane optical phonon frequencies at

Γ. In unstrained graphene, the two in-plane optical phonon branches are degenerate at Γ because of rotation symmetry. By singling out a specific direction, uniaxial strain breaks this rotation symmetry and as a result the two phonon branches split. Since the Raman peak corresponding to one-phonon, defect-free processes is due to the emission of an in-plane optical phonon at Γ, the G peak splits into two peaks corresponding to the two now non-degenerate phonon branches. The difference between these two phonon frequencies is proportional to the applied strain and for low strain ($<$0.7% [Mohiuddin *et al.* (2009)]), the peak splitting is too small compared to the width of the phonons to be observable. Only for higher strain values does one record a clearly visible peak splitting, as seen in Fig. 3.11a.

For the two-phonon and/or defect-assisted peaks, uniaxial strain can also cause observable peak splitting. The mechanism, however, is a different one than that for the G peak. The observed peak splitting of, for example, the $2D$ peak has its roots in a breaking of the threefold rotation symmetry of the resonance surface of the electron and phonon wave vectors. Recall that the phonons mostly contributing to the peak are determined from resonance conditions derived in the previous section. These resonance conditions can be fulfilled or approximately fulfilled by a whole set of wave vectors for the electrons and phonons, which, however, obey the lattice symmetry. By breaking the threefold rotation symmetry of the lattice by applying uniaxial strain, the shape of the resonance surfaces in the Brillouin zone is distorted and this leads to different phonon wave vectors that satisfy them, which correspond to phonon frequencies that are a bit separated from each other. This leads to the observation of a splitting in the two-phonon and/or defect-assisted bands, as has been measured, for example, for the $2D$ peak [Yoon *et al.* (2011)] (compare Fig. 3.11b) As in the case of the G peak, the applied uniaxial strain has to be large enough to observe such a splitting with the threshold value being roughly in the same range as for the G peak.

It should be noted that, for a low-quality graphene sheet and/or an unfortunate choice of substrate, the intrinsic width of the Raman peaks due to the electron and phonon decay widths can be too large to allow a peak splitting to be observed. Instead one will only measure a further broadening of the peaks (compare for instance [Mohiuddin *et al.* (2009)], where no $2D$ peak splitting was observed, and [Yoon *et al.* (2011)], where it was observed).

Biaxial strain, on the other hand, does not lead to a splitting of the Raman peaks. Due to the hexagonal lattice structure, biaxial strain leads mostly only to a uniform stretching or contraction of the unit cell and does

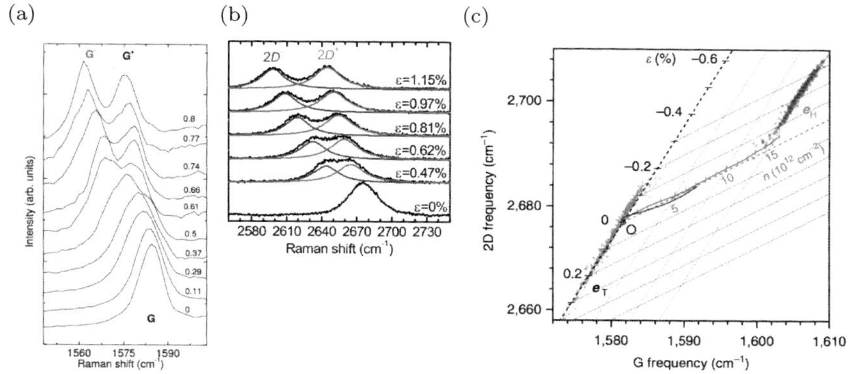

Fig. 3.11: **Effects of strain on the Raman G and $2D$ peaks.** (a,b) Influence of uniaxial strain on the Raman G (a) and $2D$ (b) peaks. Shown are experimentally recorded Raman spectra for increasing amounts of uniaxial strain, leading to a splitting of the G and $2D$ peaks. (Figures taken from [Mohiuddin *et al.* (2009)] and [Yoon *et al.* (2011)], respectively.) (c) Illustration of the separation of doping and strain-induced shifts of the Raman G and $2D$ peak positions as presented in [Lee *et al.* (2012)]. The two clusters of points represent measurements of the positions of the G and $2D$ peaks on various spots of two different samples. The dashed lines represent axes of slopes 0.7 and 2.2, corresponding to the linear relation between shifts of the $2D$ and G peaks due to doping or strain, respectively. By projecting a point in the (ω_G, ω_{2D})-plane onto these to axes, one can gain estimates for the amount of charge carrier doping and strain in the sample. The scaling of the axes is valid for hole doping and uniaxial strain, respectively. (Figure taken from [Lee *et al.* (2012)].) (For color reference, go to page 486.)

not break the rotation symmetry in a major way. Therefore, both mechanisms described in the previous paragraph do not apply in the case of biaxial strain and the Raman bands do not split. However, the mechanism that leads to a shift of the peaks remains applicable. As one would naively expect, the change of the peak positions for a given amount of strain is indeed much larger for biaxial strain than for uniaxial strain [Lee *et al.* (2012)].

Finally, we want to point out that while strain and doping both lead to a shift of the positions of the various peaks, it has recently been demonstrated that it is possible to disentangle these two sources of peak shifts [Lee *et al.* (2012)]. By focusing on the positions of G and $2D$ peaks, one can exploit the fact that the two peak shifts are linearly correlated, both for doping and strain as the source of the peak shifts. The ratio of the two shifts, however, depends on the source of the shifts: For hole doping, the ratio of the $2D$ and G peak shifts is roughly $\Delta\omega_{2D}/\Delta\omega_G|_{\text{hole dop.}} \sim 0.7$. For electron doping, the slope is much slower and deviates soon from the linear trend. For strain, the $2D$ peak, as being related to a two-phonon Raman process, is subject to much larger changes of the position compared to the G peak, with the experimentally extracted ratio of the peak shifts given by $\Delta\omega_{2D}/\Delta\omega_G|_{\text{strain}} \sim 2.2$. Under the assumption that the doping- and strain-induced shift are simply additive and not correlated, one can project the coordinates of a point in the ω_G-ω_{2D} plane onto the axes of slopes 0.7 and 2.2 and separate the doping- and strain-induced shifts (compare Fig. 3.11c), whereupon both can be converted to a value for the charge carrier concentration or the strain, respectively, using theoretical predictions [Pisana *et al.* (2007); Mohiuddin *et al.* (2009); Yoon *et al.* (2011)].

3.4 Conclusions

This chapter has given a glimpse at the rich variety of phenomena that occur in Raman spectroscopy of graphene. Due to the linear band structure of graphene, there are always electron-hole pairs in resonance with the incoming or outgoing laser light. This leads to the relatively large Raman cross section and the strong sensitivity of the Raman spectrum on the sample parameters. A full quantitative explanation of Raman peak positions, peak widths, and intensities requires the somewhat lengthy formalism of time-dependent perturbation theory. However, we have seen that kinematic considerations and a careful look at the matrix elements and energy denominators allow in many cases an intuitive understanding of the Raman spectrum without explicit calculations.

The G line position is given by the E_{2g} phonon frequency at Γ and thus it does not disperse as a function of the laser energy. Its position and width are sensitive measures for doping and strain. At the same time it remains largely unaffected by defects and by the number of layers. The $2D$ line, by contrast, disperses as a function of the laser wave length. Its splitting into subpeaks (sometimes only seen as an increased broadening of the line) is, in particular, used as a convenient measure for the number of layers in the sample. Its position is also affected by strain, but only moderately by doping. The intensity of defect related peaks is a direct measure of the defect concentration in the sample.

One of the main challenges for quantitative calculations is still the absolute position and the width of the $2D$ line (and thus also of the D line). The difficulties are twofold: (i) a fully *ab initio* calculation of the highest optical phonon branch around K is still missing (only the linear slope at K has been determined quantitatively using the GW approximation). (ii) The width is a result of both the effect of $\mathbf{k}$-space integration and also depends on the various broadenings of intermediate electron and hole states as well as on the phonon frequency.

For the time being, *ab initio* calculations of the Raman spectrum of graphene (such as, for example, the calculation presented in [Herziger *et al.* (2014)]) still rely on the input of phonon and electron lifetimes as semi-empirical parameters. All the lifetimes (due to electron-electron, electron-phonon, and phonon-phonon interaction) can, in principle, be calculated, but calculations remain quite expensive in terms of computing time. With the increasing available computing power and recent progress in code development for lifetime calculations, we are optimistic that a fully *ab initio* calculation of the Raman spectrum of graphene can be achieved in the near future.

Acknowledgments

We acknowledge support by the National Research Fund, Luxembourg (projects FNR10196198 and FNR7490149). Furthermore, we acknowledge stimulating discussions with Franscecso Mauri, Michele Lazzeri and Matteo Calandra.

Bibliography

[1] Ando, T. (2006). Anomaly of optical phonon in monolayer graphene, *J. Phys. Soc. Jpn.* **75**, 12.

[2] Basko, D. M. (2008). Theory of resonant multiphonon raman scattering in graphene, *Phys. Rev. B* **78**, 12, p. 125418.

[3] Basko, D. M. (2009). Calculation of the raman g peak intensity in monolayer graphene: role of ward identities, *New. J. Phys.* **11**, 9, p. 095011.

[4] Basko, D. M., Piscanec, S., and Ferrari, A. C. (2009). Electron-electron interactions and doping dependence of the two-phonon raman intensity in graphene, *Phys. Rev. B* **80**, 16, p. 165413.

[5] Berciaud, S., Li, X., Htoon, H., Brus, L. E., Doorn, S. K., and Heinz, T. F. (2013). Intrinsic line shape of the raman 2d-mode in freestanding graphene monolayers, *Nano Lett.* **13**, 8, pp. 3517–3523.

[6] Calizo, I., Balandin, A. A., Bao, W., Miao, F., and Lau, C. N. (2007). Temperature dependence of the raman spectra of graphene and graphene multilayers, *Nano Lett.* **7**, 9, pp. 2645–2649.

[7] Cançado, L. G., Jorio, A., Ferreira, E. H. M., Stavale, F., Achete, C. A., Capaz, R. B., Moutinho, M. V. O., Lombardo, A., Kulmala, T. S., and Ferrari, A. C. (2011). Quantifying defects in graphene via raman spectroscopy at different excitation energies, *Nano Lett.* **11**, 8, pp. 3190–3196.

[8] Cançado, L. G., Jorio, A., and Pimenta, M. A. (2007). Measuring the absolute raman cross section of nanographites as a function of laser energy and crystallite size, *Phys. Rev. B* **76**, 6, p. 064304.

[9] Chen, C.-F., Park, C.-H., Boudouris, B. W., Horng, J., Geng, B., Girit, C., Zettl, A., Crommie, M. F., Segalman, R. A., Louie, S. G., and Wang, F. (2011). Controlling inelastic light scattering quantum pathways in graphene, *Nature* **471**, 7340, pp. 617–620.

[10] Das, A., Pisana, S., Chakraborty, B., Piscanec, S., Saha, S. K., Waghmare, U. V., Novoselov, K. S., Krishnamurthy, H. R., Geim, A. K., Ferrari, A. C., and Sood, A. K. (2008). Monitoring dopants by raman scattering in an electrochemically top-gated graphene transistor, *Nat. Nanotechnol.* **3**, 4, pp. 210–215.

[11] Elias, D. C., Gorbachev, R. V., Mayorov, A. S., Morozov, S. V., Zhukov,

A. A., Blake, P., Ponomarenko, L. A., Grigorieva, I. V., Novoselov, K. S., Guinea, F., and Geim, A. K. (2011). Dirac cones reshaped by interaction effects in suspended graphene, *Nat. Phys.* **7**, 9, pp. 701–704.

[12] Ferrari, A. C. (2007). Raman spectroscopy of graphene and graphite: disorder, electron–phonon coupling, doping and nonadiabatic effects, *Solid State Commun.* **143**, 1, pp. 47–57.

[13] Ferrari, A. C. and Basko, D. M. (2013). Raman spectroscopy as a versatile tool for studying the properties of graphene, *Nat. Nanotechnol.* **8**, 4, pp. 235–246.

[14] Ferrari, A. C., Meyer, J. C., Scardaci, V., Casiraghi, C., Lazzeri, M., Mauri, F., Piscanec, S., Jiang, D., Novoselov, K. S., Roth, S., and Geim, A. K. (2006). Raman spectrum of graphene and graphene layers, *Phys. Rev. Lett.* **97**, 18, p. 187401.

[15] Ferreira, E. H. M., Moutinho, M. V. O., Stavale, F., Lucchese, M. M., Capaz, R. B., Achete, C. A., and Jorio, A. (2010). Evolution of the raman spectra from single-, few-, and many-layer graphene with increasing disorder, *Phys. Rev. B* **82**, 12, p. 125429.

[16] Fetter, A. and Walecka, J. (1971). *Quantum Theory of Many-Particle Physics* (McGraw-Hill, New York).

[17] Forster, F., Molina-Sanchez, A., Engels, S., Epping, A., Watanabe, K., Taniguchi, T., Wirtz, L., and Stampfer, C. (2013). Dielectric screening of the kohn anomaly of graphene on hexagonal boron nitride, *Phys. Rev. B* **88**, 8, p. 085419.

[18] Graf, D., Molitor, F., Ensslin, K., Stampfer, C., Jungen, A., Hierold, C., and Wirtz, L. (2007). Spatially resolved raman spectroscopy of single-and few-layer graphene, *Nano Lett.* **7**, 2, pp. 238–242.

[19] Grüneis, A., Attaccalite, C., Pichler, T., Zabolotnyy, V., Shiozawa, H., Molodtsov, S., Inosov, D., Koitzsch, A., Knupfer, M., Schiessling, J., Follath, R., Weder, R., Rudolf, P., Wirtz, L., and Rubio, A. (2008). Electron-electron correlation in graphite: a combined angle-resolved photoemission and first-principles study, *Phys. Rev. Lett.* **100**, 3, p. 037601.

[20] Hedin, L. (1965). New method for calculating the one-particle green's function with application to the electron-gas problem, *Phys. Rev.* **139**, 3A, p. A796.

[21] Herziger, F., Calandra, M., Gava, P., May, P., Lazzeri, M., Mauri, F., and Maultzsch, J. (2014). Two-dimensional analysis of the double-resonant 2d raman mode in bilayer graphene, *Phys. Rev. Lett.* **113**, 18, p. 187401.

[22] Hwang, E. H. and Das Sarma, S. (2008). Single-particle relaxation time versus transport scattering time in a two-dimensional graphene layer, *Phys. Rev. B* **77**, 19, p. 195412.

[23] Kohn, W. (1959). Image of the fermi surface in the vibration spectrum of a metal, *Phys. Rev. Lett.* **2**, 9, p. 393.

[24] Lazzeri, M., Attaccalite, C., Wirtz, L., and Mauri, F. (2008). Impact of the electron-electron correlation on phonon dispersion: Failure of lda and gga dft functionals in graphene and graphite, *Phys. Rev. B* **78**, 8, p. 081406.

[25] Lee, J. E., Ahn, G., Shim, J., Lee, Y. S., and Ryu, S. (2012). Optical sepa-

ration of mechanical strain from charge doping in graphene, *Nat. Commun.* **3**, p. 1024.

[26] Mahan, G. D. (2000). *Many-particle physics* (Springer Science & Business Media).

[27] Malard, L. M., Pimenta, M. A., Dresselhaus, G., and Dresselhaus, M. S. (2009). Raman spectroscopy in graphene, *Phys. Rep.* **473**, 5, pp. 51–87.

[28] Mañes, J. L. (2007). Symmetry-based approach to electron-phonon interactions in graphene, *Phys. Rev. B* **76**, 4, p. 045430.

[29] Maultzsch, J., Reich, S., Thomsen, C., Requardt, H., and Ordejón, P. (2004). Phonon dispersion in graphite, *Phys. Rev. Lett.* **92**, 7, p. 075501.

[30] Mohiuddin, T. M. G., Lombardo, A., Nair, R. R., Bonetti, A., Savini, G., Jalil, R., Bonini, N., Basko, D. M., Galiotis, C., Marzari, N., Novoselov, K. S., Geim, A. K., and Ferrari, A. C. (2009). Uniaxial strain in graphene by raman spectroscopy: G peak splitting, grüneisen parameters, and sample orientation, *Phys. Rev. B* **79**, 20, p. 205433.

[31] Neumann, C., Reichardt, S., Venezuela, P., Drögeler, M., Banszerus, L., Schmitz, M., Watanabe, K., Taniguchi, T., Mauri, F., Beschoten, B., Rotkin, S. V., and Stampfer, C. (2015). Raman spectroscopy as probe of nanometre-scale strain variations in graphene, *Nat. Commun.* **6**, p. 8462.

[32] Onida, G., Reining, L., and Rubio, A. (2002). Electronic excitations: density-functional versus many-body green's-function approaches, *Rev. Mod. Phys.* **74**, 2, p. 601.

[33] Peskin, M. E. and Schroeder, D. V. (1995). *An Introduction to Quantum Field Theory* (Westview Press).

[34] Pisana, S., Lazzeri, M., Casiraghi, C., Novoselov, K. S., Geim, A. K., Ferrari, A. C., and Mauri, F. (2007). Breakdown of the adiabatic born–oppenheimer approximation in graphene, *Nat. Mater.* **6**, 3, pp. 198–201.

[35] Piscanec, S., Lazzeri, M., Mauri, F., Ferrari, A. C., and Robertson, J. (2004). Kohn anomalies and electron-phonon interactions in graphite, *Phys. Rev. Lett.* **93**, 18, p. 185503.

[36] Reich, S., Thomsen, C., and Maultzsch, J. (2004). *Carbon Nanotubes: Basic Concepts and Physical Properties* (Wiley-VCH).

[37] Saito, R., Dresselhaus, G., and Dresselhaus, M. S. (1998). *Physical Properties of Carbon Nanotubes* (World Scientific).

[38] Thomsen, C. and Reich, S. (2000). Double resonant raman scattering in graphite, *Phys. Rev. Lett.* **85**, 24, p. 5214.

[39] Venezuela, P., Lazzeri, M., and Mauri, F. (2011). Theory of double-resonant raman spectra in graphene: Intensity and line shape of defect-induced and two-phonon bands, *Phys. Rev. B* **84**, 3, p. 035433.

[40] Yoon, D., Son, Y.-W., and Cheong, H. (2011). Strain-dependent splitting of the double-resonance raman scattering band in graphene, *Phys. Rev. Lett.* **106**, 15, p. 155502.

[41] Yu, P. Y. and Cardona, M. (2010). *Fundamentals of Semiconductors*, 4th edn. (Springer).

Part 2

Nonlinear Optical Properties

Chapter 4

Microscopic View on the Ultrafast Carrier Dynamics in Graphene

Ermin Malic[1,*], Torben Winzer[2], Faris Kadi[2], and Andreas Knorr[2]

[1]*Department of Physics,*
Chalmers University of Technology,
SE-41296 Gothenburg, Sweden

[2]*Institut für Theoretische Physik,*
Nichtlineare Optik und Quantenelektronik,
Technische Universität Berlin,
10623 Berlin, Germany

[*]*ermin.malic@chalmers.se*

We present a theoretical study on the ultrafast carrier dynamics in graphene including direct comparison to recent experimental pump-probe results. Based on graphene Bloch equations, we resolve the Coulomb- and phonon-driven dynamics of optically excited carriers in time, momentum, and angle. Our calculations reveal how the optically generated anisotropic non-equilibrium carrier distribution relaxes towards its initial thermal distribution. The main steps of this dynamics include a Coulomb-dominated ultrafast carrier thermalization on a timescale of tens of femtoseconds, a phonon-driven scattering in all momentum directions giving rise to an isotropic distribution in the first hundred femtosecond, and a slower phonon-assisted carrier cooling occurring on a picosecond timescale. We reveal the characteristic features of the carrier dynamics in the near infrared as well as below the optical phonon energy and in the terahertz region close to the Dirac point. In particular, we emphasize the crucial role of Auger scattering processes that give rise to a technologically promising carrier multiplication in graphene.

4.1 Introduction and Motivation

Graphene's exceptional gapless and linear electronic band structure close to the Dirac points gives rise to fascinating ultrafast phenomena, which are of great interest both for fundamental research as well as for technological applications [Malic *et al.* (2013)]. In particular, carrier scattering channels bridging the valence and the conduction band can change the charge carrier density resulting in a carrier multiplication (CM) [Winzer *et al.* (2010)], cf. Fig. 4.1(a). This is a promising phenomenon for graphene-based photodetecting devices.

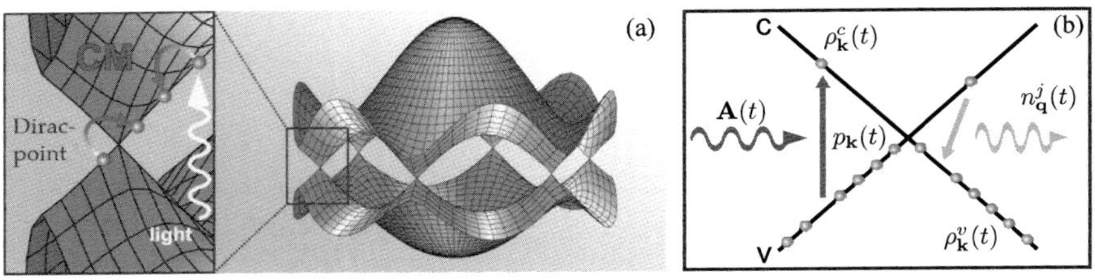

Fig. 4.1 **Band structure and relaxation channels.** (a) Band structure of graphene in the entire Brillouin zone exhibiting six Dirac points, where conduction and valence band cross. The resulting lack of a band gap and the linear bands give rise to new relaxation channels for excited carriers in graphene, cf. the zoom-in into the region around a Dirac point. These Auger processes bridge the valence and the conduction band and can lead to a technologically promising carrier multiplication. (b) Illustration of microscopic quantities necessary to describe the dynamics of optically excited carriers. Figure adapted from Ref. [Malic *et al.* (2013)]. (For color reference, go to page 487.)

To microscopically access the dynamics of non-equilibrium carriers in graphene, we have developed a theoretical approach based on the density matrix formalism [Malic *et al.* (2013, 2011a)]. The core of the approach are graphene Bloch equations allowing us to temporally and spectrally resolve the dynamics of electrons in different regimes. We take into account all relevant many-particle interactions on a consistent microscopic footing. This includes (i) carrier-light interaction responsible for the generation of a non-equilibrium carrier population, (ii) carrier-carrier interaction responsible for the ultrafast thermalization of the excited system, and (iii) carrier-phonon interaction responsible for energy equalization between the electronic and the lattice system. Modeling differential transmission spectra based on graphene Bloch equations, we are able to directly compare our results to current pump-probe experiments.

The chapter is structured as followed: We start with a brief introduction of the theoretical framework. Then, we separately discuss single relaxation mechanisms, before we study the full carrier dynamics in the infrared regime, below the optical phonon energy, and in the THz spectral range. Finally, we analyze several interesting graphene-specific features of carrier dynamics including anisotropic carrier relaxation and the generation of a significant carrier multiplication.

4.2 Theoretical Framework

The results presented in this chapter are based on calculations within the density matrix formalism [Haug and Koch (2004); Rossi and Kuhn (2002); Knorr *et al.* (1996); Kira and Koch (2006)] offering microscopic access to time-, momentum-, and angle-resolved carrier and phonon dynamics in a non-equilibrium situation. The calculated microscopic quantities are: (i) microscopic polarization $p_{\mathbf{k}} = \langle a^+_{v\mathbf{k}} a_{c\mathbf{k}} \rangle$, (ii) carrier occupation probability $\rho^\lambda_{\mathbf{k}} = \langle a^+_{\lambda\mathbf{k}} a_{\lambda\mathbf{k}} \rangle$ in the state $\mathbf{k}$ within the band $\lambda = v, c$ (denoting the valence and the conduction band), and (iii) phonon number $n^j_{\mathbf{q}} = \langle b^+_{j\mathbf{q}} b_{j\mathbf{q}} \rangle$ in the mode j and with the momentum $\mathbf{q}$. Here, we have expressed the quantities in the formalism of second quantization introducing creation and annihilation operators for electrons $(a^+_{\lambda\mathbf{k}}, a_{\lambda\mathbf{k}})$ and phonons $(b^+_{j\mathbf{q}}, b_{j\mathbf{q}})$. Figure 4.1(b) illustrates these quantities for graphene focusing on the physics around the Dirac point: First, an optical pulse described by the vector potential $\mathbf{A}(\mathbf{t})$ is applied to optically excite the carrier system: by lifting electrons from the valence into the conduction band, a non-equilibrium carrier distribution is generated. The microscopic polarization $p_{\mathbf{k}}$ is a measure for the transition probability between the two bands. The optical excitation and the subsequent scattering dynamics changes the occupation probabilities $\rho^c_{\mathbf{k}}$ and $\rho^v_{\mathbf{k}}$ in the involved bands. The carrier scattering with phonons transfers the optically injected energy to the lattice system and changes the phonon number $n^j_{\mathbf{q}}$ beyond the bath approximation. Therefore, the temporal evolution of $\rho^\lambda_{\mathbf{k}}(t), p_{\mathbf{k}}(t)$, and $n^j_{\mathbf{q}}(t)$ contains the relevant information needed to understand the dynamics in the non-equilibrium system.

The temporal evolution of these microscopic quantities is obtained by exploiting the Heisenberg equation of motion [Haug and Koch (2004)]

$$i\hbar \frac{d}{dt}\mathcal{O}(t) = [\mathcal{O}(t), H]_- = \mathcal{O}(t)H - H\mathcal{O}(t) \qquad (4.1)$$

with an exemplary operator $\mathcal{O}$ and the many-particle Hamilton operator H. In this work, H consists of the (i) free-carrier and phonon contribution H_0, (ii) carrier-field interaction $H_{\mathrm{c,f}}$ describing optical excitation of graphene into a non-equilibrium, (iii) carrier-carrier interaction $H_{\mathrm{c,c}}$ accounting for Coulomb-induced intra- and interband scattering, and (iv) carrier-phonon interaction $H_{\mathrm{c,p}}$ denoting phonon-assisted intra- and intervalley processes. The non-interaction part H_0 reads $H_0 = \sum_{\mathbf{l}} \varepsilon_{\mathbf{l}}\, a_{\mathbf{l}}^{+} a_{\mathbf{l}} \; + \sum_{\mathbf{u}} \hbar\omega_{\mathbf{u}} \left(b_{\mathbf{u}}^{+} b_{\mathbf{u}} + \frac{1}{2}\right)$ with the electronic single-particle energy $\varepsilon_{\mathbf{l}}$ and the phonon energy $\hbar\omega_{\mathbf{u}}$, where $\mathbf{l} = (\mathbf{k}, \lambda)$ and $\mathbf{u} = (\mathbf{q}, j)$ are a compound indices containing the electron (phonon) momentum $\mathbf{k}(\mathbf{q})$, the electronic band λ, and the phonon mode j. The carrier-field interaction is expressed within the radiation gauge [Scully and Zubairy (1997)] as $H_{\mathrm{c,f}} = i\hbar \frac{e_0}{m_0} \sum_{\mathbf{l}_1,\mathbf{l}_2} \boldsymbol{M}_{\mathbf{l}_1,\mathbf{l}_2} \cdot \boldsymbol{A}(t)\, a_{\mathbf{l}_1}^{+} a_{\mathbf{l}_2}$ with the elementary charge e_0, the free electron mass m_0, and the vector potential $\mathbf{A}(\mathbf{t})$. The strength of this carrier-light coupling is given by the optical matrix element $\mathbf{M}_{\mathbf{l}_1,\mathbf{l}_2}$ that will be discussed below together with other coupling elements appearing in the Hamilton operator. The carrier-carrier interaction reads $H_{\mathrm{c,c}} = \frac{1}{2} \sum_{\mathbf{l}_1,\mathbf{l}_2,\mathbf{l}_3,\mathbf{l}_4} V_{\mathbf{l}_3,\mathbf{l}_4}^{\mathbf{l}_1,\mathbf{l}_2}\; a_{\mathbf{l}_1}^{+} a_{\mathbf{l}_2}^{+} a_{\mathbf{l}_4} a_{\mathbf{l}_3}$ including the Coulomb matrix element $V_{\mathbf{l}_3,\mathbf{l}_4}^{\mathbf{l}_1,\mathbf{l}_2}$. Finally, the last contribution of the Hamilton operator describes the interaction between carriers and phonons and reads $H_{\mathrm{c,p}} = \sum_{\mathbf{l}_1,\mathbf{l}_2} \sum_{\mathbf{u}} \left(g_{\mathbf{u}}^{\mathbf{l}_1,\mathbf{l}_2}\, a_{\mathbf{l}_1}^{+} a_{\mathbf{l}_2} b_{\mathbf{u}} + g_{\mathbf{u}}^{\mathbf{l}_1,\mathbf{l}_2 *}\, a_{\mathbf{l}_2}^{+} a_{\mathbf{l}_1} b_{\mathbf{u}}^{+}\right)$ with the carrier-phonon matrix elements $g_{\mathbf{u}}^{\mathbf{l}_1,\mathbf{l}_2}$. The Hamilton operator becomes graphene-specific by inserting electron and phonon dispersion relations as well as the coupling elements. These are calculated with tight-binding wave functions in the nearest-neighbor approximation [Malic *et al.* (2013)]. This approach is known to be a good approximation for graphene close to the Dirac point [Reich *et al.* (2002)]. Note that the theory does not rely on fitting parameters. The appearing tight-binding parameter is fixed according to DFT calculations at the beginning of the calculation and is not changed anymore. The dielectric background is introduced by a static screening of the Coulomb potential.

Treating all many-particle interactions on the same microscopic footing within the second-order Born-Markov approximation, we obtain graphene Bloch equations [Malic *et al.* (2013, 2011a)]:

$$\dot{p}_{\mathbf{k}}(t) = i\Delta\omega_{\mathbf{k}} p_{\mathbf{k}}(t) - i\Omega_{\mathbf{k}}^{vc}(t)\big(\rho_{\mathbf{k}}^{c}(t) - \rho_{\mathbf{k}}^{v}(t)\big) + \mathcal{U}_{\mathbf{k}}(t) - \gamma_{\mathbf{k}}(t) p_{\mathbf{k}}(t), \tag{4.2}$$

$$\dot{\rho}_{\mathbf{k}}^{\lambda}(t) = \pm 2\mathrm{Im}\big(\Omega_{\mathbf{k}}^{vc*}(t) p_{\mathbf{k}}(t)\big) + \Gamma_{\lambda,\mathbf{k}}^{\mathrm{in}}(t)\big(1 + \rho_{\mathbf{k}}^{\lambda}(t)\big) - \Gamma_{\lambda,\mathbf{k}}^{\mathrm{out}}(t) \rho_{\mathbf{k}}^{\lambda}(t), \tag{4.3}$$

$$\dot{n}_{\mathbf{q}}^{j}(t) = \Gamma_{j,\mathbf{q}}^{\mathrm{em}}(t)\big(1 + n_{\mathbf{q}}^{j}(t)\big) - \Gamma_{j,\mathbf{q}}^{\mathrm{abs}}(t) n_{\mathbf{q}}^{j}(t) - \gamma_j\big(n_{\mathbf{q}}^{j}(t) - n_B\big) \tag{4.4}$$

with the energy difference $\hbar\Delta\omega_{\mathbf{k}} = (\varepsilon^v_{\mathbf{k}} - \varepsilon^c_{\mathbf{k}})$, the Rabi frequency $\Omega^{vc}_{\mathbf{k}}(t) = i\frac{e_0}{m_0}\mathbf{M}^{vc}_{\mathbf{k}} \cdot \mathbf{A}(t)$ describing the optical excitation of graphene, the Bose-Einstein distribution n_B denoting the equilibrium distribution of phonons, and the phonon lifetime γ_j^{-1}. The electronic band structure around the Dirac points reads $\varepsilon^{\lambda}_{\mathbf{k}} = \sigma_\lambda v_F k$ with $\sigma_c = 1$ and $\sigma_v = -1$. The gradient $v_F = -\gamma_0 a_0\sqrt{3}/2 \approx 0.605\,\text{eVnm}$ corresponds to a carrier velocity of 0.92 nm/fs, which is approximately 300 times slower than the vacuum light velocity. The equations include all relevant two-particle relaxation channels including Coulomb- and phonon-induced intra- and interband as well as intra- and intervalley scattering processes. The time- and momentum-dependent scattering rates $\Gamma_{\lambda\mathbf{k}}(t) = \Gamma^{\text{cc}}_{\lambda\mathbf{k}}(t) + \Gamma^{\text{cp}}_{\lambda\mathbf{k}}(t)$ entering the equation for the carrier occupation $\rho^{\lambda}_{\mathbf{k}}(t)$ describe the strength of the carrier-carrier (cc) and carrier-phonon (cp) scattering processes. For the phonon occupation, we obtain the corresponding emission and absorption rates $\Gamma^{\text{em/abs}}_{j,\mathbf{q}}$ [Malic *et al.* (2011a)]. The many-particle scattering also contributes to diagonal ($\gamma_{\mathbf{k}}(t)$) and off-diagonal dephasing ($\mathcal{U}_{\mathbf{k}}(t)(t)$) of the microscopic polarization $p_{\mathbf{k}}(t)$. More details on the theoretical approach can be found in Ref. [Malic *et al.* (2011a)].

4.3 Relaxation Channels in Non-equilibrium

By solving graphene Bloch equations, we can track the way of optically excited carriers towards their equilibrium distribution. To thoroughly understand this relaxation dynamics, we first separately study the single relaxation channels including (i) carrier-light interaction optically inducing a non-equilibrium, (ii) carrier-carrier interaction giving rise to carrier thermalization and under certain conditions carrier multiplication, and (iii) carrier-phonon interaction accounting for carrier cooling and orientational relaxation. In the next section, we will study the full carrier dynamics combining these relaxation channels and explicitly considering their interplay.

4.3.1 *Carrier-light interaction*

It is important to treat the carrier-light interaction on the same microscopic footing as the many-particle scattering processes. The simple assumption of a non-equilibrium distribution as initial condition would neglect carrier-carrier and carrier-phonon scattering taking place during the time scale of the excitation pulse.

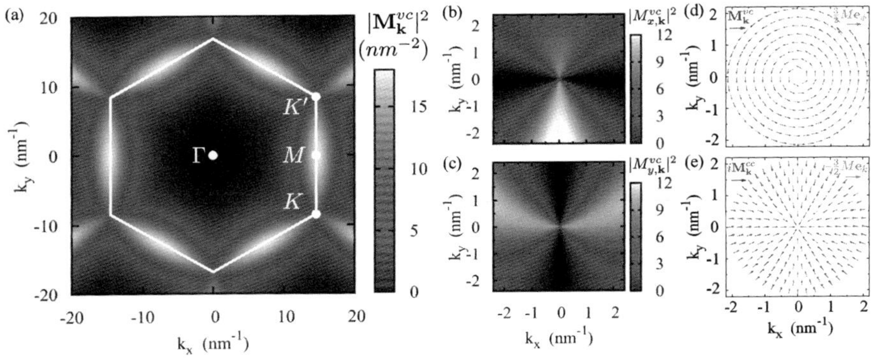

Fig. 4.2 **Optical matrix element.** (a) Surface plot showing the square of the absolute value of the optical interband matrix element in the momentum space. (b) - (c) Corresponding x- and y-components around the K' point. (d) - (e) The interband and the intraband matrix element (blue arrows) can be approximated by a pure rotational and radial fields (red arrows), respectively. Figure adapted from Ref. [Winzer (2013)]. (For color reference, go to page 487.)

4.3.1.1 *Optical matrix element*

We start the section with a discussion of the optical matrix element that contains all symmetry-imposed selection rules and determines the strength of the coupling between the carrier system and an external optical field. It corresponds to the expectation value of the momentum operator $\mathbf{p} = -i\hbar\nabla$. Throughout this book chapter, we refer the expression optical matrix element to

$$\boldsymbol{M}^{\lambda\lambda'}_{\mathbf{k},\mathbf{k}'} = \langle \Psi^{\lambda}(\mathbf{k},\mathbf{r})|\nabla|\boldsymbol{\Psi}^{\lambda'}(\mathbf{k}',\mathbf{r})\rangle \,. \tag{4.5}$$

Since the momentum of light is negligibly small, we focus on direct optical transitions with $\mathbf{k} = \mathbf{k}'$.

Inserting tight-binding wave functions [Reich *et al.* (2002)] and exploiting the symmetry of the $2p_z$-orbitals, we obtain an analytic expression for the interband optical matrix element [Grüneis *et al.* (2003); Malić *et al.* (2006)]

$$\mathbf{M}^{vc}_{\mathbf{k}} = \frac{M}{|e(\mathbf{k})|}\mathrm{Re}\left[e^{*}(\mathbf{k})\sum_{i=1}^{3} e^{i\mathbf{k}\mathbf{b}_i}\frac{\mathbf{b}_i}{|\mathbf{b}_i|}\right] \tag{4.6}$$

with the constant $M \approx 3.7\,\mathrm{nm}^{-1}$ and the tight-binding nearest-neighbor sum $e(\mathbf{k}) = \sum_{i=1}^{3} e^{i\mathbf{k}\mathbf{b}_i}$ accounting for the graphene lattice symmetry ($\mathbf{b_i}$ are vectors connecting the neighboring carbon atoms). For the intraband optical matrix element $\mathbf{M}_{\mathbf{k}}^{\lambda\lambda}$, replace Re by iIm in Eq. (4.6). The optical matrix element fulfills the symmetry relations $\mathbf{M}_{\mathbf{k}}^{vc} = \mathbf{M}_{\mathbf{k}}^{cv}$ and $\mathbf{M}_{\mathbf{k}}^{vv} = -\mathbf{M}_{\mathbf{k}}^{cc}$. Optical properties of graphene are determined by the optical interband matrix element [Malic *et al.* (2011a)], which is shown in Fig. 4.2 (a). While the maximum carrier-light coupling is found at the M point resulting in a pronounced peak in absorption spectra [Yang *et al.* (2009); Malic *et al.* (2011a); Mak *et al.* (2011); Chae *et al.* (2011)], it vanishes at the the Γ point manifesting an optical selection rule for graphene. For excitations with linearly polarized pulses, the projection of the matrix element to the polarization orientation is the crucial quantity. Therefore, we show in Figs. 4.2 (b) and (c) the x- and the y-component of the optical interband matrix, respectively. We find that in the momentum space the x-component $M_{x,\mathbf{k}}^{v,c}$ is the largest in the y direction, while $M_{y,\mathbf{k}}^{vc}$ is the most pronounced in the x direction.

Considering the optical matrix element as a vector field close to the Dirac point, it can be shown that $\mathbf{M}_{\mathbf{k}}^{vc}$ corresponds to an approximately pure rotational field, while the intraband matrix element is a radial field close to the Dirac point [Winzer (2013)]

$$\mathbf{M}_{\mathbf{k}}^{vc} = \frac{3}{2} M \mathbf{e}_{\phi} \quad \text{and} \quad \mathbf{M}_{\mathbf{k}}^{cc} = i\frac{3}{2} M \mathbf{e}_{k}. \tag{4.7}$$

In Figs. 4.2(d) and (e), the tight-binding optical matrix element from Eq. (4.6) and its simplified expression from Eq. (4.7) are directly compared in the vicinity of the K point. The direction of rotation of the interband matrix element is opposite at the K' point.

4.3.1.2 *Optical excitation*

A non-equilibrium carrier distribution is generated by applying an optical pulse described by a Gaussian-shaped vector potential

$$\mathbf{A}(t) = A_0\, \hat{\mathbf{e}}_{\mathcal{P}}\, e^{-\frac{t^2}{2\sigma_t^2}} \cos(\omega_L t), \tag{4.8}$$

where A_0 is the amplitude with the dimension fsV/nm, $\hat{\mathbf{e}}_{\mathcal{P}} = (\cos\phi_{\mathcal{P}}, \sin\phi_{\mathcal{P}})$ is the polarization unit vector, σ_t is the pulse duration, and $\hbar\omega_L$ is the excitation energy. The studies presented in this section will be performed with an energy of $\hbar\omega_L = 1.5\,\text{eV}$ and a full-width-half-maximum of $\sigma_{\text{FWHM}} = 10\,\text{fs}$ corresponding to a pulse duration of $\sigma_t = \sigma_{\text{FWHM}}/(2\sqrt{2\ln(2)}) \approx 5]fs$.

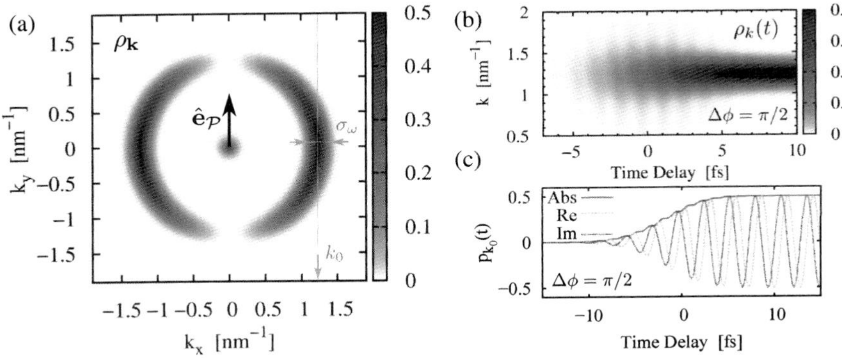

Fig. 4.3 **Optically excited carrier distribution.** (a) Highly anisotropic non-equilibrium carrier distribution after optical excitation with a pulse characterized by an energy of 1.5 eV and a temporal duration of 10 fs. (b) Time- and momentum resolved carrier occupation during the excitation process in direction of the strongest carrier-light interaction ($\Delta\phi = \pi/2$). (c) Temporal evolution of the microscopic polarization at $\mathbf{k} = (k_0, \phi_{\mathcal{P}} + \pi/2)$. Figure taken from Ref. [Winzer (2013)].

Focusing on the solution of the free-particle graphene Bloch equations, i.e. taking into account only carrier-light interaction, we obtain for a linearly polarized optical pulse with a polarization in y direction $\hat{\mathbf{e}}_{\mathcal{P}} = \hat{\mathbf{e}}_y$ a non-equilibrium carrier distribution, cf. Fig. 4.3. Remarkably, we observe a maximum carrier occupation in k_x direction, while no carriers can be found in the direction parallel to the pulse polarization. The highly anisotropic carrier distribution can be traced back to the anisotropy of the optical matrix element, cf. Fig. 4.2. A successive integration of the graphene Bloch equation in the lowest order of excitation results in $\dot{p}_{\mathbf{k}} \propto |\mathbf{M}^{vc}_{\mathbf{k}} \cdot \hat{\mathbf{e}}_{\mathcal{P}}|$ and $\dot{\rho}^{\lambda}_{\mathbf{k}} \propto |\mathbf{M}^{vc}_{\mathbf{k}} \cdot \hat{\mathbf{e}}_{\mathcal{P}}|^2$ [Malic *et al.* (2010)]. Exploiting the symmetry of the matrix element, the angle dependence of the excited carrier distribution follows the relation $\dot{\rho}^{\lambda}_{\mathbf{k}} \propto \sin^2(\Delta\phi)$, where $\Delta\phi = \phi_{\mathbf{k}} - \phi_{\mathcal{P}}$ is the angle between electron wave vector with $\mathbf{k} = (k, \phi_{\mathbf{k}})$ and pulse polarization, cf. Fig. 4.3(a).

Furthermore, the excited distribution is energetically centered around $k_0 \approx 1.25\,\mathrm{nm}^{-1}$ corresponding to the excitation energy $\hbar\omega_L = 2\nu_F k_0$. The width of the carrier distribution reflects the spectral width of the optical pulse $\sigma_\omega = \sigma_t^{-1}$. The carrier population close to the Dirac point corresponds to the initial thermal Fermi distribution at room temperature (present before the pulse). Figure 4.3(b) shows the time-resolved carrier occupation in direction of the strongest carrier-light interaction during the excitation process. The oscillatory build-up of the population directly reflects the frequency of the applied optical pulse in second order $2\omega_L$, which arises from the treatment beyond the rotating wave approximation [Malic (2008)]. In contrast, the microscopic polarization $p_{\mathbf{k}}$ is proportional to the exciting field in first order giving rise to resonant oscillations at k_0, cf. Fig 4.3(c).

Besides the excitation energy $\hbar\omega_L$ and the pulse width σ_t, the pump fluence $\mathcal{E}_{pf}$ plays an important role for understanding the carrier dynamics. It is a measure for the excitation strength and is defined by the time-integrated intensity of the pulse per area [Diels *et al.* (1996)]. Assuming that the pulse duration σ_t is significantly longer than a single cycle duration $\tau = 2\pi/\omega$, the electric field can be expressed as $\mathbf{E}(t) = -\partial_t \mathbf{A}(t) \approx \omega_L A_0 \hat{\mathbf{e}}_0 \exp\left(-\frac{t^2}{2\sigma_t^2}\right)\sin(\omega_L t)$. The intensity is defined via the Poynting vector by $I(t) = |\mathbf{S}(t)| = \varepsilon_0 c |\mathbf{E}(t)|^2$ [Jackson (1998)]. Within the same approximation ($\sigma_t \gg \tau$), the intensity can be replaced by its average $\hat{I}(t)$ over a single cycle duration $\hat{I}(t) = \frac{1}{\tau}\int_{t-\tau/2}^{t+\tau/2} dt' |\mathbf{S}(t')| \approx \frac{\omega_L^2 A_0^2 \varepsilon_0 c}{2} e^{-\frac{t^2}{\sigma_t^2}}$. Finally, the pump fluence is given by the integral over the intensity

$$\mathcal{E}_{pf} = \int_{-\infty}^{+\infty} dt \hat{I}(t) = A_0^2 \omega_L^2 \sigma_t \varepsilon_0 \sqrt{\pi} c/2. \tag{4.9}$$

It is proportional to the pulse duration σ_t and to the squared amplitude A_0^2 and excitation frequency ω_L^2.

4.3.2 *Carrier-carrier interaction*

Due to the weak screening in the atomically thin graphene, the Coulomb interaction plays a crucial role for the relaxation dynamics of optically excited carriers.

4.3.2.1 *Coulomb matrix element*

The Coulomb matrix element determines the strength of the Coulomb interaction and reads in second quantization [Haug and Koch (2004)]

$$V_{\mathbf{cd}}^{\mathbf{ab}} = \langle \Psi_{\mathbf{a}}(\mathbf{r})\Psi_{\mathbf{b}}(\mathbf{r}')|V(\mathbf{r}-\mathbf{r}')|\Psi_{\mathbf{d}}(\mathbf{r}')\Psi_{\mathbf{c}}(\mathbf{r})\rangle = V_q \mathcal{V}_q^2 f_{\mathbf{ac}} f_{\mathbf{bd}} \delta_{\mathbf{k}_a+\mathbf{k}_b,\mathbf{k}_c+\mathbf{k}_d}. \tag{4.10}$$

Here, $q = |\mathbf{k}_c - \mathbf{k}_a|$ is the distance between the interacting carriers in the momentum space, $V_q = \frac{1}{L^2}\frac{e_0^2}{2\varepsilon_0 q}$ is the Fourier transform of the two-dimensional Coulomb potential, and ε_0 is the vacuum permittivity. The Coulomb matrix element can be analytically calculated by inserting tight-binding wave functions $\Psi_{\mathbf{i}}$ including $2p_z$-orbitals with an effective atomic number Z^* [Malic *et al.* (2011a)]. The integral over $2p_z$-orbitals results in a graphene-specific momentum-dependent function $\mathcal{V}_q = \left((qa_B/Z^*)^2 + 1\right)^{-3}$ modifying the Coulomb potential. For processes requiring a large momentum transfer q, this function causes an effective decay of the Coulomb potential $V_{\mathbf{cd}}^{\mathbf{ab}}$ with q^{-13}. However, for all processes taking place within one Dirac valley the impact of $\mathcal{V}_q$ is negligibly small, since $qa_B \ll Z^*$ and therefore $\mathcal{V}_q \approx 1$. The appearing Kronecker $\delta_{\mathbf{k}_a+\mathbf{k}_b,\mathbf{k}_c+\mathbf{k}_d}$ ensures the conservation of momentum for all Coulomb-induced processes. Furthermore, the tight-binding wave functions give rise to a graphene-specific form factor

$$f_{\mathbf{ac}} = \frac{1}{2}\left(1 + \sigma_{\lambda_a\lambda_c}\frac{e^*(\mathbf{k}_a)e(\mathbf{k}_c)}{|e(\mathbf{k}_a)||e(\mathbf{k}_c)|}\right) \approx \frac{1}{2}\left(1 + \sigma_{\lambda_\mathbf{a}\lambda_\mathbf{c}}e^{-i(\phi_c-\phi_a)}\right), \quad (4.11)$$

where the abbreviation $\sigma_{\lambda_\mathbf{a}\lambda_\mathbf{c}} = \sigma_{\lambda_\mathbf{a}}\sigma_{\lambda_\mathbf{c}}$ (with $\sigma_c = 1$ and $\sigma_v = -1$) has been introduced. Within the linear approximation close to the Dirac point, the form factor describes the angle dependence of each Coulomb process, where ϕ_i is the orientation of $\mathbf{k}_i$. Here, the Coulomb matrix element becomes maximal for parallel processes within a band ($\phi_c - \phi_a = 0$ in Eq. (4.11)), whereas it vanishes for anti-parallel electron wave vectors ($\phi_c - \phi_a = \pi$), i.e. for back-scattering [Ando *et al.* (1998)]. For interband processes the situation is opposite.

The presence of carriers in the graphene sheet as well as in the substrate screens the pure Coulomb potential $V_\mathbf{q}$. Assuming that the substrate is located only on one side of the graphene layer, the corresponding screening can be considered by an effective dielectric background constant $\varepsilon_{sub} = (\varepsilon_{r,sub} + \varepsilon_{r,vac})/2$ resulting in $V_\mathbf{q}^{sub} = V_\mathbf{q}/\varepsilon_{sub}$ [Ando (2006)]. The intrinsic screening $\varepsilon(\mathbf{q})$ can be consistently derived within the density matrix formalism from higher-order correlations [Verdenhalven *et al.* (2013)]. The result corresponds to the static limit of the Lindhard formula [Jiang *et al.* (2007)]

$$\varepsilon(\mathbf{q}) = 1 - 2V_\mathbf{q}^{sub}\sum_{\lambda\lambda'\mathbf{k}}\frac{\rho_{\mathbf{k}-\mathbf{q}}^{\lambda'} - \rho_\mathbf{k}^\lambda}{\varepsilon_{\mathbf{k}-\mathbf{q}}^{\lambda'} - \varepsilon_\mathbf{k}^\lambda}\left|\int d\mathbf{r}\Psi_\lambda^*(\mathbf{k})e^{-i\mathbf{q}\cdot\mathbf{r}}\Psi_{\lambda'}(\mathbf{k})\right|^2. \quad (4.12)$$

and the screened Coulomb potential reads $W_\mathbf{q} = V_\mathbf{q}/(\varepsilon_{sub}\varepsilon(\mathbf{q}))$.

The probability of a Coulomb scattering event depends on the initial and final state of the two involved carriers, i.e. on four sets of quantum numbers $\mathbf{i} = (\mathbf{k}_i, \lambda_i)$. However, energy and momentum conservation

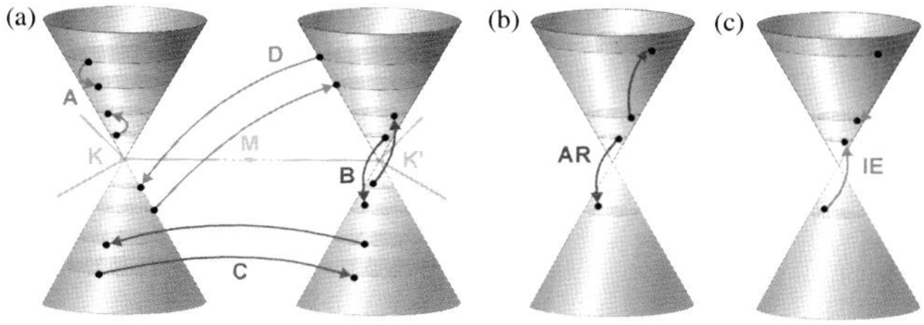

Fig. 4.4 **Coulomb-induced scattering channels** Schematic illustration of carrier-carrier scattering channels in graphene including (a) intra- (A,C) and interband (B,D) as well as intra- (A,B) and intervalley (C,D) processes. (b) Auger recombination (AR) and (c) impact excitation (IE). Figure adapted from Ref. [Malic *et al.* (2011a)].

considerably reduces the allowed phase space. The momentum conservation ensured by the Kronecker delta appearing in the Coulomb matrix element reduces the number of independent momentum vectors to three: $(\mathbf{k_a}, \mathbf{k_b}, \mathbf{k_c}, \mathbf{k_d}) \to (\mathbf{k}, \mathbf{k}', \mathbf{k}+\mathbf{q}, \mathbf{k}'-\mathbf{q})$. Due to the energy conservation appearing as delta functions in Coulomb-induced scattering rates in the graphene Bloch equations, the momentum space can be restricted to the area around the Dirac points (characterized by a linear dispersion) after an optical excitation in the near-infrared or below. Furthermore, we distinguish intraband ($\lambda_a = \lambda_c$ and $\lambda_b = \lambda_d$), interband ($\lambda_a \neq \lambda_c$ and $\lambda_b \neq \lambda_d$), and Auger scattering processes, cf. Fig. 4.4. In the case of Auger scattering, one electron changes the band, whereas the second remains within its initial band. Their appearance is a direct consequence of the linear and gapless band structure of graphene. In contrast, in conventional semiconductors they are energetically forbidden or at least extremely inefficient. Since they change the carrier density, these processes differ qualitatively from all other Coulomb-induced scattering channels. Auger recombination (AR) and the inverse process of impact excitation (IE) are distinguished, cf. Figs. 4.4(b) and (c). While AR annihilates an electron-hole pair, IE increases the number of optically excited carriers leading to the technologically promising process of carrier multiplication, cf. Section 4.6 for a more detailed discussion.

4.3.2.2 *Coulomb-induced relaxation dynamics*

Coulomb-induced scattering has a crucial impact on the relaxation dynamics of non-equilibrium carriers. Figure 4.5 illustrates the temporal evolution of the angle-averaged carrier distribution $\rho_k = \int_0^{2\pi} \rho(k,\phi)\,d\phi$ during

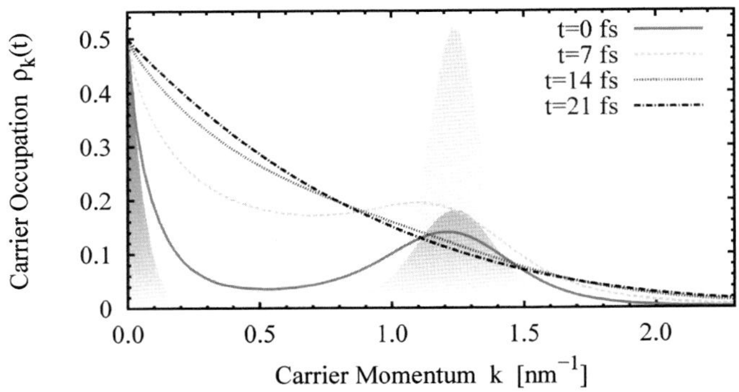

Fig. 4.5 **Coulomb-induced relaxation dynamics.** Time- and momentum-resolved purely Coulomb-driven dynamics of the angle averaged carrier distribution during and after an excitation pulse with a duration of $\sigma_t = 7\,\text{fs}$. Carrier-carrier scattering gives rise to an ultrafast carrier thermalization resulting in a hot, spectrally broad Fermi distribution. For comparison, the red and green shaded areas illustrate the carrier distribution at the center (0 fs) and after the optical pulse (7 fs) without carrier-carrier scattering. Figure taken from Ref. [Winzer (2013)].

and immediately after an optical excitation with an energy of 1.5 eV (corresponding to a momentum $k_0 = 1.25\,\text{nm}^{-1}$). The excited non-equilibrium becomes redistributed on a femtosecond timescale resulting in a hot, spectrally broad Fermi distribution reached after approximately 20 fs. A significant part of the relaxation dynamics occurs already during the ultrashort optical pulse (7 fs). The shaded areas in Fig. 4.5 show the carrier occupation without taking into account carrier-carrier scattering demonstrating their crucial importance for the relaxation dynamics. In particular, Auger scattering processes turn out to be significantly important, as discussed in Section 4.6.

4.3.3 *Carrier-phonon interaction*

To reach the initial carrier distribution characterized by a Fermi function at room temperature, band-to-band recombination and energy redistribution processes between the carrier and the lattice system are required. In contrast to Coulomb-induced processes, carrier-phonon scattering transfers energy from the excited carrier to the lattice system via intraband cooling and interband recombination.

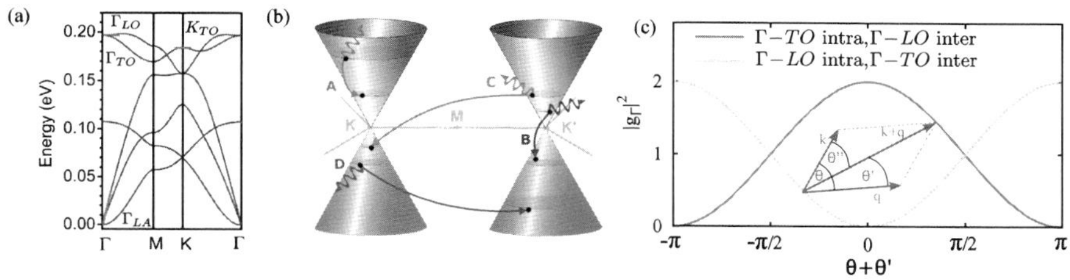

Fig. 4.6 **Carrier-phonon scattering in graphene.** (a) Phonon dispersion of graphene along the high-symmetry lines $\Gamma - M - K - \Gamma$. The red lines mark the phonon modes that are important for the relaxation dynamics. (b) Phonon-induced scattering channels including intravalley intra- (A) and interband scattering (B) induced by Γ-phonons as well as intervalley inter- (C) and intraband scattering (D) induced by K-phonons. For phonon absorption processes, all arrows need to be inverted. (c) Angle dependence of the matrix element for optical Γ phonons. The inset illustrates the definition of the scattering angles θ, θ', and θ'' appearing in the carrier-phonon matrix elements. Figure (a) adapted from Ref. [Sanders *et al.* (2013)], (b) from [Malic *et al.* (2011a)], and (c) from [Malic *et al.* (2013)]

4.3.3.1 *Carrier-phonon matrix element*

The probability for carrier-phonon scattering is determined by the electron and phonon dispersion relation and the carrier-phonon matrix element. The phonon dispersion is shown in Fig. 4.6(a) illustrating optical and acoustic phonons along high-symmetry lines within the Brillouin zone of graphene [Maultzsch *et al.* (2004); Sanders *et al.* (2013)]. The dispersion of optical phonons in graphene is characterized by two kinks (Kohn anomalies) at the Γ and the K point [Piscanec *et al.* (2004)]. The corresponding Γ-E_{2g} and K-A_1' phonon modes exhibit strong electron-phonon coupling. The doubly degenerate E_{2g} mode splits into an upper longitudinal optical (LO) and a lower transverse optical (TO) mode. The energy of optical phonons around these high-symmetry points can be approximated by the Einstein model assuming approximately constant energies with $\hbar\omega_{\Gamma LO} = 198\,\text{meV}$, $\hbar\omega_{\Gamma TO} = 192\,\text{meV}$, and $\hbar\omega_{KTO} = 162\,\text{meV}$, cf. red lines in Fig. 4.6(a) [Maultzsch *et al.* (2004)]. The momentum transfer during a phonon-induced scattering process is either very small ($q \approx 0$) corresponding to the intravalley scattering via Γ phonons or very large ($q \approx \frac{4\pi}{\sqrt{3}a_0}$) corresponding to the intervalley scattering between K and K' points, cf. Fig. 4.6(b).

S. Piscanec and co-workers have shown that the slope of the kinks in the phonon dispersion relation is directly proportional to the square of the coupling element allowing their experimental estimation [Piscanec *et al.*

(2004)]. The electron-phonon coupling elements for the ΓLO, ΓTO, and the K mode read [Piscanec *et al.* (2004)]

$$|g_{\mathbf{q}\Gamma\mathbf{j}}^{\mathbf{k}\lambda\lambda'}|^2 = \frac{1}{N}\tilde{g}_\Gamma^2\left(1 + c_{\Gamma j}^{\lambda\lambda'}\cos(\theta+\theta')\right), \tag{4.13}$$

$$|g_{\mathbf{qKj}}^{\mathbf{k}\lambda\lambda'}|^2 = \frac{1}{N}\tilde{g}_K^2\left(1 + c_{Kj}^{\lambda\lambda'}\cos(\theta'')\right) \tag{4.14}$$

with j denoting the LO or TO phonon mode with $c_{\Gamma TO}^{\lambda\lambda'} = c_{\Gamma LO}^{\lambda\lambda} = c_{KTO}^{\lambda\lambda} = -1$ and $c_{\Gamma TO}^{\lambda\lambda} = c_{\Gamma LO}^{\lambda\lambda'} = c_{KTO}^{\lambda\lambda'} = +1$ with $\lambda \neq \lambda'$. Furthermore, $\tilde{g}_{KTO}^2 = 0.0994\,\mathrm{eV}^2$ for KTO phonons, $\tilde{g}_\Gamma^2 = 0.0405\,\mathrm{eV}^2$ for ΓLO and ΓTO phonons, and the scattering angles are defined by $\theta = \angle(\mathbf{k},\mathbf{q})$, $\theta' = \angle(\mathbf{k}+\mathbf{q},\mathbf{q})$, and $\theta'' = \angle(\mathbf{k},\mathbf{k}+\mathbf{q})$, cf. the inset in Fig. 4.6(c). The factor N corresponds to the number of graphene unit cells with $N = \frac{A}{A_{\mathrm{uc}}} = \frac{2A}{a_0\sqrt{3}}$. The graphene area A cancels in the graphene Bloch equations after performing the q-sum.

The carrier-phonon matrix elements do not depend on the magnitude of the momentum transfer q of the involved phonons, but they exhibit a characteristic angle dependence for each phonon mode. The transversal intraband and longitudinal interband coupling become maximal for $\theta + \theta' = 0$, cf. Fig. 4.6(c). This condition is fulfilled either for parallel scattering ($\mathbf{k} \uparrow\uparrow \mathbf{q}$) or if $q < k$ and $\mathbf{k} \uparrow\downarrow \mathbf{q}$ are simultaneously fulfilled. In the first case the electron ends up in an energetically higher state requiring absorption of a phonon. Thus, this process is proportional to the corresponding phonon population $n_{\mathbf{q}}^j$, which only contributes significantly in the presence of hot phonons. In contrast, the anti-parallel case with $q < k$ corresponds to forward scattering along the Dirac cone via emission of a phonon, which typically provides the highest probability ($\propto n_{\mathbf{q}}^j + 1$). The strongest transversal interband and l ongitudinal intraband coupling is obtained for $\theta + \theta' = \pm\pi$, cf. Fig. 4.6(c).

The dispersion of acoustic phonons is approximately linear close to the Γ point with $\hbar\omega_{j,q} = \nu_j q$. We take into account only the strongest ΓLA phonon mode with $\nu_{\Gamma LA} = 13.2\,\mathrm{meVnm} \approx \nu_F/46$ [Hwang and Das Sarma (2008a); Tse and Das Sarma (2009)] allowing for continuous energy transfer to the electron system in contrast to optical phonons. Due to $\nu_{\Gamma LA} < v_F$ the emission or absorption of an acoustic phonon only accounts for electronic intraband processes, whereas interband processes are energetically forbidden [Tse and Das Sarma (2009)]. To obtain the corresponding carrier-phonon matrix element, we follow the approach of Das Sarma and co-workers and obtain for phonon-induced intraband scattering [Tse and Das Sarma (2009)]

$$|g_{\mathbf{q}\Gamma\mathbf{LA}}^{\mathbf{k}\lambda\lambda}|^2 = \frac{1}{A}\frac{q\,D^2\hbar^2}{\rho_M\nu_{\Gamma\mathrm{LA}}}\left(1 - \cos(\theta'')\right) \tag{4.15}$$

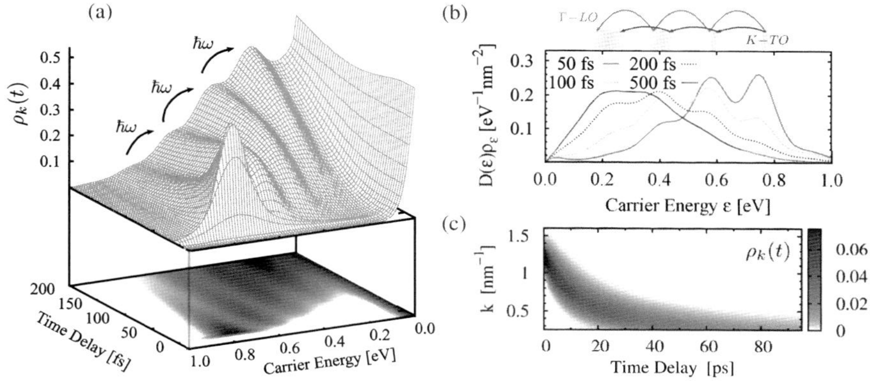

Fig. 4.7 **Phonon-induced relaxation dynamics.** (a) Time- and energy-resolved purely phonon-driven dynamics of the angle averaged carrier distribution during the first 200 fs after optical excitation at $\hbar\omega_L = 1.5\,\text{eV}$. Due to the strict energy conservation the carrier relaxation occurs in discrete Markov steps of the optical phonon energies. (b) Carrier distribution for selected time delays considering the density of states. Multiple emission of optical phonons with different energies gives rise to a broadening of the optically generated non-equilibrium distribution on a femtosecond timescale. (c) Scattering with acoustic phonons is much slower taking place on a hundred picosecond timescale. Figure taken from Ref. [Winzer (2013)].

with the deformation potential $D = 16$ eV and the graphene mass density $M = 7.6 \cdot 10^{-8}\,\text{gcm}^{-2}$. Note that in contrast to optical phonons, the coupling directly depends on the absolute value of the momentum q.

4.3.3.2 *Phonon-induced relaxation dynamics*

Phonon-induced scattering plays a crucial role for the relaxation dynamics of non-equilibrium carriers, since it connects the excited carrier system with the lattice enabling an energy relaxation. Due to the strict energy conservation within the applied Markov approximation in graphene Bloch equations, the scattering occurs in discrete steps of the optical phonon energies resulting in a phonon cascade from the excitation momentum towards the Dirac point, cf. Fig. 4.7(a). The first scattering processes takes place immediately after the applied pulse, while the second and the third relaxation steps appear flatter and delayed, since they require significant carrier occupation at the energies $\hbar\omega_L/2 - \hbar\omega_j$ and $\hbar\omega_L/2 - 2\hbar\omega_j$, respectively. In each relaxation step, a phonon is emitted transferring energy from the carrier to the lattice system. Figure 4.7(b) shows the dynamics of the carrier population multiplied with the linear density of states. Here, the area below the curves corresponds to the carrier density that is conserved as long as no phonon-induced carrier recombination takes place. Multiple emission of optical phonons with different energy leads to a broadening of the carrier distribution, as illustrated in Fig. 4.7(b) Since optical phonons are much more efficient, the carrier dynamics induced by acoustic phonons is shown separately in Fig. 4.7(c). Induced by arbitrary small energy transfers, the non-equilibrium carrier distribution at the excitation energy shifts slowly towards the Dirac point on a timescale of hundred picoseconds, cf. Fig. 4.7(c).

4.4 Carrier Relaxation Dynamics

Having understood the single Coulomb- and phonon-induced relaxation channels, now we investigate the full carrier dynamics taking into account both relaxation mechanisms. Since they compete with each other (e.g. Auger- and phonon-induced recombination processes), the combined dynamics will not correspond to the sum of single contributions. Furthermore, depending on the excitation regime (carrier density, temperature, excitation energy), either carrier-carrier or carrier-phonon scattering can be the dominant relaxation mechanism. In this section, we will discuss the dynamics in three different experimentally accessible excitation regimes: near-infrared, below optical phonon energy, and terahertz (THz).

4.4.1 *Carrier dynamics in the near-infrared*

Here, we generate a non-equilibrium distribution in the near-infrared by apply an optical Gaussian-shaped pulse characterized by an energy of $\hbar\omega_L = 1.5\,\text{eV}$, a full-width-half-maximum of $\sigma_{\text{FWHM}} = 10\,\text{fs}$, and a fluence of $\mathcal{E}_{pf} = 1\,\mu\text{Jcm}^{-2}$. Figure 4.8 shows temporally and spectrally resolved angle-averaged carrier distribution corresponding to the occupation probability ρ_k weighted with the density of states $D(\omega)$. For the time $t = 0\,\text{fs}$ corresponding to the center of the excitation pulse (red line), we observe a non-equilibrium distribution centered at $\varepsilon = 0.75\,\text{eV}$ and the thermal background population at room temperature in the low energy tail. Then, the non-equilibrium electrons are redistributed to energetically lower states via combined carrier-carrier and carrier-phonon scattering aiming finally at their initial equilibrium distribution via carrier cooling and

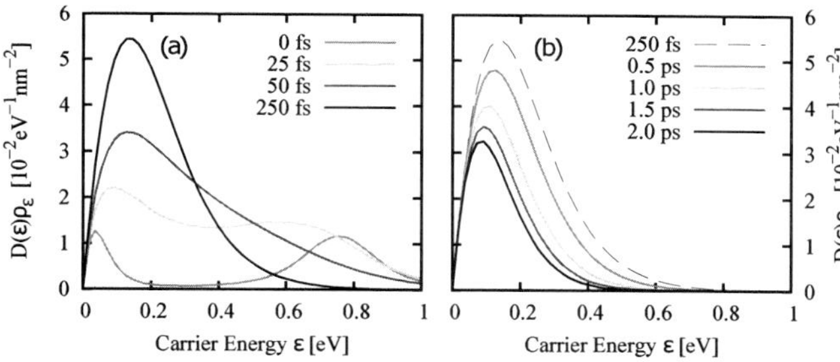

Fig. 4.8 **Thermalization and cooling of carriers.** Energy-resolved carrier distribution for selected times after the optical excitation illustrating (a) the initial ultrafast process of establishing a hot thermalized carrier distribution and (b) the phonon-induced carrier cooling taking place on a slower picosecond timescale. Figure taken from Ref. [Winzer (2013)].

carrier recombination. We observe that the carrier dynamics is characterized by two processes occurring on different time scales. First, the initial non-equilibrium distribution becomes thermalized resulting in a hot, spectrally broad quasi-equilibrium in the first hundred femtoseconds, cf. Fig. 4.8(a). Second, phonon-induced scattering redistributes energy from the excited carrier system to the lattice on a picosecond timescale resulting in a narrowing of the carrier distribution, cf. Fig. 4.8(b).

The process of thermalization and carrier cooling cannot be strictly separated in time, since carrier-phonon scattering contributes to both the thermalization and the energy dissipation. Nevertheless, high-resolution pump-probe experiments measuring the differential transmission signal (DTS) have found two distinct decay times with τ_1 typically in the range of 100 fs and τ_2 typically around 1 ps and represent typical thermalization and cooling times [Dawlaty *et al.* (2008); George *et al.* (2008); Sun *et al.* (2008); Plochocka *et al.* (2009); Choi *et al.* (2009); Kumar *et al.* (2009); Newson *et al.* (2009); Shang *et al.* (2010); Wang *et al.* (2010); Huang *et al.* (2010); Breusing *et al.* (2011); Winnerl *et al.* (2011); Strait *et al.* (2011); Brida *et al.* (2013); Gierz *et al.* (2013); Johannsen *et al.* (2013); Plötzing *et al.* (2014); Gierz *et al.* (2015)] .To model these experiments, we approximate the DTS $[\Delta T/T_0(\tau,\omega)]$ by the pump-induced change in the carrier occupation [Kadi *et al.* (2014)]

$$\Delta T/T_0(\tau,\omega) \propto \frac{|\mathbf{M}^{vc}_{k_0}\cdot\hat{\mathbf{e}}_A|^2 e_0\hbar^2}{8\pi m_o \nu_F^2}\left[\Delta\rho_{k_0}(\tau,\omega)-\Delta\rho_{k_0}(-\infty,\omega)\right] \tag{4.16}$$

evaluated at the excited and probed state $k_0 = \hbar\omega_L/2v_F$ with $\Delta\rho_{k_0} =$

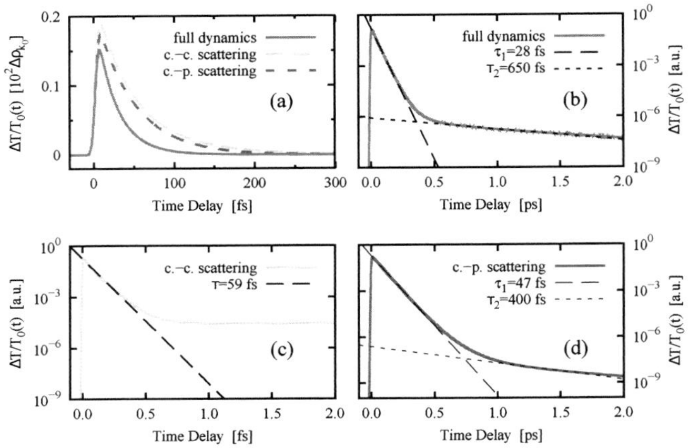

Fig. 4.9 **Differential transmission spectra.** (a) Temporal evolution of the DTS considering the full carrier dynamics (red solid), only carrier-carrier scattering (green long dashed), and only carrier-phonon scattering (blue short dashed). The separate logarithmic plots (b)-(d) demonstrate the bi- or mono-exponential decay of the DTS depending on the relaxation mechanism. Figure taken from Ref. [Winzer (2013)].

$\rho^c_{k_0} - \rho^v_{k_0}$. Figure 4.9(a) shows the differential transmission spectrum after optical excitation and the subsequent relaxation via carrier-carrier and carrier-phonon scattering. The initial sharp increase of the signal accounts for the generation of carriers during the applied pulse. The increased carrier occupation leads to an absorption bleaching and results in an enhanced transmission. After the pulse, the carrier relaxation is responsible for the decrease of the DT signal. Our calculations show a clear bi-exponential decay, cf. the logarithmic plot in Figure 4.9(b). The extracted time constants of $\tau_1 = 28\,\text{fs}$ and $\tau_2 = 650\,\text{fs}$ are in good agreement with experimental studies [Dawlaty *et al.* (2008); Breusing *et al.* (2011)]. A third decay time is expected at larger times around 300 ps stemming from the carrier cooling via acoustic phonons [Winnerl *et al.* (2011)]. The comparison of the temporal evolution of the DTS and the carrier population (cf. Fig. 4.8) reveals that the first exponential decay component τ_1 reflects the ultrafast carrier thermalization, while the second decay is governed by the phonon-induced carrier cooling. The latter process is slowed down due to the increased Pauli blocking for a thermalized carrier distribution and a heated phonon system. Our theoretical approach allows to investigate the contribution of carrier-carrier and carrier-phonon scattering separately, cf. Figs. 4.9(c) and (d). In the case of purely Coulomb-induced dynamics, the DTS offers a single exponential decay of $\tau_1^{(cc)} = 59\,\text{fs}$. After establishing a hot thermalized distribution, the dynamics comes to a hold, since no energy relaxation can take place. In contrast, the purely phonon-induced DTS exhibits a bi-exponential decay with $\tau_1^{(cp)} = 47\,\text{fs}$ and $\tau_2^{(cp)} = 400\,\text{fs}$ revealing that phonons contribute both to carrier thermalization and cooling. Interestingly, the full dynamics is not just the sum of the single relaxation processes, since $\tau_2 \neq \tau_2^{(cp)}$ and $1/\tau_1 \neq 1/\tau_1^{(cc)} + 1/\tau_1^{(cp)}$ reflecting an interplay between Coulomb- and phonon-induced processes.

4.4.1.1 *Phonon-assisted intraband absorption*

Most pump-probe studies exhibit a zero-crossing after the initial decay in the DTS [Newson *et al.* (2009); Plochocka *et al.* (2009); Sun *et al.* (2008, 2010); Winnerl *et al.* (2011); Shang *et al.* (2010); Breusing *et al.* (2011)]. So far, this feature could not be reproduced in Fig. 4.9. For some of the experiments, the negative DTS component can be clearly traced back to the predominant intraband absorption. This is clearly the case, when the photon energy is smaller than twice the value of the Fermi energy [Sun *et al.* (2008); Winnerl *et al.* (2011); Sun *et al.* (2010)]. However, for all other cases, where negative DT components occur for photon energies much larger than the Fermi energy, the underlying mechanism has been controversially discussed in literature. In particular, pump-induced band structure renormalization for transient [Breusing *et al.* (2011)] or in the long time limit [Roberts *et al.* (2013)] at the M-point as well as intraband absorption [Malard *et al.* (2013)] have been suggested as explanations for the observed negative DT signals.

Our microscopic calculations reveal that phonon-assisted intraband absorption processes can explain the experimentally observed zero-crossing from positive to negative DT signals. To include these additional relaxation channels, we extend the Bloch equations by taking into account the interplay between carrier-phonon and carrier-light interactions. When deriving an equation of motion describing the temporal evolution of the carrier occupation ρ_k, phonon-assisted quantities appear [Malic *et al.* (2011a)]. Their dynamics has been determined only by carrier-phonon matrix elements, so far. Now, we also include the coupling of phonon-assisted quantities to the carrier-light Hamilton operator resulting in an additional contribution in ρ_k describing phonon-assisted optical transitions, cf. Fig. 4.10(a) [Kadi *et al.* (2014)] . Since the experimentally observed negative DTS occurs clearly after the optical excitation, we focus on investigating the interaction of the probe pulse with the pump-induced non-equilibrium carrier occupation $\rho_{\mathbf{k}}^{\lambda,0}(t)$. We apply a perturbative approach assuming $\rho_{\mathbf{k}}^{\lambda}(t) = \rho_{\mathbf{k}}^{\lambda,0}(t) + \delta\rho_{\mathbf{k}}^{\lambda}(t)$ with $\delta\rho_{\mathbf{k}}^{\lambda}(t)$ describing a small perturbation in the carrier occupation induced by the probe pulse. The perturbation is in driven by phonon-assisted intraband absorption processes. Solving the equation for $\delta\rho_{\mathbf{k}}^{\lambda}(t)$, we obtain

$$\delta\rho_{\mathbf{k}}^{\lambda}(\omega,\tau) = \frac{\Gamma_{\lambda,\mathbf{k}}^{\mathrm{in,A}}(\omega,\tau)(1-\rho_{\mathbf{k}}^{\lambda,0}(\tau)) - \Gamma_{\lambda,\mathbf{k}}^{\mathrm{out,A}}(\omega,\tau)\rho_{\mathbf{k}}^{\lambda,0}(\tau)}{i\omega+\tilde{\gamma}} \tag{4.17}$$

with a damping constant $\tilde{\gamma}$ and the field-assisted scattering rates $\Gamma_{\lambda,\mathbf{k}}^{\mathrm{in/out,A}}$, which are responsible for the phonon-induced optical intraband absorption. This results in an additional contribution to differential transmission spectra [Kadi *et al.* (2014); Kadi (2015)]

$$\Delta T/T_0(\tau,\omega)|_{\mathrm{intra}} \propto \frac{i}{A\omega A_0^t}\sum_{\lambda\mathbf{k}} \mathbf{M}_{\mathbf{k}}^{\lambda\lambda}\cdot\hat{\mathbf{e}}_{\mathbf{A}}\mathrm{Im}\left[\rho_{\mathbf{k}}^{\lambda}(\tau,\omega) - \rho_{\mathbf{k}}^{\lambda}(-\infty,\omega)\right] \tag{4.18}$$

with A_0^t as the amplitude of the test pulse and with A as the graphene area that cancels after performing the sum over $\mathbf{k}$. This contribution accounting for probe-pulse-induced phonon-assisted intraband absorption processes adds up to the DTS contribution stemming from interband transitions that have been discussed in the previous section, cf. Eq. (4.16). While the latter gives rise to absorption bleaching and a positive DTS, the new contribution leads to an increased absorption after the excitation. Depending on the interplay with the interband contribution, an overall negative DT signal can be reached in certain excitation regimes.

Figure 4.10(b) shows a differential transmission spectrum after an optical excitation with a pulse characterized by a photon energy of 1.5 eV, a full-width-half maximum of 50 fs and a pump fluence of 8 μJcm^{-2}. Taking into account only interband transitions, we observe an entirely positive DT signal (orange line) exhibiting a bi-exponential decay. However, when including phonon-assisted intraband absorption processes, the DTS exhibits a distinct zero-crossing after approximately 300 fs followed by a subsequent recovery towards zero on a ps timescale. The intraband absorption does not change the first decay rate τ_1, whereas the second decay τ_2 is considerably slowed down, since it is now limited by slower carrier recombination processes that are responsible for the recovery of the DT signal. Our results are in a very good agreement with experimental data [Winnerl *et al.* (2013)]. Figures 4.10(c) and (d) show a direct theory-experiment comparison illustrating a series of DT spectra obtained for different pump fluences covering the range of 2 - 24 μJcm^{-2}. We observe similar timescales for the

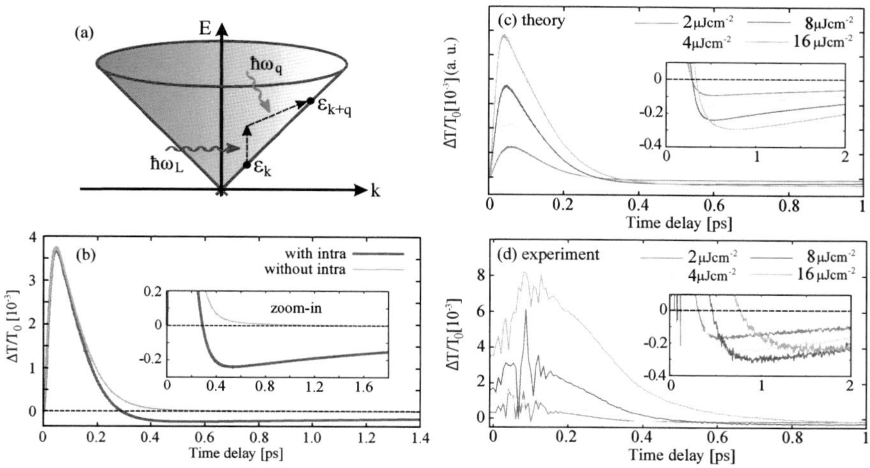

Fig. 4.10 **DTS including intraband absorption.** (a) Schematic illustration of phonon-assisted intraband absorption. (b) DTS with and without intraband absorption processes. Transient differential transmission spectrum for different pump fluences including the phonon-assisted intraband absorption processes. (c)-(d) Direct comparison between theory and experiment illustrating DTS for different pump fluences. Figure adapted from [Kadi *et al.* (2014)].

two decay components τ_1 and τ_2, the same trend for the zero-crossing, and the fluence-dependent increase of positive and negative DTS components.

At high fluences, the intraband absorption is enhanced due to the larger number of available carriers in the conduction band. At the same time, increasing the fluence gives rise to a more efficient Pauli blocking, which increases the absorption bleaching and leads to stronger positive DT signals. The ratio between the maximal negative and positive DTS decreases with the fluence both in experiment and theory reaching a maximal value in the range of 10% in the weak excitation regime. The decrease can be traced back to the predominant role of the Pauli blocking at high fluences prevailing over the increased efficiency of intraband absorption. This also explains the observation both in theory and experiment that the zero crossing occurs at larger delay times with the increasing fluence, cf. the insets of Figs. 4.10(c) and (d). The appearing quantitative deviations with respect to the more pronounced temporal shift of the zero-crossing for increasing fluences might be due to the neglected effects of Coulomb-induced intraband absorption processes and energy renormalization.

4.4.2 Carrier dynamics below the optical phonon energy

The constant optical phonon energy of about 200 meV presents a distinguished energy scale for carrier-optical-phonon scattering. Tuning the optical excitation below this particular energy allows us to strongly suppress scattering with optical phonons, cf. Fig. 4.11(a). As a result, the much slower acoustic phonons for excitations in the near-infrared might become now the predominant energy dissipating process.

To address this open question, we model pump-probe experiments for excitations above (245 meV) and below (72 meV) the energy of optical phonons. The obtained pump-induced change of carrier occupation in the excited states $\Delta\rho_{k_0}$ is directly compared to experimental differential transmission spectra $\Delta T/T_0$ for both excitation energies, cf. Figs. 4.11(b) and (c). We find a qualitatively excellent agreement between theory and experiment: First of all, both theory and experiment reveal a clearly slowed-down carrier dynamics compared to the excitation in the infrared. This can be traced back to the reduced phase space closer to the Dirac point, i.e. the number of available scattering partners decreases with the carrier energy. Second, the drastically slower dynamics for excitations below the energy of optical phonons, i.e. in the case of 72 meV, is due to the strong Pauli blocking of intraband scattering with optical phonons. More specifically, we observe both in experiment and theory a bi-exponential decay for 245 meV, whereas at 72 meV the decay is mono-exponential. For the 245 meV excitation, the fast decay time in the sub-picosecond range is ascribed to Coulomb-induced thermalization, in particular dominated by the processes of impact excitation accounting for an increase in the carrier density. As a result, the energy of the excitation pulse is redistributed to an enhanced number of carriers resulting in a cooling of the electronic system on a timescale of some tens of picoseconds. In the case of 72 meV, only the slower decay component can be observed and is ascribed to the phonon-induced cooling (enabled by the preceding thermalization via carrier-carrier scattering).

Surprisingly, we find that optical phonons play the predominant role even for the 72 meV excitation, cf. the dashed red line in Fig. 4.11(b) illustrating the DT signal for excitation at 72 meV without taking into account scattering with optical phonons. We find a much weaker decay demonstrating the importance of optical phonons. At first sight, this is unintuitive, because at low photon energies, these channels should be strongly suppressed. However, carrier-carrier intraband scattering efficiently thermalizes the

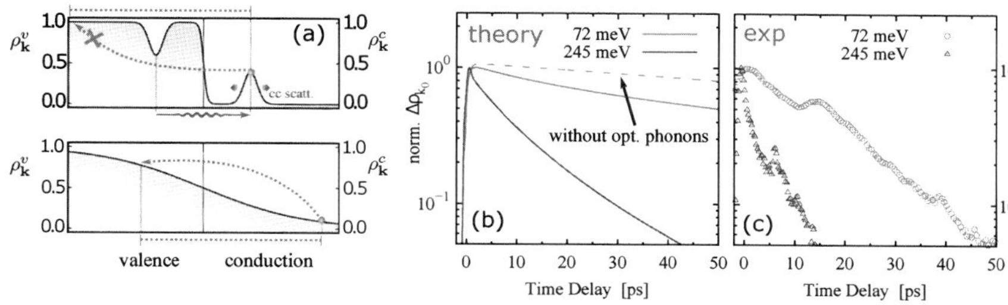

Fig. 4.11 **Differential transmission below and above the optical phonon energy.** (a) For excitations below the optical phonon energies (72 meV), direct scattering with optical phonons is strongly suppressed by Pauli blocking (top). Coulomb-induced broadening of the non-equilibrium distribution opens up the relaxation pathway via optical phonon emission (bottom). (b) Theoretically evaluated pump-induced change of occupation in the excited states $\Delta\rho_{k_0}$ and (c) experimentally measured differential transmission spectra $\Delta T/T_0$ at 72 meV (red) and 245 meV (blue). The red dashed line shows the DTS without taking into account optical phonons. Note that the small secondary peaks appearing in the experimental data are artifacts from the experimental setup. Figure adapted from [Winzer (2013)].

non-equilibrium carriers toward a spectrally broad thermalized distribution, cf. Fig. 4.11(a). As a result, a significant number of carriers can be found at states far above the excitation energy enabling intraband scattering via optical phonons.

The scattering with acoustic phonons is much weaker and occurs as a further decay constant in the range of 300 ps at both excitation energies in theory and experiment (not shown). A quantitative comparison between experiment and theory shows that the decay times in theory are by a factor of about 2 larger than in experiment. The discrepancy may be caused by distinct processes which are not considered in the model, such as scattering with impurities [Hwang and Das Sarma (2008b)] or Z-phonons [Kusminskiy *et al.* (2009)]. Furthermore, non-Markovian effects may play a role close to the Dirac point.

4.4.3 *Carrier dynamics in the THz region*

While the carrier dynamics in the near-infrared and below the optical phonon energy has been well understood in literature, the relaxation channels very close to the Dirac point (few meV range) have still not been entirely uncovered. To address this interesting question, we model pump-probe experiments, where the pump pulse excites carriers in the near-infrared region and the probe pulse tests the carrier occupation in the THz regime. Recent experiments reveal that the optically induced THz conductivity can be either positive or negative. The positive component has been attributed to enhanced free-carrier intraband absorption [George *et al.* (2008); Strait *et al.* (2011)], while the negative component is still unclear. It has been ascribed to different many-particle mechanisms, such as stimulated THz emission [Docherty *et al.* (2012)], enhanced carrier scattering with optical phonons, surface optical phonons or charge impurities [Jnawali *et al.* (2013); Shi *et al.* (2014); Frenzel *et al.* (2013)], and carrier heating [Tielrooij *et al.* (2013); Jensen *et al.* (2014); Frenzel *et al.* (2014)]. So far, simple phenomenological Drude models have been applied to understand the THz features. Furthermore, disorder-assisted electron-phonon scattering processes, the so-called supercollisions, have been proposed to account for the experimentally observed cooling times that are much faster than expected in the THz region with a vanishing density of states [Song *et al.* (2012)].

Here, we show that it is of crucial importance to describe the THz carrier dynamics on microscopic footing. Our calculations based on graphene Bloch equations show that Drude models have severe shortcomings in the THz range. Furthermore, recent experiments on different samples characterized by different degrees of disorder suggest that supercollisions might not be the appropriate explanation for some of the THz experiments. To address the THz dynamics and model optical-pump/THz-probe differential transmission, we extend the Bloch equations by including the drift term $\mathbf{E} \cdot \nabla_{\mathbf{k}} \rho_{\mathbf{k}}^{\lambda}$ expressing the field-induced intraband transitions. The latter are crucial in the THz regime, where interband transitions are strongly suppressed by Pauli blocking.

It is numerically extremely challenging to temporally and spectrally resolve the carrier dynamics from the near-infrared down to the THz regime. Furthermore, in the optical regime, drift terms contribute beyond the rotating wave approximation. Assuming a weak THz test pulse, the carrier occupation $\rho_{\mathbf{k}}^{\lambda}(t)$ is treated perturbatively by assuming $\rho_{\mathbf{k}}^{\lambda}(t) = \rho_{\mathbf{k}}^{\lambda,0}(t) + \delta\rho_{\mathbf{k}}^{\lambda}(t)$ with pump pulse-induced occupation $\rho_{\mathbf{k}}^{\lambda,0}(t)$ and the small occupation change due to the $\delta\rho_{\mathbf{k}}^{\lambda}(t)$ THz test pulse. Furthermore, we exploit the fact that after the optical excitation in the near-infrared, a hot Fermi distribution is reached within the first tens of femtoseconds due to the ultrafast carrier-carrier and carrier-phonon scattering. The subsequent dynamics is characterized by the temporal evolution of the temperature $T_{\lambda}(t)$ and the Fermi level $\varepsilon_F^{\lambda}(t)$ for electrons and holes. As a result, calculating $T_{\lambda}(t)$ and $\varepsilon_F^{\lambda}(t)$, we have access to the pump pulse-induced occupation $\rho_{\mathbf{k}}^{\lambda,0}(t)$. The dynamics of the test-pulse-induced change of the carrier occupation $\delta\rho^{\lambda}_{\mathbf{k}}(t)$ is calculated from the linearized graphene Bloch equations yielding [Mihnev *et al.* (2016); Kadi (2015)]

$$\frac{d}{dt}\delta\rho^{\lambda}_{\mathbf{k}}(t) = -\frac{e_0}{\hbar}\mathbf{E}\cdot\nabla_{\mathbf{k}}\rho^{\lambda,0}_{\mathbf{k}}(t) - \Gamma^{0}_{\lambda\mathbf{k}}(t)\,\delta\rho^{\lambda}_{\mathbf{k}}(t), \tag{4.19}$$

where $\Gamma^{0}_{\lambda\mathbf{k}}(t) = \Gamma^{\text{in},0}_{\lambda\mathbf{k}}(t) + \Gamma^{\text{out},0}_{\lambda\mathbf{k}}(t)$ are time-dependent scattering rates describing the purely pump-induced dynamics from optical pump pulses. Inserting the numerically evaluated Fourier transform of $\delta\rho^{\lambda}_{\mathbf{k}}(t)$, the DTS can be obtained via Eq. 4.18. The contribution of interband transitions to the DTS can be neglected due to the strong Pauli blocking in the THz range stemming from doping or the fast relaxing pump-induced carriers.

Figure 4.12 shows a direct comparison between our theoretical calculations and recently performed optical-pump/THz-probe experiments performed on highly doped ($E_F \approx 100 - 400\,\text{meV}$) and quasi-undoped ($E_F \approx 10\,\text{meV}$) graphene samples for different substrate temperatures and a fixed pump fluence [Mihnev *et al.* (2016)]. Both in theory and experiment, we observe an opposite behavior depending on the doping: While in highly doped graphene samples, the DTS is positive for all pump-probe delay times (corresponding to a pump-induced decrease of the THz absorption), we find an entirely negative DTS for quasi-undoped graphene (corresponding to an increased THz absorption after optical excitation). Furthermore, we find an approximately mono-exponential decay of the DT signal for highly doped graphene. The corresponding relaxation time is in the range of $1 - 2\,\text{ps}$ with sample-to-sample variation within $20 - 30\%$ indicating that disorder-assisted electron-phonon processes are not the dominant cooling mechanism in the investigated samples. The relaxation time is independent of the substrate temperature [cf. Figs. 4.13(a), (b)] and weakly increases with the applied pump fluence (not shown). The negligible temperature dependence can be traced back to the insensitivity of the THz absorption to the substrate temperature at least up to 300 K for highly doped graphene samples with $E_F = 100 - 400\,\text{meV}$.

The situation is more complicated for the quasi-undoped case, where the dynamics strongly depends on the substrate temperature. Here, the DTS shows a change from a fast mono-exponential decay at 300 K to a slower bi-exponential decay at cryogenic temperatures. The fast relaxation time at room temperature is in the range of $4-7\,\text{ps}$, which is slower compared to the case of highly doped graphene, where carrier-carrier scattering is more efficient due to the increased number of scattering partners. The two decay

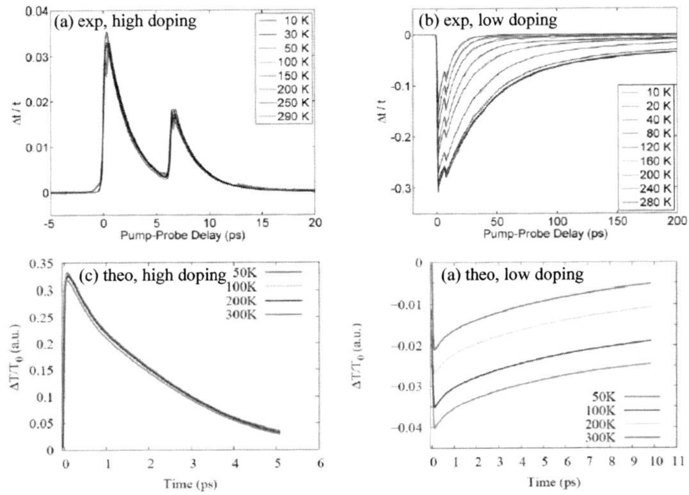

Fig. 4.12 **Optical-pump/THz-probe differential transmission spectra.** Direct comparison between experimental [(a), (b)] and theoretical [(c), (d)] differential THz transmission at the energy of the THz probe pulse for a highly doped [(a), (c)] and quasi-undoped undoped [(b), (d)] graphene sample with a Fermi energy of $E_F \approx 100-400$ meV and $E_F \approx 10$ meV, respectively. The spectra are shown for different substrate temperatures and for a fixed pump fluence. The secondary peak appearing in experimental data is due to a round-trip reflection of the optical pump pulse inside the substrate. Both theory and experiment shows the same qualitative features: The DT signal is positive for all time delays and shows a mono-exponential decay in the case of a highly doped graphene sample. In contrast, for quasi-undoped graphene the DTS is entirely negative and evolves from fast mono-exponential relaxation at room temperature to a slower bi-exponential relaxation at cryogenic temperatures. Figure adapted from [Mihnev *et al.* (2016)].

components at very low temperatures are ascribed to two distinct cooling mechanisms. The fast decay on a timescale of tens of picoseconds is due to the combined effect of carrier-carrier and carrier-optical-phonon scattering in the absence of disorder. In contrast to the case of highly doped graphene, the relaxation time strongly decreases with the substrate temperature, cf. Figs. 4.13(c), (d). Our calculations reveal that this dependence can be traced back to the temperature dependence of the equilibrium transmission T_0 rather than to the dynamic non-equilibrium part of the differential transmission.

The applied microscopic theory can be reduced to the widely used standard Drude model by assuming constant time- and momentum-independent

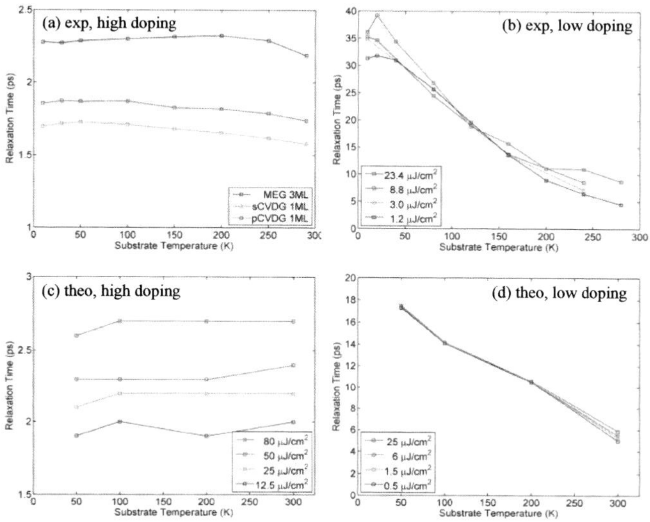

Fig. 4.13 **THz carrier relaxation times.** Direct comparison of experimentally measured [(a), (b)] and theoretically predicted [(c), (d)] relaxation times extracted from differential transmission spectra. The relaxation times are shown as a function of the substrate temperature at a fixed pump fluence for a highly doped [(a), (c)] and quasi-undoped [(b), (d)] graphene samples. Theory and experiment are in good agreement showing a strong/weak temperature dependence for undoped/doped graphene. Figure adapted from [Mihnev *et al.* (2016)].

scattering rates in Eq. (4.19). To investigate the validity of the Drude model that is often applied as a phenomenological basis for the interpretation of THz experiments, we calculate the differential transmission expected from such a model as a function of the pump-induced transient carrier temperature $T(t)$ and Fermi energy $E_F(t)$, cf. Fig. 4.14. We compare the results with solutions of the full graphene Bloch equations for disorder-free highly doped graphene with $E_F = 300\,\text{meV}$ evaluated at a low and a high fluence of $12.5\,\mu\text{Jcm}^{-2}$ and $80\,\mu\text{Jcm}^{-2}$, respectively, cf. the red and the blue line in Fig. 4.14. We find that in contrast to the experiment and fully microscopic calculations the Drude model predicts a negative DTS for very high carrier temperatures that can be reached at early relaxation times for high fluences. Assuming a widely used approximation that the pump pulse only heats the carrier system leaving the Fermi level unchanged, the Drude model predicts a negative DTS for all times, cf. the dashed black line in Fig. 4.14. As a result, the standard Drude model is not sufficient enough to

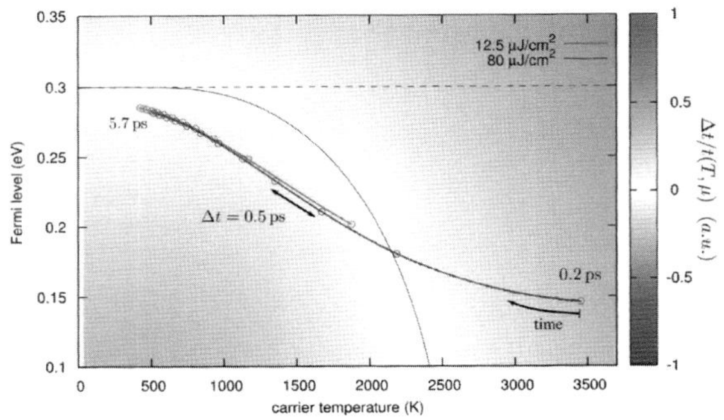

Fig. 4.14 **Differential THz transmission according to the Drude model.** DTS as expected from a Drude model for an initial temperature of 300 K and a Fermi level of 300 meV plotted as a function of the transient carrier temperature T and the Fermi level E_F. The solid black line separates regions of positive and negative DTS. The dashed black line illustrates the expected DTS for a constant Fermi level. The red (blue) line shows the path through the T-E_F-map for a fixed pump fluence of 12.5 μJcm^{-2} (80 μJcm^{-2}) obtained from the solution of the full graphene Bloch equations for the pump-induced dynamics. The first circle on the line shows the carrier system 0.2 ps after the pump pulse, when the carrier distribution can be extrapolated by a Fermi distribution. The time delay between the circles is 0.5 ps. Figure taken from Ref. [Mihnev *et al.* (2016)].

explain the THz dynamics in graphene. A complete understanding requires the solution of graphene Bloch equations including the explicitly time- and momentum-dependent scattering rates.

4.5 Orientational Carrier Dynamics

So far, we have focused on the time- and energy-resolved dynamics of an angle-averaged non-equilibrium carrier population $\rho_k = \int_0^{2\pi} \rho(k, \phi)\, d\phi$, since the aim was to reveal the relaxation channels responsible for carrier thermalization and cooling. Having in mind the anisotropy of the optical matrix element (cf. Fig. 4.2) resulting in an initial optically excited highly anisotropic carrier distribution (cf. Fig. 4.3), it is of great interest to study the angle-dependent relaxation dynamics. Therefore, we now distinguish Coulomb- and phonon-induced scattering processes along and across the Dirac cone, i.e. we resolve the carrier distribution in angle ϕ. Scattering along the cone occurs when the momenta $\mathbf{k_i}$ and $\mathbf{k_f}$ of the initial and the final electron state are parallel to each other. They bring excited electrons quickly into the vicinity of the Dirac point without a momentum transfer

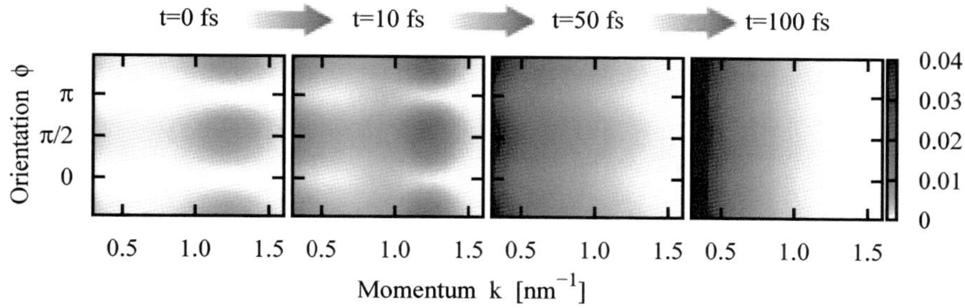

Fig. 4.15 **Angle-resolved orientational carrier dynamics.** Surface plot illustrating the temporal evolution of the optically excited carrier occupation $\rho_{k,\phi}(t)$ as a function of the momentum k and the angle ϕ (denoting the direction in the momentum space with respect to the pulse polarization). The initially highly anisotropic carrier distribution centered at the excitation momentum of $1.25\,\text{nm}^{-1}$ equilibrates via carrier-carrier and carrier-phonon scattering within 100 fs in both carrier momentum k and angle ϕ. Figure taken from Ref. [Winzer (2013)].

across the cone. As a result, these channels conserve the initial anisotropy and are the driving force for the ultrafast thermalization of carriers. In contrast, scattering processes across the cone (orientational relaxation) redistribute carriers along different directions in the momentum space leading to an isotropic carrier distribution.

4.5.1 *Orientational carrier dynamics in the near-infrared*

First, we focus on the impact of the orientational relaxation in the near-infrared regime (with excitation energy of $\hbar\omega_L = 1.5\,\text{eV}$). Fig. 4.15 illustrates the initial phase of the momentum- and angle-resolved carrier dynamics accounting for Coulomb- and phonon-induced scattering processes. The angle ϕ denotes the direction in the momentum space with respect to the linear pulse polarization, i.e. $\phi = \angle(\mathbf{k}, \hat{\mathbf{e}}_{\mathcal{P}})$. At the time $t = 0\,\text{fs}$ corresponding to the center of the excitation pulse, we observe a strongly anisotropic non-equilibrium carrier distribution with maxima centered at $k_0 = \hbar\omega_L/2v_F \approx 1.25\,\text{nm}^{-1}$ and in the orientation perpendicular to the pulse polarization, i.e. $\phi = \pi/2$ ($\pm 3\pi/2$). Ultrafast scattering takes place already during the excitation pulse (with a width of 10 fs) giving rise to a broadening of the carrier occupation in angle and momentum. However, the scattering occurs predominately directly towards the Dirac cone leaving the states parallel to the pulse polarization, i.e. at $\phi = 0$ and π, unoccupied. This changes after approximately 10 fs, the states slightly below

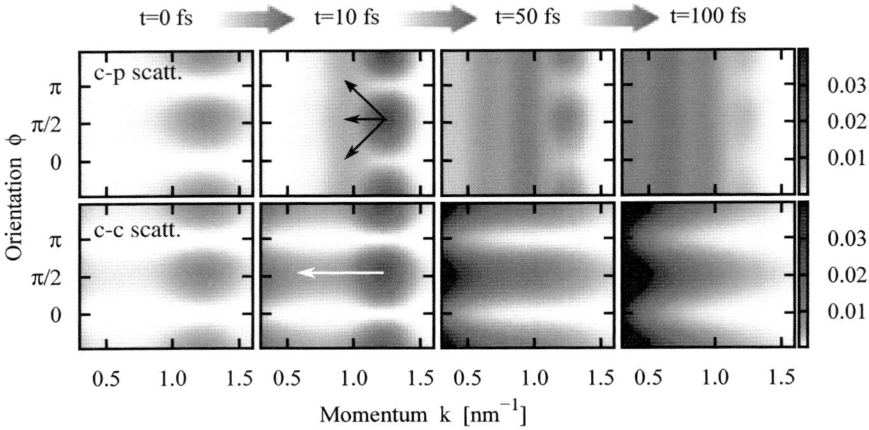

Fig. 4.16 **Relaxation mechanism for orientational dynamics.** Momentum- and angle-resolved carrier dynamics due to purely phonon-induced (top) and purely Coulomb-induced relaxation (bottom). Carrier-phonon scattering gives rise to a highly efficient orientational relaxation across the Dirac cone, whereas carrier-carrier scattering is mainly responsible for an ultrafast momentum redistribution along the Dirac cone. Figure taken from Ref. [Winzer (2013)].

the excitation energy become filled in all directions forming a strip, cf. Fig. 4.15. This orientational relaxation proceeds on an ultrafast timescale resulting in a nearly completely isotropic carrier distribution after 100 fs [Malic *et al.* (2011b, 2012)].

To investigate which particular relaxation channel is responsible for the efficient orientational relaxation, we separately calculate the carrier dynamics induced only by carrier-phonon and carrier-carrier scattering, respectively. Figure 4.16 illustrates a qualitatively different behavior depending on the scattering mechanism. After 10 fs, the isotropic strip of occupied states below the excitation from the full dynamics can clearly be assigned to the phonon-induced dynamics, cf. Fig. 4.16 (top). In particular, the distance between the strip and the excitation state k_0 corresponds approximately to an optical phonon energy. As a result, the initially anisotropic carrier distribution becomes isotropic close to the excitation energy already after the emission of a single phonon. The subsequent cascade of phonon emissions leads to an isotropic distribution after approximately 100 fs. A further study has shown that ΓLO phonons predominantly contribute to scattering across the Dirac cone and hence to an isotropic distribution. This can be understood by the angle-dependence of the corresponding

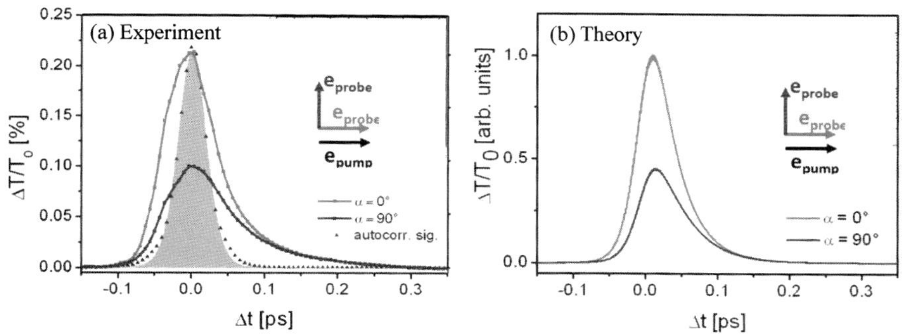

Fig. 4.17 **Polarization-dependent differential transmission.** Direct comparison between (a) experimentally measured and (b) theoretically predicted differential transmission spectra for parallel (red) and perpendicular (blue) polarized pump and probe pulses, cf. the inset. The DT signal strongly depends on the relative polarization in the first 100 fs after optical excitation demonstrating the presence of an anistropic carrier distribution. Figure adapted from Ref. [Mittendorff *et al.* (2014)].

carrier-phonon matrix elements [cf. Eq. (4.13)] that are highly efficient for intraband scattering all over the cone. In contrast, carrier-carrier scattering mainly occurs towards the Dirac point conserving the initial anisotropy to a large extent, cf. Fig. 4.16 (bottom). After approximately 100 fs, the carrier occupation can be approximated by hot Fermi distributions with different temperatures depending on the direction in the momentum space. This can be traced back to the angle-dependence of the Coulomb matrix element preferring parallel scattering along the Dirac cone, cf. Eqs. (4.10) and (4.11). As a result, the Coulomb-induced orientational relaxation is less efficient leaving the carrier distribution anisotropic on a timescale of few hundreds of femtoseconds.

Finally, we investigate the impact of the anisotropic carrier distribution on the experimentally accessible differential transmission. As discussed earlier, the DTS can be approximated by the pump-induced change of the carrier occupation $\Delta\rho_{\mathbf{k}_0}$ in the optically excited state $\mathbf{k}_0 = (k_0, \phi)$, cf. Eq. (4.16). To uncover the influence of anisotropy, we model polarization-dependent pump-probe experiments, i.e. we vary the angle between the polarization of the pump and the probe pulse. In very good qualitative and quantitative agreement with the experiment, our theory predicts an amplitude ratio slightly above 2:1 for parallel and perpendicular polarization, cf. Fig. 4.17. The observed polarization dependence is a direct evidence for an anisotropic carrier distribution. For times larger than 150 fs,

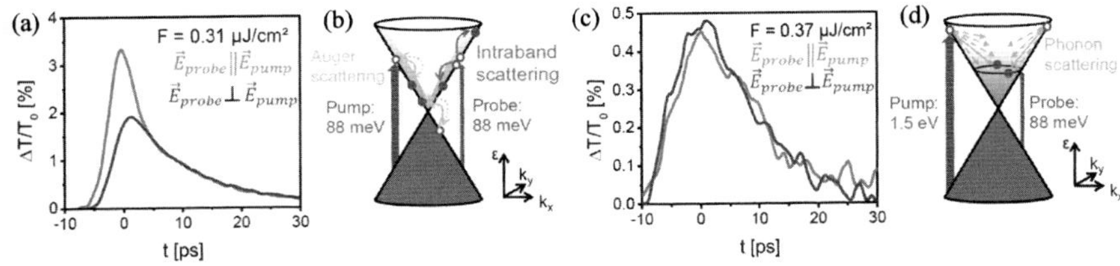

Fig. 4.18 **Polarization-dependent single- and two-color pump-probe experiments.** Differential transmission for parallel and perpendicular polarized pump and probe pulses for (a) single-color and (c) two-color pump-probe experiments at a similar pump fluence of $0.31\,\mu\text{Jcm}^{-2}$ and $0.37\,\mu\text{Jcm}^{-2}$, respectively. The energy of the probe pulse is with 88 meV well below the energy of optical phonons. The pump pulse energy is either 88 meV or 1.5 eV. (b), (d) Schematic illustration of the excitation conditions and the predominant relaxation channels for single- and two-color pump-probe experiments, respectively. Figure adapted from Ref. [Koenig-Otto *et al.* (2016)].

the two polarization configuration become indistinguishable indicating that isotropy has been reached on this timescale. All observations follow the theoretical predictions in [Malic *et al.* (2012)].

4.5.2 *Orientational dynamics below optical phonon energy*

Now, we use the gained insights from the previous section to tune the carrier dynamics in such a way that it is entirely determined by Coulomb scattering. To reach this goal, we apply an optical pulse with an energy of 88 meV that is well below the optical phonon energy. To ensure that also scattering from the high-energy tails of the thermalized carrier distribution with optical phonons is negligible, the study is carried out at low temperatures and low pump fluences. Modeling degenerate polarization-dependent pump-probe experiments reveals that under these excitation conditions the anisotropic carrier distribution persists for several picoseconds, i.e. we observe a difference in the DT signal for parallel and perpendicular polarized pump and probe pulses on this timescale, cf. Figs. 4.18(a) and (b). This is in strong contrast to the excitation in the near-infrared where isotropy is reached already after 150 fs, cf. Fig. 4.17(a).

Having eliminated scattering with optical phonons and having in mind that acoustic phonons occur on a timescale of few hundreds of picoseconds, Coulomb-induced scattering processes present the only efficient relaxation channel. However, Coulomb interaction prefers parallel scattering along the Dirac cone preserving the optically induced anisotropy in the momentum-dependent carrier distribution, cf. Fig. 4.16 (bottom). According to Eqs.

(4.10) and (4.11), we can approximate the intraband Coulomb matrix element V as $|V|^2 \propto (1+\cos\phi_1)(1+\cos\phi_2)$ with $\phi_{1,2}$ as the angle between the momenta of initial and final state of two scattering electrons. As a resulting, the scattering is most efficient for parallel processes, i.e. $\phi_i = 0$. They become less important with increasing scattering angle and vanish for backscattering ($\phi_i = \pi$). As a result, the low efficiency for Coulomb-induced scattering across the Dirac cone accounts for the long-lived anisotropy in the excitation regime, where scattering with optical phonons is strongly suppressed.

Modeling polarization-dependent two-color pump-probe experiment at the same probe conditions, however with the pump pulse exciting the carrier system at 1.5 eV, we find a completely different behavior: The distribution probed at the energy of 88 meV is isotropic for all delay times, cf. Fig. 4.18(c). The reason is that the excitation at near-infrared allows coupling with optical phonons, which efficiently scatter across the Dirac cone redistributing the non-equilibrium carriers in the entire momentum, cf. Fig. 4.18(d). Once the carriers reach the probed state at 88 meV, their distribution is already entirely isotropic. Our theoretical calculations reproduce well the qualitative features observed in the experiment (not shown).

4.6 Carrier Multiplication

The most fascinating consequence of the exceptional gapless and linear band structure of graphene on its carrier dynamics is the possibility of carrier multiplication (CM), i.e. generation of multiple electron-hole pairs by absorption of a single photon via Auger-induced scattering. This is an ultrafast many-particle phenomenon that is of great interest both for fundamental physics as well as technological applications. Auger processes are specific Coulomb-driven scattering events, where one electron bridges the valence and the conduction band, while the other involved electron remains in the same band, cf. Figs. 4.4(b) and (c). As a result, Auger scattering changes the overall charge carrier density. We distinguish Auger recombination (AR) from its inverse process of impact excitation (IE). The latter leads to an increase of the carrier density and is responsible for the appearance of CM, cf. Fig. 4.19(a). However, CM takes place only, if IE prevails over AR. The efficiency of both processes can be estimated depending on the carrier distribution and the resulting Pauli blocking. After the optical injection of carriers, the intraband scattering of excited carriers far above the Dirac point has the same efficiency for IE and AR, since the states below

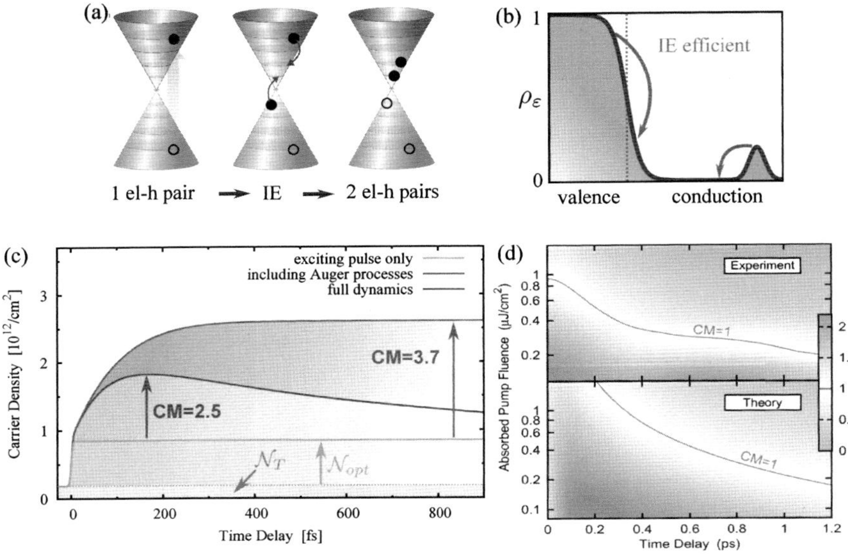

Fig. 4.19 **Carrier multiplication in graphene.** (a) Schematic illustration of the impact excitation (IE), where the absorption of a single photon gives rise to a generation of multiple carrier-hole pairs. (b) In optically excited system, IE can be very efficient, while the inverse process of Auger recombination is strongly suppressed via Pauli blocking resulting from largely filled valence and empty conduction band states close to the Dirac point. (c) Temporal evolution of the carrier density accounting for all relaxation channels (blue line), only carrier-light interaction (green line), and only Coulomb scattering (red). IE leads to a multiplication of the initially excited carrier density $\mathcal{N}_{opt}$ by a factor of 3.7. Phonon-induced recombination counteracts the multiple charge carrier generation and reduces the CM factor to 2.5. Figures (a), (b) adapted from [Winzer and Malić (2012)], (c) taken from [Winzer (2013)], and (d) taken from Ref. [Plötzing *et al.* (2014)]. (For color reference, go to page 487.)

and above are nearly empty right after the optical excitation. In contrast, the probability of the assisting interband process close to the Dirac point, is much higher for IE scaling with $\rho^v(1-\rho^c) \approx 1$, since $\rho^v \approx 1$ and $\rho^c \approx 0$, cf. Fig. 4.19(b). As a result, IE is significantly favored by Pauli blocking during the initial carrier dynamics.

Solving graphene Bloch equations, we can calculate the temporal evolution of the carrier density at different excitation conditions addressing the question when an optimal CM can be reached. Our calculations reveal a significant increase of the carrier density after optical excitation, cf. Fig. 4.19(c). To understand the underlying processes, we subsequently switch on different relaxation channels. Accounting only for electron-light interaction, the applied ultrashort optical pulse lifts electrons from the valence into the conduction band resulting in an optically induced carrier density $\mathcal{N}_{opt}$, which remains constant after the pulse (green line). Considering additionally Coulomb-induced scattering channels including Auger processes results in a significant increase of the carrier density even after the pulse has been switched off (red line). The many-particle-induced carrier generation can be quantified by a carrier multiplication factor CM, which is defined as the total number of generated electron-hole pairs per absorbed photon

$$CM = \frac{\mathcal{N} - \mathcal{N}_T}{\mathcal{N}_{opt}}, \tag{4.20}$$

where $\mathcal{N}$ is the total carrier density and $\mathcal{N}_T$ is the thermal carrier background. In the case of a purely Coulomb-driven dynamics, we obtain a carrier multiplication factor of $CM = 3.7$. However, to complete the picture considering a full carrier dynamics, we also include phonon-induced relaxation channels (blue line) directly competing with Auger processes. An excited electron can scatter to an energetically lower state by emitting a phonon and reducing the efficiency for Auger recombination. Furthermore, phonon-induced interband scattering can directly change the carrier density. Since recombination processes involving the emission of phonons are predominant, phonons reduce the carrier density directly counteracting the effect of CM. Nevertheless, taking the full dynamics into account we still observe a maximum carrier multiplication of approximately $CM = 2.5$ that persists on a picosecond timescale, cf. the blue line in Fig. 4.19(c).

Since Auger processes are sensitive to the excited carrier density, we have investigated the temporal evolution of CM as a function of the absorbed pump fluence, cf. Fig. 4.19(d). Depending on delay time and the pump fluence, we can clearly distinguish regions with $CM > 1$ (blue) and $CM < 1$ (red). We find the largest CM values at low pump fluences in the range of $0.1\,\mu\text{Jcm}^{-2}$ persisting on a timescale of several picoseconds. In contrast, in the strong excitation regime, Pauli blocking prefers AR over IE resulting in a decrease of the carrier density and thus in $CM < 1$. At intermediate pump fluences in the range of $1\,\mu\text{Jcm}^{-2}$, we observe a smaller carrier multiplication on a much shorter timescale [Winzer and Malić (2012)]. Here, IE still prevails over AR, however the asymmetry between these two processes due to Pauli blocking is restricted to a shorter time range, since the number of scattering partners is increased accelerating the relaxation dynamics and leading to a faster equilibration between the IE and AR processes. Our results are in very good agreement with a recent time-dependent ARPES study [Gierz *et al.* (2015)] as well as with a high-resolution multi-color pump-probe study confirming the theoretically predicted trend for different excitation regimes [Plötzing *et al.* (2014)], cf. Fig. 4.19(d). Recently, Auger processes have been shown to also dominate the carrier dynamics in Landau-quantized graphene [Wendler *et al.* (2014); Mittendorff *et al.* (2015)]

4.6.1 *Carrier multiplication in doped graphene samples*

So far, we have investigated the appearance of carrier multiplication for undoped graphene. However, most realistic graphene samples are doped. Therefore, we investigate now the influence of an extrinsic carrier density on the dynamics of optically excited carriers. In the presence of doping, we have to reconsider the definition of charge carriers. For n-doping, we can either count electrons above the Dirac point or above the Fermi level (for holes accordingly in p-doped samples). For optical measurements, probing vertical carrier transitions the Dirac point is the more appropriate reference. Then, the carrier density reads

$$n_{\mathrm{CM}}(T, E_F) = \frac{\sigma_s \sigma_v}{L^2} \sum_{\lambda=e,h;\,\mathbf{k}} \rho^{\lambda}_{\mathbf{k}}, \tag{4.21}$$

where σ_s (σ_v) denotes the spin (valley) degeneracy and L^2 the graphene area. In contrast to optical measurements, for electric transport phenomena hot electrons around the Fermi level play the crucial role. Thus, charge carriers are defined with respect to the Fermi level, i.e. for n-doped graphene the upper Dirac cone is split into $\rho^{h,c}_{\mathbf{k}} = 1 - \rho^{c}_{\mathbf{k}}$ for $k < k_F$ and $\rho^{e,c}_{\mathbf{k}} = \rho^{c}_{\mathbf{k}}$ for $k > k_F$ with the Fermi momentum k_F. The bottom cone remains unaffected with $\rho^{h,v}_{\mathbf{k}} \equiv \rho^{h}_{\mathbf{k}} = 1 - \rho^{v}_{\mathbf{k}}$. The according *hot* carrier density n_{hCM} is given by

$$n_{\mathrm{hCM}} = \frac{\sigma_s \sigma_v}{L^2} \left(\sum_{\mathbf{k}} \rho^{h}_{\mathbf{k}} + \sum_{\mathbf{k}}^{k<k_F} \rho^{h,c}_{\mathbf{k}} + \sum_{\mathbf{k}}^{k>k_F} \rho^{e,c}_{\mathbf{k}} \right). \tag{4.22}$$

Note that symmetric results are obtained for n- and p-doped graphene. For undoped graphene, both definitions of carrier density and carrier multiplication are equivalent for symmetry reasons.

The definition of the charge carrier density has a direct impact on the definition of the carrier multiplication. Having the Dirac point as the reference, carrier multiplication occurs due to Auger scattering bridging the valence and the conduction band. In the following, we label this process as carrier multiplication (CM). Having the Fermi level as the reference, carrier multiplication occurs via Coulomb-induced intraband scattering bridging

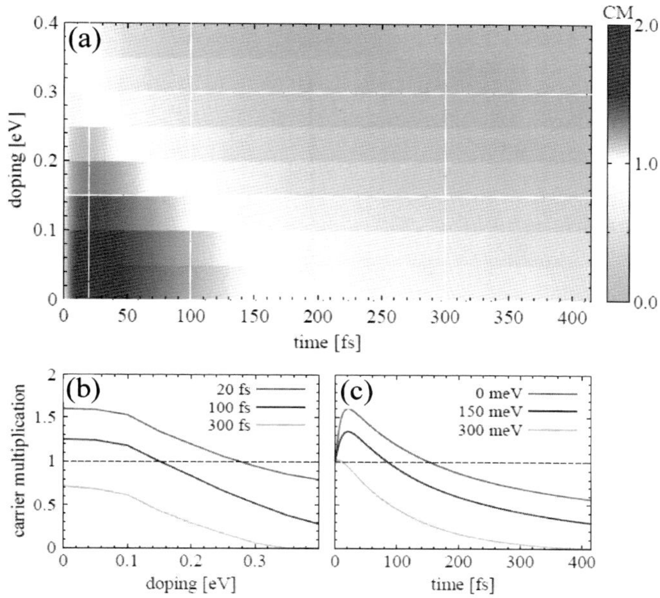

Fig. 4.20 **Doping-dependent carrier multiplication (CM).** (a) Temporal evolution of the doping-dependent CM for a fixed absorbed pump fluence $\varepsilon_{\text{abs}} = 0.3\,\mu\text{Jcm}^{-2}$. (b) Doping-dependent CM for three fixed time delays and (c) the temporal evolution of CM for different doping levels. Figure taken from [Kadi *et al.* (2015)].

the states below and above the Fermi level, cf. the inset in Fig. 4.21(a). According to literature [Tielrooij *et al.* (2013)], we label this process as hot carrier multiplication (hCM).

To include doping effects, we extend the graphene Bloch equations by deriving equations of motion for both electron and hole occupations $\rho_{\mathbf{k}}^{e,c}, \rho_{\mathbf{k}}^{h,v}$. Furthermore, we include dynamic screening of the Coulomb potential accounting for the increased carrier density. Our calculations reveal that for n-doped graphene, electron relaxation is, as expected, faster resulting in a thermalized hot Fermi distribution already after 30 fs, while for p-doped graphene the dynamics of holes is more efficient. This can be traced back to the increased number of scattering partners for electrons or holes depending on the type of doping. Furthermore, we find that doping reduces the efficiency of CM, cf. Fig. 4.20(a). Considering an absorbed pump fluence of $\varepsilon_{\text{abs}} = 0.3\mu J$ cm^{-2}, the highest CM factor of 1.7 is obtained for undoped graphene, cf. Fig. 4.20(b). It remains on a timescale of

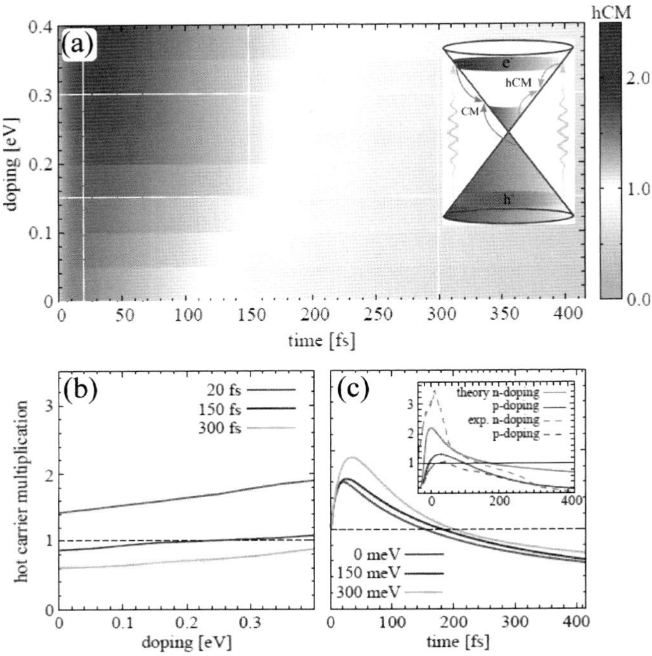

Fig. 4.21 **Doping-dependent hot carrier multiplication (hCM).** The same investigation as in Fig. 4.20 for the case of hot carrier multiplication. The inset in (a) illustrates the processes responsible for CM and hCM in n-doped graphene. The inset in (f) shows a direct comparison between theoretically predicted (solid lines) and experimentally measured (dashed lines) data [Johannsen *et al.* (2015)] for: (i) n-doped graphene ($E_F = 380\,\text{meV}$) with an absorbed pump fluence of $\varepsilon_{\text{abs}} = 0.5\,\mu\text{Jcm}^{-2}$ and (ii) p-doped graphene ($E_F = 240\,\text{meV}$) with a fluence of $\varepsilon_{\text{abs}} = 1.5\,\mu\text{Jcm}^{-2}$. Figure adapted from [Kadi *et al.* (2015)].

up to 150 fs for undoped graphene and becomes significantly shorter with increasing doping, cf. Fig. 4.20(c). For doping values higher than 300 meV, CM completely vanishes. This behavior can be traced back to the strongly efficient impact excitation at low doping, which is a result of the large gradient in carrier occupation around the Dirac point. This changes for highly doped graphene presenting optimal conditions for the inverse process of Auger recombination.

In contrast, our calculations predict a clear increase of the hCM with doping, cf. Fig. 4.21(a). We obtain hCM factors of up to approximately 2 occurring on a timescale of 200 fs for highly doped graphene with $E_F = 300\,\text{meV}$. There is nearly a linear dependence between hCM and doping, cf.

Fig. 4.20(b). The intraband Coulomb scattering around the Fermi energy is responsible for the appearance of hCM. With increasing doping, the spectral region for these *intraband Auger processes* shifts to higher density of states. As a result, they become generally more efficient. The probability for *intraband IE processes* (cf. the inset in Fig. 4.21(a)) is given by $\rho^{h,c}(1-\rho^{e,c})$, which is initially large compared to the probability for *intraband AR processes* scaling with $\rho^{e,c}(1-\rho^{h,c})$. The initial strong imbalance between IE and AR gives rise to the observed pronounced hCM.

Applying our microscopic approach, we can explain the recently performed time-resolved ARPES measurements on n- and p-doped graphene [Johannsen *et al.* (2015)]. A direct comparison of theoretically predicted and experimentally obtained values for hot carrier multiplication is shown in the inset of Fig. 4.21(c). The theory captures well all features observed in the experiment, such as the clearly higher hCM for n-doped graphene. This difference is not due to the type of doping, but can rather be explained by the differences in the applied pump fluence and the doping value of the investigated graphene samples. HCM increases almost linearly with doping [cf. Fig. 4.21(c)] and becomes less efficient at high pump fluences [similarly to the case of CM, cf. Fig. 4.19(d)]. As a result, the n-doped graphene sample shows a much more pronounced hCM, since its Fermi level is significantly higher and since the experiment has been performed at a clearly smaller pump fluence compared to the p-doped graphene sample.

4.7 Conclusion and Outlook

We have presented an extensive microscopic view on different aspects of the ultrafast carrier dynamics in graphene. Our approach is based on the density matrix formalism with graphene Bloch equations in its core. It includes optical excitation of a non-equilibrium distribution as well as Coulomb- and phonon-induced relaxation channels on the same microscopic footing offering a microscopic access to time-, momentum-, and angle-resolved relaxation dynamics.

Our calculations reveal that the optical excitation generates a highly anisotropic non-equilibrium distribution. Efficient carrier-carrier and carrier-phonon scattering processes quickly redistribute carriers resulting in a hot thermalization distribution on a timescale of 100 fs. In particular, scattering with ΓLO phonons also performs an ultrafast orientational relaxation all over the Dirac cone destroying the initial anisotropy, whereas Coulomb-induced channels preferably scatter co-linearly preserving the anisotropy. On a slower sub-picosecond timescale, carrier cooling occurs via carrier-phonon scattering, where again optical phonons play the crucial role. As a result, we predict a bi-exponential decay in differential transmission spectra in excellent agreement with high-resolution pump-probe experiments. Furthermore, we reveal that phonon-assisted intraband scattering processes are responsible for the observed negative components in differential tr ansmission spectra.

For excitations below the optical phonon energy, the dynamics considerably slows down, since scattering with optical phonons becomes strongly suppressed by Pauli blocking. Surprisingly, we still observe how optical phonons significantly contribute if the pump fluence is high enough. This can be explained by the carrier-carrier scattering building a spectrally broad thermalized distribution. Here, the probability to find carriers clearly above the excitation energy is high opening the relaxation channel with optical phonons. In contrast, at low pump fluences only Coulomb-induced processes determine the dynamics resulting in a quick thermalization however in a long-lived anisotropy.

Investigating the carrier dynamics in the terahertz region very close to the Dirac point, many-particle-assisted intraband absorption processes become crucial. In agreement with recent optical-pump/THz probe experiments, we show an entirely negative differential transmission for undoped graphene samples, whereas in the case of high doping there is a positive DTS. The behavior is found for different samples characterized by different degrees of disorder suggesting that disorder-induced supercollisions might not play the crucial role. We show that it is of crucial importance to describe the THz carrier dynamics on microscopic footing and that widely used phenomenological Drude models have severe shortcomings in the THz range. We note that for a full description of the THz dynamics in coherently coupled graphene stacks, a combination of many-particle-induced scattering (discussed in this chapter) and radiative coupling processes (discussed in the chapter ?? by Micheal Wörner and Thomas Elsaesser) needs to be taken into account.

Finally, in contrast to conventional structures, we have found Auger processes to be very important for understanding the carrier dynamics in graphene. Due to the linear and gapless band structure, they become extremely efficient depending on the excitation regime, they can result in a considerable carrier multiplication. In agreement with recent experiments, we have found maximal CM of about 2 in the low-excitation regime persisting on a timescale of a few picoseconds. Furthermore, we have explained the doping dependence of CM and the so-called hot carrier multiplication referring to the generation of hot carriers with respect to the Fermi level.

As outlook, more studies are needed to reveal the THz dynamics close to the Dirac point, in particular the direct competition between radiative decay and many-particle scattering processes needs to be investigated on microscopic footing. Furthermore, the numerically challenging non-Markovian treatment of the carrier dynamics should be performed to study the importance of quantum mechanical memory effects in the first femtoseconds after an optical excitation. It would be also interesting to study the change of the relaxation dynamics in the presence of an electrical field - a situation that is of crucial importance for graphene-based optoelectronic devices.

Bibliography

[1] Ando, T. (2006). Screening effect and impurity scattering in monolayer graphene, *Journal of the Physical Society of Japan* **75**, 7, p. 074716, doi: 10.1143/JPSJ.75.074716, `http://jpsj.ipap.jp/link?JPSJ/75/074716/`.

[2] Ando, T., Nakanishi, T., and Saito, R. (1998). Berry's phase and absence of back scattering \ in carbon nanotubes, *Journal of the Physical Society of Japan* **67**, 8, pp. 2857–2862, doi:10.1143/JPSJ.67.2857, `http://jpsj.ipap.jp/link?JPSJ/67/2857/`.

[3] Breusing, M., Kuehn, S., Winzer, T., Malić, E., Milde, F., Severin, N., Rabe, J. P., Ropers, C., Knorr, A., and Elsaesser, T. (2011). Ultrafast nonequilibrium carrier dynamics in a single graphene layer, *Physical Review B* **83**, 15, p. 153410, doi:10.1103/PhysRevB.83.153410, `http://link.aps.org/doi/10.1103/PhysRevB.83.153410`.

[4] Brida, D., Tomadin, A., Manzoni, C., Kim, Y. J., Lombardo, A., Milana, S., Nair, R. R., Novoselov, K. S., Ferrari, A. C., Cerullo, G., and Polini, M. (2013). Ultrafast collinear scattering and carrier multiplication in graphene, *Nature Commun.* **4**, p. 1987, `http://dx.doi.org/10.1038/ncomms2987`.

[5] Chae, D., Utikal, T., Weisenburger, S., Giessen, H., and v. Klitzing K. (2011). Excitonic fano resonance in free-standing graphene, *Nano Lett.* **11**, p. 1379.

[6] Choi, H., Borondics, F., Siegel, D., Zhou, S., Martin, M. C., Lanzara, A., and Kaindl, R. A. (2009). Broadband electromagnetic response and ultrafast dynamics of few-layer epitaxial graphene, *Appl. Phys. Lett.* **94**, 17, p. 172102.

[7] Dawlaty, J. M., Shivaraman, S., Chandrashekhar, M., Rana, F., and Spencer, M. G. (2008). Measurement of ultrafast carrier dynamics in epitaxial graphene, *Applied Physics Letters* **92**, 4, pp. 042116–042116–3, doi: doi:10.1063/1.2837539, `http://apl.aip.org/resource/1/applab/v92/i4/p042116_s1`.

[8] Diels, J.-C. P., Rudolph, W. P., Diels, J.-C., and Rudolph, W. (1996). *Ultrashort Laser Pulse Phenomena: Fundamentals, Techniques, and Applications on a Femtosecond Time Scale*, 1st edn. (Academic Press), ISBN 0122154924.

[9] Docherty, C. J., Lin, C.-T., Joyce, H. J., Nicholas, R. J., Herz, L. M., Li, L.-J., and Johnston, M. B. (2012). Extreme sensitivity of graphene

photoconductivity to environmental gases, *Nat Commun* **3**, p. 1228, doi: 10.1038/ncomms2235, `http://dx.doi.org/10.1038/ncomms2235`.

[10] Frenzel, A. J., Lui, C. H., Fang, W., Nair, N. L., Herring, P. K., Jarillo-Herrero, P., Kong, J., and Gedik, N. (2013). Observation of suppressed terahertz absorption in photoexcited graphene, *Applied Physics Letters* **102**, 11, 113111, doi:http://dx.doi.org/10.1063/1.4795858, `http://scitation.aip.org/content/aip/journal/apl/102/11/10.1063/1.4795858`.

[11] Frenzel, A. J., Lui, C. H., Shin, Y. C., Kong, J., and Gedik, N. (2014). Semiconducting-to-metallic photoconductivity crossover and temperature-dependent drude weight in graphene, *Phys. Rev. Lett.* **113**, p. 056602, doi:10.1103/PhysRevLett.113.056602, `http://link.aps.org/doi/10.1103/PhysRevLett.113.056602`.

[12] George, P. A., Strait, J., Dawlaty, J., Shivaraman, S., Chandrashekhar, M., Rana, F., and Spencer, M. G. (2008). Ultrafast optical-pump terahertz-probe spectroscopy of the carrier relaxation and recombination dynamics in epitaxial graphene, *Nano Letters* **8**, 12, pp. 4248–4251, doi:10.1021/nl8019399, `http://dx.doi.org/10.1021/nl8019399`.

[13] Gierz, I., Calegari, F., Aeschlimann, S., Cervantes, M. C., Cacho, C., Chapman, R. T., Springate, E., Link, S., Starke, U., Ast, C. R., and Cavalleri, A. (2015). Tracking primary thermalization events in graphene with photoemission at extreme timescales, *Phys. Rev. Lett.* **115**, p. 086803.

[14] Gierz, I., Petersen, J. C., Mitrano, M., Cacho, C., Turcu, I. C. E., Springate, E., Stöhr, A., Köhler, A., Starke, U., and Cavalleri, A. (2013). Snapshots of non-equilibrium dirac carrier distributions in graphene, *Nature Materials* **12**, 12, pp. 1119–1124, doi:10.1038/nmat3757.

[15] Grüneis, A., Saito, R., Samsonidze, G. G., Kimura, T., Pimenta, M. A., Jorio, A., Filho, A. G. S., Dresselhaus, G., and Dresselhaus, M. S. (2003). Inhomogeneous optical absorption around the k point in graphite and carbon nanotubes, *Physical Review B* **67**, 16, p. 165402, doi:10.1103/PhysRevB.67.165402, `http://link.aps.org/doi/10.1103/PhysRevB.67.165402`.

[16] Haug, H. and Koch, S. W. (2004). *Quantum Theory of the Optical and Electronic Properties of Semiconductors* (World Scientific), ISBN 9789812387561.

[17] Huang, L., Hartland, G. V., Chu, L.-Q., Luxmi, Feenstra, R. M., Lian, C., Tahy, K., and Xing, H. (2010). Ultrafast transient absorption microscopy studies of carrier dynamics in epitaxial graphene, *Nano Letters* **10**, 4, pp. 1308–1313, doi:10.1021/nl904106t, `http://dx.doi.org/10.1021/nl904106t`.

[18] Hwang, E. H. and Das Sarma, S. (2008a). Acoustic phonon scattering limited carrier mobility in two-dimensional extrinsic graphene, *Physical Review B* **77**, 11, p. 115449, doi:10.1103/PhysRevB.77.115449, `http://link.aps.org/doi/10.1103/PhysRevB.77.115449`.

[19] Hwang, E. H. and Das Sarma, S. (2008b). Single-particle relaxation time versus transport scattering time in a two-dimensional graphene layer, *Physical Review B* **77**, 19, p. 195412, doi:10.1103/PhysRevB.77.195412, `http://link.aps.org/doi/10.1103/PhysRevB.77.195412`.

[20] Jackson, J. D. (1998). *Classical Electrodynamics Third Edition*, 3rd edn. (Wiley), ISBN 047130932X.

[21] Jensen, S. A., Mics, Z., Ivanov, I., Varol, H. S., Turchinovich, D., Koppens, F. H. L., Bonn, M., and Tielrooij, K. J. (2014). Competing ultrafast energy relaxation pathways in photoexcited graphene, *Nano Letters* **14**, 10, pp. 5839–5845, doi:10.1021/nl502740g, pMID: 25247639.

[22] Jiang, J., Saito, R., Samsonidze, G. G., Jorio, A., Chou, S. G., Dresselhaus, G., and Dresselhaus, M. S. (2007). Chirality dependence of exciton effects in single-wall carbon nanotubes: Tight-binding model, *Physical Review B* **75**, 3, p. 035407, doi:10.1103/PhysRevB.75.035407, http://link.aps.org/doi/10.1103/PhysRevB.75.035407.

[23] Jnawali, G., Rao, Y., Yan, H., and Heinz, T. F. (2013). Observation of a transient decrease in terahertz conductivity of single-layer graphene induced by ultrafast optical excitation, *Nano Letters* **13**, 2, pp. 524–530, doi:10.1021/nl303988q, http://dx.doi.org/10.1021/nl303988q, http://dx.doi.org/10.1021/nl303988q, pMID: 23330567.

[24] Johannsen, J. C., Ulstrup, S., Cilento, F., Crepaldi, A., Zacchigna, M., Cacho, C., Turcu, I. C. E., Springate, E., Fromm, F., Raidel, C., Seyller, T., Parmigiani, F., Grioni, M., and Hofmann, P. (2013). Direct view of hot carrier dynamics in graphene, *Physical Review Letters* **111**, 2, p. 027403, doi:10.1103/PhysRevLett.111.027403, http://link.aps.org/doi/10.1103/PhysRevLett.111.027403.

[25] Johannsen, J. C., Ulstrup, S., Crepaldi, A., Cilento, F., Zacchigna, M., Miwa, J. A., Cacho, C., Chapman, R. T., Springate, E., Fromm, F., Raidel, C., Seyller, T., King, P. D. C., Parmigiani, F., Grioni, M., and Hofmann, P. (2015). Tunable carrier multiplication and cooling in graphene, *Nano Letters* **15**, 1, pp. 326–331, doi:10.1021/nl503614v, pMID: 25458168.

[26] Kadi, F. (2015). *Relaxation Dynamics in Graphene*, Ph.D. thesis, Technische Universität Berlin.

[27] Kadi, F., Winzer, T., Knorr, A., and Malic, E. (2015). Impact of doping on the carrier dynamics in graphene, *Sci. Rep.* **5**, p. 16841.

[28] Kadi, F., Winzer, T., Malic, E., Knorr, A., Göttfert, F., Mittendorff, M., Winnerl, S., and Helm, M. (2014). Microscopic Description of Intraband Absorption in Graphene: The Occurrence of Transient Negative Differential Transmission, *Phys. Rev. Lett.* **113**, p. 035502, doi:10.1103/PhysRevLett.113.035502, http://link.aps.org/doi/10.1103/PhysRevLett.113.035502.

[29] Kira, M. and Koch, S. (2006). Many-body correlations and excitonic effects in semiconductor spectroscopy, *Progress in Quantum Electronics* **30**, 5, pp. 155–296, doi:10.1016/j.pquantelec.2006.12.002, http://www.sciencedirect.com/science/article/pii/S0079672706000280.

[30] Knorr, A., Hughes, S., Stroucken, T., and Koch, S. W. (1996). Theory of ultrafast spatio-temporal dynamics in semiconductor heterostructures , *Chemical Physics* **210**, pp. 27 – 47, doi:http://dx.doi.org/10.1016/0301-0104(96)00120-6, http://www.sciencedirect.com/science/

article/pii/0301010496001206, confined Excitations in Molecular and Semiconductor Nanostructures.

[31] Koenig-Otto, J. C., Mittendorff, M., Winzer, T., Malic, E., Knorr, A., Berger, C., de Heer, W. A., Pashkin, A., Schneider, H., Helm, M., and Winnerl, S. (2016). Slow non-collinear coulomb scattering in the vicinity of the dirac point in graphene, *Phys. Rev. Lett.* **117**, 0847401.

[32] Kumar, S., Anija, M., Kamaraju, N., Vasu, K. S., Subrahmanyam, K. S., Sood, A. K., and Rao, C. N. R. (2009). Femtosecond carrier dynamics and saturable absorption in graphene suspensions, *Applied Physics Letters* **95**, 19, pp. 191911–191911–3, doi:doi:10.1063/1.3264964, http://apl.aip.org/resource/1/applab/v95/i19/p191911_s1.

[33] Kusminskiy, V. S., Campbell, D. K., and Castro Neto, A. H. (2009). Lenosky's energy and the phonon dispersion of graphene, *Physical Review B* **80**, 3, p. 035401, doi:10.1103/PhysRevB.80.035401, http://link.aps.org/doi/10.1103/PhysRevB.80.035401.

[34] Mak, K. F., Shan, J., and Heinz, T. F. (2011). Seeing many-body effects in single- and few-layer graphene: Observation of two-dimensional saddle-point excitons, *Physical Review Letters* **106**, 4, p. 046401, doi:10.1103/PhysRevLett.106.046401, http://link.aps.org/doi/10.1103/PhysRevLett.106.046401.

[35] Malard, L. M., Mak, K. F., Neto, A. H. C., Peres, N. M. R., and Heinz, T. F. (2013). Observation of intra- and inter-band transitions in the transient optical response of graphene, *New Journal of Physics* **15**, 1, p. 015009, http://stacks.iop.org/1367-2630/15/i=1/a=015009.

[36] Malic, E. (2008). *Many-particle theory of optical properties in low-dimensional nanostructures*, Ph.D. thesis, Technische Universität Berlin, Berlin.

[37] Malić, E., Hirtschulz, M., Milde, F., Knorr, A., and Reich, S. (2006). Analytical approach to optical absorption in carbon nanotubes, *Physical Review B* **74**, 19, p. 195431, doi:10.1103/PhysRevB.74.195431, http://link.aps.org/doi/10.1103/PhysRevB.74.195431.

[38] Malic, E., Richter, M., Hartmann, G., Gomis-Bresco, J., Woggon, U., and Knorr, A. (2010). Analytical description of gain depletion and recovery in quantum dot optical amplifiers, *New Journal of Physics* **12**, 6, p. 063012, doi:10.1088/1367-2630/12/6/063012, http://iopscience.iop.org/1367-2630/12/6/063012.

[39] Malic, E., Weber, C., Richter, M., Atalla, V., Klamroth, T., Saalfrank, P., Reich, S., and Knorr, A. (2011a). Microscopic model of the optical absorption of carbon nanotubes functionalized with molecular spiropyran photoswitches, *Physical Review Letters* **106**, 9, p. 097401, doi:10.1103/PhysRevLett.106.097401, http://link.aps.org/doi/10.1103/PhysRevLett.106.097401.

[40] Malic, E., Winnerl, S., and Knorr, A. (2013). *Graphene and Carbon Nanotubes: Ultrafast Optics and Relaxation Dynamics*, 1st edn. (Wiley-VCH), ISBN 3527411615.

[41] Malic, E., Winzer, T., Bobkin, E., and Knorr, A. (2011b). Microscopic theory of absorption and ultrafast many-particle kinetics in graphene, *Physical Review B* **84**, 20, p. 205406, doi:10.1103/PhysRevB.84.205406, http://link.aps.org/doi/10.1103/PhysRevB.84.205406.

[42] Malic, E., Winzer, T., and Knorr, A. (2012). Efficient orientational carrier relaxation in optically excited graphene, *Applied Physics Letters* **101**, 21, pp. 213110–213110-3, doi:doi:10.1063/1.4767356, http://apl.aip.org/resource/1/applab/v101/i21/p213110_s1.

[43] Maultzsch, J., Reich, S., Thomsen, C., Requardt, H., and Ordejón, P. (2004). Phonon dispersion in graphite, *Physical Review Letters* **92**, 7, p. 075501, doi:10.1103/PhysRevLett.92.075501, http://link.aps.org/doi/10.1103/PhysRevLett.92.075501.

[44] Mihnev, M. T., Kadi, F., Divin, C. J., Winzer, T., Lee, S., Liu, C., Zhong, Z., Berger, C., de Heer, W. A., Malic, E., Knorr, A., and Norris, T. B. (2016). Microscopic origins of the terahertz carrier relaxation and cooling dynamics in graphene, *Nature Commun.* **7**, 11617 (2016).

[45] Mittendorff, M., Wendler, F., Malic, E., Knorr, A., Orlita, M., Potemski, M., Berger, C., de Heer, W. A., Schneider, H., Helm, M., and Winnerl, S. (2015). Carrier dynamics in Landau-quantized graphene featuring strong Auger scattering, *Nature Phys.* **11**, pp. 75–81, http://dx.doi.org/10.1038/nphys3164.

[46] Mittendorff, M., Winzer, T., Malic, E., Knorr, A., Berger, C., de Heer, W. A., Schneider, H., Helm, M., and Winnerl, S. (2014). Anisotropy of excitation and relaxation of photogenerated charge carriers in graphene, *Nano Letters* **14**, 3, pp. 1504–1507, doi:10.1021/nl404730y, pMID: 24559191.

[47] Newson, R. W., Dean, J., Schmidt, B., and van Driel, H. M. (2009). Ultrafast carrier kinetics in exfoliated graphene and thin graphite films, *Optics Express* **17**, 4, pp. 2326–2333, doi:10.1364/OE.17.002326, http://www.opticsexpress.org/abstract.cfm?URI=oe-17-4-2326.

[48] Piscanec, S., Lazzeri, M., Mauri, F., Ferrari, A. C., and Robertson, J. (2004). Kohn anomalies and electron-phonon interactions in graphite, *Physical Review Letters* **93**, 18, p. 185503, doi:10.1103/PhysRevLett.93.185503, http://link.aps.org/doi/10.1103/PhysRevLett.93.185503.

[49] Plochocka, P., Kossacki, P., Golnik, A., Kazimierczuk, T., Berger, C., de Heer, W. A., and Potemski, M. (2009). Slowing hot-carrier relaxation in graphene using a magnetic field, *Physical Review B* **80**, 24, p. 245415, doi:10.1103/PhysRevB.80.245415, http://link.aps.org/doi/10.1103/PhysRevB.80.245415.

[50] Plötzing, T., Winzer, T., Malic, E., Neumaier, D., Knorr, A., and Kurz, H. (2014). Experimental Verification of Carrier Multiplication in Graphene, *Nano Lett.* **14**, 9, pp. 5371–5375, doi:10.1021/nl502114w, http://dx.doi.org/10.1021/nl502114w, http://dx.doi.org/10.1021/nl502114w, pMID: 25144320.

[51] Reich, S., Maultzsch, J., Thomsen, C., and Ordejón, P. (2002). Tight-binding description of graphene, *Physical Review B* **66**, 3, p. 035412, doi:10.1103/PhysRevB.66.035412, http://link.aps.org/doi/10.1103/PhysRevB.66.035412.

[52] Roberts, A. T., Binder, R., Kwong, N. H., Golla, D., Cormode, D., Leroy, B. J., Everitt, H. O., and Sandhu, A. (2013). Optical characterization of electron-phonon interactions at the saddle point in graphene, *Phys. Rev. Lett.* **112**, p. 187401.

[53] Rossi, F. and Kuhn, T. (2002). Theory of ultrafast phenomena in photoexcited semiconductors, *Reviews of Modern Physics* **74**, 3, pp. 895–950, doi:10.1103/RevModPhys.74.895, `http://link.aps.org/doi/10.1103/RevModPhys.74.895`.

[54] Sanders, G. D., Nugraha, A. R. T., Sato, K., Kim, J.-H., Kono, J., Saito, R., and Stanton, C. J. (2013). Theory of coherent phonons in carbon nanotubes and graphene nanoribbons, *Journal of Physics: Condensed Matter* **25**, 14, p. 144201, doi:10.1088/0953-8984/25/14/144201, `http://iopscience.iop.org/0953-8984/25/14/144201`.

[55] Scully, M. O. and Zubairy, M. S. (1997). *Quantum Optics*, 1st edn. (Cambridge University Press), ISBN 0521435951.

[56] Shang, J., Luo, Z., Cong, C., Lin, J., Yu, T., and Gurzadyan, G. G. (2010). Femtosecond UV-pump/visible-probe measurements of carrier dynamics in stacked graphene films, *Applied Physics Letters* **97**, 16, pp. 163103–163103–3, doi:doi:10.1063/1.3504704, `http://apl.aip.org/resource/1/applab/v97/i16/p163103_s1`.

[57] Shi, S.-F., Tang, T.-T., Zeng, B., Ju, L., Zhou, Q., Zettl, A., and Wang, F. (2014). Controlling graphene ultrafast hot carrier response from metal-like to semiconductor-like by electrostatic gating, *Nano Letters* **14**, 3, pp. 1578–1582, doi:10.1021/nl404826r, pMID: 24564302.

[58] Song, J. C. W., Reizer, M. Y., and Levitov, L. S. (2012). Disorder-assisted electron-phonon scattering and cooling pathways in graphene, *Physical Review Letters* **109**, 10, p. 106602, doi:10.1103/PhysRevLett.109.106602, `http://link.aps.org/doi/10.1103/PhysRevLett.109.106602`.

[59] Strait, J. H., Wang, H., Shivaraman, S., Shields, V., Spencer, M., and Rana, F. (2011). Very slow cooling dynamics of photoexcited carriers in graphene observed by optical-pump terahertz-probe spectroscopy, *Nano Letters* **11**, 11, pp. 4902–4906, doi:10.1021/nl202800h, `http://dx.doi.org/10.1021/nl202800h`.

[60] Sun, D., Divin, C., Berger, C., de Heer, W. A., First, P. N., and Norris, T. B. (2010). Spectroscopic measurement of interlayer screening in multilayer epitaxial graphene, *Physical Review Letters* **104**, 13, p. 136802, doi:10.1103/PhysRevLett.104.136802, `http://link.aps.org/doi/10.1103/PhysRevLett.104.136802`.

[61] Sun, D., Wu, Z.-K., Divin, C., Li, X., Berger, C., de Heer, W. A., First, P. N., and Norris, T. B. (2008). Ultrafast relaxation of excited dirac fermions in epitaxial graphene using optical differential transmission spectroscopy, *Physical Review Letters* **101**, 15, p. 157402, doi:10.1103/PhysRevLett.101.157402, `http://link.aps.org/doi/10.1103/PhysRevLett.101.157402`.

[62] Tielrooij, K. J., Song, J. C. W., Jensen, S. A., Centeno, A., Pesquera, A., Zurutuza Elorza, A., Bonn, M., Levitov, L. S., and Koppens, F. H. L. (2013). Photoexcitation cascade and multiple hot-carrier generation in graphene,

Nature Physics **9**, p. 248, doi:10.1038/nphys2564, http://www.nature.com/nphys/journal/vaop/ncurrent/full/nphys2564.html#/access.

[63] Tse, W.-K. and Das Sarma, S. (2009). Energy relaxation of hot dirac fermions in graphene, *Physical Review B* **79**, 23, p. 235406, doi:10.1103/PhysRevB.79.235406, http://link.aps.org/doi/10.1103/PhysRevB.79.235406.

[64] Verdenhalven, E., Binder, R., Knorr, A., and Malić, E. (2013). Derivation of the screened bloch equations and application to carbon nanostructures, *Chemical Physics* **413**, pp. 3–10, doi:10.1016/j.chemphys.2012.07.013, http://www.sciencedirect.com/science/article/pii/S0301010412002819.

[65] Wang, H., Strait, J. H., George, P. A., Shivaraman, S., Shields, V. B., Chandrashekhar, M., Hwang, J., Rana, F., Spencer, M. G., Ruiz-Vargas, C. S., and Park, J. (2010). Ultrafast relaxation dynamics of hot optical phonons in graphene, *Appl. Phys. Lett.* **96**, 8, 081917.

[66] Wendler, F., Knorr, A., and Malic, E. (2014). Carrier multiplication in graphene under Landau quantization, *Nature Commun.* **5**, p. 3703, http://dx.doi.org/10.1038/ncomms4703.

[67] Winnerl, S., Göttfert, F., Mittendorff, M., Schneider, H., Helm, M., Winzer, T., Malic, E., Knorr, A., Orlita, M., Potemski, M., Sprinkle, M., Berger, C., and Heer, W. A. d. (2013). Time-resolved spectroscopy on epitaxial graphene in the infrared spectral range: relaxation dynamics and saturation behavior, *Journal of Physics: Condensed Matter* **25**, 5, p. 054202, doi:10.1088/0953-8984/25/5/054202, http://iopscience.iop.org/0953-8984/25/5/054202.

[68] Winnerl, S., Orlita, M., Plochocka, P., Kossacki, P., Potemski, M., Winzer, T., Malic, E., Knorr, A., Sprinkle, M., Berger, C., de Heer, W. A., Schneider, H., and Helm, M. (2011). Carrier relaxation in epitaxial graphene photoexcited near the dirac point, *Physical Review Letters* **107**, 23, p. 237401, doi:10.1103/PhysRevLett.107.237401, http://link.aps.org/doi/10.1103/PhysRevLett.107.237401.

[69] Winzer, T. (2013). *Ultrafast Carrier Relaxation Dynamics in Graphene*, Ph.D. thesis, Technische Universität Berlin.

[70] Winzer, T., Knorr, A., and Malic, E. (2010). Carrier multiplication in graphene, *Nano Letters* **10**, 12, pp. 4839–4843, doi:10.1021/nl1024485, http://dx.doi.org/10.1021/nl1024485.

[71] Winzer, T. and Malić, E. (2012). Impact of auger processes on carrier dynamics in graphene, *Physical Review B* **85**, 24, p. 241404, doi:10.1103/PhysRevB.85.241404, http://link.aps.org/doi/10.1103/PhysRevB.85.241404.

[72] Yang, L., Deslippe, J., Park, C.-H., Cohen, M. L., and Louie, S. G. (2009). Excitonic effects on the optical response of graphene and bilayer graphene, *Physical Review Letters* **103**, 18, p. 186802, doi:10.1103/PhysRevLett.103.186802, http://link.aps.org/doi/10.1103/PhysRevLett.103.186802.

Chapter 5

Theory of Optical Nonlinearities in Graphene

Jin Luo Cheng[1,2], Nathalie Vermeulen[1] and John E. Sipe[2,*]
[1]*Brussels Photonics Team (B-PHOT), Department of Applied Physics and Photonics (IR-TONA), Vrije Universiteit Brussel, Pleinlaan 2, 1050 Brussel, Belgium*
[2]*Department of Physics and Institute for Optical Sciences, University of Toronto, 60 St. George Street, Toronto, Ontario, Canada M5S 1A7*
**sipe@physics.utoronto.ca*

The third order optical nonlinearity of graphene is theoretically investigated using the semiconductor Bloch equations under the independent particle approximation. The electronic states are described by a two band tight binding model, and the scattering is described by phenomenological parameters. For nominally weak optical fields, perturbation theory at zero temperature gives analytic expressions for the third order conductivities under the linear dispersion approximation around the Dirac points. These expressions show many singularities and divergences, which indicates a complex nonlinear optical behavior of graphene. We illustrate the dependence of several nonlinear optical effects, such as third harmonic generation, Kerr effects and two photon absorption, and parametric frequency conversion, on the relaxation parameters and temperature. The analytic results from perturbation theory, with the approximation employed, are verified by numerical solutions. The latter also allow us to discuss the effects of saturation on the optical nonlinearity. The results of the independent particle theory presented here are smaller than the values extracted from experiments by at least two orders of magnitude.

5.1 Introduction

Graphene shows unique and remarkable optical properties: The linear optical response is characterized by a universal conductivity [Nair *et al.* (2008)] $\sigma_0 = e^2/4\hbar$ at wavelengths from the mid-infrared to the visible, with a monolayer absorbance of about 2.3% [Mak *et al.* (2008)]. Thus while a single layer of graphene is highly transparent, considered as a monolayer it is a highly absorbing material, and saturated absorption is easy to achieve [Bao *et al.* (2009)]. Although the graphene crystal has center-of-inversion symmetry, it can exhibit second-order nonlinear responses arising from interface effects [Dean and van Driel (2009, 2010); Bykov *et al.* (2012); An *et al.* (2013, 2014); Sipe *et al.* (1987)], nonuniformity of optical field [Mikhailov (2007); Mikhailov and Ziegler (2008); Glazov (2011); Mikhailov (2011); Glazov and Ganichev (2014)], the presence of natural curvature fluctuations of suspended graphene [Lin *et al.* (2014)], or the presence of dc currents [Wu *et al.* (2012); Bykov *et al.* (2012); Avetissian *et al.* (2012); An *et al.* (2013, 2014); Cheng *et al.* (2014a, 2015b)]. Nevertheless, the first nonvanishing nonlinear susceptibility arising in pristine graphene is the third order susceptibility [Glazov and Ganichev (2014)], and that is the focus of this chapter. The third order nonlinear optical response shows a complex dependence on the measurement method, light frequency, and perhaps even the sample preparation. The measurements on different nonlinear effects, such as parametric frequency conversion (PFC, also four-wave mixing), third harmonic generation (THG), and Kerr effects and two photon absorption (corresponding to a nonlinear refractive index), indicate that the effective response coefficients are in general orders of magnitude larger than those of normal insulators, semiconductors, and metals. The measured effective susceptibilities of graphene are listed in Table 5.1.

In moving towards any application, an important step is understanding the optical nonlinearities of graphene, as well as their dependence on the chemical potential, the temperature, and sample quality. For most cases, the field $\boldsymbol{E}(t)$ to which electrons respond only causes a small variation in the motion of the electrons, and thus can be treated as a uniform weak perturbation. Then the optical response of graphene can be described by

$$J^d(t) = \int \frac{d\omega}{2\pi} e^{-i\omega t} \sigma^{(1);da}(\omega) E^a_\omega + \int \frac{d\omega_1 d\omega_2 d\omega_3}{(2\pi)^3} e^{-i(\omega_1+\omega_2+\omega_3)t} \times \sigma^{(3);dabc}(\omega_1,\omega_2,\omega_3) E^a_{\omega_1} E^b_{\omega_2} E^c_{\omega_3} + \cdots . \tag{5.1}$$

Here $J^d(t)$ is the sheet charge current along direction d (x or y), which can be converted to an effective bulk current density $J^d(t)\delta(z)$ for the

Table 5.1 Measured nonlinear susceptibility $\chi^{(3)}$, nonlinear refractive index n_2, and saturation light intensity I_{sat}.

type	λ (μm)	$\chi^{(3)}$ (m^2/V^2)		Reference
PFC	$\lambda_s = 1.179$ $\lambda_p = 0.969$	2×10^{-15}		[Hendry *et al.* (2010)]
THG	$\lambda = 1.72$	0.4×10^{-16}		[Kumar *et al.* (2013)]
THG	$\lambda = 1.55$	4.2×10^{-15}		[Säynätjoki *et al.* (2013)]
THG	$\lambda = 0.79$	$\sim 10^{-19}$		[Hong *et al.* (2013)]
		n_2 (m^{2}W^{-1})	I_{sat} (GW/cm^2)	
Kerr	$\lambda = 1.55$	10^{-11}	0.074	[Zhang *et al.* (2012)]
Kerr	$\lambda = 1.55$	10^{-17}		[Gu *et al.* (2012)]
Kerr	$\lambda = 0.532$	2.5×10^{-13}		[Wu *et al.* (2011)]
Kerr	$\lambda = 0.78 - 1.1$		1-10	[Yang *et al.* (2011)]

graphene positioned at $z = 0$; $\sigma^{(1);da}(\omega)$ is the linear conductivity tensor, $\sigma^{(3);dabc}(\omega_1, \omega_2, \omega_3)$ the third order conductivity tensor, and $\boldsymbol{E}_\omega$ is the Fourier component of $\boldsymbol{E}(t)$, defined by

$$\boldsymbol{E}(t) = \int \frac{d\omega}{2\pi} \boldsymbol{E}_\omega e^{-i\omega t} . \tag{5.2}$$

To compare with the susceptibility value for bulk materials, the effective susceptibilities of graphene can be extracted as

$$\chi_{\text{eff}}^{(1);da}(\omega) = \frac{\sigma^{(1);da}(\omega)}{-i\omega\epsilon_0 d_{\text{gr}}} , \tag{5.3}$$

$$\chi_{\text{eff}}^{(3);dabc}(\omega_1, \omega_2, \omega_3) = \frac{\sigma^{(3);dabc}(\omega_1, \omega_2, \omega_3)}{-i(\omega_1 + \omega_2 + \omega_3)\epsilon_0 d_{\text{gr}}} , \tag{5.4}$$

with $d_{\text{gr}} = 3.3$Å an effective thickness of the graphene monolayer.

This chapter focuses on the theory of graphene's nonlinear optical response at the level of the independent particle approximation, with the inclusion of phenomenological relaxation parameters. For light in the infrared to the visible, most optical transitions occur around the Dirac points, the electronic band structure can be well approximated by the linear dispersion approximation, and then analytic expressions for $\sigma^{(3);dabc}(\omega_1, \omega_2, \omega_3)$ can be obtained at zero temperature, even with the inclusion of phenomenological relaxation parameters. As shown in Section 5.4.1, the full expressions are complex, and show many divergences. However, in the relaxation-free limit and at zero temperature, they can be greatly simplified for two physical situations: (1) For undoped graphene, a simple expression is found

$$\sigma^{(3);xxxx}(\omega_1, \omega_2, \omega_3) = \frac{\pi\sigma_3}{\hbar^4(\omega_1 + \omega_2)(\omega_2 + \omega_3)(\omega_3 + \omega_1)(\omega_1 + \omega_2 + \omega_3)} , \tag{5.5}$$

with the constant $\sigma_3 \equiv \sigma_0(\hbar v_F e)^2/\pi$ with v_F the Fermi velocity. (2) In doped graphene when $\hbar\omega_i \ll |\mu|$ with μ the chemical potential, the contributions mainly come from the electron movement around the Fermi surface driven by the low frequency electric field, and the expression becomes

$$\sigma^{(3);xxxx}(\omega_1, \omega_2, \omega_3) \approx \frac{i\sigma_3}{2|\mu|\hbar^3\omega_1\omega_2\omega_3} \tag{5.6}$$

In the following, a quantum theory based on the semiconductor Bloch equations (SBEs) is presented for the third order nonlinear conductivities of graphene.

5.2 Electronic States in a Two-band Tight Binding Model

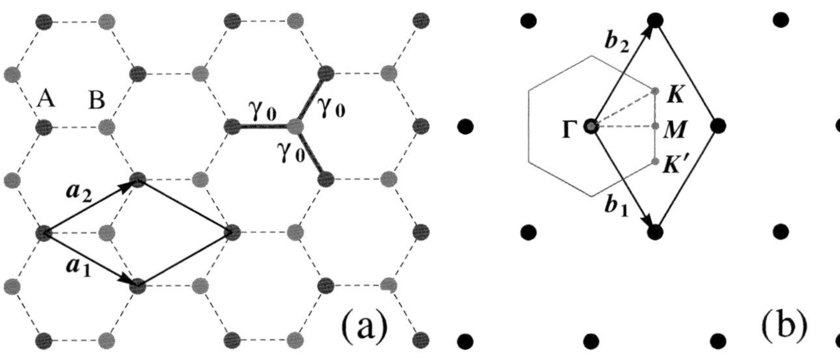

Fig. 5.1 Illustration of the graphene lattice. (a) The lattice in real space. The two lattice sites are noted by A and B, the primitive vectors are $\boldsymbol{a}_1$ and $\boldsymbol{a}_2$, and the nearest neighbor hopping parameter is γ_0. (b) The lattice in reciprocal space. The primitive reciprocal vectors are $\boldsymbol{b}_1$ and $\boldsymbol{b}_2$. The purple hexagon is the first Brillouin zone; Γ, $\boldsymbol{K}$, $\boldsymbol{K}'$, and $\boldsymbol{M}$ are the high symmetry points.

Graphene is a two dimensional material formed by carbon atoms in a honey-comb lattice, as shown in Fig. 5.1 (a). The crystal lattice can be generated by two primitive lattice vectors, as

$$\boldsymbol{a}_1 = a_0\left(\frac{\sqrt{3}}{2}\hat{\boldsymbol{x}} - \frac{1}{2}\hat{\boldsymbol{y}}\right), \quad \boldsymbol{a}_2 = a_0\left(\frac{\sqrt{3}}{2}\hat{\boldsymbol{x}} + \frac{1}{2}\hat{\boldsymbol{y}}\right), \tag{5.7}$$

where $a_0 = 2.46$Å is the lattice constant. In each unit cell, there are two inequivalent carbon atoms, at the sites A and B. The reciprocal lattice, which is formed by all vectors $\boldsymbol{G}$ which satisfy $e^{i\boldsymbol{a}_1\cdot\boldsymbol{G}} = e^{i\boldsymbol{a}_2\cdot\boldsymbol{G}} = 1$, is shown in Fig. 5.1 (b). The reciprocal lattice can be generated by the primitive reciprocal vectors

$$\boldsymbol{b}_1 = \frac{2\pi}{a_0}\left(\frac{1}{\sqrt{3}}\hat{\boldsymbol{x}} - \hat{\boldsymbol{y}}\right), \quad \boldsymbol{b}_2 = \frac{2\pi}{a_0}\left(\frac{1}{\sqrt{3}}\hat{\boldsymbol{x}} + \hat{\boldsymbol{y}}\right). \tag{5.8}$$

The first Brillouin zone is illustrated as a purple hexagon in Fig. 5.1 (b).

For low photon energy excitation, the important bands are associated with the π and π^* orbitals of molecular physics, and result from a tight-binding model employing carbon $2p_z$ orbitals $\phi(\boldsymbol{r} - \boldsymbol{R}_{nm\alpha})$, which can be chosen to be orthogonal to each other, with each being well-localized around atom position $\boldsymbol{R}_{nm\alpha} = n\boldsymbol{a}_1 + m\boldsymbol{a}_2 + \boldsymbol{\tau}_\alpha$, with n, m being integers and $\alpha = A, B$ standing for the two inequivalent sites; $\boldsymbol{\tau}_A = 0$ and $\boldsymbol{\tau}_B = (\boldsymbol{a}_1 + \boldsymbol{a}_2)/3$. The matrix elements of the unperturbed electron Hamiltonian H_0 between these orbitals can be approximated as

$$\begin{aligned}
&\int d\boldsymbol{r}\phi^*(\boldsymbol{r} - \boldsymbol{R}_{n_1m_1\alpha_1})H_0\phi(\boldsymbol{r} - \boldsymbol{R}_{n_2m_2\alpha_2}) \\
&\quad = \gamma_0\delta_{\alpha_1,A}\delta_{\alpha_2,B}\big(\delta_{n_1,n_2}\delta_{m_1,m_2} + \delta_{n_1,n_2}\delta_{m_1,m_2+1} + \delta_{m_1,m_2}\delta_{n_1,n_2+1}\big) \\
&\qquad + \{n_1m_1\alpha_1 \leftrightarrow n_2m_2\alpha_2\}\,.
\end{aligned} \tag{5.9}$$

In this model, the energy reference is set as the onsite energy of each atom, and only the nearest neighbor coupling $\gamma_0 = 2.7$ eV is considered [Castro Neto *et al.* (2009)], as shown in Fig. 5.1 (a).

The Bloch states can be written as

$$\psi_{s\boldsymbol{k}}(\boldsymbol{r}) = c^s_{A\boldsymbol{k}}\Phi_{A\boldsymbol{k}}(\boldsymbol{r}) + c^s_{B\boldsymbol{k}}\Phi_{B\boldsymbol{k}}(\boldsymbol{r})\,, \tag{5.10}$$

where $s = +(-)$ are band indexes for $\pi^*(\pi)$ bands, and $\Phi_{\alpha\boldsymbol{k}}(\boldsymbol{r}) = (2\pi)^{-1}\sum_{nm} e^{i\boldsymbol{k}\cdot\boldsymbol{R}_{nm}}\phi(\boldsymbol{r} - \boldsymbol{R}_{nm} - \boldsymbol{\tau}_\alpha)$ are tight binding basis functions. Using Eq. (5.9) the Schrödinger equation for the $c^s_{\alpha\boldsymbol{k}}$ is

$$\gamma_0\begin{pmatrix} 0 & f_{\boldsymbol{k}} \\ f^*_{\boldsymbol{k}} & 0 \end{pmatrix}\begin{pmatrix} c^s_{A\boldsymbol{k}} \\ c^s_{B\boldsymbol{k}} \end{pmatrix} = \varepsilon_{s\boldsymbol{k}}\begin{pmatrix} c^s_{A\boldsymbol{k}} \\ c^s_{B\boldsymbol{k}} \end{pmatrix}. \tag{5.11}$$

Here $f_{\boldsymbol{k}} = 1 + e^{-i\boldsymbol{k}\cdot\boldsymbol{a}_1} + e^{-i\boldsymbol{k}\cdot\boldsymbol{a}_2}$ is the structure factor. The amplitudes of the eigenfunctions identified by Eq. (5.11) are then found to be

$$\begin{pmatrix} c^s_{A\boldsymbol{k}} \\ c^s_{B\boldsymbol{k}} \end{pmatrix} = \frac{1}{\sqrt{2}}\begin{pmatrix} s\frac{f_{\boldsymbol{k}}}{|f_{\boldsymbol{k}}|} \\ 1 \end{pmatrix}, s = \pm, \tag{5.12}$$

with eigenvalues

$$\varepsilon_{s\boldsymbol{k}} = s\gamma_0|f_{\boldsymbol{k}}|\,. \tag{5.13}$$

Then the Bloch functions $\psi_{s\boldsymbol{k}}(\boldsymbol{r})$ can be rewritten as

$$\psi_{s\boldsymbol{k}}(\boldsymbol{r}) = \frac{1}{2\pi} e^{i\boldsymbol{k}\cdot\boldsymbol{r}} u_{s\boldsymbol{k}}(\boldsymbol{r}) \,,$$
$$u_{s\boldsymbol{k}}(\boldsymbol{r}) = \sum_{nm} e^{i\boldsymbol{k}\cdot(\boldsymbol{R}_{nm}-\boldsymbol{r})} \sum_{\alpha} \phi(\boldsymbol{r} - \boldsymbol{R}_{nm} - \boldsymbol{\tau}_\alpha) c^s_{\alpha\boldsymbol{k}} \,.$$

where $u_{s\boldsymbol{k}}(\boldsymbol{r}) = u_{s\boldsymbol{k}}(\boldsymbol{r} + \boldsymbol{R}_{nm})$ for any integer n, m is the periodic part.

With the application of an external homogeneous electric field $\boldsymbol{E}(t)$, the single particle Hamiltonian under the dipole interaction becomes

$$H = H_0 + H_{eR} \,, \quad H_{eR} = -e\boldsymbol{E}(t) \cdot \boldsymbol{r} \,, \tag{5.14}$$

with the electron charge $e = -|e|$. Because the operator $\boldsymbol{r}$ is not periodic, its matrix elements are not well defined in the Bloch states [Blount (1962)]. However, when $\boldsymbol{r}$ appears only in commutator $[\boldsymbol{r}, A]$ with a periodic operator A, the commutator is periodic. With the matrix elements of A defined as

$$\int d\boldsymbol{r} \psi^*_{s_1\boldsymbol{k}_1}(\boldsymbol{r}) A \psi_{s_2\boldsymbol{k}_2}(\boldsymbol{r}) = \delta(\boldsymbol{k}_1 - \boldsymbol{k}_2) A_{s_1 s_2 \boldsymbol{k}_2} \,,$$

the matrix elements of the commutator are well defined and

$$\int d\boldsymbol{r} \psi^*_{s_1\boldsymbol{k}_1}(\boldsymbol{r}) [\boldsymbol{r}, A] \psi_{s_2\boldsymbol{k}_2}(\boldsymbol{r}) = \delta(\boldsymbol{k}_1 - \boldsymbol{k}_2) \left\{ [\boldsymbol{\xi}_{\boldsymbol{k}_2}, A_{\boldsymbol{k}_2}]_{s_1 s_2} + i \boldsymbol{\nabla}_{\boldsymbol{k}} A_{s_1 s_2 \boldsymbol{k}_2} \right\} . \tag{5.15}$$

The $\boldsymbol{\xi}_{s_1 s_2 \boldsymbol{k}}$ are the Berry connections [Blount (1962); Aversa and Sipe (1995); Sipe and Shkrebtii (2000)],

$$\boldsymbol{\xi}_{s_1 s_2 \boldsymbol{k}} = i \int_{\text{cell}} \frac{d\boldsymbol{r}}{\mathcal{A}_{\text{cell}}} u^*_{s_1\boldsymbol{k}}(\boldsymbol{r}) \boldsymbol{\nabla}_{\boldsymbol{k}} u_{s_2\boldsymbol{k}}(\boldsymbol{r}) \tag{5.16}$$

and are then found to be

$$\boldsymbol{\xi}_{s_1 s_2 \boldsymbol{k}} = \begin{pmatrix} c^{s_1}_{A\boldsymbol{k}} \\ c^{s_1}_{B\boldsymbol{k}} \end{pmatrix}^\dagger \left[\begin{pmatrix} \boldsymbol{\tau}_A & 0 \\ 0 & \boldsymbol{\tau}_B \end{pmatrix} + i \boldsymbol{\nabla}_{\boldsymbol{k}} \right] \begin{pmatrix} c^{s_s}_{A\boldsymbol{k}} \\ c^{s_s}_{B\boldsymbol{k}} \end{pmatrix} . \tag{5.17}$$

Here $\mathcal{A}_{\text{cell}}$ is the area of one unit cell, and we have used an approximation $\int d\boldsymbol{r} \phi(\boldsymbol{r}) \boldsymbol{r} \phi(\boldsymbol{r} - \boldsymbol{R}_{nm\alpha}) = 0$. Substituting the eigenfunctions in Eq. (5.12) we get

$$\boldsymbol{\xi}_{ss\boldsymbol{k}} = \frac{1}{2}(\boldsymbol{\tau} - \boldsymbol{\chi}_{\boldsymbol{k}}) \,, \quad \boldsymbol{\xi}_{s\bar{s}\boldsymbol{k}} = \frac{1}{2}(\boldsymbol{\tau} + \boldsymbol{\chi}_{\boldsymbol{k}}) \,, \tag{5.18}$$

with $\boldsymbol{\chi}_{\boldsymbol{k}} = \text{Im}\,[\boldsymbol{\nabla}_{\boldsymbol{k}} f_{\boldsymbol{k}} / f_{\boldsymbol{k}}]$, and $\bar{s} = -(+)$ for $s = +(-)$.

Taking A in Eq. (5.15) as the Hamiltonian H, the matrix elements of the velocity operator

$$\boldsymbol{v} = \frac{1}{i\hbar}[\boldsymbol{r}, H] = \frac{1}{i\hbar}[\boldsymbol{r}, H_0] \tag{5.19}$$

can be obtained. Its matrix elements are diagonal with respect to $\boldsymbol{k}$ as

$$\int d\boldsymbol{r}\psi^*_{s_1\boldsymbol{k}_1}(\boldsymbol{r})\boldsymbol{v}\psi_{s_2\boldsymbol{k}_2}(\boldsymbol{r}) = \boldsymbol{v}_{s_1s_2\boldsymbol{k}_1}\delta(\boldsymbol{k}_1-\boldsymbol{k}_2)\,, \tag{5.20}$$

with

$$\boldsymbol{v}_{s_1s_2\boldsymbol{k}} = i\omega_{s_1s_2\boldsymbol{k}}\boldsymbol{\xi}_{s_1s_2\boldsymbol{k}} + \frac{1}{\hbar}\boldsymbol{\nabla}_{\boldsymbol{k}}\varepsilon_{s_1\boldsymbol{k}}\delta_{s_1s_2}\,, \tag{5.21}$$

and $\hbar\omega_{s_1s_2\boldsymbol{k}} = \varepsilon_{s_1\boldsymbol{k}} - \varepsilon_{s_2\boldsymbol{k}}$.

Approximations around the Dirac points

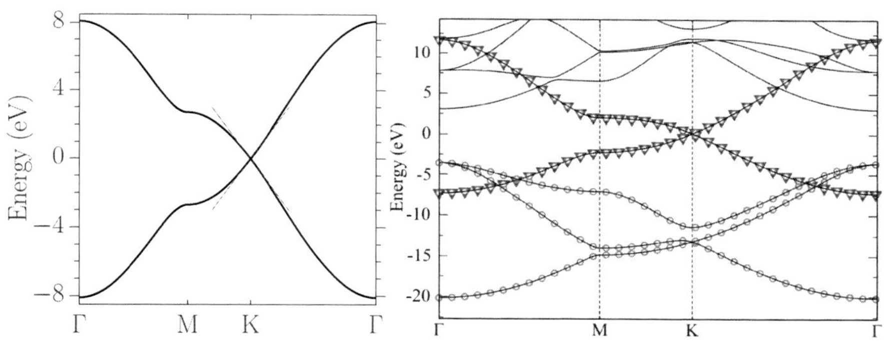

Fig. 5.2 Band structure of graphene. (left) Bands generated from Eq. (5.13) (Thick black curves) and from Eq. (5.23) (thin blue lines); (right) DFT bands (solid curves), with triangle symbols indicate the bands from carbon $2p_z$ orbitals. The right diagram is copied from [Marzari *et al.* (2012)].

From the two band tight binding model given above, these two bands intersect at the Dirac points $\mathbf{K}$ and $\mathbf{K}'$ in the Brillouin zone, (see Fig. 5.1 (b)),

$$\boldsymbol{K} = \frac{2\pi}{a_0}\left(\frac{1}{\sqrt{3}}\hat{\boldsymbol{x}} + \frac{1}{3}\hat{\boldsymbol{y}}\right)\,, \quad \boldsymbol{K}' = \frac{2\pi}{a_0}\left(\frac{1}{\sqrt{3}}\hat{\boldsymbol{x}} - \frac{1}{3}\hat{\boldsymbol{y}}\right)\,. \tag{5.22}$$

Both points satisfy $f_{\boldsymbol{K}} = f_{\boldsymbol{K}'} = 0$. For undoped or lightly doped graphene, the Fermi energy is close to the Dirac points, and the most effective optical transitions are around the Dirac points. It is usual to approximate the electronic dispersion relation as linear, *i.e.*,

$$\begin{aligned}
\varepsilon_{s\boldsymbol{K}+\boldsymbol{k}} &\approx s\hbar v_F k\,, & \varepsilon_{s\boldsymbol{K}'+\boldsymbol{k}} &\approx s\hbar v_F k\,,\\
\boldsymbol{v}_{ss\boldsymbol{K}+\boldsymbol{k}} &\approx s v_F\hat{\boldsymbol{k}}\,, & \boldsymbol{v}_{ss\boldsymbol{K}'+\boldsymbol{k}} &\approx s v_F\hat{\boldsymbol{k}}\,,\\
\boldsymbol{v}_{s\bar{s}\boldsymbol{K}+\boldsymbol{k}} &\approx is v_F\hat{\boldsymbol{k}}\times\hat{\boldsymbol{z}}\,, & \boldsymbol{v}_{s\bar{s}\boldsymbol{K}'+\boldsymbol{k}} &\approx -is v_F\hat{\boldsymbol{z}}\times\hat{\boldsymbol{k}}\,,\\
\boldsymbol{\xi}_{s\bar{s}\boldsymbol{K}+\boldsymbol{k}} &\approx \tfrac{1}{2k}\hat{\boldsymbol{k}}\times\hat{\boldsymbol{z}}\,, & \boldsymbol{\xi}_{s\bar{s}\boldsymbol{K}'+\boldsymbol{k}} &\approx -\tfrac{1}{2k}\hat{\boldsymbol{k}}\times\hat{\boldsymbol{z}}\,,
\end{aligned} \tag{5.23}$$

with the Fermi velocity $v_F = \sqrt{3}a_0\gamma_0/(2\hbar)$ and the direction of $\hat{\boldsymbol{k}} = \boldsymbol{k}/k$. These approximations follow immediately from Eqs. (5.13), (5.21), and (5.16). The band structure from Eq. (5.13) and the linear dispersion approximation from Eq. (5.23) are illustrated in the left diagram of Fig. 5.2. Compared to a DFT band structure (right diagram of Fig. 5.2), these two bands are isolated from other bands approximately within the energy range $[-3, 3]$ eV. For energies beyond this range, the excitation corresponds to promotion of electrons from the sp^2 orbitals, which is not discussed in this chapter.

5.3 Optical Response to a Uniform Electric Field: Semiconductor Bloch Equation Approach

In the independent particle approximation, the system can be described by the single particle density matrix operator $\hat{\rho}(t)$. In our two band model, it has matrix elements

$$\int d\boldsymbol{r}\psi^*_{s_1\boldsymbol{k}}(\boldsymbol{r})\hat{\rho}(t)\psi_{s_2\boldsymbol{k}}(\boldsymbol{r}) = \rho_{s_1s_2\boldsymbol{k}}(t)\delta(\boldsymbol{k}_1 - \boldsymbol{k}_2)\,. \tag{5.24}$$

The diagonal terms $\rho_{ss\boldsymbol{k}}$ describe the population of band s at wave vector $\boldsymbol{k}$, and the off-diagonal terms $\rho_{s\bar{s}\boldsymbol{k}}$ describe the optical polarization. The areal current density is then calculated by

$$\boldsymbol{J} = 2e\int \frac{d\boldsymbol{k}}{(2\pi)^2}\mathrm{Tr}[\boldsymbol{v}_{\boldsymbol{k}}\rho_{\boldsymbol{k}}(t)] \tag{5.25}$$

where the integral is over one Brillouin zone, $\boldsymbol{v}_{\boldsymbol{k}}$ denotes the matrix with elements $\boldsymbol{v}_{s_1s_2\boldsymbol{k}}$, $\rho_{\boldsymbol{k}}(t)$ denotes the matrix with elements $\rho_{s_1s_2\boldsymbol{k}}(t)$, and the prefactor 2 is from the spin degeneracy.

When there is no electric field, the system is governed by the single particle Hamiltonian H_0 and stays at its thermal equilibrium states $\rho^{(0)}_{s_1s_2\boldsymbol{k}} = n_{s_1\boldsymbol{k}}\delta_{s_1s_2}$. Here $n_{s\boldsymbol{k}} = [1+e^{(\varepsilon_{s\boldsymbol{k}}-\mu)/(k_BT)}]^{-1}$ is the Fermi-Dirac distribution at the chemical potential μ and the temperature T, and k_B is Boltzmann's constant. Assuming that the electric field is turned on adiabatically, the evolution of the density matrix follows the semiconductor Bloch equations

$$\hbar\frac{\partial\rho_{\boldsymbol{k}}(t)}{\partial t} = -i[\mathcal{E}_{\boldsymbol{k}} - e\boldsymbol{E}(t)\cdot\boldsymbol{\xi}_{\boldsymbol{k}}, \rho_{\boldsymbol{k}}(t)] - e\boldsymbol{E}(t)\cdot\boldsymbol{\nabla}_{\boldsymbol{k}}\rho_{\boldsymbol{k}} + \hbar\frac{\partial\rho_{\boldsymbol{k}}(t)}{\partial t}\bigg|_{scat}\,. \tag{5.26}$$

Here $\mathcal{E}_{\boldsymbol{k}}$ denotes the matrix with elements $\mathcal{E}_{s_1s_2\boldsymbol{k}} = \delta_{s_1s_2}\varepsilon_{s_1\boldsymbol{k}}$ and $\boldsymbol{\xi}_{\boldsymbol{k}}$ denotes the matrix with elements $\boldsymbol{\xi}_{s_1s_2\boldsymbol{k}}$. Because of $\boldsymbol{\xi}_{++\boldsymbol{k}} = \boldsymbol{\xi}_{--\boldsymbol{k}}$, the diagonal terms of $\boldsymbol{\xi}_{\boldsymbol{k}}$ do not have an effect on the commutator of Eq. (5.26). Then following Aversa and Sipe [Aversa and Sipe (1995)], the interband transitions are given by terms involving $\boldsymbol{r_k} = \boldsymbol{\xi}_{+-\boldsymbol{k}} = \boldsymbol{\xi}_{-+\boldsymbol{k}}$, and the intraband transitions given by those involving the derivatives. The interband and intraband transitions are illustrated in Fig. 5.3

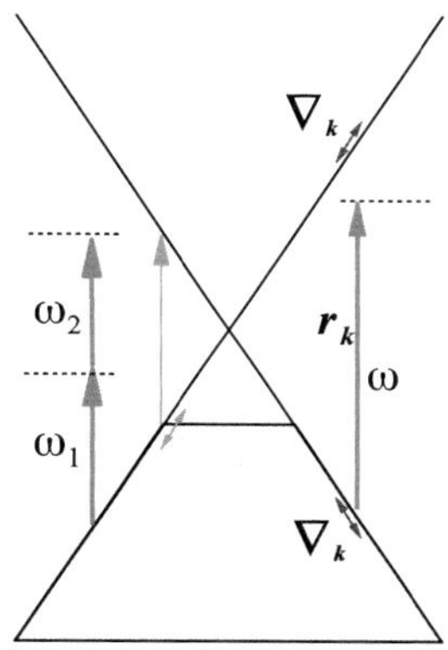

Fig. 5.3 Illustration of interband and intraband transitions in graphene. The vertical arrows indicate the interband transition, with edges connecting two bands. The oblique double arrows indicate the intraband transition. The dashed lines indicate virtual states. The gray arrows show the resonant transition occurring at the Fermi surface.

The last term in Eq. (5.26) describes the effect of scattering. In a real system, the relaxation can be induced by many aspects, such as the carrier-impurity interaction, the carrier-carrier interaction, and the carrier-phonon interaction. To strictly include all these interactions, a complicated theory [Malic *et al.* (2011)] has to be adopted, which is difficult to solve even numerically. A widely used approximation for the scattering term is the relaxation time approximation, with different relaxation times to characterize each transition process. These relaxation times can depend on the electron state energy, momentum, band index, doping level, and the external fields. However, at optical frequencies the main effect of the relaxation processes on the optical properties is to ameliorate the divergences associated with physical resonances [Zhang and Voss (2011); Cheng *et al.* (2014a); Mikhailov (2014); Cheng *et al.* (2015b); Mikhailov (2016)], and so the relaxation processes can be modeled by phenomenological constants, which in some cases can be extracted from experiments. In such a way we model the intraband (interband) relaxation process by a parameter

$\Gamma_{ss} = \Gamma_i$ ($\Gamma_{s\bar{s}} = \Gamma_e$), and then

$$\hbar \frac{\partial \rho_{s_1 s_2 \boldsymbol{k}}(t)}{\partial t}\bigg|_{\rm scat} = -\Gamma_{s_1 s_2}\left[\rho_{s_1 s_2 \boldsymbol{k}}(t) - \rho^0_{s_1 s_2 \boldsymbol{k}}\right]. \tag{5.27}$$

Equations (5.26) and (5.27) can be solved by different methods, such as the perturbation method, the rotating wave approximation, a moving frame strategy [Cheng *et al.* (2014b)], and also numerical simulation [Avetissian *et al.* (2013a,b); Cheng *et al.* (2015a); Semnani *et al.* (2016); Margulis *et al.* (2016)]. In this chapter we focus on the perturbation solution and the numerical simulation.

Other approaches

Besides the independent particle approach where the scattering is described phenomenologically, there are other approaches to consider the optical response to a uniform field: (1) The velocity gauge approach [Malic *et al.* (2011); Zhang and Voss (2011); Margulis *et al.* (2014, 2016)]. Here the light matter interaction is introduced by using the vector potential interaction $\boldsymbol{p}\cdot\boldsymbol{A}$ instead of the strategy used here. Strictly speaking, it is connected to our approach here by a gauge transformation of the electric field. However, it has been shown that this approach can lead to many apparent divergences, especially in the limit of zero frequency, which can only be eliminated by a careful use of various sum rules [Sipe and Shkrebtii (2000)]. The semiconductor equation derived from this interaction is basically equivalent to the solution of Eq. (5.26) in a moving frame. Additional attention is required for this equivalence if finite relaxation parameters are included. (2) The uniform limit of a non-uniform field response [Mikhailov (2014, 2016)]. The response is with respect to a spatially varying field $\boldsymbol{E}_\omega e^{-i\omega t} e^{i\boldsymbol{q}\cdot\boldsymbol{r}} + c.c$, and taking the limit $\boldsymbol{q} \to 0$. It can be shown that this approach is exactly the same as our approach when this limit is taken. (3) Different phenomenological relaxation processes [Al-Naib *et al.* (2014, 2015)]. Under strong Terahertz excitation, the electrons can be significantly thermalized, and the relaxation may bring the system to a quasi-equilibrium state instead of a thermal equilibrium state.

5.4 Perturbative Solution of SBEs

For a weak electric field $\boldsymbol{E}(t)$, a perturbative solution of Eq. (5.26) can be obtained up to the third order of the electric field as

$$\rho_{s_1 s_2 \boldsymbol{k}}(t) = \rho^{(0)}_{\boldsymbol{k}} + \rho^{(1)}_{s_1 s_2 \boldsymbol{k}}(t) + \rho^{(2)}_{s_1 s_2 \boldsymbol{k}}(t) + \rho^{(3)}_{s_1 s_2 \boldsymbol{k}}(t) + \cdots, \tag{5.28}$$

where $\rho_{\boldsymbol{k}}^{(j)}$ satisfies

$$\hbar\frac{\partial\rho_{\boldsymbol{k}}^{(j)}(t)}{\partial t} = -i[\mathcal{E}_{\boldsymbol{k}},\rho_{\boldsymbol{k}}^{(j)}(t)] - \Gamma_{s_1s_2}^{(j)}\rho_{s_1s_2\boldsymbol{k}}^{(j)}(t) + ie\boldsymbol{E}(t)\cdot[\boldsymbol{\xi}_{\boldsymbol{k}},\rho_{\boldsymbol{k}}^{(j-1)}] - e\boldsymbol{E}(t)\cdot\boldsymbol{\nabla}_{\boldsymbol{k}}\rho_{\boldsymbol{k}}^{(j-1)}. \quad (5.29)$$

Here the relaxation parameters are further specified for each perturbation order by $\Gamma_{ss}^{(j)} = \Gamma_i^{(j)}$ and $\Gamma_{s\bar{s}}^{(j)} = \Gamma_e^{(j)}$. Furthermore, $\rho_{\boldsymbol{k}}^{(j)}(t)$ for $j = 1, 2, 3$ are solved in a form

$$\begin{aligned}
\rho_{s_1s_2\boldsymbol{k}}^{(1)}(t) &= \int\frac{d\omega_3}{2\pi}(-e)E_{\omega_3}^{c}e^{-i\omega_3 t}\mathcal{P}_{s_1s_2\boldsymbol{k}}^{(1);c}(\omega_3)\\
\rho_{s_1s_2\boldsymbol{k}}^{(2)}(t) &= \int\frac{d\omega_2 d\omega_3}{(2\pi)^2}(-e)^2E_{\omega_2}^{b}E_{\omega_3}^{c}e^{-i(\omega_2+\omega_3)t}\mathcal{P}_{s_1s_2\boldsymbol{k}}^{(2);bc}(\omega_2,\omega_3)\\
\rho_{s_1s_2\boldsymbol{k}}^{(3)}(t) &= \int\frac{d\omega_1 d\omega_2 d\omega_3}{(2\pi)^3}(-e)^3E_{\omega_1}^{a}E_{\omega_2}^{b}E_{\omega_3}^{c}e^{-i(\omega_1+\omega_2+\omega_3)t}\\
&\quad\times\mathcal{P}_{s_1s_2\boldsymbol{k}}^{(3);abc}(\omega_1,\omega_2,\omega_3),
\end{aligned}$$

with

$$\mathcal{P}_{s_1s_2\boldsymbol{k}}^{(1);c}(\omega_3) = \frac{[\xi_{\boldsymbol{k}}^{c},\rho_{\boldsymbol{k}}^{(0)}]_{s_1s_2} + i\frac{\partial}{\partial k_c}\rho_{s_1s_2\boldsymbol{k}}^{(0)}}{\hbar\omega_3 - \hbar\omega_{s_1s_2\boldsymbol{k}} + i\Gamma_{s_1s_2}^{(1)}}, \quad (5.30)$$

$$\mathcal{P}_{s_1s_2\boldsymbol{k}}^{(2);bc}(\omega_2,\omega_3) = \frac{[\xi_{\boldsymbol{k}}^{b},\mathcal{P}_{\boldsymbol{k}}^{(1);c}(\omega_3)]_{s_1s_2} + i\frac{\partial}{\partial k_b}\mathcal{P}_{s_1s_2\boldsymbol{k}}^{(1);c}(\omega_3)}{\hbar\omega_2 + \hbar\omega_3 - \hbar\omega_{s_1s_2\boldsymbol{k}} + i\Gamma_{s_1s_2}^{(2)}}, \quad (5.31)$$

$$\mathcal{P}_{s_1s_2\boldsymbol{k}}^{(3);abc}(\omega_1,\omega_2,\omega_3) = \frac{[\xi_{\boldsymbol{k}}^{a},\mathcal{P}_{\boldsymbol{k}}^{(2);bc}(\omega_2,\omega_3)]_{s_1s_2} + i\frac{\partial}{\partial k_b}\mathcal{P}_{s_1s_2\boldsymbol{k}}^{(2);c}(\omega_2,\omega_3)}{\hbar\omega_1 + \hbar\omega_2 + \hbar\omega_3 - \hbar\omega_{s_1s_2\boldsymbol{k}} + i\Gamma_{s_1s_2}^{(3)}}. \quad (5.32)$$

The first order terms are

$$\mathcal{P}_{ss\boldsymbol{k}}^{(1);c}(\omega_3) = \frac{i}{\nu_3}\frac{\partial n_{s\boldsymbol{k}}}{\partial k_c}, \quad \mathcal{P}_{s\bar{s}\boldsymbol{k}}^{(1);c}(\omega_3) = \frac{-sr_{\boldsymbol{k}}^{c}\Delta n_{\boldsymbol{k}}}{\vartheta_3 - s\epsilon_{\boldsymbol{k}}}. \quad (5.33)$$

Here $\nu_3 \equiv \hbar\omega_3 + i\Gamma_i^{(1)}$, $\vartheta_3 \equiv \hbar\omega_3 + i\Gamma_e^{(1)}$, $\Delta n_{\boldsymbol{k}} = n_{+\boldsymbol{k}} - n_{-\boldsymbol{k}}$ and $\epsilon_{\boldsymbol{k}} = \varepsilon_{+\boldsymbol{k}} - \varepsilon_{-\boldsymbol{k}}$. The second order terms are

$$\begin{aligned}
\mathcal{P}_{ss\boldsymbol{k}}^{(2);bc}(\omega_2,\omega_3) &= \frac{1}{\nu_0}\left[-\frac{1}{\nu_3}\frac{\partial^2 n_{s\boldsymbol{k}}}{\partial k_b\partial k_c} + sr_{\boldsymbol{k}}^{b}r_{\boldsymbol{k}}^{c}\Delta n_{\boldsymbol{k}}\left(\frac{1}{\vartheta_3+\epsilon_{\boldsymbol{k}}} + \frac{1}{\vartheta_3-\epsilon_{\boldsymbol{k}}}\right)\right],\\
\mathcal{P}_{s\bar{s}\boldsymbol{k}}^{(2);bc}(\omega_2,\omega_3) &= \frac{-is}{\vartheta_0 - s\epsilon_{\boldsymbol{k}}}\left[\frac{\partial}{\partial k_b}\left(\frac{r_{\boldsymbol{k}}^{c}\Delta n_{\boldsymbol{k}}}{\vartheta_3 - s\epsilon_{\boldsymbol{k}}}\right) + \frac{r_{\boldsymbol{k}}^{b}}{\nu_3}\frac{\partial\Delta n_{\boldsymbol{k}}}{\partial k_c}\right].
\end{aligned} \quad (5.34)$$

Here $\nu_0 \equiv \hbar\omega_0 + i\Gamma_i^{(2)}$ and $\vartheta_0 \equiv \hbar\omega_0 + i\Gamma_e^{(2)}$ with $\omega_0 = \omega_2 + \omega_3$. The third order terms are

$$\begin{aligned}
\mathcal{P}_{ss\boldsymbol{k}}^{(3);abc}(\omega_1,\omega_2,\omega_3) &= \frac{1}{\nu\nu_0\nu_3}P_{1;s\boldsymbol{k}}^{abc} + \frac{1}{\nu\nu_0}P_{2;s\boldsymbol{k}}^{abc}(\vartheta_3) \\
&\quad + \frac{1}{\nu\nu_3}P_{3;s\boldsymbol{k}}^{abc}(\vartheta_0) + \frac{1}{\nu}P_{4;s\boldsymbol{k}}^{abc}(\vartheta_0,\vartheta_3)\,, \\
\mathcal{P}_{s\bar{s}\boldsymbol{k}}^{(3);abc}(\omega_1,\omega_2,\omega_3) &= \frac{1}{\nu_0\nu_3}P_{5;s\boldsymbol{k}}^{abc}(\vartheta) + \frac{1}{\nu_0}P_{6;s\boldsymbol{k}}^{abc}(\vartheta,\vartheta_3) \\
&\quad + \frac{1}{\nu_3}P_{7;s\boldsymbol{k}}^{abc}(\vartheta,\vartheta_0) + P_{8;s\boldsymbol{k}}^{abc}(\vartheta,\vartheta_0,\vartheta_3)\,, \qquad (5.35)
\end{aligned}$$

Here $\nu \equiv \hbar\omega + i\Gamma_i^{(3)}$ and $\vartheta \equiv \hbar\omega + i\Gamma_e^{(3)}$ with $\omega \equiv \omega_1 + \omega_2 + \omega_3$. The terms $P_{i;s\boldsymbol{k}}^{abc}$ with $i = 1, \cdots, 4$ are related to the populations at band s and are given by

$$\begin{aligned}
P_{1;s\boldsymbol{k}}^{abc} &= -i\frac{\partial^3 n_{s\boldsymbol{k}}}{\partial k_a \partial k_b \partial k_c}\,, \\
P_{2;s\boldsymbol{k}}^{abc}(\vartheta_3) &= is\frac{\partial}{\partial k_a}\left[r_{\boldsymbol{k}}^b r_{\boldsymbol{k}}^c \Delta n_{\boldsymbol{k}}\left(\frac{1}{\vartheta_3+\epsilon_{\boldsymbol{k}}} + \frac{1}{\vartheta_3-\epsilon_{\boldsymbol{k}}}\right)\right], \\
P_{3;s\boldsymbol{k}}^{abc}(\vartheta_0) &= isr_{\boldsymbol{k}}^a r_{\boldsymbol{k}}^b\left(\frac{1}{\vartheta_0+\epsilon_{\boldsymbol{k}}} + \frac{1}{\vartheta_0-\epsilon_{\boldsymbol{k}}}\right)\frac{\partial \Delta n_{\boldsymbol{k}}}{\partial k_c}\,, \\
P_{4;s\boldsymbol{k}}^{abc}(\vartheta_0,\vartheta_3) &= isr_{\boldsymbol{k}}^a\left[\frac{1}{\vartheta_0+\epsilon_{\boldsymbol{k}}}\frac{\partial}{\partial k_b}\left(\frac{r_{\boldsymbol{k}}^c\Delta n_{\boldsymbol{k}}}{\vartheta_3+\epsilon_{\boldsymbol{k}}}\right) + \frac{1}{\vartheta_0-\epsilon_{\boldsymbol{k}}}\frac{\partial}{\partial k_b}\left(\frac{r_{\boldsymbol{k}}^c\Delta n_{\boldsymbol{k}}}{\vartheta_3-\epsilon_{\boldsymbol{k}}}\right)\right],
\end{aligned}$$

while the terms $P_{i;s\boldsymbol{k}}^{abc}$ with $i = 5, \cdots, 8$ are related to the interband polarization and are given by

$$\begin{aligned}
P_{5;s\boldsymbol{k}}^{abc}(\vartheta) &= s\frac{r_{\boldsymbol{k}}^a}{\vartheta - s\epsilon_{\boldsymbol{k}}}\frac{\partial^2 \Delta n_{\boldsymbol{k}}}{\partial k_b \partial k_c}\,, \\
P_{6;s\boldsymbol{k}}^{abc}(\vartheta,\vartheta_3) &= -2s\frac{r_{\boldsymbol{k}}^a r_{\boldsymbol{k}}^b r_{\boldsymbol{k}}^c \Delta n_{\boldsymbol{k}}}{\vartheta - s\epsilon_{\boldsymbol{k}}}\left(\frac{1}{\vartheta_3+\epsilon_{\boldsymbol{k}}} + \frac{1}{\vartheta_3-\epsilon_{\boldsymbol{k}}}\right), \\
P_{7;s\boldsymbol{k}}^{abc}(\vartheta,\vartheta_0) &= \frac{s}{\vartheta - s\epsilon_{\boldsymbol{k}}}\frac{\partial}{\partial k_a}\left(\frac{r_{\boldsymbol{k}}^b}{\vartheta_0 - s\epsilon_{\boldsymbol{k}}}\frac{\partial \Delta n_{\boldsymbol{k}}}{\partial k_c}\right), \\
P_{8;s\boldsymbol{k}}^{abc}(\vartheta,\vartheta_0,\vartheta_3) &= \frac{s}{\vartheta - s\epsilon_{\boldsymbol{k}}}\frac{\partial}{\partial k_a}\left[\frac{1}{\vartheta_0 - s\epsilon_{\boldsymbol{k}}}\frac{\partial}{\partial k_b}\left(\frac{r_{\boldsymbol{k}}^c\Delta n_{\boldsymbol{k}}}{\vartheta_3 - s\epsilon_{\boldsymbol{k}}}\right)\right].
\end{aligned}$$

In all these expressions, each $\boldsymbol{r}_{\boldsymbol{k}}$ stands for an interband transition process, and each derivative stands for an intraband transition process.

5.4.1 *Third order conductivity around the Dirac points*

With the perturbative solution of $\mathcal{P}_{\boldsymbol{k}}^{(3);abc}(\omega_1,\omega_2,\omega_3)$, the third order current can be calculated using Eq. (5.25), as

$$J^{(3);d}(t) = 2e\int \frac{d\boldsymbol{k}}{(2\pi)^2}\sum_{s_1s_2} v^d_{s_2s_1\boldsymbol{k}}\rho^{(3)}_{s_1s_2\boldsymbol{k}}(t)\,. \tag{5.36}$$

Generally, the perturbative third order current is also given as

$$J^{(3);d}(t) = \int \frac{d\omega_1 d\omega_2 d\omega_3}{(2\pi)^3} e^{-i\omega t}\widetilde{\sigma}^{(3);dabc}(\omega_1,\omega_2,\omega_3)E^a_{\omega_1}E^b_{\omega_2}E^c_{\omega_3}\,, \tag{5.37}$$

where $\omega = \omega_1+\omega_2+\omega_3$ and $\widetilde{\sigma}^{(3);dabc}(\omega_1,\omega_2,\omega_3)$ is the third order conductivity, which is not yet symmetrized with respect to the electric field components, and is given as

$$\begin{aligned}
&\widetilde{\sigma}^{(3);dabc}(\omega_1,\omega_2,\omega_3)\\
&= i\sigma_3\left[\frac{\mathcal{S}_1^{dabc}}{\nu\nu_0\nu_3} + \frac{\mathcal{S}_2^{dabc}(\vartheta_3)}{\nu\nu_0} + \frac{\mathcal{S}_3^{dabc}(\vartheta_0)}{\nu\nu_3} + \frac{\mathcal{S}_4^{dabc}(\vartheta_0,\vartheta_3)}{\nu} + \frac{\mathcal{S}_5^{dabc}(\vartheta)}{\nu_0\nu_3}\right.\\
&\qquad\left. + \frac{\mathcal{S}_6^{dabc}(\vartheta,\vartheta_3)}{\nu_0} + \frac{\mathcal{S}_7^{dabc}(\vartheta,\vartheta_0)}{\nu_3} + \mathcal{S}_8^{dabc}(\vartheta,\vartheta_0,\vartheta_3)\right],
\end{aligned} \tag{5.38}$$

Here $\mathcal{S}_i^{dabc}$ are the contributions from $P_{i;\boldsymbol{k}}^{(3);abc}$. The symmetrization of the terms $\widetilde{\sigma}^{(3);ijkl}(\omega_j,\omega_k,\omega_l)$ can be performed following the standard convention of nonlinear optics [Boyd (2008)] by permuting the indices (jkl) to arrive at the nonlinear conductivity $\sigma^{(3);dabc}(\omega_1,\omega_2,\omega_3)$, as

$$\begin{aligned}
&\sigma^{(3);dabc}(\omega_1,\omega_2,\omega_3)\\
&= \frac{1}{6}\left[\widetilde{\sigma}^{(3);dabc}(\omega_1,\omega_2,\omega_3) + \widetilde{\sigma}^{(3);dbca}(\omega_2,\omega_3,\omega_1) + \widetilde{\sigma}^{(3);dcab}(\omega_3,\omega_1,\omega_2)\right.\\
&\quad\left. + \widetilde{\sigma}^{(3);dacb}(\omega_1,\omega_3,\omega_2) + \widetilde{\sigma}^{(3);dcba}(\omega_3,\omega_2,\omega_1) + \widetilde{\sigma}^{(3);dbac}(\omega_2,\omega_1,\omega_3)\right].
\end{aligned}$$

The quantities $\mathcal{S}_i^{dabc}$, $\widetilde{\sigma}^{(3);dabc}$, and $\sigma^{(3);dabc}$ are all fourth order tensors. Neglecting the optical response in the z direction, there are in all 8 nonzero components for the D_{6h} symmetry of graphene, among which three are independent; they are

$$\begin{aligned}
&\sigma^{(3);xxyy} = \sigma^{(3);yyxx}, \quad \sigma^{(3);xyxy} = \sigma^{(3);yxyx}, \quad \sigma^{(3);xyyx} = \sigma^{(3);yxxy},\\
&\sigma^{(3);xxxx} = \sigma^{(3);yyyy} = \sigma^{(3);xxyy} + \sigma^{(3);xyxy} + \sigma^{(3);xyyx}.
\end{aligned}$$

Using the nonzero independent components, the third order current in Eq. (5.37) can be written as

$$\boldsymbol{J}^{(3)}(t) = 3\int \frac{d\omega_1 d\omega_2 d\omega_3}{(2\pi)^3} e^{-i\omega t}\sigma^{(3);xxyy}(\omega_1,\omega_2,\omega_3)\boldsymbol{E}_{\omega_1}\boldsymbol{E}_{\omega_2}\cdot\boldsymbol{E}_{\omega_3}\,. \tag{5.39}$$

In the following, we write the independent nonzero components of fourth rank tensors as column vectors, ordering the independent components of a fourth rank tensor T^{dabc} as $T = \begin{bmatrix} T^{xxyy} \\ T^{xyxy} \\ T^{xyyx} \end{bmatrix}$.

For the optical response to visible or infrared light, the dominant optical transitions are only around the Dirac points, and all quantities can be well approximated by those in Eq. (5.23). Then the integration in Eq. (5.36) can be worked out analytically, with the use of partial fractions. By employing the fixed vectors

$$A_1 = \begin{bmatrix} -3 \\ 1 \\ 1 \end{bmatrix}, A_2 = \begin{bmatrix} 1 \\ -3 \\ 1 \end{bmatrix}, A_3 = \begin{bmatrix} 1 \\ 1 \\ -3 \end{bmatrix}, A_0 = \begin{bmatrix} 1 \\ 1 \\ 1 \end{bmatrix}, \tag{5.40}$$

where note $A_0 = -(A_1 + A_2 + A_3)$, we can present the analytic expressions for the different components of $\widetilde{\sigma}^{(3);dabc}$ appearing in Eq. (5.38) at zero temperature, as

$$\mathcal{S}_1 = \frac{1}{|\mu|} A_0 \,, \tag{5.41}$$

$$\mathcal{S}_2(\vartheta_3) = \mathcal{G}_\mu(\vartheta_3) \frac{A_0}{\vartheta_3^2} - \frac{1}{|\mu|} \frac{A_0}{\vartheta_3} \,, \tag{5.42}$$

$$\mathcal{S}_3(\vartheta_0) = \mathcal{H}_\mu(\vartheta_0) \frac{A_3}{\vartheta_0} - \frac{1}{|\mu|} \frac{A_3}{\vartheta_0} \,, \tag{5.43}$$

$$\begin{aligned} \mathcal{S}_4(\vartheta_0, \vartheta_3) = &-\mathcal{G}_\mu(\vartheta_3) \frac{\vartheta_3 A_2 + \vartheta_2 A_3}{\vartheta_2^2 \vartheta_3^2} + \mathcal{G}_\mu(\vartheta_0) \frac{(\vartheta_0 + \vartheta_2) A_2 + \vartheta_2 A_3}{\vartheta_0^2 \vartheta_2^2} \\ &-\mathcal{H}_\mu(\vartheta_0) \frac{A_2}{\vartheta_0 \vartheta_2} + \frac{1}{|\mu|} \frac{A_3}{\vartheta_0 \vartheta_3} \,, \end{aligned} \tag{5.44}$$

$$\mathcal{S}_5(\vartheta) = \mathcal{H}_\mu(\vartheta) \frac{A_0}{\vartheta} + \mathcal{I}_\mu(\vartheta) A_1 - \frac{1}{|\mu|} \frac{A_0}{\vartheta} \,, \tag{5.45}$$

$$\mathcal{S}_6(\vartheta, \vartheta_3) = -\mathcal{G}_\mu(\vartheta_3) \frac{\vartheta A_0}{\vartheta_3^2(\vartheta^2 - \vartheta_3^2)} + \mathcal{G}_\mu(\vartheta) \frac{\vartheta_3 A_0}{\vartheta^2(\vartheta^2 - \vartheta_3^2)} + \frac{1}{|\mu|} \frac{A_0}{\vartheta \vartheta_3} \,, \tag{5.46}$$

$$\begin{aligned} \mathcal{S}_7(\vartheta, \vartheta_0) = &\mathcal{H}_\mu(\vartheta_0) \left(\frac{A_2}{\vartheta_1^2} - \frac{A_3}{\vartheta_0 \vartheta_1} \right) + \mathcal{H}_\mu(\vartheta) \left(\frac{A_3}{\vartheta \vartheta_1} - \frac{A_2}{\vartheta_1^2} \right) \\ &-\mathcal{I}_\mu(\vartheta) \frac{A_2}{\vartheta_1} + \frac{1}{|\mu|} \frac{A_3}{\vartheta \vartheta_0} \,, \end{aligned} \tag{5.47}$$

and

$$\begin{aligned}
&\mathcal{S}_8(\vartheta,\vartheta_0,\vartheta_3)\\
&= \mathcal{G}_\mu(\vartheta_3)\left[\frac{A_2}{(\vartheta-\vartheta_3)\vartheta_2^2\vartheta_3}+\frac{\vartheta^2\vartheta_2+\vartheta_3^3+\vartheta\vartheta_3(-3\vartheta_0+2\vartheta_3)}{(\vartheta-\vartheta_3)^3\vartheta_2^2\vartheta_3^2}A_3\right]\\
&+\mathcal{G}_\mu(\vartheta_0)\left[-\frac{\vartheta_0\vartheta_1+\vartheta_1\vartheta_2-\vartheta_0\vartheta_2}{\vartheta_0^2\vartheta_1^2\vartheta_2^2}A_2-\frac{\vartheta_1\vartheta_2-\vartheta_0^2-\vartheta_0\vartheta_2}{\vartheta_1^2\vartheta_0^2\vartheta_2^2}A_3\right]\\
&+\mathcal{G}_\mu(\vartheta)\left[-\frac{1}{\vartheta\vartheta_1^2(\vartheta-\vartheta_3)}A_2-\frac{5\vartheta^2+\vartheta_3(\vartheta_0+\vartheta_3)-\vartheta(3\vartheta_0+4\vartheta_3)}{\vartheta\vartheta_1^2(\vartheta-\vartheta_3)^3}A_3\right]\\
&+\mathcal{H}_\mu(\vartheta_0)\left(\frac{A_2}{\vartheta_0\vartheta_1\vartheta_2}-\frac{A_3}{\vartheta_1^2\vartheta_2}\right)+\mathcal{H}_\mu(\vartheta)\frac{4\vartheta^2-3\vartheta\vartheta_0-2\vartheta\vartheta_3+\vartheta_0\vartheta_3}{\vartheta\vartheta_1^2(\vartheta-\vartheta_3)^2}A_3\\
&+\mathcal{I}_\mu(\vartheta)\frac{A_3}{\vartheta_1(\vartheta-\vartheta_3)}-\frac{1}{|\mu|}\frac{A_3}{\vartheta\vartheta_0\vartheta_3}. \qquad (5.48)
\end{aligned}$$

where $\vartheta_2=\vartheta_0-\vartheta_3$, $\vartheta_1=\vartheta-\vartheta_0$, and

$$\mathcal{G}_\mu(\vartheta)=\ln\left|\frac{2|\mu|+\vartheta}{2|\mu|-\vartheta}\right|+i\left(\pi+\arctan\frac{\vartheta_r-2|\mu|}{\vartheta_i}-\arctan\frac{\vartheta_r+2|\mu|}{\vartheta_i}\right), \qquad (5.49)$$

$$\mathcal{H}_\mu(\vartheta)=\frac{1}{2|\mu|-\vartheta}+\frac{1}{2|\mu|+\vartheta}, \qquad (5.50)$$

$$\mathcal{I}_\mu(\vartheta)=\frac{1}{(2|\mu|+\vartheta)^2}-\frac{1}{(2|\mu|-\vartheta)^2}, \qquad (5.51)$$

where $\vartheta=\vartheta_r+i\vartheta_i$ is a complex argument with ϑ_r and ϑ_i being real numbers.

The expressions in Eqs. (5.41-5.48) are written into terms of $\mathcal{G}_\mu$, $\mathcal{H}_\mu$, $\mathcal{I}_\mu$, and $|\mu|^{-1}$. Then the unsymmetrized conductivity can be reorganized as

$$\begin{aligned}\widetilde{\sigma}^{(3);dabc}(\omega_1,\omega_2,\omega_3)&=\widetilde{\sigma}_A^{(3);dabc}(\omega_1,\omega_2,\omega_3)+\widetilde{\sigma}_B^{(3);dabc}(\omega_1,\omega_2,\omega_3)\\&\quad+\widetilde{\sigma}_C^{(3);dabc}(\omega_1,\omega_2,\omega_3) \qquad (5.52)\end{aligned}$$

where $\widetilde{\sigma}_A$ includes all terms involving $\mathcal{G}_\mu$, $\widetilde{\sigma}_B$ includes all terms involving $\mathcal{H}_\mu$ and $\mathcal{I}_\mu$, and the remainder, $\widetilde{\sigma}_C$, includes all terms proportional to $|\mu|^{-1}$. Similar separations are also used for the symmetrized conductivity $\sigma^{(3);dabc}$. The term $\widetilde{\sigma}_C$ can be simplified to yield

$$\widetilde{\sigma}_C^{(3)}(\omega_1,\omega_2,\omega_3)=\frac{\sigma_3}{|\mu|}\left(\frac{A_0}{\nu_0}-\frac{A_3}{\vartheta_0}\right)\frac{(\Gamma_e^{(3)}-\Gamma_i^{(3)})(\Gamma_e^{(1)}-\Gamma_i^{(1)})}{\nu\vartheta\nu_3\vartheta_3}. \qquad (5.53)$$

5.4.2 *Intraband and interband contributions*

The term in $\widetilde{\sigma}^{(3);dabc}(\omega_1, \omega_2, \omega_3)$ that contains only contributions from the formal intraband component (derivatives) of Eq. (5.26) is the term proportional to $\mathcal{S}_1$; it varies with $|\mu|$ as $|\mu|^{-1}$, which is qualitatively different than the variation as $|\mu|$ of the corresponding Drude term in the linear conductivity. Although the $|\mu|^{-1}$ dependence can be associated with the intraband contribution, yet as $|\mu| \to 0$ the contribution to $\widetilde{\sigma}^{(3);dabc}(\omega_1, \omega_2, \omega_3)$ involving only the formal interband component (terms involving $\boldsymbol{r_k}$) of Eq. (5.26), which is proportional to $\mathcal{S}_6$, also becomes important; while it includes contributions from $\mathcal{G}_\mu(\theta_3)$ and $\mathcal{G}_\mu(\theta)$, there is also a term proportional to $|\mu|^{-1}$. The formally "mixed" terms, $\mathcal{S}_j$, with $j = 2, 3, 4, 5, 7, 8$, also provide terms proportional to $|\mu|^{-1}$. The summation of all these terms, all formally involving different proportions of interband and intraband contributions, gives Eq. (5.53); the $|\mu|^{-1}$ behavior in $\widetilde{\sigma}_C^{(3);dabc}(\omega_1, \omega_2, \omega_2)$ cannot be associated with motion that is just formally intraband. This is evidence for the blurring of the distinction between intraband and interband motion as $|\mu| \to 0$, which might have been expected physically.

5.4.3 *Relaxation free limit*

The results for $\widetilde{\sigma}^{(3);dabc}(\omega_1, \omega_2, \omega_3)$ show a complicated dependence on the ω_j, on the $\Gamma_{i/e}^{(j)}$, and on μ. For general frequencies ω_i and chemical potential μ for which divergences do not arise, the limit of $\widetilde{\sigma}^{(3);dabc}(\omega_1, \omega_2, \omega_3)$ can be safely taken as $\Gamma_{i/e}^{(j)} \to 0$. Obviously, in that case $\widetilde{\sigma}_C^{(3);dabc}(\omega_1, \omega_2, \omega_3) \to 0$, and even though $\widetilde{\sigma}_B^{(3);dabc}(\omega_1, \omega_2, \omega_3)$ does not go to zero, its symmetrized counterpart $\sigma_B^{(3);dabc}(\omega_1, \omega_2, \omega_3) \propto \Gamma_i^{(j)}$ does. Thus in general, the relaxation free limit only includes the term $\widetilde{\sigma}_A^{(3);dabc}(\omega_1, \omega_2, \omega_3)$, and recovers the results of Cheng *et al.* [Cheng *et al.* (2014b)], where the calculation was performed using an independent particle theory with infinitesimal relaxation parameters, but based on a different analytic approach. For those conductivities evaluated at frequencies where divergences occur, it is not surprising that the relaxation-free limits can depend on the order in which the relaxation parameters are taken to vanish; the response in such a situation strongly depends on the scattering, and phenomenological approximations may be too simplistic. For example, taking ω_{comb} to be a possible frequency combination appearing in the expression in Eqs. (5.41-5.48), resonant transitions (real or virtual) occur as $|\hbar\omega_{\text{comb}}| = 2|\mu|$. Then the function $\mathcal{H}_\mu$ or $\mathcal{I}_\mu$ becomes $\mathcal{H}_\mu(\omega_{\text{comb}} + i\hbar^{-1}\Gamma_e) \propto \Gamma_e^{-1}$ or $\mathcal{I}_\mu(\omega_{\text{comb}} + i\hbar^{-1}\Gamma_e) \propto \Gamma_e^{-2}$ respectively, and then $\sigma_B^{(3);dabc} \propto \Gamma_i \Gamma_e^{-1}$; its limit depends on the sequence of limits of $\Gamma_i \to 0$ and $\Gamma_e \to 0$, and so there seems to be no single well-defined relaxation free limit within this phenomenological theory.

It is also interesting to consider two further limits: (i) the limit of no doping $|\mu| \to 0$ and (ii) small frequencies $\hbar\omega_i \ll |\mu|$. Both cases give $\sigma^{(3);xxyy} = \sigma^{(3);xyxy} = \sigma^{(3);xyyx} = \sigma^{(3);xxxx}/3$ and the results of $\sigma^{(3);xxxx}$ are given in Eqs. (5.5) and (5.6), respectively.

In Eq. (5.6) the conductivity is proportional to $|\mu|^{-1}$. As expected, this result recovers the relaxation free, semiclassical result [Mikhailov (2007)] of a perturbative response divergent as $|\mu|^{-1}$, for $\hbar\omega_i \ll |\mu|$, and the result associated in that calculation with purely intraband motion. However, the recovery arises in a nontrivial way, because $\widetilde{\sigma}_C^{(3)}(\omega_1, \omega_2, \omega_3)$ vanishes in the relaxation-free limit, and the chemical potential μ only appears in $\mathcal{G}_\mu$ of $\widetilde{\sigma}_A^{(3);dabc}$. A careful check shows that Eq. (5.6) is exactly the same as the contribution proportional to the $\mathcal{S}_1$ term, and agrees with the relaxation free results [Mikhailov (2007)] involving only intraband motion. While the physically appropriate result of purely intraband motion is recovered in this limit as it should be, the connection to formally intraband, interband, and mixed responses in $\sigma^{(3);dabc}(\omega_1, \omega_2, \omega_3)$ is less than direct.

5.4.4 *Divergences and limits*

There are a number of divergences that arise in the nonlinear conductivity coefficients. Some of them are only apparent: For example, there seem to be divergences when $\vartheta - \vartheta_3 = 0$, but a careful collection of terms shows that even in the absence of relaxation $\lim_{\delta\to 0} \widetilde{\sigma}^{(3);dabc}(-\omega, \omega + \delta, \omega_3)$ is finite. Some of the divergences are of course real: There are divergences for $2|\mu| \pm \vartheta = 0$ in the functions $\mathcal{G}(\vartheta)$, $\mathcal{H}(\vartheta)$, and $\mathcal{I}(\vartheta)$, which lead to divergences in $\widetilde{\sigma}^{(3);dabc}(\omega_1, \omega_2, \omega_3)$. These are associated with interband optical transitions, and for nonvanishing relaxation they occur at frequencies removed from the real axis; we will see how some of them affect the structure of $\sigma^{(3);dabc}(\omega_1, \omega_2, \omega_3)$ later. There are also divergences associated with $\vartheta + \vartheta_3 = 0$. In the absence of relaxation these occur when $\omega_1 + \omega_2 + 2\omega_3 = 0$, and lead to a divergence in $\widetilde{\sigma}^{(3);dabc}(-\omega_2 - 2\omega_3 + \delta, \omega_2, \omega_3)$ as δ^{-1}. A special case of these is when $\vartheta = 0$ and $\vartheta_j = 0$ for $j = 0, 1, 2$, or 3. Some of the associated conductivity terms, such as $\sigma^{(3);dabc}(-\omega, \omega, \omega)$ and $\sigma^{(3);dabc}(-\omega, -\omega, 2\omega)$ will be mentioned in Section 5.5.3.

All of these divergences only occur at complex frequencies in the presence of relaxation, and have their analogs in gapped systems. Of a different nature are the divergences that arise as $|\mu| \to 0$. While in the absence of relaxation and taking into account only intraband motion, the third order nonlinear response coefficient that can be extracted from the full nonlinear response is divergent [Mikhailov (2007)] as $|\mu|^{-1}$, one might hope that in the presence of relaxation this would be ameliorated. Yet in general it is not. A more sophisticated theory is in order to address the limit $|\mu| \to 0$. Note that this divergence only exists in the perturbation theory, not in a full solution of Eq. (5.26).

5.4.5 *Finite temperature*

Generalizing a zero temperature calculation to finite temperature usually introduces two new features: The initial electron distribution is modified, and the relaxation rates are changed. The latter is implicit in the choice of relaxation parameters. In a perturbative calculation with analytic results, the former can be taken into account in the following simple way: Explicitly displaying the chemical potential and temperature dependence, we write $n_{s\mathbf{k}}(\mu, T)$ for the electron distribution at equilibrium, and $\sigma^{(3)}(\mu, T)$ for the nonlinear conductivity. By using

$$n_{s\mathbf{k}}(\mu, T) = \int_{-\infty}^{\infty} dx F_\mu(x, T) \frac{\partial}{\partial x} n_{s\mathbf{k}}(x, 0) \tag{5.54}$$

with $F_\mu(x, T) = [1 + e^{\beta(x-\mu)}]^{-1}$, the conductivity at finite temperature can be related to the zero temperature conductivity via

$$\begin{aligned} \sigma^{(3)}(\mu, T) &= \int_{-\infty}^{\infty} dx F_\mu(x, T) \frac{\partial}{\partial x} \sigma^{(3)}(x, 0) \\ &= \beta \int_{-\infty}^{\infty} dx F_\mu(x, T) \left[1 - F_\mu(x, T)\right] \sigma^{(3)}(x, 0) \,. \end{aligned} \tag{5.55}$$

Here the second line is obtained by using the partial integration and the condition $\sigma^{(3)}(x \to \pm\infty, 0) = 0$. Because $F_\mu(x, T)[1 - F_\mu(x, T)]$ is a pulse function located at $x = \mu$ with a width of the order of the thermal energy, the conductivity at finite temperature T can be obtained by averaging the zero temperature values over the chemical potential in an energy window with a width of the order of magnitude of the thermal energy. In Eqs. (5.41) to (5.48), the chemical potential appears in the function $\mathcal{G}_\mu$ in $\widetilde{\sigma}_A$, functions $\mathcal{H}_\mu$ and $\mathcal{I}_\mu$ in $\widetilde{\sigma}_B$, and $|\mu|^{-1}$ in $\widetilde{\sigma}_C$. Therefore, the conductivity at finite temperature is determined by applying Eq. (5.55) to these quantities. At zero temperature, the functions $\mathcal{G}_\mu(\epsilon + i\Gamma)$, $\mathcal{H}_\mu(\epsilon + i\Gamma)$, and $\mathcal{I}_\mu(\epsilon + i\Gamma)$ show fine structures around $\epsilon = \pm 2|\mu|$ in an energy range that is related to the value of Γ. At finite temperature, their counterparts are calculated from Eq. (5.55), and their fine structures are smeared out, in a similar way as when using a large effective Γ. However, the term $\widetilde{\sigma}_C^{(3);dabc}$, which involves $|\mu|^{-1}$, remains divergent at finite temperature due to the singularity of the integrand in Eq. (5.55) at $x = 0$; this divergence is not affected by the relaxation. Based on Eq. (5.53), where this term is nonzero only at $\Gamma_i^{(j)} \neq \Gamma_e^{(j)}$, the divergence shows that either the perturbation theory or the assumption of unequal intraband and interband relaxation times in undoped graphene is not adequate, and more realistic treatments of the scattering and temperature are required. Nonetheless, from a full numerical solution of Eq. (5.26) and (5.27) [Cheng *et al.* (2015a)], we find that contributions from the $|\mu|^{-1}$ term only give a small contribution to the total conductivity at finite temperature. In the following, we consider $\Gamma_i^{(j)} = \Gamma_i$, $\Gamma_e^{(j)} = \Gamma_e$, and $\Gamma_i = \Gamma_e$.

When the chemical potential μ and the frequencies $\{\omega_i\}$ are chosen to be away from resonant transitions, the conductivity is a smooth function of the chemical potential. Considering that the thermal energy $k_B T$ is only about $\sim$ 25.8 meV at room temperature, the conductivity at room temperature is close to the value at zero temperature away from resonant transitions. However, around resonant transitions where the conductivity diverges, the effects of finite temperature can be important.

5.5 Nonlinear Phenomena: Perturbative Results

In this section, we will discuss the third order conductivities associated with different nonlinear processes, including third harmonic generation, Kerr effects and two photon absorption, and parametric frequency conversion.

5.5.1 *Third harmonic generation*

The third order conductivity for third harmonic generation satisfies $\sigma^{(3);xxyy} = \sigma^{(3);xyxy} = \sigma^{(3);xyyx} = \sigma^{(3);xxxx}/3$. In the limit of no relaxation and at zero temperature, it can be written as

$$\sigma^{(3);xxyy}(\omega,\omega,\omega) = \frac{i\sigma_3}{144(\hbar\omega)^4}\Big[17\mathcal{G}_\mu(\hbar\omega + i0^+) - 64\mathcal{G}_\mu(2\hbar\omega + i0^+) + 45\mathcal{G}_\mu(3\hbar\omega + i0^+)\Big]. \tag{5.56}$$

In Fig. (5.4) the spectra of $\sigma^{(3);xxxx}(\omega,\omega,\omega)$ at $|\mu| = 0.3$ eV for zero and room temperature are plotted. At zero temperature, the relaxation-free results are given as the thick (red) curves in Fig. 5.4 (a). This figure shows the

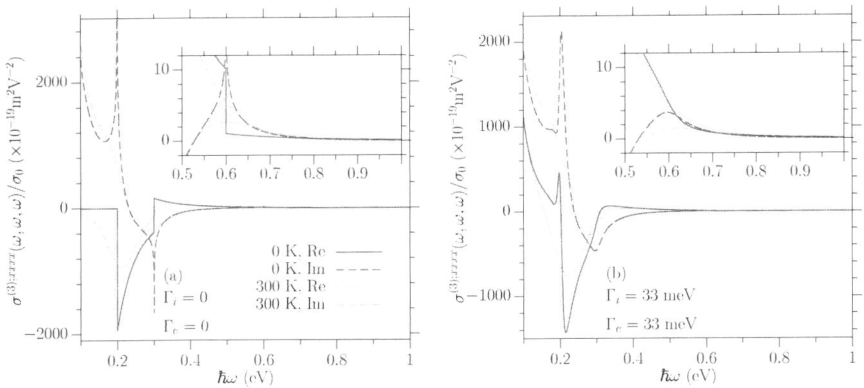

Fig. 5.4 Spectra of $\sigma^{(3);xxxx}(\omega,\omega,\omega)$ at zero (thick red curves) and room (thin blue curves) temperatures for different relaxation parameters: (a) $\Gamma_i = \Gamma_e = 0$, (b) $\Gamma_i = \Gamma_e = 33$ meV. The real (imaginary) parts of the conductivity are given by the solid (dashed) curves; we have taken $|\mu| = 0.3$ eV. The insets focus on results in the region $[0.5, 1]$ eV. The diagram is from [Cheng *et al.* (2015b)]. (For color reference, go to page 488.)

step function of the real parts and the logarithmic divergence of the imaginary parts at three resonant photon energies $\hbar\omega = 0.2$, 0.3, and 0.6 eV, which are induced by the function $\mathcal{G}_\mu$ and correspond to the resonant transitions for which the chemical potential gap $2|\mu|$ matches the energies of three photons, two photons, and one photon, respectively [Cheng *et al.* (2014b)]. With relaxation rates $\Gamma_i = \Gamma_e = 33$ meV included, the conductivity is a smooth function of ω, and plotted in Fig. 5.4 (b). It shows the following effects: (i) the divergent peaks of the imaginary parts of the conductivity in Fig. 5.4(a) become finite and broadened, (ii) the step functions of the real parts become continuous, (iii) the real parts become finite as $\hbar\omega < 2|\mu|/3 = 0.2$ eV, and increase rapidly with decreasing frequency. They receive contributions not only from intraband transitions, describing Drude-like effects, but also from the interband transitions due to the linear dispersion relation of graphene (for example, see the prefactor $(\hbar\omega)^{-4}$ in Eq. (5.56)).

At room temperature, the conductivities for these two cases look very similar to each other, and the fine structures caused by the resonant transitions are smeared out. Temperature affects the conductivity by smearing and lowering the peaks caused by functions $\mathcal{G}$, $\mathcal{H}$, and $\mathcal{I}$, which has an effect similar to increasing the value of relaxation parameter Γ_e.

It is possible to compare with the THG experiments listed in Table 5.1. Setting $\hbar\omega = 0.72$ eV, as in the experiment by Kumar *et al.* [Kumar *et al.*

(2013)], the effective susceptibility without the inclusion of relaxation at zero temperature can be estimated as $|\chi_{\text{eff}}^{(3);xxxx}| = 3.25 \times 10^{-19}\text{m}^2/\text{V}^2$ for $|\mu| \to 0$, and $6 \times 10^{-19}\text{m}^2/\text{V}^2$ for $|\mu| = 0.3$ eV. Both are much smaller than the experimental values. The small value is partly induced by cancellations of each $\mathcal{G}_\mu$ term in Eq. (5.56). From the inset of Fig. 5.4 we see that the results do not change too much at room temperature, or with the inclusion of reasonable finite relaxation parameters. The discrepancy between theory and experiment will be partly addressed in Section 5.6.2.

5.5.2 *Kerr effects and two photon absorption*

The third order conductivity $\sigma^{(3);xxxx}(\omega, \omega, -\omega)$ is usually treated as an intensity dependent correction to the linear conductivity $\sigma^{(1);xx}(\omega)$ to give an effective one $\sigma_{\text{eff}}^{(1);xx}(\omega) = \sigma^{(1);xx}(\omega) + 3\sigma^{(3);xxxx}(\omega, \omega, -\omega)|E_\omega^x|^2$.

In the limit of no relaxation and at zero temperature, it has been shown [Cheng *et al.* (2014b)] that $\sigma^{(3);dabc}(-\omega, \omega, \omega)$ has many divergences, and its behavior in the neighborhood of equal frequencies can be written as

$$\sigma^{(3);dabc}(-\omega, \omega + \delta_1, \omega + \delta_2) = \frac{\mathcal{T}_1^{dabc}(\omega)}{\delta_1\delta_2} + \frac{\mathcal{T}_2^{dabc}(\omega; \delta_2)}{\delta_1} + \frac{\mathcal{T}_2^{dabc}(\omega; \delta_1)}{\delta_2} + \mathcal{T}_3^{dabc}(\omega; \delta_1, \delta_2)\,, \tag{5.57}$$

where $\mathcal{T}_1^{dabc}$, $\mathcal{T}_2^{dabc}$, and $\mathcal{T}_3^{dabc}$ are all smooth functions of δ_1 and δ_2. The strength of the singularities is determined by $\mathcal{T}_1^{dabc}$ and $\mathcal{T}_2^{dabc}$, which are real functions and are only nonzero when the photon energy is greater than that for the onset of one-photon absorption ($\hbar\omega > 2|\mu|$). For fixed photon energy $\hbar\omega$, the appearance of the divergence as μ decreases from $2|\mu| > \hbar\omega$ to $2|\mu| \le \hbar\omega$ indicates that it is associated with the existence of electrons (holes) at the $\boldsymbol{k}$ where one-photon absorption is possible. Physically, at these $\boldsymbol{k}$ the third order correction to one-photon absorption would lead to the perturbative description of the saturation, but in the absence of relaxation that correction diverges, as it would for an inhomogeneously broadened collection of two-level systems. At zero temperature, the sharp Fermi surface can strictly exclude such electrons (holes) for $2|\mu| > \hbar\omega$. However, at finite temperature thermal fluctuations will always place some electrons (holes) at $\boldsymbol{k}$ where one-photon absorption can occur, and so the divergence in the third order response will exist for any photon energy.

But for nonzero Γ_i and Γ_e, $\sigma^{(3);dabc}(-\omega, \omega + \delta_1, \omega + \delta_2)$ is a smooth function of real ω, δ_1, and δ_2. For a pulse response when the energy broadening of the pulse is less than the relaxation energies, it is reasonable to set

$\delta_1 = \delta_2 = 0$, and then the conductivity can be written as

$$\sigma^{(3);dabc}(-\omega,\omega,\omega) = \frac{\mathcal{L}_1^{dabc}(\omega)}{\Gamma_i\Gamma_e} + \frac{\mathcal{L}_2^{dabc}(\omega;\Gamma_e)}{\Gamma_i} + \frac{\mathcal{L}_3^{dabc}(\omega;\Gamma_i,\Gamma_e)}{\Gamma_e} + \mathcal{L}_4^{dabc}(\omega;\Gamma_i,\Gamma_e)\,, \quad (5.58)$$

Fig. 5.5(a) gives the relaxation-free calculation, which is done as $\lim_{\Gamma_i=\Gamma_e\to 0}\sigma^{(3);dabc}(-\omega,\omega,\omega)$. Three regimes are apparent: (1) $\hbar\omega < |\mu|$, in which both one- and two-photon absorption are absent, and the real part of the conductivity is zero. The imaginary part at low photon energy scales as $(\hbar\omega)^{-3}$. At $\hbar\omega = |\mu|$, the real part shows a step function, while the imaginary part shows a logarithmic divergence. (2) $|\mu| < \hbar\omega < 2|\mu|$, in which two-photon absorption is present but one-photon absorption is still absent. The real part of the conductivity here scales as $(\hbar\omega)^{-4}$. Around $\hbar\omega = 2|\mu|$, the imaginary part shows a divergence $(\hbar\omega - 2|\mu|)^{-2}$. For frequencies satisfying $\hbar\omega < 2|\mu|$, if the graphene is subject to a Gaussian pulse sufficiently narrow in frequency, the nonlinear current induced will still have a shape that is approximately Gaussian. (3) $\hbar\omega > 2|\mu|$, where both two- and one-photon absorption are present. The imaginary part of the conductivity diverges as $(\hbar\omega - 2|\mu|)^{-2}$ around $\hbar\omega = 2|\mu|$, and the real part diverges for the entire region $\hbar\omega > 2|\mu|$. At finite temperature and in the absence of relaxation, the real part diverges for *any* photon energy $\hbar\omega$. As we discussed after Eq. (5.57), the divergence of the real part of $\sigma^{(3);dabc}(-\omega,\omega,\omega)$ is induced by the existence of electrons (holes) at the $\boldsymbol{k}$ where one-photon absorption occurs; at zero temperature, these electrons (holes) only exist when the chemical potential $|\mu| < \hbar\omega/2$, while at finite temperature, they exist at any chemical potential due to thermal fluctuations.

Figure 5.5(b) gives the results with relaxation parameters $\Gamma_i = \Gamma_e = 33$ meV. Relaxation affects the conductivity in a complex way, but there are some qualitative features that can be identified: (i) In the neighborhood of the divergences that arise in the relaxation-free calculation, including the divergent regime $\hbar\omega > 2\,|\mu|$ and the special frequency $\hbar\omega = |\mu|$, both the real and imaginary parts of the conductivity are lowered and are everywhere finite. (ii) For the relaxation parameters used here, the real part of the conductivity is negative for $\hbar\omega > 2|\mu|$. Because of the presence in this frequency range of one-photon absorption, which is always positive, the two-photon absorption processes indicated by the real part of $\sigma^{(3);dabc}(-\omega,\omega,\omega)$ can be understood as a correction to the simple linear prediction of the absorption. In fact, we can find a range of electric fields large enough, *i.e.* $|E_\omega^x|^2 > -\mathrm{Re}[\sigma^{(1);xx}(\omega)]/\mathrm{Re}[3\sigma^{(3);xxxx}(\omega,\omega,-\omega)]$,

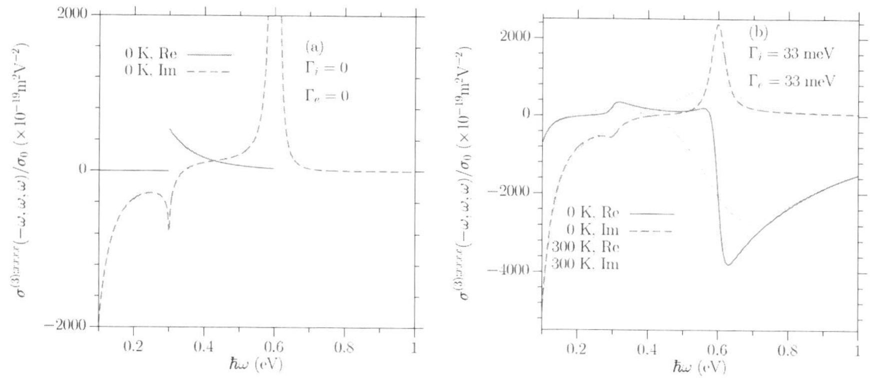

Fig. 5.5 Spectra of $\sigma^{(3);xxxx}(-\omega, \omega, \omega)$ at zero (thick red curves) and room (thin blue curves) temperatures with different relaxation parameters: (a) $\Gamma_i = \Gamma_e = 0$, (b) $\Gamma_i = \Gamma_e = 33$ meV. The chemical potential is $|\mu| = 0.3$ eV. The real (imaginary) parts of the conductivity are given by the solid (dashed) curves. In figure (a) the real part (solid curve) of $\sigma^{(3);xxxx}(-\omega, \omega, \omega)$ diverges for all $\hbar\omega > 2|\mu|$. The diagram is from [Cheng *et al.* (2015b)]. (For color reference, go to page 488.)

so that $\sigma_{\text{eff}}^{da}(\omega)$ is negative; for a field anywhere near or above this strength the perturbative result is naturally suspect. (iii) Even for frequencies in the range $|\mu| < \hbar\omega < 2|\mu|$, where only two-photon absorption is present in the absence of relaxation, the real part of the nonlinear conductivity $\sigma^{(3);dabc}(-\omega, \omega, \omega)$ can be negative. Yet in the presence of relaxation the linear conductivity $\sigma^{(1);xx}(\omega)$ acquires a real part in this frequency range, and the real part of $\sigma_{\text{eff}}^{xx}(\omega)$, for example, is always positive for small enough electric field amplitudes, indicating absorption. However, these results indicate the sensitivity to the relaxation parameters of both the third order conductivity, and its interplay with the first order conductivity, and a more sophisticated description of the scattering is clearly in order. At room temperature, the peaks or divergences are further broadened. For a given frequency ω, the regime $\hbar\omega > 2|\mu|$ always contributes to a finite temperature calculation due to the average over the chemical potential. The absolute values of the real part of the conductivity in the regime $\hbar\omega < 2|\mu|$ also significantly increase.

The results of experimental studies are often described in terms of an effective bulk nonlinear refractive index n_{2b} and an effective bulk nonlinear absorption coefficient β_b, or equivalently a saturation intensity I_{sat} for cases

$\hbar\omega \geq 2|\mu|$ where one-photon absorption is present. They are connected by

$$1 + \frac{\sigma_{\text{eff}}^{(1);xx}(\omega)}{-i\omega\epsilon_0 d_{\text{gr}}} = \left(n_b + i\frac{c}{2\omega}\alpha_b\right)^2 , \tag{5.59}$$

$$n_b = n_{0b} + n_{2b} I , \quad \alpha_b = \alpha_{0b} + \beta_b I , \tag{5.60}$$

with $I = 2\epsilon_0 n_0 c |E_\omega|^2$ the light intensity. The values of n_{0b} and α_{0b} can be obtained by setting $E_\omega = 0$. For weak light intensity, n_{2b} and β_b can be extracted by expanding both sides of Eq. (5.59) with respect to E_ω, as

$$n_{2b} + i\frac{c}{2\omega}\beta_b \approx \frac{3}{4\epsilon_0 c} \frac{\sigma^{(3);xxx}(-\omega,\omega,\omega)}{\sigma^{(1);xx}(\omega) - i\omega\epsilon_0 d_{\text{gr}}} . \tag{5.61}$$

The saturation intensity is defined from the saturation absorption, which is phenomenologically written as

$$\alpha_b = \frac{\alpha_{0b}}{1 + I/I_{\text{sat}}} . \tag{5.62}$$

At weak light intensity, comparing the expansion of the right hand side with the second equation of Eq. (5.60) gives

$$I_{\text{sat}} = -\alpha_{0b}/\beta_b . \tag{5.63}$$

Although these effective bulk quantities can indeed be defined for graphene, their inclusion in the calculation of the optical response typically is not trivial because their effect is not small. For example, the imaginary part of the left hand side of Eq. (5.59) is usually not a small quantity, thus α_{0b} and n_{0b} have to be calculated carefully [del Coso and Solis (2004)]. Experiments usually measure the absorbance α instead of the absorption coefficient α_b. For isolated graphene, the absorbance of normally incident light is proportional to $\alpha \propto \text{Re}[\sigma_{\text{eff}}^{xx}(\omega)]$. Beyond weak field limits, the effective linear conductivity may be written as $\sigma_{\text{eff}}^{(1);xx}(\omega) = \sigma^{(1);xx}(\omega) + 3\sigma_{nl}(\omega)|E_\omega^x|^2$. Then using the quantities α, α_0, and β in Eq. (5.62) yields

$$\text{Re}[\sigma_{nl}(\omega)] = -\frac{\text{Re}[\sigma^{(1);xx}(\omega)]}{3E_{\text{sat}}^2} \frac{1}{1 + (|E_\omega|/E_{\text{sat}})^2} , \tag{5.64}$$

where E_{sat} is the saturation field. The weak field limit gives $\sigma_{nl}(\omega) = \sigma^{(3);xxxx}(\omega,\omega,-\omega)$, and then

$$E_{\text{sat}} = \sqrt{-\frac{\text{Re}[\sigma^{(1);xx}(\omega)]}{3\text{Re}[\sigma^{(3);xxxx}(-\omega,\omega,\omega)]}} , \tag{5.65}$$

The saturation intensity in air can then be calculated as $I_{\text{sat}} = 2\epsilon_0 c E_{\text{sat}}^2$.

In the experiment by Zhang *et al.* [Zhang *et al.* (2012)], the photon energy was $\hbar\omega = 0.8$ eV. Assuming $\mu \in [0.4, 0.7]$ eV, we calculate $n_2 \sim 10^{-11}\text{cm}^2/\text{W}$, much less than their measured value. However, if the doping level were much lower the experiment could have fallen in the divergent regime identified above, where we would not expect the perturbation approach adopted here to be sufficient.

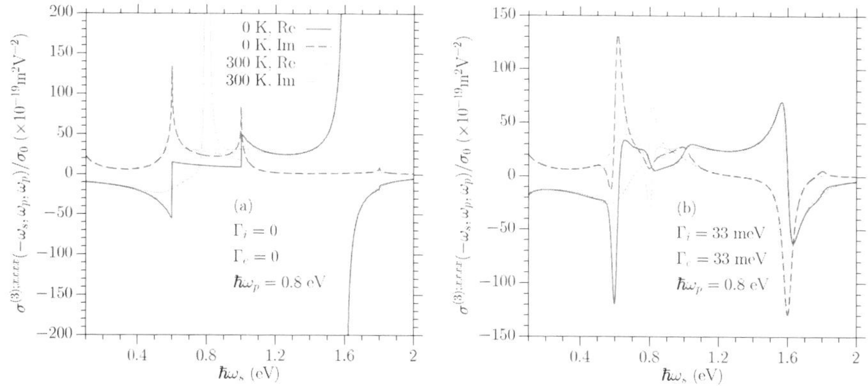

Fig. 5.6 Spectra of $\sigma^{(3);xxxx}(-\omega_s, \omega_p, \omega_p)$ for different relaxation parameters at zero (thick red curves) and room temperature (thin blue curves). Solid (dashed) curves give their real (imaginary) parts separately. In the calculation, the chemical potential is $|\mu| = 0.5$ eV, and the pump photon energy is $\hbar\omega_p = 0.8$ eV. The diagram is from [Cheng *et al.* (2015b)]. (For color reference, go to page 488.)

5.5.3 *Parametric frequency conversion*

Third order nonlinearities can lead to the appearance of new frequencies via parametric frequency conversion, which is described by $\sigma^{(3);dabc}(-\omega_s, \omega_p, \omega_p)$. Here ω_p is the frequency of a strong pump field, and ω_s is the signal frequency converted by interaction with the pump to an idler frequency $\omega_i = 2\omega_p - \omega_s$. Possible resonant transitions occur as any of the frequencies $|\omega_p|$, $|\omega_s|$, $|\omega_p - \omega_s|$, $|2\omega_p|$, or $|2\omega_p - \omega_s|$ equal $2|\mu|/\hbar$. In Fig. 5.6 we plot the dependence of $\sigma^{(3);xxxx}(-\omega_s, \omega_p, \omega_p)$ on ω_s for different relaxation parameters at $\hbar\omega_p = 0.8$ eV and $|\mu| = 0.5$ eV.

At zero temperature, the calculations show peaks/step functions for resonant transitions at $\hbar\omega_{s1} = 2\hbar\omega_p - 2|\mu| = 0.6$ eV, $\hbar\omega_{s2} = 2|\mu| = 1.0$ eV, or $\hbar\omega_{s3} = 2|\mu| + \hbar\omega_p = 1.8$ eV, both with and without the inclusion of relaxation. Around these resonant transitions, the behavior of the conductivity can be analyzed as follows:

(1) Around $\omega_s = \omega_{s1} + \delta\omega$, the idler photon energy $\hbar\omega_i = 2|\mu| - \hbar\delta\omega$ is close to the onset of one-photon absorption. By taking $\vartheta = \hbar\omega_i + i\Gamma_e = 2|\mu| - \hbar\delta\omega + i\Gamma_e$, the conductivity as $\delta\omega \to 0$ is determined by functions $\mathcal{G}_\mu(\vartheta) \sim \ln(\hbar\delta\omega - i\Gamma_e)$, $\mathcal{H}_\mu(\vartheta) \sim (\hbar\delta\omega - i\Gamma_e)^{-1}\Gamma_i$, and $\mathcal{I}_\mu(\vartheta) \sim (\hbar\delta\omega - i\Gamma_e)^{-2}\Gamma_i$. In the relaxation-free limit, only $\mathcal{G}_\mu(\vartheta)$ contributes a logarithmic divergence to the imaginary part, and a step change in the real part [in Fig. 5.6 (a)] for nonzero $\delta\omega$. With the inclusion of relaxation characterized by $\Gamma_i = \Gamma_e = 33$ meV, the resonances are smeared out.

(2) Around $\omega_s = \omega_{s2} + \delta\omega$, the signal frequency is close to the onset of one-photon absorption. For non-resonant transitions in a usual semiconductor, ω_{s_2} and ω_{s_1} are interchangeable in the expression for the parametric frequency conversion conductivity [Boyd (2008)]; here in graphene they yield asymmetric peaks because the resonant transitions dominate. In the limit of no relaxation, the conductivity shows a logarithmic divergence that is easily smeared out by the inclusion of small relaxation parameters.

(3) Around $\omega_s = \omega_{s3} + \delta\omega$: By taking $\vartheta = -\hbar\omega_s + \hbar\omega_p + i\Gamma_e = -2|\mu| - \hbar\delta\omega + i\Gamma_e$, the conductivity as $\delta\omega \to 0$ is determined by functions $\mathcal{G}_\mu(\vartheta) \sim \ln(\hbar\delta\omega - i\Gamma_e)$ and $\mathcal{H}_\mu(\vartheta) \sim (\hbar\delta\omega - i\Gamma_e)^{-1}\Gamma_i$. In the limit of no relaxation, $\mathcal{G}_\mu$ gives a small peak.

(4) At finite temperature there is a further smearing of the peaks around the resonances.

Besides these resonant transitions, two singularities are apparent: (i) the singularity around $\omega_s = 2\omega_p$, which corresponds to two-color coherent current injection. The singularity is not determined by the behavior of the functions $\mathcal{G}$, $\mathcal{H}$, $\mathcal{I}$, but by the coefficients that premultiply them in Eqs. (5.41) to (5.48). Around $\omega_s = 2\omega_p$ we put $\omega_s = 2\omega_p + \delta\omega$ and find

$$\sigma^{(3);xxxx}(-2\omega_p + \delta\omega, \omega_p, \omega_p) = \frac{1\text{eV}}{\hbar\delta\omega + i\Gamma_i} i\sigma_{inj}^{(3);xxxx}(\omega_p) + \sigma_R^{(3);xxxx}(\omega_p)\,. \tag{5.66}$$

Two-color coherent current injection requires both one-photon absorption (for ω_s) and two photon absorption (for ω_p), *i.e.*, $\hbar\omega_p > |\mu|$. The parameters we have adopted in Fig. 5.6 fulfill this criterion, and thus the singularity appears. Finite temperatures do not qualitatively affect this singularity because it is not related to the chemical potential. (ii) The strong response around $\omega_s = \omega_p$ is related to the third order correction to one-photon absorption, and only appears at finite temperature. Since $|\mu| < \hbar\omega_p < 2|\mu|$, at zero temperature only two-photon absorption is present and there is no one-photon absorption. However, at finite temperature thermal fluctuations will place electrons where one-photon absorption can occur, and the third order correction to that will lead, in the absence of relaxation, to a divergent result as discussed in Section 5.5.2; in the presence of relaxation the result will not be divergent but very large, describing the saturation of the one-photon absorption at the level of the third order response.

From the laser intensity used in the experiement of Hendry *et al.* [Hendry *et al.* (2010)], theoretical values of the nonlinear coefficients can be estimated: At zero temperature and without dampings, Eq. (5.5) gives

$\sigma_0^{-1}\sigma^{(3)}(-\omega_s, \omega_p, \omega_p) = 1.6 \times 10^{-18}$ m^2/V^2 or $\chi^{(3)} = 1.5i \times 10^{-17}$ m^2/V^2. Again, the theoretical value is much smaller than the experimental value. A full calculation from Eqs. (5.38) and (5.41)-(5.48) shows that the results are not a sensitive function of the chemical potential for $|\mu| < 0.3$ eV; therefore at room temperature, the damping free calculation yields about the same result as the zero temperature calculation. Even if relaxation parameters are included, the results for $|\chi^{(3)}|$ at room temperature for $|\mu| < 0.3$ eV differ from the zero temperature results by less than 5% for parameters $(\Gamma_i, \Gamma_e) = (33, 33)$ meV.

5.6 Numerical Solution of SBEs

The analytic perturbative results have been discussed in the preceding sections. However, as pointed out, there are discrepancies between experimental results and theoretical predictions. Using the appropriate experimental parameters, the susceptibility values obtained by present theories are orders of magnitude smaller than measured values [Cheng *et al.* (2014b, 2015b)]. To perform the perturbative calculation, two approximations have been used: the electric fields are considered sufficiently weak, and the dispersion relation is approximated as linear. The first is the key assumption for perturbation theory. But there are a host of resonant transitions in graphene, and it is not easy to identify exactly how small an electric field needs be to be properly considered "weak". The second approximation is used because for most of the experiments the optical transitions occur around the Dirac points, where the linear dispersion approximation works very well for the energy spectrum. However, this may be not adequate for the third order nonlinearity calculation.

In this section, by numerically solving the SBEs in doped graphene, these two approximations are investigated by considering the dependence of the optical response on the chemical potential [Cheng *et al.* (2015a)]: We investigate the validity of the perturbative treatment for nominally weak fields, and we study how saturation can affect the nonlinearity when the fields become strong.

5.6.1 *Weak fields*

To make the numerical simulation easier, the electric field is chosen as a Gaussian pulse $\boldsymbol{E}(t) = \hat{\boldsymbol{x}} E_0 e^{-t^2/\Delta_c^2} e^{-i\omega_c t} + c.c.$, which is centered at frequency ω_c with a duration Δ_c. The numerical calculation is performed for $T = 300$ K, $\Delta_c = 100$ fs, $\hbar\omega_c = 0.6$ eV, and $\Gamma_i = \Gamma_e = 33$ meV. The optical response is investigated at different frequencies ω_c and $3\omega_c$ for its dependence on the electric field amplitude E_0 and the chemical potential μ. The details about the separation of the nonlinear parts of the response and the procedure to fit the nonlinear coefficients are presented by Cheng *et al.* [Cheng *et al.* (2015a)].

The comparison with the analytic results is performed for $E_0 = 10^6$ V/m. For the coefficients of both the third harmonic generation and the Kerr effects, the agreement is very good, and this suggests that for such fields the perturbative results presented above are reliable. It also indicates that the usual Dirac point approximations adopted in the perturbative calculation, involving the linear dispersion relation and the form of the matrix elements, do not introduce any significant errors in calculating the linear and nonlinear optical response of doped graphene at incident photon energies around $\hbar\omega = 0.6$ eV.

5.6.2 *Strong fields*

For strong fields, there are two different regimes depending on the chemical potential: (i) $2|\mu| < \hbar\omega_c$. Here one photon absorption exists and carriers can be injected from the "−" band to the "+" band. Stronger electric fields inject more carriers. If the electrons have a finite lifetime in the states into which they are injected, their injection prevents the effectiveness of further absorption. For strong incident fields the nonlinear response at ω_c and $3\omega_c$ can be expected to be characterized by third harmonic generation and terms that behave phenomenologically as Kerr effects and two-photon absorption. A strict perturbative description requires high orders contribution, which can be collected to give a field dependent third order nonlinearity coefficients $\sigma_{nl}(\omega_c)$ [as defined above Eq. (5.64)] and $\sigma_{\rm THG}(\omega_c)$. (ii) $2|\mu| > \hbar\omega_c$. When two photon absorption exists, *i.e.*, $2|\mu| > \hbar\omega_c > |\mu|$, two photon absorption can inject carriers, but it is less efficient than one photon absorption. Thus saturation requires higher electric fields. When two photon absorption is absent, *i.e.*, $|\mu| > \hbar\omega_c$, the carrier injection process occurs in a higher order absorption process, which is much less efficient. Here we briefly discuss how the saturation affects the optical nonlinearity in case (i).

Figure 5.7(a) gives the dependence of the numerically determined $\sigma_{nl}(\omega_c)$ as a function of E_0 for two different chemical potentials $\mu = 0$ and 0.2 eV. The real part of $\sigma_{nl}(\omega_c)$ is fitted to an expression $-\sigma_f/[3(E_{\rm sat}^2 + E_0^2)]$

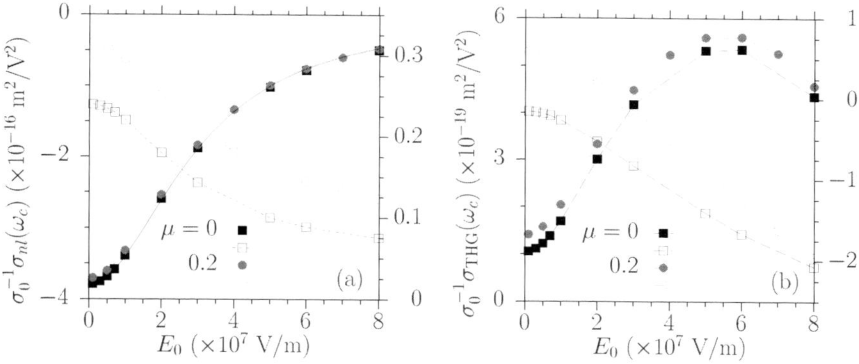

Fig. 5.7 (color online) Saturation effect on the coefficients of Kerr effects and two photon absorption (a) and third harmonic generation (b). Here the filled symbols are the real part with the values at the left y-axis, and the hollow symbols are the imaginary part with the values at the right y-axis. The dashed lines are drawn to guide the eye. The solid lines are fitting curves (see the text). The diagram is from [Cheng *et al.* (2015a)].

with two parameters σ_f and $E_{\rm sat}$. The fittings (shown as solid curves) are very good for $\mathrm{Re}[\sigma_f]/\sigma_0 \approx -1$ and -1 for $|\mu| = 0$ and 0.2 eV, respectively, with the saturation fields the same in both cases as $E_{\rm sat} \approx 3 \times 10^7$ V/m. Comparing the fitted form with Eq. (5.64), and noting that the linear conductivities are given by $\sigma^{(1);xx}(\omega_c)/\sigma_0 = 1$ and $0.96 - 0.11i$, respectively, for our parameter sets, the closeness of the fitted σ_f with the real part of these linear conductivity $\sigma^{(1);xx}(\omega_c)$ indicates that the saturation can indeed be attributed to linear absorption. Further, by using the numerical values of $\sigma_{nl}(\omega_c)/\sigma_0$ at a weak field value $E_0 = 10^6$ V/m, which for our two parameter sets are both -3.7×10^{-16} m^2/V^2, Eq. (5.65) leads to both saturation fields as 3×10^7 V/m, which are very close to the fitted values. The field dependence of $\mathrm{Im}[\sigma_{nl}(\omega_c)]$, which at least in the weak field limit can be related to the real part of the nonlinear response via nonlinear Kramers-Kronig relations, varies in a more complicated way.

The saturation field can also be estimated from only the linear absorption coefficients. Physically, the saturation effect occurs when the injected electron density from one-photon absorption is comparable to the density of states in the region of $\boldsymbol{k}$ space where the electrons are injected. The injected electron density is $\hbar/\Gamma_i \xi^{xx}(\omega_c) E_m^2$ with one-photon absorption coefficient [Cheng *et al.* (2014b)] $\xi^{xx}(\omega_c) = 2\mathrm{Re}[\sigma^{(1);xx}(\omega_c)]/(\hbar\omega_c)$ and the critical field amplitude E_m, while the total available states are estimated as those satisfying $-\Gamma_e \le \varepsilon_{+\boldsymbol{k}} - \varepsilon_{-\boldsymbol{k}} - \hbar\omega_c \le \Gamma_e$. Then the critical field amplitude E_m is estimated as

$$E_m \approx \sqrt{\frac{2\Gamma_i\Gamma_e}{\pi}\frac{\sigma_0}{\text{Re}[\sigma^{(1);xx}(\omega_c)]}\frac{\hbar\omega_c}{\hbar|e|v_F}}\,. \tag{5.67}$$

This can be used to find approximate values of $E_m \sim 2.8 \times 10^7$ V/m for those two parameter sets. They are close to the fitted saturation field.

We now turn from $\sigma_{nl}(\omega_c)$ to $\sigma_{\text{THG}}(\omega_c)$. We find that saturation can significantly affect THG, as shown in Fig. 5.7(b). Both the real and imaginary parts of $\sigma_{\text{THG}}(\omega_c)$ depend strongly on the electric field. The imaginary part even changes its sign from positive to negative values with increasing electric field, while the real part shows peaks around $E_0 = 5 \times 10^7$ V/m. Compared to the values at $E_0 = 10^6$ V/m, these peak absolute values are about 5 times larger. In the measurement of THG [Säynätjoki *et al.* (2013); Kumar *et al.* (2013); Hong *et al.* (2013)] the procedure used to prepare the samples would indicate the chemical potentials should be very low; thus saturation may well occur and the effective THG coefficients $\sigma_{\text{THG}}(\omega_c)$ may be above their perturbative values. Physically, the peak is induced by the competing contributions to the THG by the carriers injected by one photon absorption and two photon absorption [Cheng *et al.* (2015a)].

5.7 Conclusion and Outlook

We investigated the third order nonlinear optical conductivity of doped graphene employing the semiconductor Bloch equations and describing the scattering phenomenologically; the electronic states were obtained from a two-band tight binding model. Two different methods were used to solve the semiconductor Bloch equations: the perturbation method for weak fields and numerical simulations for strong fields.

Perturbative analyses play a central role in nonlinear optics. Even in the independent-particle approximation, with relaxation treated phenomenologically, the calculated response tensors indicate the nonlinear optical effects that are allowed, and point to where resonances can lead to an interesting dependence on time and frequency. It is a starting point for more realistic calculations. When focusing on the optical transition around the Dirac points of graphene, where the widely used linear dispersion relation is a good approximation, an analytic expression for general third order conductivities was obtained by solving the semiconductor Bloch equations perturbatively at zero temperature, taking different relaxation parameters for interband and intraband optical transitions. The nonlinear conductivities at finite temperature were obtained by an appropriate integration over the chemical potential. The conductivities show a complicated dependence on photon energy, chemical potential, and the relaxation parameters. We discussed in detail third harmonic generation, Kerr effects and two-photon absorption, and parametric frequency conversion.

Comparison of these results with experiments is difficult, since in many of the reported experiments the graphene samples have not been characterized in the linear regime, and neither the relaxation parameters nor even the chemical potential have been identified. Results for some nonlinear response coefficients, such as that describing the Kerr effect, are predicted by our calculations to be so sensitive to these parameters that we cannot hazard a comparison of theory to experiment. Yet our results for third harmonic generation and parametric frequency conversion are insensitive enough to these parameters that we can conclude our results are about two orders of magnitude smaller than those extracted from experiments [Cheng *et al.* (2014b)], even with the adoption of reasonable relaxation parameters. We speculated on the causes of this disagreement; it is of course early days for both detailed experimental and theoretical studies of such nonlinear effects in graphene. But these disagreements may persist, and a long and difficult journey may be necessary to understand the details of the full nonlinear optical response of graphene. Even so we can expect that, as in the study of the nonlinear optical response of other materials, the kind of perturbative calculation we have presented here will provide a useful port of embarkation, paving the way to a better physical insight in the complex nonlinear optical response of graphene.

Even with the inclusion of relaxation we found that the perturbative approach itself is problematic at vanishing chemical potential, as might be expected from a similar result in the limit that only intraband contributions are considered [Mikhailov (2007)], except in the special case that either first- or third-order interband and intraband relaxation rates are set equal. The perturbative approach adopted is unproblematic for doped graphene at zero temperature, but is a concern at finite temperature, since thermal fluctuations always place some electrons or holes near the Dirac points. Yet numerical calculations of the full semiconductor Bloch equations indicate that the contribution of such electrons to the full optical response is small, so this effect does not afflict our results.

The numerical calculations also confirmed the correctness of the perturbative results at weak electric fields, the use of the linear dispersion approximation, and the consideration of optical transitions only around the Dirac points. With an increase in the electric field amplitude, both the effective Kerr coefficient and the THG coefficients show dependences on the field strength, which can be attributed to saturation effects. For $\hbar\omega_c > 2|\mu|$ where one-photon absorption exists, the saturation effects can be characterized by a saturation field. The amplitude of the effective third harmonic generation coefficient can increase up to 5 times as the electric field changes from 10^6 V/m to 8×10^7 V/m. However, compared to the two orders of magnitude difference between the values from the perturbation calculation and experiments [Cheng *et al.* (2014b, 2015b)], this small factor indicates the discrepancies between theory and expeirments may be attributed to one of a number of possible limitations in the theoretical description: (1) the two band tight binding model is too simple to describe the correct optical nonlinearity of graphene, and a full band structure calculation may be required, (2) neglected thermal effects induced by temperature change and gradients may play a role in the response, and (3) the inclusion of realistic scattering and many-body effects may be required even for qualitative agreement with experiments.

Bibliography

[1] Al-Naib, I., Sipe, J. E., and Dignam, M. M. (2014). High harmonic generation in undoped graphene: Interplay of inter- and intraband dynamics, *Phys. Rev. B* **90**, 24, p. 245423, doi:10.1103/physrevb.90.245423, `http://dx.doi.org/10.1103/PhysRevB.90.245423`.

[2] Al-Naib, I., Sipe, J. E., and Dignam, M. M. (2015). Nonperturbative model of harmonic generation in undoped graphene in the terahertz regime, *New J. Phys.* **17**, 11, p. 113018, doi:10.1088/1367-2630/17/11/113018, `http://dx.doi.org/10.1088/1367-2630/17/11/113018`.

[3] An, Y. Q., Nelson, F., Lee, J. U., and Diebold, A. C. (2013). Enhanced optical second-harmonic generation from the current-biased graphene/sio 2 /si(001) structure, *Nano Lett.* **13**, 5, pp. 2104–2109, doi:10.1021/nl4004514, `http://dx.doi.org/10.1021/nl4004514`.

[4] An, Y. Q., Rowe, J. E., Dougherty, D. B., Lee, J. U., and Diebold, A. C. (2014). Optical second-harmonic generation induced by electric current in graphene on si and SiC substrates, *Phys. Rev. B* **89**, 11, p. 115310, doi:10.1103/physrevb.89.115310, `http://dx.doi.org/10.1103/PhysRevB.89.115310`.

[5] Aversa, C. and Sipe, J. E. (1995). Nonlinear optical susceptibilities of semiconductors: Results with a length-gauge analysis, *Phys. Rev. B* **52**, pp. 14636–14645, doi:10.1103/PhysRevB.52.14636, `http://link.aps.org/doi/10.1103/PhysRevB.52.14636`.

[6] Avetissian, H. K., Avetissian, A. K., Mkrtchian, G. F., and Sedrakian, K. V. (2012). Multiphoton resonant excitation of fermi-dirac sea in graphene at the interaction with strong laser fields, *J. Nanophoton.* **6**, 1, p. 061702, doi: 10.1117/1.jnp.6.061702, `http://dx.doi.org/10.1117/1.JNP.6.061702`.

[7] Avetissian, H. K., Mkrtchian, G. F., Batrakov, K. G., Maksimenko, S. A., and Hoffmann, A. (2013a). Multiphoton resonant excitations and high-harmonic generation in bilayer graphene, *Phys. Rev. B* **88**, 16, p. 165411, doi:10.1103/physrevb.88.165411, `http://dx.doi.org/10.1103/PhysRevB.88.165411`.

[8] Avetissian, H. K., Mkrtchian, G. F., Batrakov, K. G., Maksimenko, S. A., and Hoffmann, A. (2013b). Nonlinear theory of graphene interaction with strong laser radiation beyond the dirac cone approximation: Co-

herent control of quantum states in nano-optics, *Phys. Rev. B* **88**, 24, p. 245411, doi:10.1103/physrevb.88.245411, `http://dx.doi.org/10.1103/PhysRevB.88.245411`.

[9] Bao, Q., Zhang, H., Wang, Y., Ni, Z., Yan, Y., Shen, Z. X., Loh, K. P., and Tang, D. Y. (2009). Atomic-layer graphene as a saturable absorber for ultrafast pulsed lasers, *Adv. Funct. Mater.* **19**, 19, pp. 3077–3083, doi:10.1002/adfm.200901007, `http://dx.doi.org/10.1002/adfm.200901007`.

[10] Blount, E. (1962). *Formalisms of Band Theory*, Vol. 13 (Elsevier), doi:10.1016/S0081-1947(08)60459-2, `http://dx.doi.org/10.1016/S0081-1947(08)60459-2`.

[11] Boyd, R. W. (2008). *Nonlinear Optics*, 3rd edn. (Academic).

[12] Bykov, A. Y., Murzina, T. V., Rybin, M. G., and Obraztsova, E. D. (2012). Second harmonic generation in multilayer graphene induced by direct electric current, *Phys. Rev. B* **85**, p. 121413, doi:10.1103/PhysRevB.85.121413, `http://link.aps.org/doi/10.1103/PhysRevB.85.121413`.

[13] Castro Neto, A. H., Guinea, F., Peres, N. M. R., Novoselov, K. S., and Geim, A. K. (2009). The electronic properties of graphene, *Rev. Mod. Phys.* **81**, 1, pp. 109–162, doi:10.1103/RevModPhys.81.109.

[14] Cheng, J. L., Vermeulen, N., and Sipe, J. E. (2014a). Dc current induced second order optical nonlinearity in graphene, *Opt. Express* **22**, 13, pp. 15868–15876, doi:10.1364/OE.22.015868, `http://www.opticsexpress.org/abstract.cfm?URI=oe-22-13-15868`.

[15] Cheng, J. L., Vermeulen, N., and Sipe, J. E. (2014b). Third order optical nonlinearity of graphene, *New J. Phys.* **16**, 5, p. 053014, doi:10.1088/1367-2630/16/5/053014, `http://dx.doi.org/10.1088/1367-2630/16/5/053014`, ; New J. Phys. **18**, 029501 (2016).

[16] Cheng, J. L., Vermeulen, N., and Sipe, J. E. (2015a). Numerical study of the optical nonlinearity of doped and gapped graphene: From weak to strong field excitation, *Phys. Rev. B* **92**, p. 235307, doi:10.1103/PhysRevB.92.235307, `http://link.aps.org/doi/10.1103/PhysRevB.92.235307`.

[17] Cheng, J. L., Vermeulen, N., and Sipe, J. E. (2015b). Third-order nonlinearity of graphene: Effects of phenomenological relaxation and finite temperature, *Phys. Rev. B* **91**, p. 235320, doi:http://dx.doi.org/10.1103/PhysRevB.91.235320, ; Phys. Rev. B **93**, 039904 (2016).

[18] Dean, J. J. and van Driel, H. M. (2009). Second harmonic generation from graphene and graphitic films, *Appl. Phys. Lett.* **95**, 26, 261910, doi:10.1063/1.3275740, `http://link.aip.org/link/?APL/95/261910/1`.

[19] Dean, J. J. and van Driel, H. M. (2010). Graphene and few-layer graphite probed by second-harmonic generation: Theory and experiment, *Phys. Rev. B* **82**, 12, p. 125411, doi:10.1103/physrevb.82.125411, `http://dx.doi.org/10.1103/PhysRevB.82.125411`.

[20] del Coso, R. and Solis, J. (2004). Relation between nonlinear refractive index and third-order susceptibility in absorbing media, *J. Opt. Soc. Am. B* **21**, 3, p. 640, doi:10.1364/josab.21.000640, `http://dx.doi.org/10.1364/JOSAB.21.000640`.

[21] Glazov, M. (2011). Second harmonic generation in graphene, *JETP Lett.* **93**, pp. 366–371, doi:10.1134/S0021364011070046, http://dx.doi.org/10.1134/S0021364011070046.

[22] Glazov, M. and Ganichev, S. (2014). High frequency electric field induced nonlinear effects in graphene, *Phys. Rep.* **535**, 3, pp. 101–138, doi:10.1016/j.physrep.2013.10.003, http://dx.doi.org/10.1016/j.physrep.2013.10.003.

[23] Gu, T., Petrone, N., McMillan, J. F., van der Zande, A., Yu, M., Lo, G. Q., Kwong, D. L., Hone, J., and Wong, C. W. (2012). Regenerative oscillation and four-wave mixing in graphene optoelectronics, *Nat. Photon.* **6**, 8, pp. 554–559, http://dx.doi.org/10.1038/nphoton.2012.147.

[24] Hendry, E., Hale, P. J., Moger, J., Savchenko, A. K., and Mikhailov, S. A. (2010). Coherent nonlinear optical response of graphene, *Phys. Rev. Lett.* **105**, p. 097401, doi:10.1103/PhysRevLett.105.097401, http://link.aps.org/doi/10.1103/PhysRevLett.105.097401.

[25] Hong, S.-Y., Dadap, J. I., Petrone, N., Yeh, P.-C., Hone, J., and Osgood, R. M., Jr. (2013). Optical third-harmonic generation in graphene, *Phys. Rev. X* **3**, p. 021014, doi:10.1103/PhysRevX.3.021014, http://link.aps.org/doi/10.1103/PhysRevX.3.021014.

[26] Kumar, N., Kumar, J., Gerstenkorn, C., Wang, R., Chiu, H.-Y., Smirl, A. L., and Zhao, H. (2013). Third harmonic generation in graphene and few-layer graphite films, *Phys. Rev. B* **87**, p. 121406, doi:10.1103/PhysRevB.87.121406, http://link.aps.org/doi/10.1103/PhysRevB.87.121406.

[27] Lin, K.-H., Weng, S.-W., Lyu, P.-W., Tsai, T.-R., and Su, W.-B. (2014). Observation of optical second harmonic generation from suspended single-layer and bi-layer graphene, *Appl. Phys. Lett.* **105**, 15, p. 151605, doi:10.1063/1.4898065, http://dx.doi.org/10.1063/1.4898065.

[28] Mak, K. F., Sfeir, M. Y., Wu, Y., Lui, C. H., Misewich, J. A., and Heinz, T. F. (2008). Measurement of the optical conductivity of graphene, *Phys. Rev. Lett.* **101**, p. 196405, doi:10.1103/PhysRevLett.101.196405, http://link.aps.org/doi/10.1103/PhysRevLett.101.196405.

[29] Malic, E., Winzer, T., Bobkin, E., and Knorr, A. (2011). Microscopic theory of absorption and ultrafast many-particle kinetics in graphene, *Phys. Rev. B* **84**, 20, p. 205406, doi:10.1103/physrevb.84.205406, http://dx.doi.org/10.1103/PhysRevB.84.205406.

[30] Margulis, V., Muryumin, E., and Gaiduk, E. (2016). Frequency dependence of optical third-harmonic generation from doped graphene, *Phys. Lett. A* **380**, 1-2, p. 304310, doi:10.1016/j.physleta.2015.10.044, http://dx.doi.org/10.1016/j.physleta.2015.10.044.

[31] Margulis, V. A., Muryumin, E. E., and Gaiduk, E. A. (2014). Theoretical calculations of nonlinear refraction and absorption coefficients of doped graphene, *J. Opt.* **16**, 12, p. 125203, doi:10.1088/2040-8978/16/12/125203, http://dx.doi.org/10.1088/2040-8978/16/12/125203.

[32] Marzari, N., Mostofi, A. A., Yates, J. R., Souza, I., and Vanderbilt, D. (2012). Maximally localized wannier functions: Theory and applications, *Rev. Mod. Phys.* **84**, 4, p. 14191475, doi:10.1103/revmodphys.84.1419, http://dx.doi.org/10.1103/RevModPhys.84.1419.

[33] Mikhailov, S. A. (2007). Non-linear electromagnetic response of graphene, *Europhys. Lett.* **79**, 2, p. 27002, http://stacks.iop.org/0295-5075/79/i=2/a=27002.

[34] Mikhailov, S. A. (2011). Theory of the giant plasmon-enhanced second-harmonic generation in graphene and semiconductor two-dimensional electron systems, *Phys. Rev. B* **84**, 4, p. 045432, doi:10.1103/physrevb.84.045432, http://dx.doi.org/10.1103/PhysRevB.84.045432.

[35] Mikhailov, S. A. (2014). Quantum theory of third-harmonic generation in graphene, *Phys. Rev. B* **90**, 24, p. 241301(R), doi:10.1103/physrevb.90.241301, http://dx.doi.org/10.1103/PhysRevB.90.241301.

[36] Mikhailov, S. A. (2016). Quantum theory of the third-order nonlinear electrodynamic effects of graphene, *Phys. Rev. B* **93**, p. 085403, doi:10.1103/PhysRevB.93.085403, http://link.aps.org/doi/10.1103/PhysRevB.93.085403.

[37] Mikhailov, S. A. and Ziegler, K. (2008). Nonlinear electromagnetic response of graphene: frequency multiplication and the self-consistent-field effects, *J. Phys. Condens. Matter* **20**, 38, p. 384204, http://stacks.iop.org/0953-8984/20/i=38/a=384204.

[38] Nair, R. R., Blake, P., Grigorenko, A. N., Novoselov, K. S., Booth, T. J., Stauber, T., Peres, N. M. R., and Geim, A. K. (2008). Fine structure constant defines visual transparency of graphene, *Science* **320**, 5881, p. 1308, doi:10.1126/science.1156965, http://www.sciencemag.org/content/320/5881/1308.abstract.

[39] Säynätjoki, A., Karvonen, L., Riikonen, J., Kim, W., Mehravar, S., Norwood, R. A., Peyghambarian, N., Lipsanen, H., and Kieu, K. (2013). Rapid large-area multiphoton microscopy for characterization of graphene, *ACS Nano* **7**, 10, pp. 8441–8446, doi:10.1021/nn4042909, http://dx.doi.org/10.1021/nn4042909.

[40] Semnani, B., Majedi, A. H., and Safavi-Naeini, S. (2016). Nonlinear quantum optical properties of graphene, *J. Opt.* **18**, 3, p. 035402, doi:10.1088/2040-8978/18/3/035402, http://dx.doi.org/10.1088/2040-8978/18/3/035402.

[41] Sipe, J., Moss, D., and van Driel, H. (1987). Phenomenological theory of optical second- and third-harmonic generation from cubic centrosymmetric crystals, *Phys. Rev. B* **35**, 3, p. 11291141, doi:10.1103/physrevb.35.1129, http://dx.doi.org/10.1103/PhysRevB.35.1129.

[42] Sipe, J. E. and Shkrebtii, A. I. (2000). Second-order optical response in semiconductors, *Phys. Rev. B* **61**, 8, pp. 5337–5352, doi:10.1103/PhysRevB.61.5337.

[43] Wu, R., Zhang, Y., Yan, S., Bian, F., Wang, W., Bai, X., Lu, X., Zhao, J., and Wang, E. (2011). Purely coherent nonlinear optical response in solution dispersions of graphene sheets, *Nano Lett.* **11**, 12, pp. 5159–5164, doi:10.1021/nl2023405, http://dx.doi.org/10.1021/nl2023405.

[44] Wu, S., Mao, L., Jones, A. M., Yao, W., Zhang, C., and Xu, X. (2012). Quantum-enhanced tunable second-order optical nonlinearity in bilayer graphene, *Nano Lett.* **12**, 4, pp. 2032–2036, doi:10.1021/nl300084j, http://

pubs.acs.org/doi/abs/10.1021/nl300084j.

[45] Yang, H., Feng, X., Wang, Q., Huang, H., Chen, W., Wee, A. T. S., and Ji, W. (2011). Giant two-photon absorption in bilayer graphene, *Nano Lett.* **11**, 7, pp. 2622–2627, doi:10.1021/nl200587h, http://dx.doi.org/10.1021/nl200587h.

[46] Zhang, H., Virally, S., Bao, Q., Ping, L. K., Massar, S., Godbout, N., and Kockaert, P. (2012). Z-scan measurement of the nonlinear refractive index of graphene, *Opt. Lett.* **37**, 11, pp. 1856–1858, doi:10.1364/OL.37.001856, http://ol.osa.org/abstract.cfm?URI=ol-37-11-1856.

[47] Zhang, Z. and Voss, P. L. (2011). Full-band quantum-dynamical theory of saturation and four-wave mixing in graphene, *Opt. Lett.* **36**, 23, p. 4569, doi:10.1364/OL.36.004569, http://dx.doi.org/10.1364/OL.36.004569.

Chapter 6

Nonlinear Optical Experiments on Graphene

Hui Zhao
Department of Physics and Astronomy,
The University of Kansas, Lawrence, Kansas 66045, USA
huizhao@ku.edu

This chapter introduces recent experimental studies on nonlinear optical properties of graphene, including third harmonic generation and saturable absorption. Basic principles, experimental designs, and examples of experimental results are presented.

6.1 Introduction

Due to the unique bandstructure of graphene, its interaction with light is strong and broadband in nature. The linear optical absorption coefficient of graphene is as high as 7×10^7 m^{-1}, making it an ideal broadband absorber [Nair *et al.* (2008); Mak *et al.* (2008)]. This property, combined with the high carrier mobilities, makes graphene a promising candidate for applications such as broadband photodetectors [Xia *et al.* (2009); Mueller *et al.* (2010); Gabor *et al.* (2011); Sun *et al.* (2012)] and modulators [Liu *et al.* (2011)]. The strong and broadband coupling of light of graphene also suggests its potential uses as a nonlinear optical material. However, since graphene is centrosymmetric, its second-order nonlinearities are only from its interface with substrate, and are rather weak [Dean and van Driel (2009)]. Hence, third-order nonlinear optical effects are the leading nonlinear responses. Recently, third-order nonlinear optical effects of graphene has been investigated in a number of experiments. For example, four-wave mixing experiment was performed on graphene samples[Hendry *et al.*

(2010)]. Third harmonic generation was observed in graphene and graphite flakes[Kumar *et al.* (2013); Hong *et al.* (2013)]. Two-photon absorption of graphene has been observed [Yang *et al.* (2011)] and used in a quantum interference scheme to inject ballistic photocurrents [Sun *et al.* (2010a)]. Furthermore, saturation absorption has been widely used to study photocarrier dynamics in graphene.

This chapter provides a concise discussion on the basic aspects of nonlinear optical experiments performed on graphene. Principles, experimental designs, and major findings on various nonlinear optical experiments will be introduced. Examples of experimental results will also be summarized on third harmonic generation and saturable absorption.

6.2 Third Harmonic Generation

6.2.1 *Third-order nonlinearities and third harmonic generation*

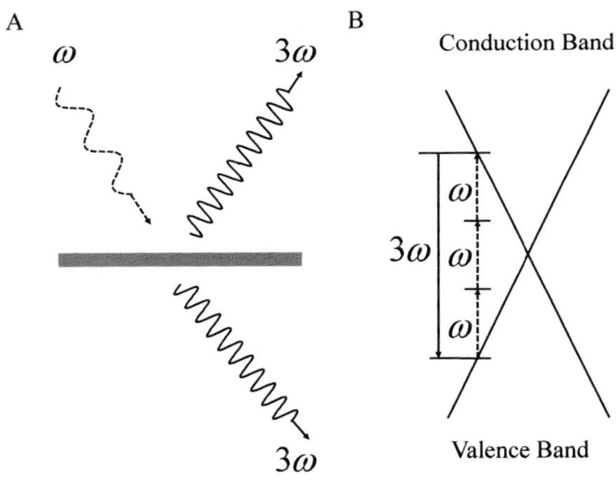

Fig. 6.1 A: Illustration of third harmonic generation process where a fundamental beam of angular frequency ω generates third harmonic radiations at 3ω through nonlinear interaction with the medium. B: Energy diagram of the third harmonic generation process where three photons combine to one photon with triple energy.

The nonlinear optical response of a material can be described by the nonlinear polarization driven by the incident light,

$$P = \varepsilon_0[\chi^{(1)}E(t) + \chi^{(2)}E^2(t) + \chi^{(3)}E^3(t) + ...], \tag{6.1}$$

where $\chi^{(2)}$ and $\chi^{(3)}$ are second- and third-order nonlinear susceptibilities, respectively. Under nonresonant conditions, the magnitudes of these nonlinear optical susceptibilities are on the order of 10^{-12} m/V and 10^{-24} m^2/V^2, respectively [Boyd (2008)]. The third-order polarization,

$$P^{(3)} = \frac{1}{4}\varepsilon_0\chi^{(3)}\mathcal{E}^3\cos 3\omega t + \varepsilon_0\chi^{(3)}\mathcal{E}^3\cos\omega t, \tag{6.2}$$

is the source of third-order nonlinear optical processes. Here, the first term oscillates at the triple frequency of the fundamental frequency, and gives rise to third harmonic generation. In this process, as illustrated in Fig. 6.1A, the fundamental beam with an angular frequency of ω generates its third harmonic beam at 3ω. From the energy point of view, three photons of the fundamental beam combine to one photon with triple photon energy (Fig. 6.1B).

6.2.2 *Experimental setup*

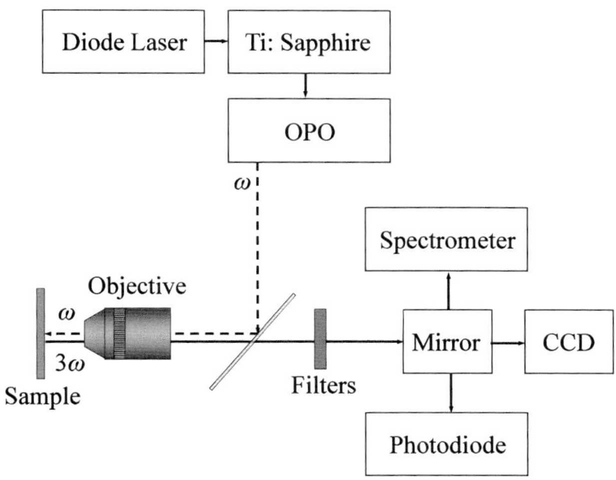

Fig. 6.2 Experimental setup to observe third harmonic generation in graphene and few-layer graphite flakes.

An optical spectroscopic system used to study third harmonic generation in graphene [Kumar *et al.* (2013)] is schematically shown in Fig. 6.2. Here, a high-power and continuous-wave diode laser was used to pump a Ti-doped sapphire laser, which generates 100-fs pulses at a repetition rate of 80 MHz with a average power of 2 W. The output is used to pump an optical parametric oscillator (OPO). The idler output of the OPO was used as the fundamental pulse (ω) of the third harmonic generation measurement. The ω pulse (dashed line in Fig. 6.2) was reflected by a dichroic beamsplitter that reflects near infrared but transmits visible light, and was focused onto the sample by a microscope objective lens with a numerical aperture of 0.4. The focused spot size is about 3.5 μm at the sample location. The third harmonic generated (3ω, solid line in Fig. 6.2) was collected in the reflection geometry by the same objective lens. A combination of color filters is used to prevent the reflected ω beam from reaching the detectors.

The experimental setup was equipped with three types of detection units, which can be selected by using an adjustable mirror. First, a silicon charge-coupled device (CCD) camera, which is connected to a video monitor, provides direct visualization of the sample and the 3ω spot. This unit was used for aligning and positioning the laser spot on the desired position of the sample, and for precisely determining the size of the laser spot. Second, a spectrometer, which is equipped with a thermoelectric-cooled silicon CCD camera, allows spectral resolution of the 3ω and detection of weak 3ω signal. Third, a single-channel silicon photodiode with an internal gain and an external lock-in amplifier was used for high-throughput detection of background-free 3ω signal.

6.2.3 *Examples of experimental results*

In the following, examples of observed third harmonic generation in graphene flakes will be presented [Kumar *et al.* (2013)]. Here, the measurement was performed with monolayer graphene and few-layer graphite flakes produced by standard mechanical exfoliation techniques, using silicon substrates with a 300-nm dioxide layer [Novoselov *et al.* (2005)]. The atomic layer numbers of the fabricated flakes were determined by a combination of optical contrast technique [Blake *et al.* (2007)] and atomic force microscopy.

Figure 6.3 shows an representative spectrum of the third harmonic signal generated by a fundamental pulse of 1720 nm. The fundamental wavelength was determined by observing its second harmonic generated using a barium borate crystal with the same setup. The central wavelength of the third harmonic observed is precisely one third of the fundamental wavelength, providing clear evidence of its origin. Furthermore, no signal was detectable under the same conditions when the laser spot was moved to the substrate,

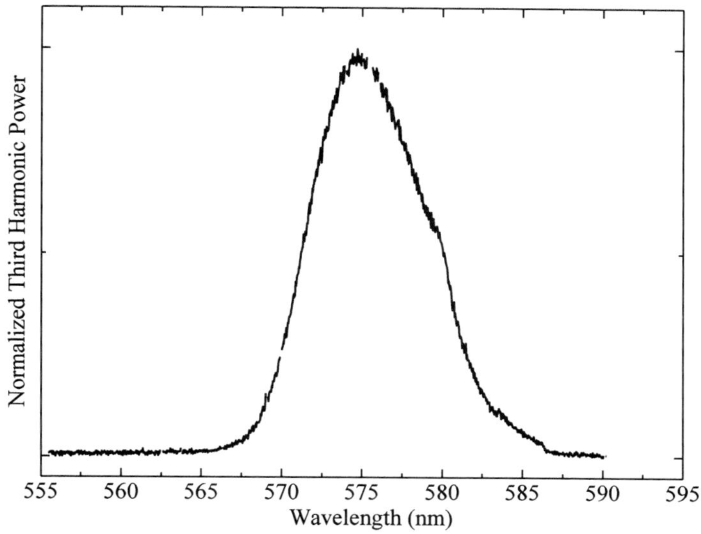

Fig. 6.3 Spectrum of third harmonic generated from a monolayer graphene flake with a fundamental pulse of 1720 nm.

confirming that the graphene flake is responsible for the third harmonic generation process. A series of polarization analysis of the observed signal showed that, for a linearly polarized fundamental pulse, the third harmonic is linearly polarized along the same direction.

In third harmonic generation, the power of the third harmonic is expected to be cubic in the fundamental power [Boyd (2008)]. To confirm that, the average power of the third harmonic beam was measured as a function of the fundamental power, as shown in Fig. 6.4. The experimental results (symbols) are well consistent with the cubic dependence (the solid curve).

The results shown in Fig. 6.4 can be used to estimate the real part of $\chi^{(3)}$. Here, bulk concept of $\chi^{(3)}$ is used to describe the strength of nonlinear response of graphene, in order to facilitate the comparison with other bulk materials. For a thin layer and ignoring the absorption of the fundamental, the third harmonic field amplitude can be written as

$$\mathcal{E}_{3\omega} = \frac{1}{4}\frac{i3\omega}{2n_{3\omega}c}d\chi^{(3)}\mathcal{E}_{\omega}^{3}, \tag{6.3}$$

where c is the speed of light in a vacuum, d is the thickness of the layer, $n_{3\omega}$ is the index of refraction of the material at 3ω [Boyd (2008)]. Using

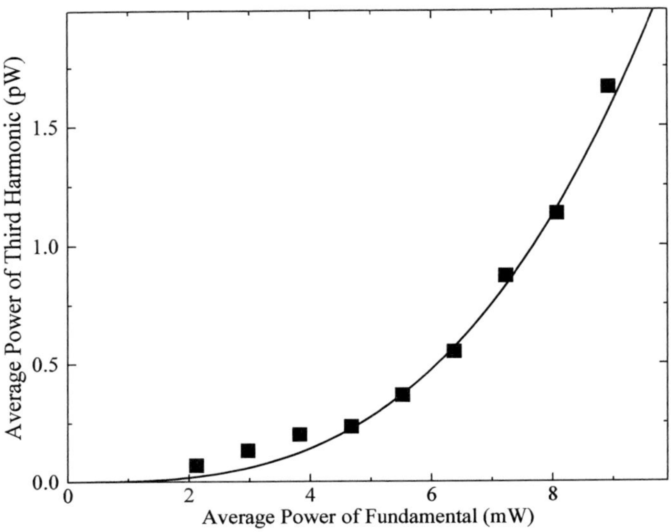

Fig. 6.4 Averaged power of third harmonic generated from a monolayer graphene flake as a function of the fundamental power. The solid curve is a cubic fit.

the fundamental Gaussian pulse of

$$I_\omega(r,t) = I_0 e^{-4\ln 2(r/w)^2} e^{-4\ln 2(t/\tau)^2}, \tag{6.4}$$

with a spot diameter w and a pulsewidth τ (both in full width at half maximum). The peak intensity, I_0, is related to the average power, $\bar{P_\omega}$, by

$$\bar{P_\omega} = F \frac{\sqrt{\pi}\tau}{\sqrt{4\ln 2}} \frac{\pi w^2}{4\ln 2} I_0, \tag{6.5}$$

where F is the repetition rate of the fundamental pulses. Using the experimental and material parameters of d = 0.33 nm, $n_\omega \approx n_{3\omega} = 2.4$, w = 3.5 μm, τ = 320 fs, λ =1720.4 nm, and F = 81 MHz, a value of $\chi^{(3)} \sim 0.4 \times 10^{-16}$ m^2/V^2 was obtained [Kumar *et al.* (2013)].

Further measurements were performed to investigate the third harmonic generation as the sample evolves from monolayer graphene to multilayer graphite. Figure 6.5 shows the generated third harmonic power from samples with different numbers of atomic layers. The solid curve is calculated by assuming a thickness independent $\chi^{(3)}$ of $\sim 0.8\times 10^{-16}$ m^2/V^2. The consistence between the data and the calculation suggests that $\chi^{(3)}$ is independent of the thickness. That is, monolayer graphene and multilayer graphite have similar third harmonic generation properties (per unit length).

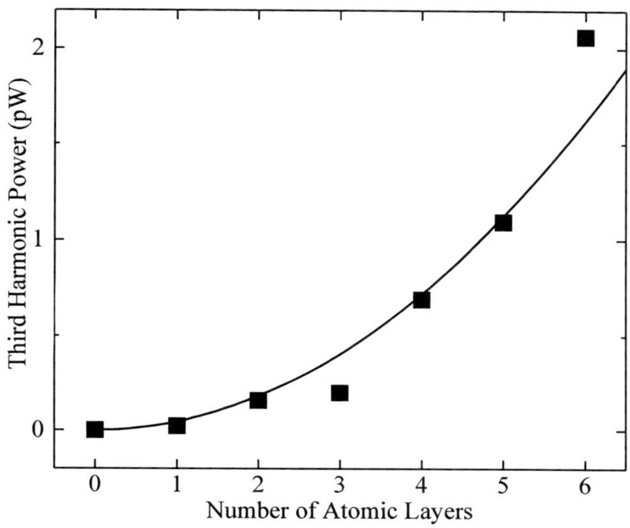

Fig. 6.5 Average power of the third harmonic beam as a function of the number of atomic layer. The solid curve is a quadratic fit.

To compare the magnitude of the third-order nonlinear optical susceptibility of graphene and thin graphite flakes with other optical materials, we note that this value is several orders of magnitude larger than transparent materials [Boyd (2008)]. This can be attributed to the enhancement originates from the resonant nature of the intermediate and final transitions of third harmonic generation, as illustrated in Fig. 6.1B. On the other hand, $\chi^{(3)}$ of graphene and graphite is comparable to metal nanoparticles with similar resonant enhancements [Boyd (2008)]. It should be pointed out that the resonant enhancement is accompanied by linear absorption of both fundamental and third harmonic beams, which are detrimental to the conversion efficiency, which is often the major considerations for applications involving third harmonic generation.

The discussions above are based on experiments performed with graphene and graphite samples that were mechanically exfoliated from graphite bulk crystals. Simultaneously, third harmonic generation was also observed from graphene samples fabricated by chemical vapor deposition, and a similar magnitude of $\chi^{(3)}$ was obtained [Hong *et al.* (2013)]. Furthermore, microscopy tools based on third harmonic generation were developed for rapid characterization of graphene and few-layer graphite [Saynatjoki *et al.* (2013)].

6.3 Saturable Absorption and Transient Absorption Spectroscopy

6.3.1 *Saturable absorption*

Saturable absorption is another important third order nonlinear optical process. In this effect, for high light intensities, the absorption coefficient of a sample becomes a function of the light intensity. In most cases, the absorption coefficient decreases with the intensity in the form of

$$\alpha = \frac{\alpha_0}{1 + I/I_S}, \tag{6.6}$$

where α_0 is the low-intensity (linear) absorption coefficient, with I_S been called saturation intensity. Recently, satuable absorption of graphene has been observed, and used as saturable absorbers in mode-locking pulsed lasers [Sun *et al.* (2010b); Bao *et al.* (2009)].

One important mechanism of saturable absorption is the phase-space state filling effect of photocarriers. In this mechanism, the photocarriers excited by the incident light occupy some of the energy states in the conduction and valence bands, and hence prevent further excitation of photocarriers to these states. Due to the finite lifetime of these carriers in these energy states, the absorption saturation effect occurs not only to the light pulse that excites these carriers, but also exist for a time-delayed secondary pulse, so long as it arrives at a time when these states are still occupied. Such as time-delayed and incoherent absorption saturation, also known as transient absorption, provides an effective way to monitor the dynamics of the photocarriers.

6.3.2 *Transient absorption spectroscopy*

In a transient absorption measurement utilizing saturable absorption, a strong pulse, called pump, is used to excite photocarriers. The differential absorption, that is, the relative change of the absorption coefficient, of the sample experienced by a time-delayed and weak probe pulse can be written as

$$\frac{\Delta\alpha}{\alpha_0} = \frac{\alpha(N) - \alpha_0}{\alpha_0} = -\frac{N}{N + N_S}, \tag{6.7}$$

where N and N_S are the carrier density and saturation density, respectively. The negative sign reflects the fact that the absorption is reduced by the population of carriers. The saturation density represents the density when the absorption was reduced by half. In most situations, the photocarrier density used in measurements is only a small fraction of the total density of states, and hence $N \ll N_S$. Under these conditions, we have $|\Delta\alpha/\alpha_0| \approx N/N_S \ll 1$.

In practice, differential absorption can only be measured through differential transmission or differential reflection. In the transmission geometry, for example, with a sample of thickness of L, the differential transmission can be written as

$$\frac{\Delta T}{T_0} = \frac{T(N) - T_0}{T_0} = e^{-\Delta\alpha L} - 1, \tag{6.8}$$

where T and T_0 are the transmission coefficients of the sample with and without the presence of photocarriers, respectively. Under the condition of $\Delta\alpha L \ll 1$ and $N \ll N_S$, we have

$$\frac{\Delta T}{T_0} = -\Delta\alpha L = \alpha_0 L \frac{N}{N_S}. \tag{6.9}$$

Hence, under these conditions, the differential transmission is proportional to the carrier density. Such a linear relation between the differential transmission signal and the photocarriers density to be monitored is highly desired. It is often confirmed by measuring the peak differential transmission signal as a function of the injected carrier density, which can be controlled by and deduced from the pump pulse fluence. In principle, however, differential transmission can still be used to monitor photocarrier dynamics even if such a linear relation cannot be assumed, so long as the relation is determined experimentally. In the measurements, the differential transmission is measured as a function of the probe delay, probe location, and probe photon energy. From these results, and by using the relation between the signal and the carrier density, one can, at least in principle, deduce the carrier density as a function of time, space, and energy after the pump excitation.

In situations where a transmission measurement is inconvenient or even impossible to perform, one can study the carrier dynamics by measuring differential reflection, defined as $\Delta R/R_0 = [R(N) - R_0]/R_0$, where R and R_0 are the reflection coefficients of the sample with and without the presence of photocarriers, respectively. Due to the complex nature of reflection, especially when multiple reflection from the substrate is involved, the relation between the differential reflection and the differential absorption can be rather complicated. However, in most cases, the differential reflection is still approximately proportional to the carrier density.

6.3.3 *Experimental setup*

Figure 6.6 shows the configuration of a typical transient absorption setup. The pump and probe pulses are incident to the sample from the same side. The pump pulse first goes through a retroreflector, which is installed on a linear stage. By moving the retroreflector along the direction of the incident beam, the travel distance, and thus the arrival time of the pump pulse at the sample, can be varied. A beamsplitter is used to send the pump pulse to a lens (usually a microscope objective lens with high numerical aperture in order to study small flakes), which focused the pump pulse to the sample. The probe pulse is reflected to the lens with by another beamsplitter, and is focused similarly. The reflection probe is collected by the same lens, and is sent to a photodiode as part of a balanced detector. A set of filters is used to prevent the reflected pump from reaching the photodiode. If the graphene sample is on a transparent substrate, the transmitted probe can also be collected by another lens, and sent to the same photodiode. In this configuration, one can choose to measure either differential reflection or differential transmission.

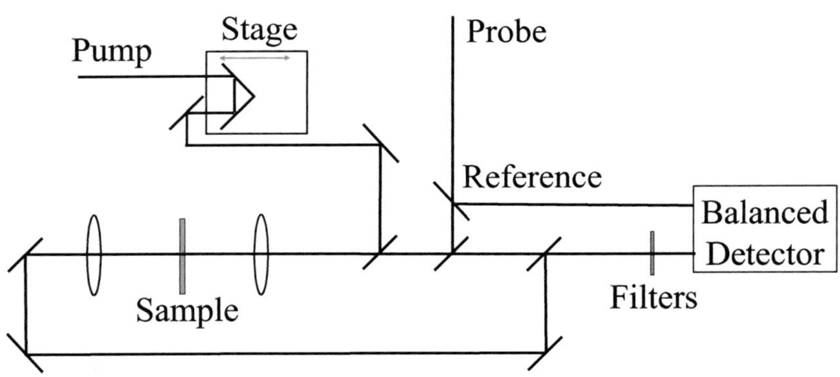

Fig. 6.6 Configuration of a typical transient absorption spectroscopy setup.

Before reaching the beamsplitter, a portion of the probe is picked up and sent to the other photodiode of the balanced detector, which is used as a reference beam. The power of this beam is carefully adjusted to match the probe power reaching the photodiode, when the pump beam is blocked. The balance detector outputs a voltage that is proportional to the difference of the optical powers received by the two photodiode. With proper alignment and adjustment, this output is zero without the pump pulse. With the pump pulse injecting photocarriers, and if the probe pulse arrives at the sample while the photocarriers still populate the energy states the probe couples, the absorption coefficient for the probe pulse is different from the linear absorption coefficient. Hence, the reflected or transmitted probe power is different from the reference, causing a nonzero output voltage of the balanced detector, which is proportional to the differential transmission or reflection. In this configuration, the differential transmission or reflection can be measured as a function of the time delay of the probe pulse with respect to the pump pulse to study the decay process of the photocarriers. When needed, the signal can also be measured as a function of the probe location, achieved by moving the probe laser spot on the sample with respect to the pump spot by tilting the beamsplitter that sends the probe to the focusing lens.

The advantage of using the balanced detector is that, the noise of the probe laser beam is equally distributed between the probe and reference beams, and is hence largely canceled since the balanced detector outputs the difference of the two. The degree of the cancellation depends on the quality of the beam alignment and the balanced detector used. Typically, the output of the balanced detector is measured with a lock-in amplifier, which is referenced to modulation of the pump intensity, to further enhance the signal-to-noise ratio.

In the following sections, examples of transient absorption measurements on graphene will be presented.

6.3.4 *Transient absorption of graphene*

The first example to be presented is on differential transmission measurements of reduced graphene oxides. Graphene oxide flakes were synthesized by using a modified Hummers' method [Hummeres and Offeman (1958)], and were spin coated on a quartz substrate to form a film. The graphene oxide film was transformed to a multilayer graphene film by thermal reduction at 1000°C [Wang *et al.* (2010b)]. The number of graphene layers is controlled to be about 50 in order to facilitate differential transmission measurements. The film absorbs about 50% of incident light in visible range.

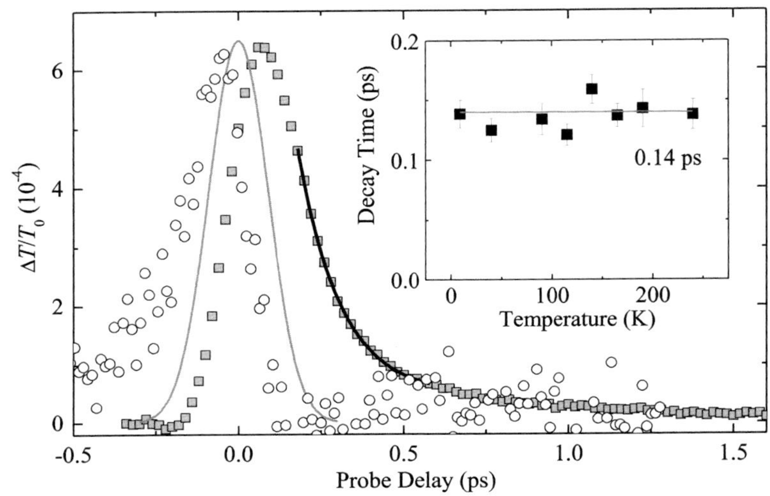

Fig. 6.7 $\Delta T/T_0$ measured from a reduced graphene oxide film at a lattice temperature of 9 K with a 750-nm pump and 810-nm probe pulses (filled squares). The solid curve over the squares is an exponential fit. The open circles are measured under the same conditions, but with the 810-nm pulse acting as the pump and 750-nm pulse the probe, and was scaled to match the height of the other data. The gray curve shows the cross-correlation of the two pulses, which represents the instrumental response time. The inset shows the deduced decay time as a function of the lattice temperature.

The filled squares in the main panel of Fig. 6.7 show the differential transmission signal measured with a probe pulse of 810-nm, with a pump pulse of 750-nm injecting a peak photocarrier density of 8.0×10^{12} cm^{-2} [Ruzicka *et al.* (2010b)]. A very short transient was observed, indicating a short lifetime of the photocarriers. In order to determine the zero probe delay, the differential transmission measurement was repeated with the same setup, but with the 750-nm pulse as the probe and the 810-nm pulse as the pump. The results are shown as the open circles in Fig. 6.7, which has been scaled to match the height of the other signal. Note that the horizontal axis of Fig. 6.7 is defined as the time lag of the 810-nm pulse with respect to the 750-nm pulse. Hence, it gives the negative of the probe delay when the 750-nm acts as the probe. The gray curve is the cross-correlation of the pump and probe pulses calculated from the pulsewidths. Clearly, the carrier dynamics persist longer than the cross-correlation.

To quantitatively describe the decay of the signal, a portion of the decay curve, from 0.2 to 0.6 ps, was fit by a single exponential function with a time constant of about 0.14 ps, as indicated as the solid curve in Fig. 6.7. By repeating the measurement at different lattice temperatures, it was found that the decay time is independent of the lattice temperature, as shown in the inset of Fig. 6.7. Note that the probe monitors population of the energy states that are about 766 meV above the Dirac points. Hence, this time constant reflects the carrier lifetime on such high energy states. It is not the recombination lifetime of photocarriers.

6.3.5 *Energy relaxation*

The ultrashort decay time of the transient absorption shown in Fig. 6.7 makes it challenging to study details of the photocarrier dynamics. Photocarriers are injected with a nonthermal energy distribution, which is close to a Gaussian distribution dictated by the spectrum of the pump pulse. The first process these carriers undergo is thermalization, during which the carrier system establish a Fermi-Dirac distribution with a well defined temperature. This thermalization process is achieved by inelastic carrier-carrier scattering, through which the carriers exchange their energy. It is known that carrier-carrier scattering is enhanced in graphene due to its linear energy dispersion to a scattering time as short as 10 fs [Xing *et al.* (2010)]. Indeed, ultrafast pump-probe [George *et al.* (2008); Wang *et al.* (2010a); Huang *et al.* (2010)] and nonlinear photoluminescence experiments [Liu *et al.* (2010); Lui *et al.* (2010); Stöhr *et al.* (2010)] have indicated that the thermalization of high density carriers in graphene is extremely fast. Therefore, carriers injected by pump pulses of about 100 fs is expected to reach a Fermi-Dirac distribution within the pulse duration. The next process the photocarriers undergo is energy relaxation, or cooling. In this process, the carrier distribution remains Fermi-Dirac, but with a temperature decreasing towards lattice temperature. This process is achieved by phonon emission of carriers. The thermalized and cooled photocarriers then recombine, radiatively or nonradiatively, to complete the life cycle of photocarriers.

A two-probe scheme was used to study the energy relaxation of hot photocarriers [Ruzicka *et al.* (2011)]. The basic idea of this approach is illustrated in the left panel of Fig. 6.8. A pump pulse of 800-nm (with

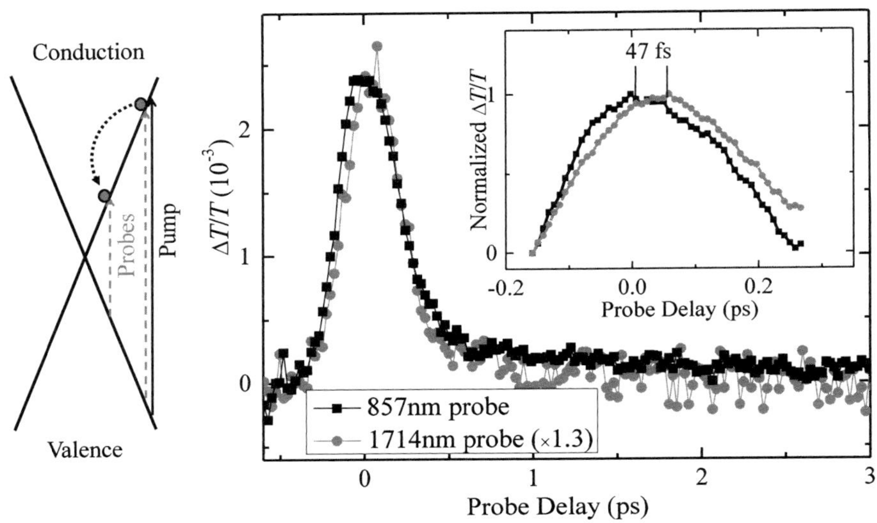

Fig. 6.8 Left: Illustration of the two-probe scheme. The photocarriers injected by a pump pulse are monitored by two probe pulses with different photon energies. Right: Differential reflection signal measured with a 857-nm (squares) and 1714-nm (circles) probes.

a photon energy of 1.550 eV, solid vertical line) inject photocarriers with an excess energy of 775 meV. Assuming that the thermalization process is much faster than the 100-fs pulse duration, the injected photocarriers form a Fermi-Dirac distribution with an average energy of 775 meV. These carriers are monitored by separately measuring differential transmission signals of two probe pulses of wavelengths of 857 and 1714 nm, respectively. According to their photon energies, they probe energy states that are 720 and 360 meV from the Dirac points, respectively (dashed vertical lines). Hence, the differential transmission of the 857-nm probe reaches a peak when the population of the energy states of 720 meV reaches a maximum. A simple analysis of the Fermi-Dirac distribution indicates that this occurs when the average energy of carriers is also 720 meV. Similarly, the differential transmission signal of the 1714-nm probe peaks when the average energy drops to 360 meV. Hence, the difference between the two peaks is the time duration for the carriers to relax their average energy from 720 to 360 meV.

The right panel of Fig. 6.8 shows the results of such a measurement on a reduced graphene oxide sample. As clearly shown in the inset, the signal of the 1714-nm probe peaks at 47 fs after the 857-nm probe. For this experiment, it is crucial to precisely control the relative arrival time of the two probe pulses. This was achieved by monitoring sum-frequency generation of the two pulses in a [110] oriented GaAs thin film. This procedure allows determination of the arrival times of the two probe pulses with an accuracy better than 5 fs. The time of 47 fs for energy relaxation of 360 meV corresponds to a rate of 8 meV/fs. Furthermore, since the major energy relaxation channel is the emission of G-mode optical phonons with an energy 0.195 eV, [Berciaud *et al.* (2010)] the energy relaxation rate measured gives an optical phonon emission time of about 25 fs. The deduced value of the optical phonon emission time is consistent with theoretical calculations. [Tse and Das Sarma (2007); Tse *et al.* (2008)]

6.3.6 *Photocarrier diffusion*

As the last example, the transient absorption measurements performed with high spatial resolution can be used to monitor the diffusion process of photocarriers in a graphene layer [Ruzicka *et al.* (2010a)].

In such a measurement, a tightly focused pump pulse injects photocarriers with a thin Gaussian distribution, dictated by the shape of the laser spot. Due to the in-plane density gradient, photocarriers diffuse in the graphene layer, causing a broadening of the distribution. In this process, the electron and hole moves as a unit due to their Coulomb attraction, which is known as ambipolar diffusion. A time-delayed and spatially scanned probe pulse is used to monitor the spatiotemporal dynamics of the photocarriers. Figure 6.9C shows an example of the measured spatial profiles of differential transmission, which is proportional to the spatial profile of the photocarrier density, with a probe delay of 80 fs. The measurement was performed with an epitaxial graphene sample.

The diffusion of photocarriers in graphene follows the diffusion equation,

$$\frac{\partial N}{\partial t} = D\nabla^2 N - \frac{N}{\tau_r}, \tag{6.10}$$

where D is the diffusion coefficient and τ_r is the lifetime of the carriers. With a Gaussian initial distribution, solution of the diffusion equations gives the evolution of the density as [Zhao *et al.* (2009)]

$$N(r,t) = N_0 \left(\frac{w_0^2}{w^2(t)}\right) e^{-4\ln(2)r^2/w^2(t) - t/\tau_r}, \tag{6.11}$$

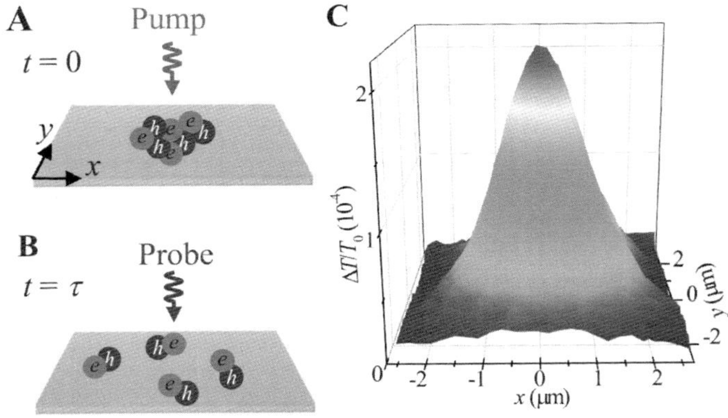

Fig. 6.9 A: A tightly focused pump laser pulse injects photocarriers with a thin Gaussian distribution across the spot. B: The injected photocarriers diffusion in the graphene layer, causing a broadening of the profile. The diffusion process is monitored by spatially resolving differential transmission of a tightly focused probe pulse. C: An example of the differential transmission profile measured with a probe delay of 80 fs, which reflects the spatial profile of photocarrier density. (For color reference, go to page 489.)

where

$$w^2(t) = w_0^2 + 16\ln(2)Dt \tag{6.12}$$

is the squared full-width of half maximum of the profile.

To determine the diffusion coefficient, the profiles along the $\hat{x}$ axis for many probe delays were acquired, as shown in Fig. 6.10A. The red curve in Fig. 6.10B shows the decay of the signal at $x = 0$. This decay reflects the movement of carriers out of the probing states of 775 meV, from which one can estimate the evolution of the carrier temperature, as shown as the blue line in Fig. 6.10B. At all probe delays, the spatial profiles of the differential transmission signal indeed remain a Gaussian shape. Figure 6.10C shows a few examples with various probe delays as labeled. From Gaussian fits to the profiles measured in the range of probe delays from 0.08 to 0.24 ps, the expansion of the squared width, $w^2(\tau) - w^2(\tau_0)$, where $w(\tau_0)$ is the width of the profile at probe delay $\tau_0 = 0.08$ ps, was obtained, and was shown in Fig. 6.10D. From a linear fit (solid line), a diffusion coefficient of $D = 1.2 \times 10^4$ cm^{2}s^{-1} was deduced.

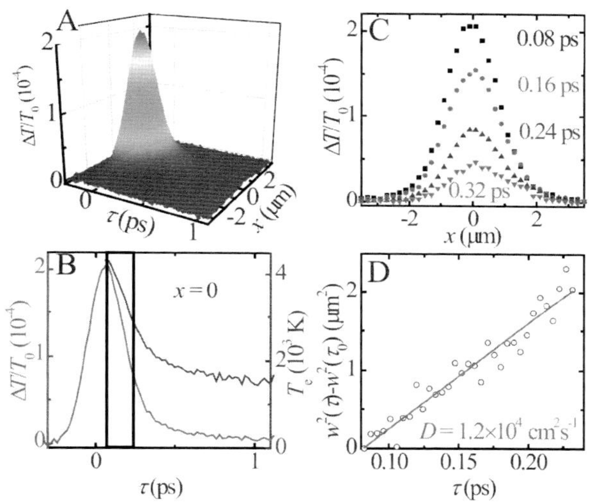

Fig. 6.10 A: The differential transmission signal measured as a function of probe position (x) and the probe delay (τ). B: A cross section of A with $x = 0$ (red line, left axis) and the deduced carrier temperature (blue line, right axis). C: Examples of the spatial profiles measured with different values of the probe delay. D: The squared width of the profile as a function of probe delay. The solid line is a linear fit with a diffusion coefficient of $D = 1.2 \times 10^4$ cm^{2}s^{-1}. (For color reference, go to page 489.)

Bibliography

[1] Bao, Q., Zhang, H., Wang, Y., Ni, Z., Yan, Y., Shen, Z. X., Loh, K. P., and Tang, D. Y. (2009). Atomic-layer graphene as a saturable absorber for ultrafast pulsed lasers, *Adv. Funct. Mater.* **19**, pp. 3077–3083.

[2] Berciaud, S., Han, M. Y., Mak, K. F., Brus, L. E., Kim, P., and Heinz, T. F. (2010). Electron and optical phonon temperatures in electrically biased graphene, *Phys. Rev. Lett.* **104**, p. 227401.

[3] Blake, P., Hill, E. W., Neto, A. H. C., Novoselov, K. S., Jiang, D., Yang, R., Booth, T. J., and Geim, A. K. (2007). Making graphene visible, *Appl. Phys. Lett.* **91**, p. 063124.

[4] Boyd, R. W. (2008). *Nonlinear Optics*, 3rd edn. (Academy Press, San Diego, USA).

[5] Dean, J. J. and van Driel, H. M. (2009). Second harmonic generation from graphene and graphitic films, *Appl. Phys. Lett.* **95**, p. 261910.

[6] Gabor, N. M., Song, J. C. W., Ma, Q., Nair, N. L., Taychatanapat, T., Watanabe, K., Taniguchi, T., Levitov, L. S., and Jarillo-Herrero, P. (2011). Hot carrier-assisted intrinsic photoresponse in graphene, *Science* **334**, p. 648.

[7] George, P. A., Strait, J., Dawlaty, J., Shivaraman, S., Chandrashekhar, M., Rana, F., and Spencer, M. G. (2008). Ultrafast optical-pump terahertz-probe spectroscopy of the carrier relaxation and recombination dynamics in epitaxial graphene, *Nano Lett.* **8**, p. 4248.

[8] Hendry, E., Hale, P. J., Moger, J., Savchenko, A. K., and Mikhailov, S. A. (2010). Coherent nonlinear optical response of graphene, *Phys. Rev. Lett.* **105**, p. 097401.

[9] Hong, S. Y., Dadap, J. I., Petrone, N., Yeh, P. C., Hone, J., and Osgood, R. M. (2013). Optical third-harmonic generation in graphene, *Phys. Rev. X* **3**, p. 021014.

[10] Huang, L., Hartland, G. V., Chu, L. Q., Luxmi, Feenstra, R. M., Lian, C., Tahy, K., and Xing, H. (2010). Ultrafast transient absorption microscopy studies of carrier dynamics in epitaxial graphene, *Nano Lett.* **10**, p. 1308.

[11] Kumar, N., Kumar, J., Gerstenkorn, C., Wang, R., Chiu, H. Y., Smirl, A. L., and Zhao, H. (2013). Third harmonic generation in graphene and few-layer graphite films, *Phys. Rev. B* **87**, p. 121406.

[12] Liu, M., Yin, X. B., Ulin-Avila, E., Geng, B. S., Zentgraf, T., Ju, L., Wang, F., and Zhang, X. (2011). A graphene-based broadband optical modulator, *Nature* **474**, p. 64.

[13] Liu, W.-T., Wu, S. W., Schuck, P. J., Salmeron, M., Shen, Y. R., and Wang, F. (2010). Nonlinear broadband photoluminescence of graphene induced by femtosecond laser irradiation, *Phys. Rev. B* **82**, p. 081408R.

[14] Lui, C. H., Mak, K. F., Shan, J., and Heinz, T. F. (2010). Ultrafast photoluminescence from graphene, *Phys. Rev. Lett.* **105**, p. 127404.

[15] Mak, K. F., Sfeir, M. Y., Wu, Y., Lui, C. H., Misewich, J. A., and Heinz, T. F. (2008). Measurement of the optical conductivity of graphene, *Phys. Rev. Lett.* **101**, p. 196405.

[16] Mueller, T., Xia, F. N. A., and Avouris, P. (2010). Graphene photodetectors for high-speed optical communications, *Nat. Photonics* **4**, p. 297.

[17] Nair, R. R., Blake, P., Grigorenko, A. N., Novoselov, K. S., Booth, T. J., Stauber, T., Peres, N. M. R., and Geim, A. K. (2008). Fine structure constant defines visual transparency of graphene, *Science* **320**, p. 1308.

[18] Novoselov, K. S., Geim, A. K., Morozov, S. V., Jiang, D., Katsnelson, M. I., Grigorieva, I. V., Dubonos, S. V., and Firsov, A. A. (2005). Two-dimensional gas of massless Dirac fermions in graphene, *Nature* **438**, pp. 197–200.

[19] Ruzicka, B. A., Kumar, N., Wang, S., Loh, K. P., and Zhao, H. (2011). Two-probe study of hot carriers in reduced graphene oxide, *J. Appl. Phys.* **109**, p. 084322.

[20] Ruzicka, B. A., Wang, S., Werake, L. K., Weintrub, B., Loh, K. P., and Zhao, H. (2010a). Hot carrier diffusion in graphene, *Phys. Rev. B* **82**, p. 195414.

[21] Ruzicka, B. A., Werake, L. K., Zhao, H., Wang, S., and Loh, K. P. (2010b). Femtosecond pump-probe studies of reduced graphene oxide thin films, *Appl. Phys. Lett.* **96**, p. 173106.

[22] Saynatjoki, A., Karvonen, L., Riikonen, J., Kim, W., Mehravar, S., Norwood, R. A., Peyghambarian, N., Lipsanen, H., and Kieu, K. (2013). Rapid large-area multiphoton microscopy for characterization of graphene, *ACS Nano* **7**, pp. 8441–8446.

[23] Stöhr, R. J., Kolesov, R., Pflaum, J., and Wrachtrup, J. (2010). Fluorescence of laser-created electron-hole plasma in graphene, *Phys. Rev. B* **82**, p. 121408R.

[24] Sun, D., Aivazian, G., Jones, A. M., Ross, J. S., Yao, W., Cobden, D., and Xu, X. D. (2012). Ultrafast hot-carrier-dominated photocurrent in graphene, *Nat. Nanotechnol.* **7**, p. 114.

[25] Sun, D., Divin, C., Rioux, J., Sipe, J. E., Berger, C., de Heer, W. A., First, P. N., and Norris, T. B. (2010a). Coherent control of ballistic photocurrents in multilayer epitaxial graphene using quantum interference, *Nano. Lett.* **10**, pp. 1293–1296.

[26] Sun, Z. P., Hasan, T., Torrisi, F., Popa, D., Privitera, G., Wang, F. Q., Bonaccorso, F., Basko, D. M., and Ferrari, A. C. (2010b). Graphene mode-locked ultratast laser, *ACS Nano* **4**, pp. 803–810.

[27] Tse, W. K. and Das Sarma, S. (2007). Phonon-induced many-body renormalization of the electronic properties of graphene, *Phys. Rev. Lett.* **99**, p. 236802.

[28] Tse, W. K., Hwang, E. H., and Das Sarma, S. (2008). Ballistic hot electron transport in graphene, *Appl. Phys. Lett.* **93**, p. 023128.

[29] W. S. Hummeres Jr and Offeman, R. E. (1958). Preparation of graphitic oxide, *J. Am. Chem. Soc.* **80**, p. 1339.

[30] Wang, H., Strait, J. H., George, P. A., Shivaraman, S., Shields, V. B., Chandrashekhar, M., Hwang, J., Rana, F., Spencer, M. G., Ruiz-Vargas, C. S., and Park, J. (2010a). Ultrafast relaxation dynamics of hot optical phonons in graphene, *Appl. Phys. Lett.* **96**, p. 081917.

[31] Wang, S. A., Ang, P. K., Wang, Z. Q., Tang, A. L. L., Thong, J. T. L., and Loh, K. P. (2010b). High mobility, printable, and solution-processed graphene electronics, *Nano Lett.* **10**, p. 92.

[32] Xia, F. N., Mueller, T., Lin, Y. M., Valdes-Garcia, A., and Avouris, P. (2009). Ultrafast graphene photodetector, *Nat. Nanotechnol.* **4**, p. 839.

[33] Xing, G., Guo, H., Zhang, X., Sum, T. C., and Huan, C. H. A. (2010). The physics of ultrafast saturable absorption in graphene, *Opt. Express* **18**, p. 4564.

[34] Yang, H. Z., Feng, X. B., Wang, Q., Huang, H., Chen, W., Wee, A. T. S., and Ji, W. (2011). Giant two-photon absorption in bilayer graphene, *Nano Lett.* **11**, pp. 2622–2627.

[35] Zhao, H., Mower, M., and Vignale, G. (2009). Ambipolar spin diffusion and D'yakonov-Perel' spin relaxation in GaAs quantum wells, *Phys. Rev. B* **79**, p. 115321.

Part 3

Terahertz Response

Chapter 7

Optical Response of Graphene under Intense Terahertz Fields

Jun Zhou[1] and Ming-Wei Wu[2,*]

[1] *Center for Phononics and Thermal Energy Science, School of Physics Science and Engineering, Tongji University, Shanghai 200092, China*

[2] *Hefei National Laboratory for Physical Sciences at Microscale and Department of Physics, University of Science and Technology of China, Hefei, Anhui 230026, China*

[*] *mwwu@ustc.edu.cn*

Owing to the recent developments of the intense terahertz radiation technology, the optical response of graphene under pulsed and continuous wave intense terahertz fields are of great interest. Due to the exotic band structure, the dynamics of carriers excited by photon of meV energy in graphene are significantly different from that in conventional semiconductors. Strong nonlinear effects are found from experimental observations and theoretical calculations. The interplay between the interband transition and the intraband transition leads to novel transport and optical properties. In the presence of continuous wave terahertz fields, the electronic structure of graphene is found to be modified due to the appearance of the Floquet states. A dynamical energy gap near the Dirac points can be opened and controlled by the terahertz fields.

7.1 Introduction

Graphene has attracted tremendous interests in the past decade because of its novel electronic properties [Castro Neto *et al.* (2009); Das Sarma *et al.* (2011)]. Its exact two-dimensional hexagonal crystal structure leads to the absence of energy gap and the linear dispersion relations near two distinguishable Dirac points noted as K and K′. This gapless band structure with linear dispersion relations, as shown in Fig. 7.1, gives rise to exotic transport and optical properties of graphene which are remarkably different from the properties of conventional semiconductors. In the case of conventional semiconductors, for instance, the interband transition occurs only when the photon energy exceeds the band gap. The case is different in graphene, according to the Pauli blocking, the interband transition occurs when the photon energy is larger than twice of the Fermi energy (E_F), which is tunable and usually in the range from zero to hundreds of meV, as shown in Fig. 7.1(a). In undoped graphene with zero E_F, the interband transition can ideally take place for arbitrary low photon energy. Moreover, zero effective mass of carriers near the Dirac points results in novel intraband transition under low energy photon excitation as shown in Fig. 7.1(b). Therefore, both the interband and the intraband transitions can be realized simultaneously when photon energy is down to the infrared or terahertz (THz) regime. The optical response of graphene under THz fields, whose photon energy is typically several meV, is worthy of investigation for both basic scientific research and practical potential applications [Bonaccorso *et al.* (2010); Vakil and Engheta (2011)].

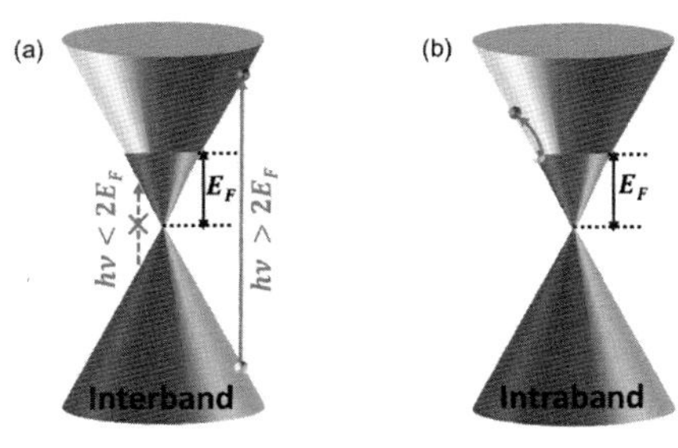

Fig. 7.1 Schematic diagrams of (a) the interband transition which occurs when $h\nu > 2E_F$ and (b) the intraband transition in the presence of optical excitation. hν is the photon energy and E_F is the Fermi energy.

This chapter gives an overview of recent progresses on the optical response of graphene under intense THz fields. Both the ultrashort pulsed THz fields and the continuous wave (cw) THz fields are discussed.

7.1.1 Models

Hamiltonian

Considering the interaction of graphene and THz field near the Dirac points, the 2×2 effective Hamiltonian of graphene near K and K′ points in the presence of an in-plane (noted as x-y plane which is parallel to graphene) THz field is [Sun and Wu (2013)]

$$\begin{aligned} H^{\mu,\mathrm{AB}}(\mathbf{k},t) &= H_0^{\mu,\mathrm{AB}}(\mathbf{k}) + H_{\mathrm{photon}}^{\mu,\mathrm{AB}}(\mathbf{k},t) \\ &= v_F \begin{pmatrix} 0 & \mu\Pi_x - i\Pi_y \\ \mu\Pi_x + i\Pi_y & 0 \end{pmatrix}. \end{aligned} \tag{7.1}$$

Here $H_0^{\mu,\mathrm{AB}}(\mathbf{k})$ is the electron effective Hamiltonian [DiVincenzo and Mele (1984)] and $H_{\mathrm{photon}}^{\mu,\mathrm{AB}}(\mathbf{k},t)$ is the electron-optical-field interaction Hamiltonian under the vector potential gauge [Peres *et al.* (2006)]. v_F is the Fermi velocity and e is the elementary charge. $\Pi_x = k_x + eA_x(t)$, $\Pi_y = k_y + eA_y(t)$ where $\mathbf{k} = (k_x, k_y)$ is the wave vector of electron and $\mathbf{A}(t) = (A_x, A_y)$ is the vector potential. The Planck constant $\hbar$ and the light speed c are set to be unit for convenience throughout this chapter. $\mu = 1$ and -1 denote K and K′ valleys, respectively. The Hilbert space is formed by A and B sublattices of hexagonal lattice. Different types of THz fields can be introduced by choosing different forms of $\mathbf{A}(t)$. The vector potential for cw THz field is [Zhou and Wu (2011)]

$$\mathbf{A}(t) = \frac{E_0}{\Omega}(-\mathbf{e}_x\cos\theta_{\mathbf{E}}\sin\Omega t + \mathbf{e}_y\sin\theta_{\mathbf{E}}\cos\Omega t). \tag{7.2}$$

The corresponding electric field is

$$\mathbf{E}(t) = E_0(\mathbf{e}_x\cos\theta_{\mathbf{E}}\cos\Omega t + \mathbf{e}_y\sin\theta_{\mathbf{E}}\sin\Omega t). \tag{7.3}$$

Here Ω and E_0 are the frequency and the strength of THz field, respectively. $\theta_{\mathbf{E}}$ describes the polarization of the field. A circularly polarized THz field is introduced by choosing $\theta_{\mathbf{E}} = \pi/4$ and linearly polarized THz fields along x- and y-directions are introduced by choosing $\theta_{\mathbf{E}} = 0$ and $\pi/2$, respectively.

Similarly, a linearly polarized pulsed THz field along x-direction is introduced by considering an electric field with Gaussian function [Sun and Wu (2013)]

$$\mathbf{E}(t) = -E_0\mathbf{e}_x\cos(\Omega t)e^{-\frac{t^2}{2\delta_t^2}}, \tag{7.4}$$

where δ_t characterizes the pulse duration. The corresponding vector potential is $\mathbf{A}(t) = -\int \mathbf{E}(t)dt$ which can be approximately written as $\mathbf{A}(t) \approx \frac{E_0}{\Omega}\mathbf{e}_x\sin(\Omega t)e^{-t^2/(2\delta_t^2)}$ for simplicity.

Starting from the Hamiltonian in Eq. (7.1) with given $\mathbf{A}(t)$, one can solve the time-dependent Schrödinger equation

$$i\frac{\partial}{\partial t}\Psi^{\mu}_{\mathbf{k}}(t) = H^{\mu,\mathrm{AB}}(\mathbf{k},t)\Psi^{\mu}_{\mathbf{k}}(t). \tag{7.5}$$

Solutions of cw THz fields

For cw THz fields with perfect periodicity, the exact solutions can be obtained by using Fourier expansions [Wright *et al.* (2009)]. Taking one step further, the Floquet states have been employed to solve the time-dependent Schrödinger equation in graphene [Zhou and Wu (2011)] in a similar way to two-dimensional electron gas in GaAs quantum wells [Cheng and Wu (2005)]. The solutions of the Schrödinger equation can be refered to as the Floquet state

$$\Psi^{\mu}_{\mathbf{k}\alpha}(t) = e^{-i\varepsilon^{\mu}_{\mathbf{k}\alpha}t}\sum_{n=-\infty}^{\infty} e^{in\Omega t}\psi^{\mu n}_{\mathbf{k}\alpha}, \tag{7.6}$$

where n is an arbitrary integer and α represents the branch index of the solutions. The eigenvalues of the Hamiltonian are noted as $\varepsilon^{\mu}_{\mathbf{k}\alpha}$ and the corresponding eigenvectors are two component spinors $\psi^{\mu n}_{\mathbf{k}\alpha} = (\psi^{\mu n+}_{\mathbf{k}\alpha}, \psi^{\mu n-}_{\mathbf{k}\alpha})^T$ where $\sigma = \pm$ denotes A and B sublattices.

$$\begin{aligned}\left(\frac{\varepsilon^{\mu}_{\mathbf{k}\alpha}}{\Omega} - n\right)\psi^{\mu n\sigma}_{\mathbf{k}\alpha} &= \frac{i\beta}{2}[(\mu\cos\theta_{\mathbf{E}} - \sigma\sin\theta_{\mathbf{E}})\psi^{\mu,n-1,-\sigma}_{\mathbf{k}\alpha} - (\mu\cos\theta_{\mathbf{E}} \\ &\quad + \sigma\sin\theta_{\mathbf{E}})\psi^{\mu,n+1,-\sigma}_{\mathbf{k}\alpha}] + \frac{v_F k}{\Omega}\mu e^{-i\mu\sigma\theta_{\mathbf{k}}}\psi^{\mu n,-\sigma}_{\mathbf{k}\alpha}. \end{aligned} \tag{7.7}$$

Here $\beta = v_F e E_0/\Omega^2$ is an important dimensionless parameter which determines the normalized quasienergy $\varepsilon^{\mu}_{\mathbf{k}\alpha}/\Omega$ and the normalized momentum $v_F\mathbf{k}/\Omega$. $\theta_{\mathbf{k}}$ is the polar angle of $\mathbf{k}$. The periodicity of the Hamiltonian in Eq. (7.1) with cw THz field consequently implies that both $\varepsilon^{\mu}_{\mathbf{k}\alpha}$ and $\varepsilon^{\mu}_{\mathbf{k}\alpha} + l\Omega$, with arbitrary integer l, are the solutions of Eq. (7.7). This feature makes the notation of quasienergies and eigenvectors simpler by dividing the quantum number α into ζ and l where ζ represents the branches of solutions. Then the reduced quasienergy $\epsilon^{\mu}_{\mathbf{k}\zeta}$ is defined in the reduced Floquet zone, $(-\Omega/2, \Omega/2]$, and arbitrary quasienergy reads $\varepsilon^{\mu}_{\mathbf{k}\alpha} = \epsilon^{\mu}_{\mathbf{k}\zeta} + l\Omega$. The corresponding eigenvectors are noted as $\psi^{\mu n\sigma}_{\mathbf{k}\alpha} = \phi^{\mu,n-l,\sigma}_{\mathbf{k}\zeta}$. There are two branches of solutions whose reduced quasienergies are of the same absolute values but opposite signs, in other words $\epsilon^{\mu}_{\mathbf{k}\zeta} = -\epsilon^{\mu}_{\mathbf{k}-\zeta}$, since the 2×2 Hamiltonian matrix in Eq. (7.1) is Hermitian with zero diagonal elements. Here $\zeta = +$ denotes the quasielectron branch with $\epsilon^{\mu}_{\mathbf{k}+} \geq 0$ and $\zeta = -$ denotes the quasihole branch with $\epsilon^{\mu}_{\mathbf{k}-} \leq 0$. The two component eigenvectors of positive branch and negative branch are related as $|\phi_{\mathbf{k}+}^{\mu n}\rangle = \sigma_y|\phi_{\mathbf{k}-}^{\mu,-n}\rangle^*$ where σ_y is the y component of the Pauli matrix.

The sidebands appear at quasienergy $\xi_{\mathbf{k}\zeta}^{\mu n} = \epsilon_{\mathbf{k}\zeta}^{\mu} - n\Omega$ for the Floquet states with reduced quasienergy $\epsilon_{\mathbf{k}\zeta}^{\mu}$. Their weights are $W_{\mathbf{k}\zeta}^{\mu n} = \langle\phi_{\mathbf{k}\zeta}^{\mu n}|\phi_{\mathbf{k}\zeta}^{\mu n}\rangle$. It is easy to testify that $\xi_{\mathbf{k}+}^{\mu n} = -\xi_{\mathbf{k}-}^{\mu m} - (n+m)\Omega$ and $W_{\mathbf{k}+}^{\mu n} = W_{\mathbf{k}-}^{\mu,-n}$. The mean energy of the Floquet states can be calculated by

$$\begin{aligned}\bar{\epsilon}_{\mathbf{k}\zeta}^{\mu} &= \frac{1}{\tau_0}\int_0^{\tau_0} dt\langle\Psi_{\mathbf{k}\zeta}^{\mu}(t)|H^{\mu,\mathrm{AB}}(\mathbf{k},t)|\Psi_{\mathbf{k}\zeta}^{\mu}(t)\rangle \\ &= \epsilon_{\mathbf{k}\zeta}^{\mu} - \sum_{n\sigma} n\Omega\phi_{\mathbf{k}\zeta}^{\mu n\sigma *}\phi_{\mathbf{k}\zeta}^{\mu n\sigma},\end{aligned} \tag{7.8}$$

where $\tau_0 = 2\pi/\Omega$ is the oscillation period of the THz field. The mean energy is used to determine the filling of the states.

The eigenvalues and eigenvectors in Eq. (7.7) are usually calculated numerically. Analytical expressions of eigenvalues and eigenvectors with proper approximations have been given by Busl *et al.* [Busl *et al.* (2012)].

Solutions for pulsed THz fields

The solutions for pulsed THz fields are more complicated than the solutions for cw THz fields because of the lack of periodicity. When the pulse duration δ_t in Eq. (7.4) is much larger than the oscillation period of the THz field τ_0, there are multiple oscillations during the pulse and the amplitude of oscillations varies slowly. Then the solutions could be approximated by the solutions for cw THz fields. When δ_t is comparable to τ_0, there are only a few oscillations during the pulse and the amplitude of oscillations changes rapidly, the solutions are different from the solutions for cw THz fields.

Considering a simple case that the THz field is polarized along x-direction, the general form of the time-dependent solutions in K valley ($\mu = 1$) can be written as [Ishikawa (2010)]

$$\Psi_{\mathbf{k}}(t) = c_1(t)\Psi_{\mathbf{k},1}(t) + c_{-1}(t)\Psi_{\mathbf{k},-1}(t), \tag{7.9}$$

where

$$\Psi_{\mathbf{k},\eta}(t) = \frac{1}{\sqrt{2}}e^{-i\eta\Theta_{\mathbf{k}}(t)}\begin{pmatrix} e^{-i\theta_v(t)/2} \\ \eta e^{i\theta_v(t)/2}\end{pmatrix}. \tag{7.10}$$

Here $\eta = \pm 1$ is the branch index, $\theta_v(t)$ denotes the directional angle of vector $\mathbf{e}_x[k_x + eA(t)] + \mathbf{e}_y k_y$. The instantaneous temporal phase is defined as

$$\Theta_{\mathbf{k}}(t) = v_F\int\sqrt{[k_x + eA(t)]^2 + k_y^2}dt. \tag{7.11}$$

Kinetic Bloch equations

Optical response of graphene under pulsed THz fields can be calculated by the Bloch equations. One simple way to establish the optical Bloch equation is introducing the interband coherence $c_1(t)c^*_{-1}(t)$ and population difference between the conduction band and the valence band $|c_1(t)|^2 - |c_{-1}(t)|^2$ where c_1 and c_{-1} are the coefficients in Eq. (7.9).

Sun and Wu [Sun and Wu (2013)] gave a set of more comprehensive kinetic Bloch equations to study the carrier dynamics and the optical response of graphene under pulsed THz fields. Based on the eigenstates of $H_0^{\mu,\mathrm{AB}}(\mathbf{k})$, the Hamiltonian in Eq. (7.1) in AB lattice space can be rewritten into the space of conduction band and valence bands

$$\begin{aligned} H_0^{\mu}(\mathbf{p}) &= v_F p \varsigma_z, \\ H^{\mu}_{\text{photon}}(\mathbf{p}) &= e v_F \mathbf{A}(t) \cdot [\mathbf{p}\varsigma_z + \mu \mathbf{e}_z \times \mathbf{p}\varsigma_y]/p, \end{aligned} \tag{7.12}$$

where $\mathbf{p}$ is the momentum of electron and ς is the Pauli matrix in the conduction and valence band space.

The matrix elements of the contour-ordered Green function $G^{\mu}(1,2)$ are $G^{\mu}_{\eta,\eta'}(1,2) = -i\langle T_C[\Psi^{\mu}_{\eta}(1)\Psi^{\mu\dagger}_{\eta'}(2)]\rangle$ [Haug and Jauho (1998)]. Here T_C is the contour-ordering operator, Ψ^{μ}_{η} and $\Psi^{\mu\dagger}_{\eta'}$ are the Fermion field operators in the Heisenberg picture. Shorthand notations of space-time points $(1) = (\mathbf{r}_1, t_1)$ and $(2) = (\mathbf{r}_2, t_2)$ are employed. The equation with $\mathbf{r}_1, t_1, \mathbf{r}_2$ and t_2 can be obtained by subtracting two differential forms of the Dyson equation which are $[i\overrightarrow{\partial}_{t_1} - H_0^{\mu}(\overrightarrow{\mathbf{p}}_1)]G^{\mu}(1,2)$ and $[-i\overleftarrow{\partial}_{t_2} - H_0^{\mu}(-\overleftarrow{\mathbf{p}}_2)]G^{\mu}(1,2)$. Then the equation can be transformed to the equation with center-of-mass and relative variables $\mathbf{R} = (\mathbf{r}_1 + \mathbf{r}_2)/2$, $\mathcal{T} = (t_1 + t_2)/2$, $\mathbf{r} = \mathbf{r}_1 - \mathbf{r}_2$, and $\tau = t_1 - t_2$,

$$\begin{aligned} i\partial_{\mathcal{T}} G^{\mu}(\tau, \mathcal{T}, \mathbf{r}) = {} & [H_0^{\mu}(\mathbf{p}_r) + H^{\mu}_{\text{photon}}(\mathbf{p}_r, \mathcal{T} + \tau/2)]G^{\mu}(\tau, \mathcal{T}, \mathbf{r}) \\ & - G^{\mu}(\tau, \mathcal{T}, \mathbf{r})[H_0^{\mu}(\overleftarrow{\mathbf{p}}_r) + H^{\mu}_{\text{photon}}(\overleftarrow{\mathbf{p}}_r, \mathcal{T} - \tau/2)]. \end{aligned} \tag{7.13}$$

Here $\mathbf{p}_r$ is the momentum operator with respect to $\mathbf{r}$. Then the gauge invariant function can be written as

$$G^{\mu}(\mathbf{k}, \tau, \mathcal{T}) = \int d\mathbf{r} \exp(-i\bar{\mathbf{k}} \cdot \mathbf{r}) G^{\mu}(\tau, \mathcal{T}, \mathbf{r}), \tag{7.14}$$

where $\bar{\mathbf{k}} = \mathbf{k} - \int_{-1/2}^{1/2} d\lambda e \mathbf{A}(\mathcal{T} + \lambda\tau)$. Substituting this transformation into Eq.(7.13) and setting $\tau = 0^-$, the equation becomes

$$\begin{aligned} & \partial_{\mathcal{T}} \rho_{\mu\mathbf{k}}(\mathcal{T}) + e\partial_{\mathcal{T}} \mathbf{A}(\mathcal{T}) \cdot \nabla_{\mathbf{k}} \rho_{\mu\mathbf{k}}(\mathcal{T}) \\ & + i[H_0^{\mu}(\bar{\mathbf{k}}) + H^{\mu}_{\text{photon}}(\bar{\mathbf{k}}, \mathcal{T}), \rho_{\mu\mathbf{k}}(\mathcal{T})] = 0. \end{aligned} \tag{7.15}$$

Here $\rho_{\mu\mathbf{k}}(\mathcal{T}) = -iG^{\mu}(\mathbf{k}, 0^{-}, \mathcal{T})$ is the local density matrix of electrons with wave vector $\mathbf{k}$. $e\partial_{\mathcal{T}}\mathbf{A}(\mathcal{T}) \cdot \nabla_{\mathbf{k}}\rho_{\mu\mathbf{k}}(\mathcal{T})$ gives the drift term since $\mathbf{E}(\mathcal{T}) = -\partial_{\mathcal{T}}\mathbf{A}(\mathcal{T})$ under the vector potential gauge. The bracket with comma means commutator. Expanding $|\bar{\mathbf{k}}|$ around $|\mathbf{k}|$ and keeping the lowest order term of $\mathbf{A}(\mathcal{T})$, the gauge invariant kinetic Bloch equations except for the scattering term and the Hartree-Fock term are obtained as follows

$$\partial_t\rho_{\mu\mathbf{k}} = \partial_t\rho_{\mu\mathbf{k}}|_{\text{coh}} + \partial_t\rho_{\mu\mathbf{k}}|_{\text{drift}} + \partial_t\rho_{\mu\mathbf{k}}|_{\text{scat}}. \tag{7.16}$$

Here $\mathcal{T}$ is changed to t for convenience. $\partial_t\rho_{\mu\mathbf{k}}|_{\text{coh}}$ is the coherent term, $\partial_t\rho_{\mu\mathbf{k}}|_{\text{drift}}$ is the drift term, and $\partial_t\rho_{\mu\mathbf{k}}|_{\text{scat}}$ is the scattering term. The diagonal elements $\rho_{\mu\mathbf{k},\eta\eta} = f_{\mu\mathbf{k},\eta}$ are the electron distribution functions in band η and the off-diagonal elements $\rho_{\mu\mathbf{k},1,-1} = \rho^{*}_{\mu\mathbf{k},-1,1}$ are the interband coherence. Both electric field acceleration of electrons and interband optical pump are considered in the drift term and the coherent term, respectively. The drift term describes the acceleration of electrons by the electric field which is

$$\partial_t\rho_{\mu\mathbf{k}}|_{\text{drift}} = e\mathbf{E}(t) \cdot \nabla_{\mathbf{k}}\rho_{\mu\mathbf{k}}. \tag{7.17}$$

The interband absorption is considered in the coherent terms

$$\partial_t\rho_{\mu\mathbf{k}}|_{\text{coh}} = -i[v_F k\varsigma_z + \Sigma^{\text{HF}}_{\mu\mathbf{k}} + H^{\mu}_{\text{pump}}, \rho_{\mu\mathbf{k}}]. \tag{7.18}$$

$H^{\mu}_{\text{pump}} = -ev_F\mu A_x \sin\theta_{\mathbf{k}}\varsigma_y$. $\Sigma^{\text{HF}}_{\mu\mathbf{k}}$ is the Hartree-Fock self-energy. By introducing the slowly varying interband-polarization $P_{\mu\mathbf{k}} = \rho_{\mu\mathbf{k},1,-1}e^{i\Omega t}$, the kinetic Bloch equations of density matrix can be rewritten to equations of carrier distributions and optical polarization

$$\begin{aligned}\partial_t f_{\mu\mathbf{k},\eta} &= \eta\text{Im}(\Omega^{*}_{\mu\mathbf{k}}P_{\mu\mathbf{k}}) + e\mathbf{E}(t) \cdot \nabla_{\mathbf{k}} f_{\mu\mathbf{k},\eta} + \partial_t f_{\mu\mathbf{k},\eta}|_{\text{scat}}, \\ \partial_t P_{\mu\mathbf{k}} &= -i\Omega_{\mu\mathbf{k}}(f_{\mu\mathbf{k},1} - f_{\mu\mathbf{k},-1})/2 - i\chi^{\mu}_{\mathbf{k}}P_{\mu\mathbf{k}} \\ &\quad + e\mathbf{E}(t) \cdot \nabla_{\mathbf{k}}P_{\mu\mathbf{k}} + \partial_t P_{\mu\mathbf{k}}|_{\text{scat}}.\end{aligned} \tag{7.19}$$

Here $\chi^{\mu}_{\mathbf{k}} = 2v_F k - \Omega + \text{Tr}[(\Sigma^{\text{HF}}_{\mu\mathbf{k}} + H^{\mu}_{\text{pump}})\varsigma_z]$ is the detuning term and the Rabi frequency is

$$\Omega_{\mu\mathbf{k}} = \mu e v_F E_0 \sin\theta_{\mathbf{k}} e^{-t^2/(2\delta_t^2)}(1 - e^{2i\Omega t})/\Omega - 2\Sigma^{\text{HF}}_{\mu\mathbf{k},1,-1}e^{i\Omega t}. \tag{7.20}$$

For high-frequency photon excitation, which is faster than the variation of distribution functions and polarization, the kinetic Bloch equations in Eq. (7.19) can be further simplified by the rotating-wave approximation.

Detailed expressions of the scattering terms including the electron-electron Coulomb, the electron-impurity, the electron-optical phonon, and the electron-remote-interfacial phonon scatterings can be found in [Sun and Wu (2013)]. The kinetic Bloch equations with all these terms can be numerically solved by the method developed by Weng *et al.* [Weng *et al.* (2004)] while the drift term is treated to the third order to suppress the noise.

7.1.2 *Intensity of THz fields*

The intensity of THz field is characterized by the dimensionless parameter $\beta = v_F e E_0/\Omega^2$ in Eq. (7.7). Generally, there are two categories of interaction between graphene and THz fields from the view of field intensity: one is weak THz fields when $\beta \ll 1$ and the other is intense THz fields when $\beta \geq 1$. For $\nu = 1$ THz, $\beta \geq 1$ is satisfied when E_0 is roughly larger than 10^4 V/m. The weak THz fields case, in which a small amount of non-equilibrium carriers are excited, has been widely studied both theoretically and experimentally. The optical response under weak THz fields is in linear regime which provides a powerful tool to probe the dynamics of electrons in graphene at femtosecond and/or picosecond time scales. The so-called optical pump-THz probe spectroscopy with weak probe THz pulse based on this feature has been reported to measure optical properties such as differential transmission [Frenzel *et al.* (2014); George *et al.* (2008); Ivanov *et al.* (2015); Jensen *et al.* (2014); Mihnev *et al.* (2015); Tielrooij *et al.* (2013)]. Moreover, the THz pump-THz probe [Winnerl *et al.* (2011)] technique with both weak pump and probe pulses has also been used to measure the differential transmission which characterizes the carrier relaxation process in graphene close to the Dirac points. By these ways, the interplay of interband and intraband transitions was studied. Sorger *et al.* [Sorger *et al.* (2015)] measured the transmission of lateral periodic patterned epitaxial graphene. Different transmissions were observed when the polarization of the THz field is parallel and is perpendicular to the patterned strip. Suzuki and Hibino [Suzuki and Hibino (2015)] found large optical anisotropy of THz field while interacting with stacked graphene nanoribbons through finite element calculations. DaSilva *et al.* [DaSilva *et al.* (2015)] found that the THz conductivity of graphene on boron nitride substrate is very sensitive to the Fermi energy. Vasilyev *et al.* [Vasilyev *et al.* (2014)] measured the strong temperature-dependent THz photoconductivity in meander-shaped graphene.

In contrast to weak THz fields, more interesting physical phenomena have been reported when the intensity of THz field is large. A large amount of non-equilibrium carriers can be generated by intense THz field through both interband and intraband processes. Then the optical properties of graphene, such as the transmission, would be significantly affected. Al-Naib *et al.* [Al-Naib *et al.* (2014, 2015)] used a tight-binding model and dynamic equations for the electron density matrix to study both the interband and the intraband dynamics in graphene under THz fields with strength up to 10^5 V/m. The harmonic generation in THz regime is studied. Satou *et al.* [Satou *et al.* (2008)] used a semiclassical kinetic equation to study the non-equilibrium carriers in undoped graphene under radiation with frequency varying from THz regime to infrared regime and field strength varying from zero to 10^4 V/m. Paul *et al.* [Paul *et al.* (2013)] have used the intense field THz spectroscopy to experimentally investigate the nonlinear carrier dynamics in graphene. An enhancement of transmission of graphene due to the heating of electrons has been observed when the THz field strength is above 10^6 V/m. Especially, when the THz field is continuous, the system consisting of graphene and intense THz field exhibits new electronic structures and density of states (DOS). A dynamical gap appears near the Dirac points [Calvo *et al.* (2011); Kibis (2010); López-Rodríguez and Naumis (2008); Oka and Aoki (2009)]. Efetov *et al.* [Fistul and Efetov (2007); Syzranov *et al.* (2008)] have proposed a new type of $p-n$ junction by utilizing the appearance of energy gap. Abergel and Chakraborty [Abergel and Chakraborty (2009)] proposed a device to generate valley polarized current using intense THz field in the presence of a transverse electric field. Besides single layer graphene, they further investigated the modification of band structure and DOS in bilayer graphene under intense THz fields [Abergel and Chakraborty (2011)].

7.2 Optical Response under cw THz Fields

The optical response of graphene under cw THz fields has attracted lots of attentions. Wright *et al.* [Wright *et al.* (2009)] reported a strong nonlinear effect in undoped graphene under cw THz field with field strength around 10^5 V/m. Fourier expansion has been used to solve the time-dependent Dirac equation. Both single frequency and triple frequency optical responses are remarkably altered by the nonlinear contribution. In the presence of a cw THz field with strength above 10^6 V/m, which can not be treated as perturbation, the system consisting of graphene and THz field is a time dependent one rather than a static one. It has been pointed out that the electronic structure of planar honeycomb lattices can be modified by a strong ac electric field [Zhang *et al.* (2009)]. Zhou and Wu [Zhou and Wu (2011)] studied the time-dependent electronic structure of the system by solving the time-dependent Schrödinger equation with the help of the Floquet theory [Shirley (1965)]. The appearance of the Floquet states is predicted and a dynamical gap is found to be opened due to cw THz field. Both theoretical and experimental results are reviewed in this section.

7.2.1 Theoretical results

Energy spectrum and DOS

The energy spectrum of the system can be numerically calculated by solving the eigenvalues of Eq. (7.7) [Zhou and Wu (2011)]. Figure 7.2 shows the energy spectrum in K valley when the THz field is circularly polarized and the normalized field strength is $\beta = 0.4$. K′ valley can be considered similarly and the superscript μ is dropped for convenience. Figure 7.2(a) shows the quasienergy $\xi^n_{\mathbf{k}\zeta}$ as a function of normalized momentum $v_F k_x/\Omega$ along x-direction. Other directions are the same since the energy spectrum is isotropic in this case. The color coding shows the weight of sidebands. Energy gaps at small momentum are found when the quasienergy is $m\Omega/2 + l\Omega$ and the momentum is $m\Omega/2v_F$ with m and l being arbitrary integers. For given momentum, all the quasienergy gaps between neighboring sidebands are the same due to the periodicity of the quasienergy spectrum. This phenomenon is attributed to the single-photon or multi-photon resonance induced by ac Stark splitting which occurs when an electromagnetic field with photon energy Ω couples two electron states with energy difference $m\Omega$. In general, the quasienergy gap originates from the m-photon resonance when the momentum satisfies $2v_F k = m\Omega$. The energy gaps decrease with the increase of momentum and finally vanish for sufficient large momentum. This means the band mixing becomes negligible and the wave function trends towards the one without THz field excitation. Selected quasienergy $\epsilon_{\mathbf{k}+}$, marked by blue crosses in Fig. 7.2(a), and the mean energy $\bar{\epsilon}_{\mathbf{k}+}$ are plotted as a function of normalized momentum in Fig. 7.2(b) in comparison with the field-free electron energy. The mean energy and the quasienergy significantly deviate from the field-free electron energy at small momentum. Such deviation becomes smaller and disappears at large momentum. Zero mean energy is reached inside the quasienergy gap, except the one at zero momentum.

In contrast to the circularly polarized THz field case, the linearly polarized THz field leads to an anisotropic energy spectrum as shown in Fig. 7.3. $\beta = 2.3$ is used in the calculations. $\theta_{\mathbf{k}}$ equals to the angle between the momentum and the THz field which is polarized along x-direction. The quasienergy gaps are closed at zero momentum for all three angles. For $\theta_{\mathbf{k}} = 0$ which means the momentum is parallel to the polarization of the THz field, all gaps are closed because the interband transition term is zero. For $\theta_{\mathbf{k}} = \pi/2$ which means the momentum is perpendicular to the THz field, only the gaps around $k_y \sim 0.5\Omega/v_F$ survive and other gaps are

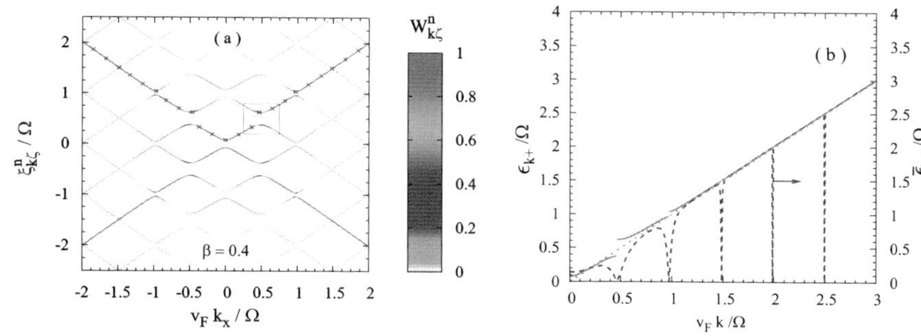

Fig. 7.2 Energy spectrum of graphene under intense circularly polarized THz field when $\beta = 0.4$. (a) Quasienergy $\xi^n_{\mathbf{k}\zeta}$ of sidebands versus the normalized momentum. The weight of sidebands are presented in color coding. (b) Selected quasienergy $\epsilon_{\mathbf{k}+}$ (red curve) and mean energy $\bar{\epsilon}_{\mathbf{k}+}$ (blue curve) versus the normalized momentum. The selected quasienergy in (b) has been marked in blue crosses in (a). The electron energy without THz field (green curve) is also presented for comparison. From [Zhou and Wu (2011)]. (For color reference, go to page 490.)

closed. For other angles like $\theta_{\mathbf{k}} = \pi/6$, most gaps are kept which is similar to the circularly polarized THz field case.

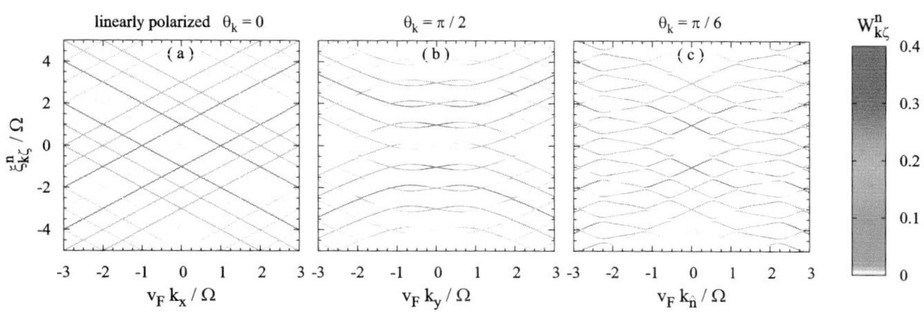

Fig. 7.3 Energy spectrum of graphene under intense linearly polarized THz field when $\beta = 2.3$. Quasienergy $\xi^n_{\mathbf{k}\zeta}$ of sidebands versus the normalized momentum along different directions. (a) $\theta_{\mathbf{k}} = 0$, (b) $\theta_{\mathbf{k}} = \pi/2$, and (c) $\theta_{\mathbf{k}} = \pi/6$. The weight of sidebands are presented in color coding. From [Zhou and Wu (2011)]. (For color reference, go to page 490.)

The time-averaged DOS is calculated from the spectral function as follows

$$D^{\text{ave}}(\omega) = g_s g_v \sum_{\mathbf{k}\zeta n} \langle \phi^n_{\mathbf{k}\zeta} | \phi^n_{\mathbf{k}\zeta} \rangle \delta(\omega - \epsilon_{\mathbf{k}\zeta} + n\Omega). \tag{7.21}$$

Here, $g_v = 2$ and $g_s = 2$ are the valley degeneracy and the spin degeneracy, respectively. Figure 7.4 shows the time-averaged DOS of both circularly and linearly polarized THz fields when $\beta = 2.3$. The very low ω part has been enlarged in the inset. There is no gap of DOS for both two cases because the quasienergy gaps at large momentum are closed for circularly polarized THz field as shown in Fig. 7.2 and the quasienergy gaps for the momentum in x-direction vanish for linearly polarized THz field.

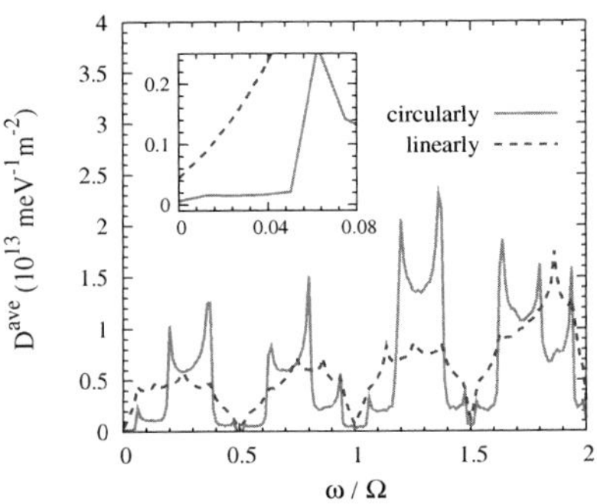

Fig. 7.4 Time-averaged DOS under circularly and linearly polarized THz fields when $\beta = 2.3$. The low ω part is in inset. From [Zhou and Wu (2011)].

Optical conductivity

The real part of optical conductivity determines the optical absorption. When a THz field of frequency ω_ℓ interacts with graphene where ℓ denotes the polarization direction, the real part of time-dependent optical conductivity can be calculated from the linear-response theory

$$\mathrm{Re}\sigma_{\ell\ell}(\mathcal{T},\omega_\ell) = -\frac{g_s g_v}{\omega_\ell}\mathrm{Im}\Pi^R_{\ell\ell}(\mathcal{T},\omega_\ell). \tag{7.22}$$

The valley index μ is dropped in Eq. (7.22) because both K and K′ valleys have same optical conductivities. $\mathrm{Im}\Pi^R_{\ell\ell}(\mathcal{T},\omega_\ell)$ is the retarded current-current correlation function in frequency space which can be calculated by using the generalized Kadanoff-Baym ansatz [Zhou and Wu (2011)]

$$\begin{aligned}\mathrm{Im}\Pi^R_{\ell\ell}(\mathcal{T},\omega_\ell) = -\pi \sum_{\mathbf{k}\zeta_1\zeta_2} \sum_{n_1 m_1 n_2 m_2} &\langle\phi^{m_2}_{\mathbf{k}\zeta_2}|\hat{j}^\ell_{\mathbf{k}}|\phi^{n_1}_{\mathbf{k}\zeta_1}\rangle\langle\phi^{m_1}_{\mathbf{k}\zeta_1}|\hat{j}^\ell_{\mathbf{k}}|\phi^{n_2}_{\mathbf{k}\zeta_2}\rangle \\ &\times (f_{\mathbf{k}\zeta_1} - f_{\mathbf{k}\zeta_2})e^{i(n_1-m_1+n_2-m_2)\Omega\mathcal{T}} \\ &\times \delta[\epsilon_{\mathbf{k}\zeta_1} - \epsilon_{\mathbf{k}\zeta_2} - (n_1+m_1-n_2-m_2)\Omega/2) + \omega_\ell]. \end{aligned} \tag{7.23}$$

Here $j^\ell_{\mathbf{k}} = ev_F\sigma_\ell$ is the current operator along ℓ direction. $f_{\mathbf{k}\zeta}$ is the Fermi distribution on the mean energy of the Floquet states which is a step function $\theta(\bar{\epsilon}_{\mathbf{k}\zeta} - E_F)$ at zero temperature. The time-averaged optical conductivity can then be calculated by substituting Eq. (7.22) and Eq. (7.23) into $\mathrm{Re}\sigma^{\mathrm{ave}}_{\ell\ell}(\omega_\ell) = \frac{1}{\tau_0}\int_0^{\tau_0} d\mathcal{T}\mathrm{Re}\sigma_{\ell\ell}(\mathcal{T},\omega_\ell)$.

For n-type graphene with large positive E_F in which the quasielectron states with quasienergy $\bar{\epsilon}_{\mathbf{k}\zeta} < E_F$ are occupied and the other states are empty, a multi-steplike structure of the real part of optical conductivity is found when a circularly polarized THz field with frequency $\omega_\ell \sim 2E_F$ is applied. This phenomenon is attributed to the sideband-modulated optical transition which is very different from the single-step structure when the THz field is not applied as shown in Fig. 7.5. There are more steps for larger field strength because more sidebands, whose quasienergies are away from $2E_F$, participate the optical transition due to their enhanced weight. In each step, there is a weak frequency dependence of optical conductivity for high Fermi energy. When the Fermi energy is low, there are dips of optical conductivity when ω_ℓ is an integer times of Ω and the field strength is strong enough. This is due to the pronounced quasienergy gaps. Similar behaviors can be found when THz field is linearly polarized.

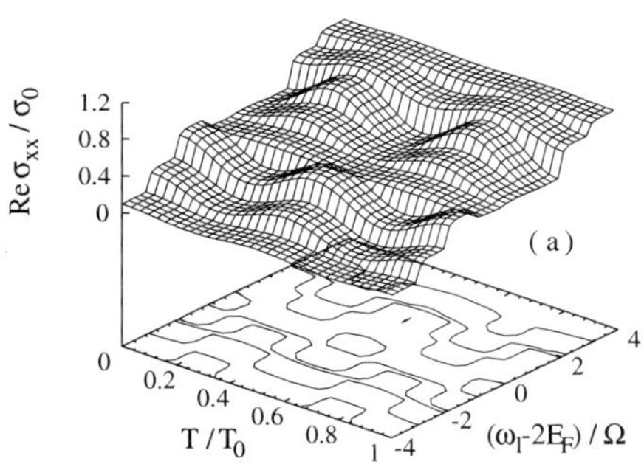

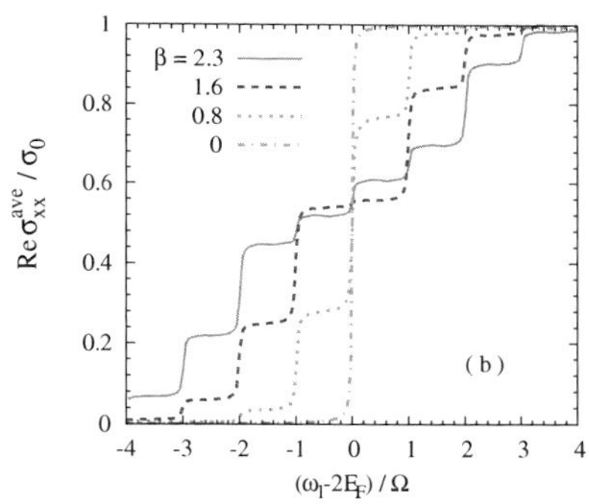

Fig. 7.5 (a) Time-dependent optical conductivity versus time and optical frequency for circularly polarized THz field when $\beta = 2.3$. (b) Time-averaged optical conductivity versus optical frequency for different β. $E_F = 6\Omega$ is used in the calculations. $\sigma_0 = \frac{e^2}{4\hbar}$ is the unit of optical conductivity. From [Zhou and Wu (2011)].

Further studies

Besides energy spectrum and optical conductivity, some other properties are also studied as listed below. The many-particle excitation in graphene under intense THz fields has been studied by Busl *et al.* [Busl *et al.* (2012)] through calculating the dynamical polarizability. They found that the plasmons exist in both undoped and doped graphene in the presence of THz fields. The plasmon frequency is tunable by THz fields. Large field results in an reduction of stability of the plasmons. Morell and Foa Torres [Morell and Foa Torres (2012)] studied the triagonal warping corrections in bilayer graphene under intense THz fields. The dynamical gaps are strongly affected by the triagonal warping in bilayer graphene. Scholz *et al.* [Scholz *et al.* (2013)] investigated the interplay between spin-orbit interactions and THz fields in graphene. They found that a circularly polarized THz field can be used not only to induce a dynamical gap, but also to close an existing gap. Finite net spin polarization can be generated in graphene by introducing THz fields. The sign and the magnitude of spin polarization can be tuned by the ratio of spin-orbit coupling parameters. López *et al.* [López *et al.* (2015)] performed a perturbation analysis on the Landau level quasienergy spectrum in graphene under intense circularly polarized THz fields. Landau level-dependent dynamical gap structures are studied.

7.2.2 *Experiments*

The Floquet states in two-dimensional systems, such as the surface of topological insulator, under intense mid-infrared cw polarized light have been experimentally observed [Wang *et al.* (2013)]. The time- and angle-resolved photoemission spectroscopy (ARPES) has been used to resolve the electronic structure. Figures 7.6 and 7.7 show the measured ARPES spectra under circularly and linearly polarized lights, respectively. The incident photon energy is 120 meV and the peak electric field strength is 2.5×10^7 V/m. The raw spectra and the spectra after subtracting the spectra before excitation are shown. For circularly polarized light excitation, the ARPES spectrum shows clear sidebands and avoided crossings along both k_x and k_y directions. Dynamical gaps are found to be open in all directions as shown in Fig. 7.6. For linearly polarized light excitation with k_x polarization, the sidebands along k_y direction can be clearly seen from Fig. 7.7. The dynamical gaps open along the direction perpendicular to the electric field and they close along the direction parallel to the electric field.

Similar experiment in graphene under intense THz fields has not been reported yet. It is believed that similar results can be obtained. The challenge for future experiment is that the photon energy of THz field is several meV which is one order of magnitude smaller than the photon energy used in [Wang *et al.* (2013)]. This implies that high resolution ARPES is required to see the fine structure of the Floquet states.

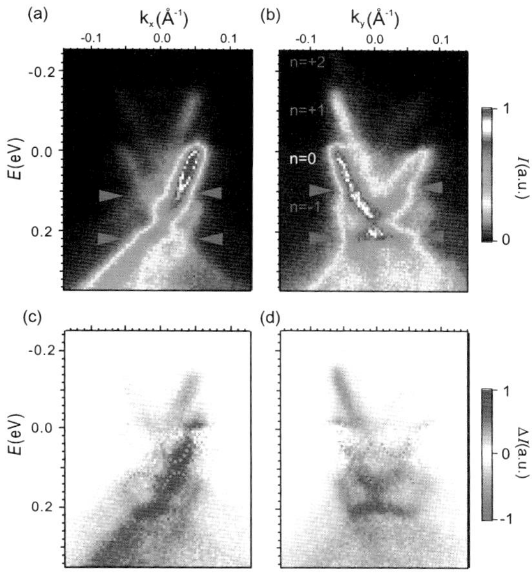

Fig. 7.6 ARPES spectra under circularly polarized light excitation. (a) and (b) show the energy spectra along k_x and k_y direction, respectively. (c) and (d) show the same spectra after subtracting the spectra without excitation. From [Wang *et al.* (2013)]. (For color reference, go to page 491.)

7.3 Optical Response under Pulsed THz Fields

Recent progresses of ultrashort intense THz radiation technology lead to a new field of high-field physics in condensed matter. The generation of ultrashort intense THz pulse with a few cycles have been reported [Hirori *et al.* (2010); Sell *et al.* (2008)]. The optical response of graphene under such ultrashort intense THz pulse is of great interest. Some theoretical and experimental works are reviewed in this section.

7.3.1 *Theoretical results*

Theoretically, strong nonlinear optical response has been reported [Mikhailov and Ziegler (2008)] by using the quasiclassical kinetic theory. They found that the average momentum of electrons in graphene decreases linearly with time while the average current remains constant for a time interval after a single rectangular pulse excitation. The strength of the field in their work is around 10^4 V/m. Stronger field with strength around 10^6 V/m has been studied [Ishikawa (2010)] by using extended optical Bloch equations derived from the time-dependent Dirac equation. The nonlinear

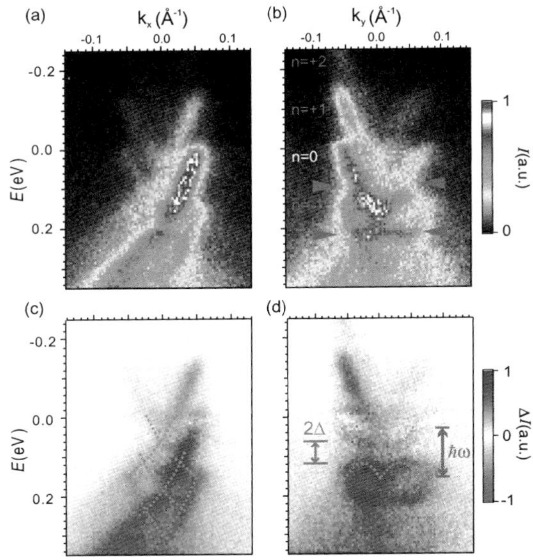

Fig. 7.7 ARPES spectra under linearly polarized light excitation. (a) and (b) show the energy spectra along k_x and k_y direction, respectively. (c) and (d) show the same spectra after subtracting the spectra without excitation. From [Wang *et al.* (2013)]. (For color reference, go to page 491.)

optical response due to the interplay between the interband dynamics and the intraband transitions has been reported. Sun and Wu [Sun and Wu (2013)] have studied the carrier dynamics when the THz field strength is up to 10^7 V/m by solving microscopic kinetic Bloch equations. Their main results are discussed as follows.

Center-of-mass drift

Sun and Wu [Sun and Wu (2013)] studied the dynamics of electrons in graphene on a SiO_2 substrate when the photon energy is $\Omega = 15$ meV (the corresponding frequency is $\nu = 3.6$ THz) and the peak electric field strength is chosen as $E_0 = 5 \times 10^5$ V/m. The oscillation period of the THz field is $\tau_0 = 276$ fs. It is larger than the twice of the full-width at half-maximum of the pump pulse which is 100 fs by choosing $\delta_t = 42.5$ fs. In other words, the pulse contains less than half cycle of electric field that means the electric field does not change direction during the pump process. This feature differs from the conventional femtosecond laser pulse which contains multiple cycles of electric field. A noticeable electron drift effect with drifted Fermi distribution after pulse excitation can then be obtained by considering the drift term which is ignored in conventional laser pulse pump. The drifted Fermi distribution under a static electric field in the presence of strong or weak electron-electron scattering, comparing to the drift effect, can be written as [Zhou and Wu (2010)]

$$f_{\mathbf{k}\eta} = \frac{1}{e^{(\eta v_F k - \mathbf{u}_\eta \cdot \mathbf{k} - E_F^\eta)/(k_B T_\eta)} + 1}, \tag{7.24}$$

$$f_{\mathbf{k}\eta} = \frac{1}{e^{(\eta v_F |\mathbf{k} - \mathbf{u}_\eta| - E_F^\eta)/(k_B T_\eta)} + 1}, \tag{7.25}$$

where $\mathbf{u}_\eta = \sum_{\mathbf{k}} \mathbf{k} f_{\mathbf{k},\eta}(t) / \sum_{\mathbf{k}} f_{\mathbf{k},\eta}(t)$ is the center-of-mass drift velocity. Figure 7.8(a) shows numerically calculated k-dependence of $\ln(f_{\mathbf{k},1}^{-1} - 1)$ along $\theta_{\mathbf{k}}$=0 and $\theta_{\mathbf{k}} = \pi$ directions for different time delays. The establishment of distribution and the electron dynamics are revealed. The minima of curves drift away from the Dirac points when $t < 25$ ps and then move back to the Dirac points when $t > 100$ ps. The curves are not linear for $t < 300$ ps that means the drifted Fermi distributions have not been established. When $t > 300$ ps, the curves become linear but the slope along $\theta_{\mathbf{k}}$=0 direction is different from the slope along $\theta_{\mathbf{k}} = \pi$ direction. Then the drifted Fermi distributions are established with the electron-electron scattering as the dominant scattering mechanism. Similar behavior of valence band can be found. Figure 7.8(b) shows the electron temperature and the drift velocity, which are obtained by fitting the distribution functions with Eq. (7.24), along different directions in both conduction band and valence band. The drift velocities and the electron temperatures along different directions trend to be the same after pulse excitation. The relative differences of the drift velocities and the enhancement of electron temperatures along different directions ($T_{\eta\theta_{\mathbf{k}}} - 300$ K) could be less than 10% after $t = 350$ fs for conduction band and after 420 fs for valence band. The drift velocities and the electron temperatures in conduction band and valence band also trend to be the same because of the interband Coulomb scattering. The calculation shows that the relative difference of the drift velocities of conduction band and valence band is less than 10% after $t = 550$ fs and the enhancement of electron temperatures of conduction band and valence band is less than 10% after $t = 350$ fs.

Figure 7.9(a) investigates the effect of impurities and photon-energy on drift velocity along the direction of electric field. The presence of impurity of density $N_i = 5 \times 10^{11}$ cm^{-2} reduces the drift velocity because of strong electron-impurity scattering which overcomes electron-phonon scattering. When the photon energy is 30 meV, the drift velocity is significantly reduced since the oscillation period of electric field is halved. Then the electric field would change direction during the pulse duration which partially

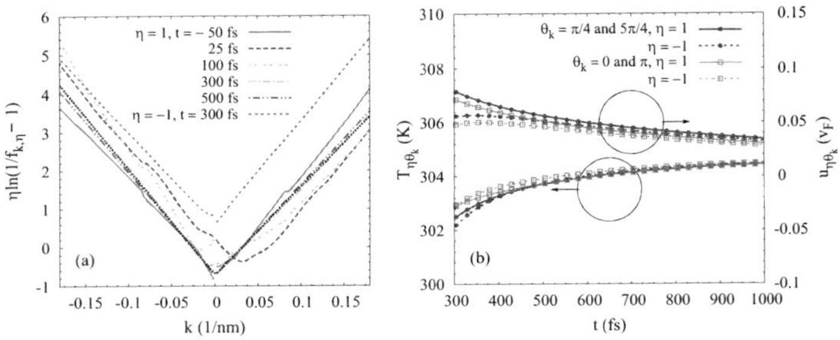

Fig. 7.8 (a) $\eta\ln(f^{-1}_{\mathbf{k},\eta}-1)$ along x- and $-x$- directions for conduction band at different time delays and for valence band at $t=300$ fs. The impurity density is zero. $T_1=302.97$ K, $v_1=0.093v_F$, and $E^1_F=18.5$ meV are used in the calculations. (b) Temporal evolution of $T_{\eta,\theta_{\mathbf{k}}}$ and $u_{\eta,\theta_{\mathbf{k}}}$ along different directions for electrons in both conduction band and valence band through fitting the drifted Fermi distribution in Eq. (7.24). From [Sun and Wu (2013)].

cancels the net momentum gained by electrons. The contributions from the pump and the Auger processes to the evolution of electron density are studied individually. The numerically calculated electron densities are shown in Fig. 7.9(b) when the pump process and the Auger process are switched off separately. In the absence of pump process, electrons can still be excited from valence band to conduction band due to interband electron-phonon scattering when electrons are heated by the drift effect of the electric field. The Auger process is found to be crucial to the dynamics of electrons. A 70% enhancement of excited electron density can be obtained at the end of pulse. The widely used rotating-wave approximation in semiconductor optics is found to be invalid for such low photon energy excitation. The fast variation of electron density during the pump process vanishes and the excited electron density decreases a lot when the rotating-wave approximation is used.

Optical transmission

After obtaining the time-dependent electron distribution functions by solving the kinetic Bloch equations, the time-dependent optical transmission for given probe-photon energy $\omega_{\rm pr}$ can be calculated as [Dawlaty *et al.* (2008)]

$$T_{\omega_{\rm pr}}(t)=\frac{1}{\left|1+\frac{N_{\rm lay}\sigma_{\omega_{\rm pr}}(t)\sqrt{\mu_0/\epsilon_0}}{1+n_{\rm ref}}\right|^2},\tag{7.26}$$

where $N_{\rm lay}$ is the number of graphene layers, $n_{\rm ref}$ is the refractive index of substrate, μ_0 is the vacuum permeability, and ϵ_0 is the vacuum permittivity.

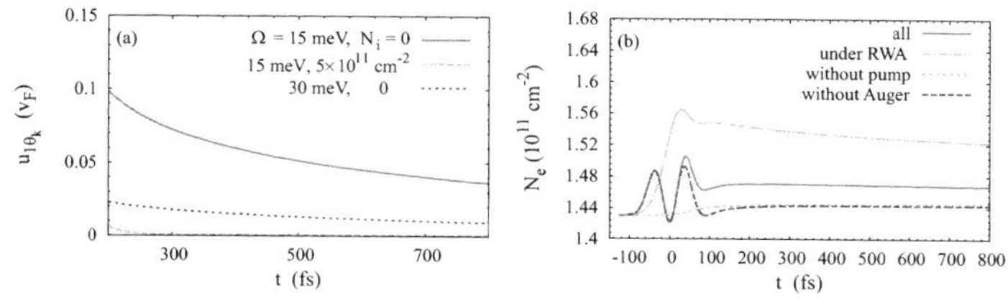

Fig. 7.9 (a) Temporal evolution of the drift velocity along x-direction for $\Omega = 15$ meV with and without impurity. The case of $\Omega = 30$ meV without impurity is also plotted. (b) Temporal evolution of the electron densities for the cases of: 1) all the processes included; 2) the rotating-wave approximation is used; 3) without pump; and 4) without Auger process. $\Omega = 15$ meV and zero impurity density are used in the calculations. From [Sun and Wu (2013)].

The time-dependent interband optical conductivity is given by [Peres *et al.* (2006); Scharf *et al.* (2013)]

$$\sigma_{\omega_{\rm pr}}(t) = -\frac{e^2}{4\pi}\int_0^{2\pi} d\theta_{\mathbf{k}_\omega} \sin^2(\theta_{\mathbf{k}_\omega})[f_{\mathbf{k}_\omega,1}(t) - f_{\mathbf{k}_\omega,-1}(t)], \tag{7.27}$$

where $|\mathbf{k}_\omega| = \omega_{\rm pr}/(2v_F)$ is the resonant point of absorption and $\theta_{\mathbf{k}_\omega}$ is the angle of $\mathbf{k}_\omega$. The differential transmission (DT) is defined as $\Delta T_{\omega_{\rm pr}}(t)/T^0_{\omega_{\rm pr}} = [T_{\omega_{\rm pr}}(t) - T^0_{\omega_{\rm pr}}]/T^0_{\omega_{\rm pr}}$ where $T^0_{\omega_{\rm pr}}$ is the transmission without pump.

Sun and Wu's [Sun and Wu (2013)] calculation results explain well the negative DT in system with high Fermi energy observed by Sun *et al.* although their experiment is not in THz regime.

7.3.2 *Experiments*

Single pulse intense THz spectroscopy

Mics *et al.* [Mics *et al.* (2015)] studied the ultrafast electrical conductivity of graphene excited by a single cycle THz pulse with peak electric field strength up to 1.2×10^7 V/m. The charge transport on picosecond timescale is found to be determined by the thermodynamic balance of thermalized electron gas in graphene. The thermal energy deposited from the THz field raises the electron temperature near-instantaneously. Both the heating and the cooling processes are on picosecond timescale. Their dynamic interplay leads to a nonlinear behavior of the ultrafast electrical conductivity.

Intense THz pump-optical probe

Tani *et al.* [Tani *et al.* (2012)] measured the near-infrared transient absorption at 800 nm to investigate the ultrafast carrier dynamics in graphene on a SiO_2 substrate excited by an intense half-circle THz pulse with peak electric field strength up to 3×10^7 V/m. Figures 7.10(a) and 7.10(b) show the temporal profiles of the normalized differential optical density (OD) of graphene at 800 nm with and without THz field excitation. The normalized differential OD is defined as $\Delta OD/OD_{ref} = (OD_{ex} - OD_{ref})/OD_{ref}$ where OD_{ex} and OD_{ref} are the OD of graphene with and without THz pulse excitation. Two different peak electric field strengths, 1.7×10^7 V/m and 3×10^7 V/m, are studied. Negative differential OD after THz field excitation is observed which means the graphene becomes more transparent. This phenomenon originates from the change of carrier distributions because the differential OD is determined by the carrier distributions at the energy of an 800 nm transition. The Boltzmann equations are used to fit the experimental data where constant relaxation times are chosen as fitting parameters. Agreement between the experimental data and the calculation results are shown in the figures. The temporal evolutions of carrier density for two peak electric field strengths calculated from the Boltzmann equations are also shown in Fig. 7.10(c) and 7.10(d). The calculation results show that the excited carrier density is much larger than the intrinsic carrier density. Stronger electric field leads to larger maximum values of differential OD as shown in Fig. 7.10(e) due to the increase of excited carrier density. A power law dependence of the maximum value on the peak electric field strength with exponent 2.5, rather than 2 for conventional interband transition, is found which indicates nonlinear coupling between THz field and graphene. Moreover, the impact ionization process is found to be promoted by intense THz fields.

Optical pump-intense THz probe

The intense THz pulse is not only used for pump but also used for probe. Hafez *et al.* [Hafez *et al.* (2015)] used intense THz field probe pulse with peak electric field strength $10^6 \sim 10^7$V/m to detect the dynamics of carriers excited by optical pump in monolayer epitaxial graphene. An enhancement of THz transmission induced by pump pulse due to the suppression of photoconductivity is measured by THz pulse with different peak electric field strengths. Figure 7.11(a) shows differential THz field $\Delta E/E_s$ of two different peak electric field strengths, where the lower one is 9×10^5 V/m and the higher one is 6.3×10^6 V/m. ΔE is the difference between the peak electric field strengths of THz probe pulses before and after optical pump and E_s is the peak of field strength transmitted through the bare substrate. The differential THz field for high peak electric field is clearly smaller than that for low peak electric field. Figure 7.11(b) shows the temporal evolution

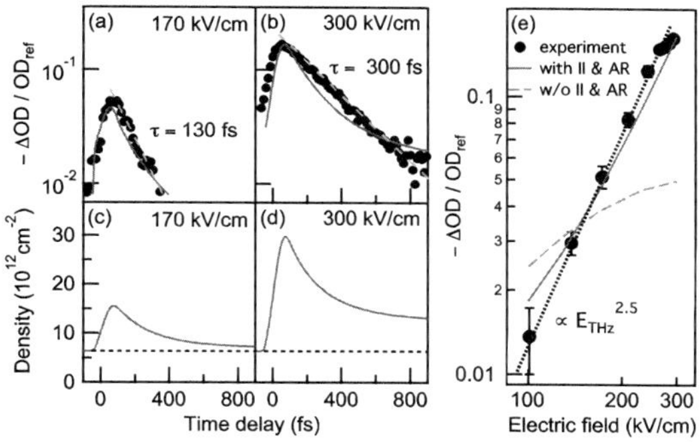

Fig. 7.10 Normalized differential OD plotted as a function of time delay with peak electric field strengths of (a) 1.7×10^7 V/m and (b) 3×10^7 V/m. The calculation results from the Boltzmann equation with fitted relaxation times are also shown in solid curves. (c) and (d) show the temporal evolution of calculated carrier density in comparison with the carrier density without excitation in dashed lines. (e) Normalized differential OD as a function of peak electric field strength. The filled circles are experimental data and the curves are the calculation results. From [Tani *et al.* (2012)].

of differential transmission $\Delta T/T_0$ while the THz field strength varies from 9×10^5 V/m to 6.3×10^6 V/m. Here T_0 is the transmission without excitation and ΔT is the enhancement of transmission after excitation. Positive ΔT means the graphene becomes transparent due to the increase of the carrier density and extra energy of the excited hot carriers. The peak of $\Delta T/T_0$ decreases with increasing THz electric field strength but increases with the pump fluence. Such nonlinear carrier dynamics is attributed to the combined effect of the optical pump fluence and the intense THz field on the carrier scattering rate.

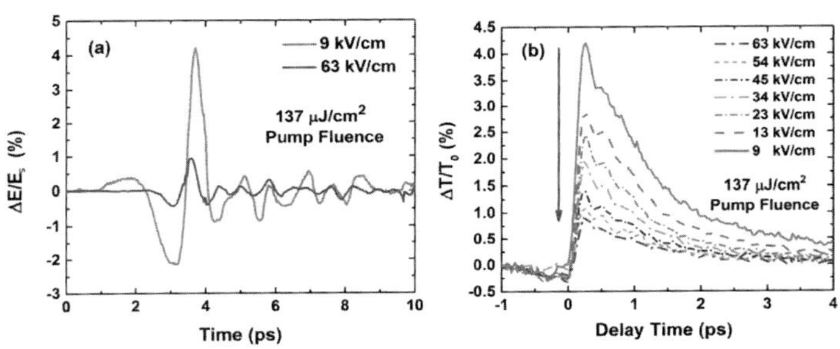

Fig. 7.11 (a) Differential THz field $\Delta E/E_s$ as a function of time delay for THz pulses with low (9×10^5 V/m) and high (6.3×10^6 V/m) peak electric field strengths. (b) Differential transmission $\Delta T/T_0$ as a function of time delay for varied intensities of probe THz pulses. The fluence of the pump pulse is fixed to 137 μJ/cm^2. From [Hafez *et al.* (2015)].

Bibliography

[1] Abergel, D. S. L. and Chakraborty, T. (2009). Generation of valley polarized current in bilayer graphene, *Appl. Phys. Lett.* **95**, 6, pp. 062107.

[2] Abergel, D. S. L. and Chakraborty, T. (2011). Irradiated bilayer graphene, *Nanotechnology* **22**, 1, pp. 015203.

[3] Al-Naib, I., Sipe, J. E., and Dignam, M. M. (2014). High harmonic generation in undoped graphene: interplay of inter- and intraband dynamics, *Phys. Rev. B* **90**, 24, pp. 245423.

[4] Al-Naib, I., Sipe, J. E., and Dignam, M. M. (2015). Nonperturbative model of harmonic generation in undoped graphene in the terahertz regime, *New J. Phys.* **17**, 11, pp. 113018.

[5] Bonaccorso, F., Sun, Z., Hasan, T., and Ferrari, A. C. (2010). Graphene photonics and optoelectronics, *Nat. Photon.* **4**, 9, pp. 611-622.

[6] Busl, M., Platero, G., and Jauho, A. -P. (2012). Dynamical polarizability of graphene irradiated by circularly polarized ac electric fields, *Phys. Rev. B* **85**, 15, pp. 155449.

[7] Calvo, H. L., Pastawski, H. M., Roche, S., and Foa Torres, L. E. F. (2011). Tuning laser-induced band gaps in graphene, *Appl. Phys. Lett.* **98**, 23, pp. 232103.

[8] Castro Neto, A. H., Guinea, F., Peres, N. M. R., Novoselov, K. S., and Geim, A. K. (2009). The electronic properties of graphene, *Rev. Mod. Phys.* **81**, 1, pp. 109.

[9] Cheng, J. L. and Wu, M. W. (2005). Intense terahertz laser fields on a two-dimensional electron gas with Rashba spin-orbit coupling, *Appl. Phys. Lett.* **86**, 3, pp. 032107.

[10] Das Sarma, S., Adam, S., Hwang, E. H., and Rossi, E. (2011). Electronic transport in two-dimensional graphene, *Rev. Mod. Phys.* **83**, 2, pp. 407.

[11] DaSilva, A. M., Jung, J., Adam, S., and MacDonald, A. H. (2015). Terahertz conductivity of graphene on boron nitride, *Phys. Rev. B* **92**, 15, pp. 155406.

[12] Dawlaty, J. M., Shivaraman, S., Chandrashekhar, M., Rana, F., and Spencer, M. G. (2008). Measurement of ultrafast carrier dynamics in epitaxial graphene, *Appl. Phys. Lett.* **92**, 4, pp. 042116; Dawlaty, J. M., Shivaraman, S., Strait, J., George, P., Chandrashekhar, M., Rana, F., Spencer,

M. G., Veksler, D., and Chen, Y. (2008). Measurement of the optical absorption spectra of epitaxial graphene from terahertz to visible, *Appl. Phys. Lett.* **93**, 13, pp. 131905.

[13] DiVincenzo, D. P. and Mele, E. J. (1984). Self-consistent effective-mass theory for intralayer screening in graphite intercalation compounds, *Phys. Rev. B* **29**, 4, pp. 1685.

[14] Frenzel, A. J., Lui, C. H., Shin, Y. C., Kong, J., and Gedik, N. (2014). Semiconducting-to-metallic photoconductivity crossover and temperature-dependent Drude weight in graphene, *Phys. Rev. Lett.* **113**, 5, pp. 056602.

[15] Fistul, M. V. and Efetov, K. B. (2007). Electromagnetic-field-induced suppression of transport through *n-p* junctions in graphene, *Phys. Rev. Lett.* **98**, 25, pp. 256803.

[16] George, P. A., Strait, J., Dawlaty, J., Shivaraman, S., Chandrashekhar, M., Rana, F., and Spencer, M. G. (2008). Ultrafast optical-pump terahertz-probe spectroscopy of the carrier relaxation and recombination dynamics in epitaxial graphene, *Nano Lett.* **8**, 12, pp. 4248-4251.

[17] Hafez, H. A., Al-Naib, I., Dignam, M. M., Sekine, Y., Oguri, K., Blanchard, F., Cooke, D. G., Tanaka, S., Komori, F., Hibino, H., and Ozaki, T. (2015). Nonlinear terahertz field-induced carrier dynamics in photoexcited epitaxial monolayer graphene, *Phys. Rev. B* **91**, 3, pp. 035422.

[18] Haug, H. and Jauho, A. -P. (1998). *Quantum Kinetics in Transport and Optics of Semiconductors*, (Springer).

[19] Hirori, H., Nagai, M., and Tanaka, K. (2010). Excitonic interactions with intense terahertz pulses in ZnSe/ZnMgSSe multiple quantum wells, *Phys. Rev. B* **81**, 8, pp. 081305(R).

[20] Ishikawa, K. L. (2010). Nonlinear optical response of graphene in time domain, *Phys. Rev. B*, **82**, 20, pp. 201402(R).

[21] Ivanov, I., Bonn, M., Mics, Z., and Turchinovich, D. (2015). Perspective on terahertz spectroscopy of graphene, *Europhys. Lett.* **111**, 6, pp. 67001.

[22] Jensen, S. A., Mics, Z., Ivanov, I., Varol, H. S., Turchinovich, D., Koppens, F. H. L., Bonn, M., and Tielrooij, K. J. (2014). Competing ultrafast energy relaxation pathways in photoexcited graphene, *Nano Lett.* **14**, 10, pp. 5839-5845.

[23] Kibis, O. V. (2010). Metal-insulator transition in graphene induced by circularly polarized photons, *Phys. Rev. B* **81**, 16, pp. 165433.

[24] López, A., Di Teodoro, A., Schliemann, J., Berche, B., and Santos, B. (2015). Laser-induced modulation of the Landau level structure in single-layer graphene, *Phys. Rev. B* **92**, 23, pp. 235411.

[25] López-Rodríguez, F. J. and Naumis, G. G. (2008). Analytic solution for electrons and holes in graphene under electromagnetic waves: gap appearance and nonlinear effects, *Phys. Rev. B* **78**, 20, pp. 201406(R).

[26] Mics, Z., Tielrooij, K. J., Parvez, K., Jensen, S. A., Ivanov, I., Feng, X. L., Müllen, K., Bonn, M., and Turchinovich, D. (2015). Thermodynamic picutre of ultrafast charge transport in graphene, *Nat. Commun.* **6**, pp. 7655.

[27] Mihnev, M. T., Wang, F., Liu, G., Rothwell, S., Cohen, P. I., Feldman, L. C., Conrad, E. H., and Norris, T. B. (2015). Evidence for bandgap opening

in buckled epitaxial graphene from ultrafast time-resolved terahertz spectroscopy, *Appl. Phys. Lett.* **107**, 17, pp. 173107.
[28] Mikhailov, S. A. and Ziegler, K. (2008). Nonlinear electromagnetic response of graphene: frequency multiplication and the self-consistent-field effects, *J. Phys.: Condens. Matter* **20**, 38, pp. 384204.
[29] Morell, E. S. and Foa Torres, L. E. F. (2012). Radiation effects on the electronic properties of bilayer graphene, *Phys. Rev. B* **86**, 12, pp. 125449.
[30] Oka, T. and Aoki, H. (2009). Photovoltaic Hall effect in graphene, *Phys. Rev. B* **79**, 8, pp. 081406(R).
[31] Paul, M. J., Chang, Y. C., Thompson, Z. J., Stickel, A., Wardini, J., Choi, H., Minot, E. D., Hou, B., Nees, J. A., Norris, T. B., and Lee, Y. S. (2013). High-field terahertz response of graphene, *New J. Phys.* **15**, pp. 085019.
[32] Peres, N. M. R., Guinea, F., and Castro Neto, A. H. (2006). Electronic properties of disordered two-dimensional carbon, *Phys. Rev. B* **73**, 12, pp. 125411.
[33] Satou, A., Vasko, F. T., and Ryzhii, V. (2008). Nonequilibrium carriers in intrinsic graphene under interband photoexcitation, *Phys. Rev. B* **78**, 11, pp. 115431.
[34] Sell, A., Leitenstorfer, A., and Huber, R. (2008). Phase-locked generation and field-resolved detection of widely tunable terahertz pulses with amplitudes exceeding 100MV/cm, *Opt. Lett.* **33**, 23, pp. 2767.
[35] Scharf, B., Perebeinos, V., Fabian, J., and Avouris, P. (2013). Effects of optical and surface polar phonons on the optical conductivity of doped graphene, *Phys. Rev. B* **87**, 3, pp. 035414.
[36] Scholz, A., López, A., and Schliemann, J. (2013). Interplay between spin-orbit interactions and a time-dependent electromagnetic field in monolayer graphene, *Phys. Rev. B* **88**, 4, pp. 045118.
[37] Shirley, J. H. (1965). Solution of the Schrödinger equation with a Hamiltonian periodic in time, *Phys. Rev.* **138**, 4B, pp. B979.
[38] Sorger, C., Preu, S., Schmidt, J., Winnerl, S., Bludov, Y. V., Peres, N. M. R., Vasilevskiy, M. I., and Weber, H. B. (2015). Terahertz response of patterned epitaxial graphene, *New J. Phys.* **17**, 5, pp. 053045.
[39] Sun, D., Wu, Z. K., Divin, C., Li, X., Berger, C., de Heer, W. A., First, P. N., Norris, T. B. (2008). Ultrafast relaxation of excited Dirac Fermions in epitaxial graphene using optical differential transmission spectroscopy, *Phys. Rev. Lett.* **101**, 15, pp. 157402.
[40] Sun, B. Y. and Wu, M. W. (2013). Microscopic theory of ultrafast dynamics of carriers photoexcited by THz and near-infrared linearly polarized laser pulses in graphene, *New J. Phys.* **15**, pp. 083038.
[41] Suzuki, S. and Hibino, H. (2015). Large optical anisotropy for terahertz light of stacked graphene ribbons with slight asymmetry, *J. Appl. Phys.* **117**, 17, pp. 174302.
[42] Syzranov, S. V., Fistul, M. V., and Efetov, K. B. (2008). Effect of radiation on transport in graphene, *Phys. Rev. B* **78**, 4, pp. 045407.

[43] Tani, S., Blanchard, F., and Tanaka, K. (2012). Ultrafast carrier dynamics in graphene under a high electric field, *Phys. Rev. Lett.* **109**, 16, pp. 166603.

[44] Tielrooij, K. J., Song, J. C. W., Jensen, S. A., Centeno, A., Pesquera, A., Zurutuza Elorza, A., Boon, M., Levitov, L. S., and Koppens, F. H. L. (2013). Photoexcitation cascade and multiple hot-carrier generation in graphene, *Nat. Phys.* **9**, 4, pp. 248-252.

[45] Vakil, A. and Engheta, N. (2011). Transformation optics using graphene, *Science* **332**, 6035, pp. 1291-1294.

[46] Vasilyev, Y. B., Vasileva, G. Y., Ivanov, Y. L., Novikov, S., and Danilov, S. N. (2014). Temperature-dependent terahertz photoconductivity in epitaxial graphene, *Appl. Phys. Lett.* **105**, 17, pp. 171105.

[47] Wang, Y. H., Steinberg, H., Jarillo-Herrero, P., and Gedik, N. (2013). Observation of Floquet-Bloch states on the surface of a topological insulator, *Science* **342**, 6157, pp. 453-457.

[48] Weng, M. Q., Wu, M. W., and Jiang, L. (2004). Hot-electron effect in spin dephasing in n-type GaAs quantum wells, *Phys. Rev. B* **69**, 24, pp. 245320.

[49] Winnerl, S., Orlita, M., Plochocka, P., Kossacki, P., Potemski, M., Winzer, T., Malic, E., Knorr, A., Sprinkle, M., Berger, C., de Heer, W. A., Schneider, H., and Helm, M. (2011). Carrier relaxation in epitaxial graphene photoexcited near the Dirac point, *Phys. Rev. Lett.* **107**, 23, pp. 237401.

[50] Wright, A. R., Xu, X. G., Cao, J. C., and Zhang, C. (2009). Strong nonlinear optical response of graphene in the terahertz regime, *Appl. Phys. Lett.* **95**, 7, pp. 072101.

[51] Zhang, W., Zhang, P., Duan, S., and Zhao, X. (2009). Quasi-energy spectra of a charged particle in planar honeycomb lattices, *New J. Phys.* **11**, pp. 063032.

[52] Zhou, Y. and Wu, M. W. (2010). Electron spin relaxation in graphene from a microscopic approach: role of electron-electron interaction, *Phys. Rev. B* **82**, 8, pp. 085304.

[53] Zhou, Y. and Wu, M. W. (2011). Optical response of graphene under intense terahertz fields, *Phys. Rev. B* **83**, 24, pp. 245436.

Chapter 8

Nonlinear Terahertz Spectroscopy on Multilayer Graphene

Michael Woerner, Thomas Elsaesser, and Klaus Reimann

Max-Born-Institut für Nichtlineare Optik und Kurzzeitspektroskopie, 12489 Berlin, Germany

The peculiar electronic band structure makes graphene highly interesting for nonlinear terahertz (THz) spectroscopy. Interband transition dipoles around the K and K' points are huge in the THz frequency range, leading to highly nonlinear effects at moderate amplitudes of THz fields. Moreover, intraband motion of electrons is intrinsically nonlinear due to the cone-like bandstructure. We discuss results from recent ultrafast THz experiments in the nonperturbative regime of light-matter interaction. Different methods including two-dimensional THz spectroscopy are applied. The generation of the third and fifth harmonic of the THz fundamental is observed in propagation experiments. Pump-probe studies with different photon energies of the pump pulses reveal induced THz absorption which decays on a time scale of a few picoseconds. THz photon echo signals are absent. A theoretical description of the experimental results requires to include combined interband/intraband transitions and currents while the decay of the pump-probe transients is a hallmark of radiative coupling in multilayer graphene.

8.1 Introduction

The two-dimensional material graphene has a peculiar electronic band structure around the K and K' points in k-space which makes it highly interesting for nonlinear THz spectroscopy. In the frequency range below 5 THz, interband transitions between lower (valence) and upper (conduction) bands display very large transition dipole moments, leading to highly nonlinear light-matter interactions at moderate amplitudes of the incident THz field. Moreover, field-driven intraband motions of electrons are expected to be highly nonlinear, since the intraband electron velocity has a constant value.

For most experiments in the THz frequency range, the electronic response of graphene is determined by the cone-like electronic band structure around the K and K' points (Dirac points) of the hexagonal Brillouin zone where the maxima of the valence and the minima of the conduction bands occur with a zero bandgap. Near these points the energy increases for the conduction band and decreases for the valence band linearly with the absolute value of the wavevector $\vec{q}$ measured from the nearest Dirac point: $E(\vec{q}) = \pm\hbar v_F|\vec{q}|$. The zero of energy is at the apex of the cone, the upper/lower sign is for the conduction/valence band, $v_F = 10^6$ m/s is the Fermi velocity [Castro Neto *et al.* (2009)]. In real space, the velocity is given by

$$\vec{v} = \hbar^{-1}\nabla_{\vec{q}}E(\vec{q}) = \pm v_F \begin{pmatrix} q_x/q \\ q_y/q \end{pmatrix}, \tag{8.1}$$

and, thus, the absolute value of the electron velocity is independent of the energy and equal to v_F. Accordingly, it is not possible to change this value with an external perturbation. This is similar to the case of elementary particles with zero rest mass, always moving with the speed of light c. Because of this similarity, electrons in graphene are often called "massless" [Novoselov *et al.* (2005)]. However, while it is not possible to change $|\vec{v}|$, a change of the *direction* of $\vec{v}$ is feasible. This fact is described by the inverse effective mass tensor m^{-1} [Ashcroft and Mermin (1976)] of the electrons in graphene, which is finite everywhere except exactly at the Dirac points. It is given by:

$$[m^{-1}(\vec{q})]_{i,j} = \frac{1}{\hbar^2}\frac{\partial^2}{\partial q_i \partial q_j}E(\vec{q}) = \frac{v_F}{\hbar q^3}\begin{pmatrix} q_y^2 & -q_x q_y \\ -q_x q_y & q_x^2 \end{pmatrix}. \tag{8.2}$$

Another consequence of graphene's band structure is the constant interband absorption over a large frequency range, having a value of $\pi\alpha = 0.023$ per layer (α: fine structure constant) [Kuzmenko *et al.* (2008); Nair *et al.* (2008); Mak *et al.* (2008)]. Interband absorption involves a transition from a valence band state at a certain $\vec{q}$ to a state in the conduction band at the same $\vec{q}$. The values of $\vec{q}$ are determined by energy conservation, i.e., the energy difference between the two bands has to be equal to the photon energy of the radiation: $h\nu = E_{\mathrm{cb}}(\vec{q}) - E_{\mathrm{vb}}(\vec{q}) = 2\hbar v_F q$. For a given frequency ν, the allowed $\vec{q}$ are on circles around the Dirac points. Thus, the combined density of states, which is proportional to the number of allowed $\vec{q}$, is proportional to q or to ν. This linear increase of the joint density of states is compensated for by the interband transition dipole $\vec{d}$, which has a magnitude inversely proportional to the transition frequency. The dipole moment $\vec{d}$ is equal to [Grüneis *et al.* (2003)]

$$\vec{d}(\vec{q}) = \frac{e}{2q^2}\begin{pmatrix} q_y \\ q_x \end{pmatrix} = \frac{ev_F}{2\pi\nu}\begin{pmatrix} q_y/q \\ -q_x/q \end{pmatrix}. \tag{8.3}$$

At low transition frequencies the interband transition dipole moments are huge (Fig. 8.1) and reach, e.g., a value of $d = e \times 80$ nm at $\nu = 2$ THz (e: elementary charge). For comparison, the interband transition dipole moment of bulk GaAs is $d = e \times 0.5$ nm [Mücke *et al.* (2001)], more than two orders of magnitude smaller. The direction of $\vec{d}$ in graphene is perpendicular to $\vec{q}$.

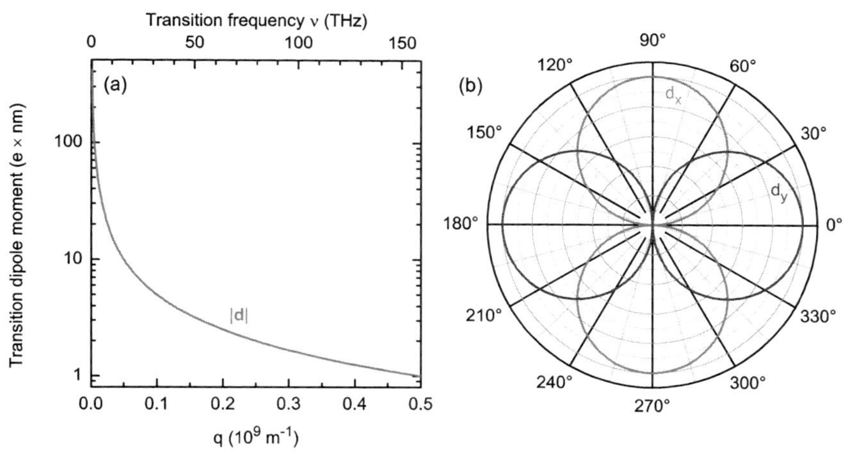

Fig. 8.1 Interband transition dipole moment $\vec{d}$. (a) $|d|$ as a function of the wavevector q in k-space as measured from the Dirac point. The upper scale gives the corresponding transition frequency ν. (b) The x and y components of $\vec{d}$ as a function of the direction of $\vec{q}$ (0° corresponds to x).

The nonlinear regime of light-matter interactions is entered at comparably small THz field amplitudes, as is evident from considering the interband Rabi frequency. At a transition frequency of $\nu = 2$ THz, the Rabi frequency $\Omega = dF/\hbar$ is equal to $2\pi\nu$ already at a field amplitude of $F = 1$ kV/cm.

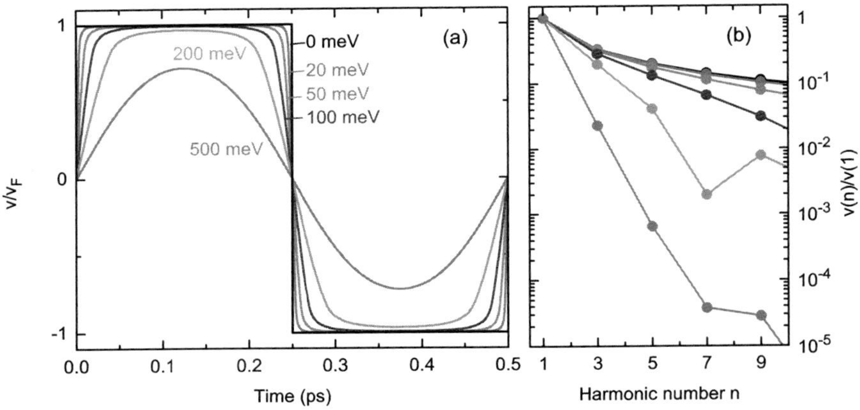

Fig. 8.2 (a) Average electron velocity in graphene with different Fermi energies E_F under the influence of an external electric field with an amplitude of 50 kV/cm and a frequency of 2 THz under the assumption of a constant electron density in the conduction band. (b) Amplitudes of the odd harmonics of the velocity relative to the fundamental.

In other words, the interaction strength is in the extreme nonlinear regime [Wegener (2005)], where perturbation theory is not applicable.

Apart from interband transitions of carriers, a THz electric field induces intraband motions if free carriers are present. Free carriers are introduced by thermal excitation at finite temperatures, doping resulting in a non-zero Fermi energy, or by interband optical excitation. For simplicity, we consider the case of a finite Fermi energy $E_F > 0$ with only one type of carriers present at zero temperature. We disregard any scattering processes and consider the number of free carriers to be constant. In this case the intraband motion in k-space is determined by the acceleration theorem,

$$\frac{d}{dt}\hbar\vec{q}(t) = -e\vec{F}(t). \tag{8.4}$$

Since there is rotational symmetry of the bandstructure around the Dirac points in the cone approximation, we assume without loss of generality that the field to be directed along x. Then, Eq. (8.4) has the solution

$$q_x(t) = q_x(0) + \frac{e}{\hbar}A(t), \quad q_y(t) = q_y(0), \quad A(t) = -\int_0^t F(t')dt'. \tag{8.5}$$

The q-interval covered by the carrier motion is determined by the extrema of $A(t)$. For a monochromatic field $F(t) = F_0 \sin(2\pi\nu t)$, the motion in k-space is sinusoidal with an amplitude of $\Delta q = eF_0/2\pi\hbar\nu$. The velocity in real space is given by $\vec{v} = \hbar^{-1}\nabla_{\vec{q}}E(\vec{q})$ and the motion in real space is not sinusoidal. For $\vec{q}(0) = 0$, the velocity in field direction v_x will be a square wave, for $q_x > 0$ it is $+v_F$, for $q_x < 0$ $-v_F$. Accordingly, the electric current should display strong contributions at the odd harmonics of ν [Mikhailov (2007); Mikhailov and Ziegler (2008)]. For a non-zero component $q_y(0)$, $v_x(t)$ will gradually change from a square wave to a sine wave as shown in Fig. 8.2 and the intensity of the higher harmonics of the current will decrease. This becomes relevant when the Fermi energy (for the case of finite temperatures, the same applies to k_BT) becomes comparable to the size of the energy excursion caused by the intraband motion, $\Delta E = eF_0v_F/2\pi\nu$, which is for $\nu = 2$ THz and $F_0 = 50$ kV/cm equal to $\Delta E = 0.4$ eV.

8.2 Single- and Multilayer Graphene

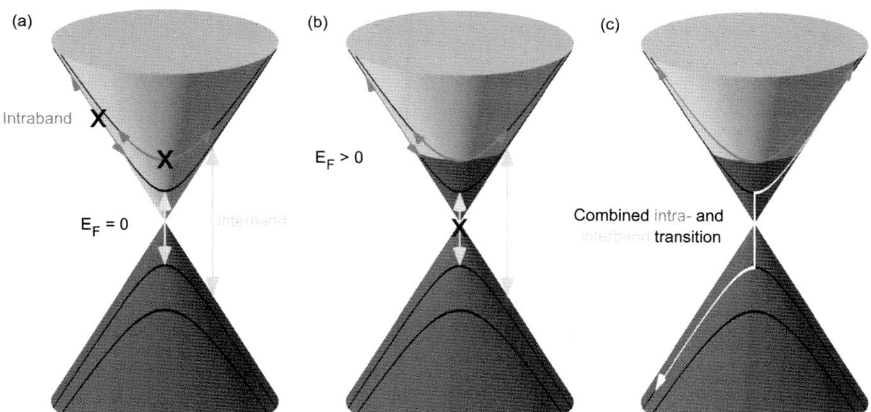

Fig. 8.3 (a) In graphene with $E_F = 0$ at zero temperature interband transitions with arbitrarily low photon energies are allowed, while intraband transitions are not possible. (b) For a finite E_F (shown is the case $E_F > 0$) interband transitions are blocked up to photon energies of $h\nu = 2|E_F|$. On the other hand, electrons in the upper band and holes in the lower band (present for $E_F < 0$) allow for intraband transitions. (c) For high THz field amplitudes the range in k-space covered by the intraband trajectories is large enough that the blocking of interband transitions by a finite E_F becomes unimportant. If an electron on such a trajectory reaches a region in k-space where the THz field is in resonance with the interband transition frequency, combined inter- and intraband trajectories will occur.

Ideal graphene has a Fermi energy $E_F = 0$, i.e., at zero temperature the lower band is completely filled and the upper band is completely empty.

This is not necessarily the case in real graphene samples. Graphene produced by exfoliating single layers from graphite [Novoselov *et al.* (2004)] usually has a negative Fermi energy around −100 meV. By introducing a gate electrode and applying a dc voltage between the electrode and the graphene layer the Fermi energy becomes tunable and it is possible to achieve a Fermi energy of zero. Unfortunately, this does not necessarily mean that the density of free carriers is zero. The electrical conductivity as a function of gate voltage shows a minimum, but in most cases maintains a non-zero value [Novoselov *et al.* (2004, 2005)]). Locally, i.e., in small areas of about 30 nm diameter, the energy position of the cone apex changes because of strain originating from corrugation of the layers. For the apex at a positive or negative finite energy, free holes or electrons are present [Martin *et al.* (2007)].

Another growth method for making graphene layers is by thermal decomposition of silicon carbide [Berger *et al.* (2004)]. Such samples are typically not single graphene layers, but multilayer graphene stacks consisting of up to several tens of layers. In structurally well-defined samples, this multilayer graphene represents a stack of electronically uncoupled individual graphene layers [Hass *et al.* (2008); Sprinkle *et al.* (2009)]. A few layers close to the SiC substrate contain a substantial concentration of free carriers , i.e., have high values of $|E_F|$, while the remaining layers have very low $|E_F|$ [Sun *et al.* (2009, 2010)]. For nonlinear measurements multilayer samples with N layers have the advantage that the nonlinear signal is N times larger than for a single layer. On the other hand, the more layers are present the more important gets radiative coupling [Stroucken *et al.* (1996); Shih *et al.* (2005); Bowlan *et al.* (2014a)], since all the layers are within a distance much less than the optical wavelength, in particular in the THz range.

Interband transitions occur at arbitrarily low frequencies for graphene with $E_F = 0$ [Fig. 8.3(a)]. Interband transitions are blocked by state-filling up to photon energies $h\nu = 2|E_F|$ for $E_F \neq 0$. On the other hand, electrons in the conduction band [Fig. 8.3(b)] or holes in the valence band give rise to intraband transitions. As a result, it is not *a priori* clear whether a finite Fermi energy will result in an increase or a decrease of the absorption in the THz range. Calculations show that finite Fermi energies lead to an increase of absorption, i.e., the increase of intraband absorption more than compensates for the decrease of interband absorption.

The graphene samples used in our experiments are grown by C-face epitaxy on SiC (from Graphene Works) and contain 40 layers. The layer stack has a total thickness of 13.2 nm, much smaller than all optical wavelengths in our experiments. This means that all layers experience the same applied electric field.

8.3 Experimental Techniques

The results presented in this article have been generated with different THz methods. First, a single THz transient of variable electric field amplitude interacts with a multilayer graphene sample and the transmitted THz field is measured in amplitude and phase by free-space electrooptic sampling. Second, two-dimensional (2D) THz spectroscopy is implemented with a pair of phase-locked THz pulses separated by a delay time τ. The nonlinear response of the sample is extracted from the total transmitted field by subtracting the transmitted field of the individual pulses as determined in independent measurement. The nonlinear signal field depends on the delay time τ and the real time t from which the nonlinear signal as a function of excitation frequency ν_τ and detection frequency ν_t is derived by a double Fourier transform. Third, an optical pump/THz probe experiment was performed in which the pump pulse photoexcites carriers and the resulting change of the THz response was probed by a THz transient, again with phase-resolved THz detection.

The experimental setup is based on a femtosecond Ti:sapphire oscillator-amplifier system, which delivers 25 fs pulses with a center wavelength of 800 nm and a pulse energy of 1 mJ at a repetition rate of 1 kHz. The 800nm pulses are split into two replica the first of which is sent into a GaSe crystal to generate THz pulses with a center frequency of 2 THz by optical rectification [Reimann (2007)]. The second replica generates either a phase-locked second THz pulse for 2D THz spectroscopy [Kuehn *et al.* (2009, 2011a,b); Junginger *et al.* (2012); Woerner *et al.* (2013); Bowlan *et al.* (2014b); Elsaesser *et al.* (2015)] or serves as excitation pulse in the pump-probe experiments. In a second pump-probe study, the 800 nm pulse was converted to the midinfrared spectral range (frequency 14.5 THz corresponding to a photon energy of 60 meV) by difference frequency generation in GaSe for near-/midinfrared pump–THz-probe measurements. In all cases, the transmitted electric field transients of the THz pulses are measured by electro-optic sampling [Wu and Zhang (1995); Leitenstorfer *et al.* (1999); Bartel *et al.* (2005)] in a ZnTe or a GaP crystal. GaP gives smaller electrooptic signals but offers a detection range extending to higher THz frequencies, in particular to higher harmonics of the 2-THz fundamental.

The linearly polarized 800 nm probe pulses for electrooptic sampling were derived from the output of the Ti:sapphire oscillator.

For 2D spectroscopy with the two phase-locked THz pulses A and B of a mutual delay τ two choppers synchronized to the pulse repetition rate allow for measuring the electric-field transients $E_{AB}(t,\tau)$ (both pulses A and B are incident on the sample), $E_A(t)$ (only pulse A), and $E_B(t,\tau)$ (only pulse B). The nonlinear signal emitted from the sample, which is proportional to the nonlinear current in the sample, is the difference [Kuehn *et al.* (2009)]

$$E_{NL}(t,\tau) = E_{AB}(t,\tau) - E_A(t) - E_B(t,\tau) \quad (8.6)$$

and depends on the delay τ and the real time t.

8.4 Experimental Results

8.4.1 *Linear THz transmission*

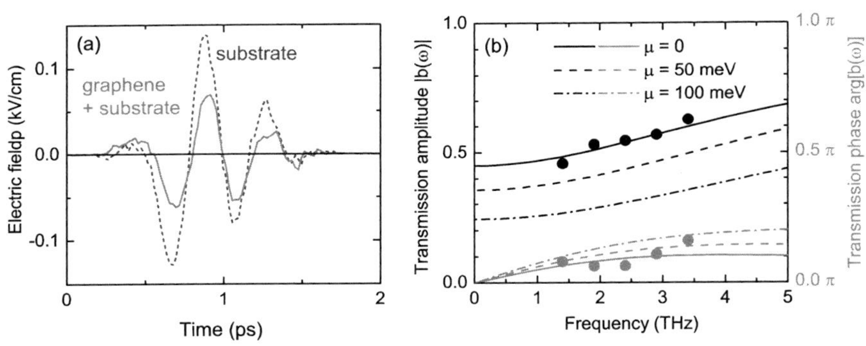

Fig. 8.4 (a) Phase-resolved linear transmission measurements of the 40-layer graphene sample at room temperature. The red solid line shows the transmitted electric field through the graphene sample on the SiC substrate, the blue dashed line the transmitted electric field through the SiC substrate alone. (b) Amplitude (left axis) and phase (right axis) of the ratio $b(\omega)$ of the Fourier transforms of the two transients. Symbols: experimental results; solid, dashed and dash-dotted lines: calculated values for different Fermi levels.

The linear THz response of the multilayer sample was determined by transmitting a single THz pulse with a low field amplitude of 0.2 kV/cm. The THz transient $E_{SG}(t)$ transmitted through the graphene stack on the SiC substrate is compared to the transient $E_S(t)$ going only through the substrate [Fig. 8.4(a)]. At 300 K the sample shows a spectrally integrated

THz transmission $\int E_{SG}(t)^2dt / \int E_S(t)^2dt$ of 0.36, at a temperature of 50 K of 0.56. Both values are below the transmission expected from quantum absorption (0.65), caused by a finite density of free carriers. An analysis of the spectrally resolved transmission data in amplitude and phase [Fig. 8.4(b)] using the Drude approach [Dawlaty *et al.* (2008); Nair *et al.* (2008); Stauber *et al.* (2008); Choi *et al.* (2009); Yan *et al.* (2011); Horng *et al.* (2011); Strait *et al.* (2011)] gives an average Fermi energy of $E_F = 24$ meV and a temperature-independent Drude scattering time of $T_s = 40$ fs. From our measurements it is not possible to determine whether $E_F > 0$ or $E_F < 0$. According to [Varchon *et al.* (2007); Sun *et al.* (2008, 2010)] we choose $E_F > 0$.

8.4.2 *Generation of THz harmonics*

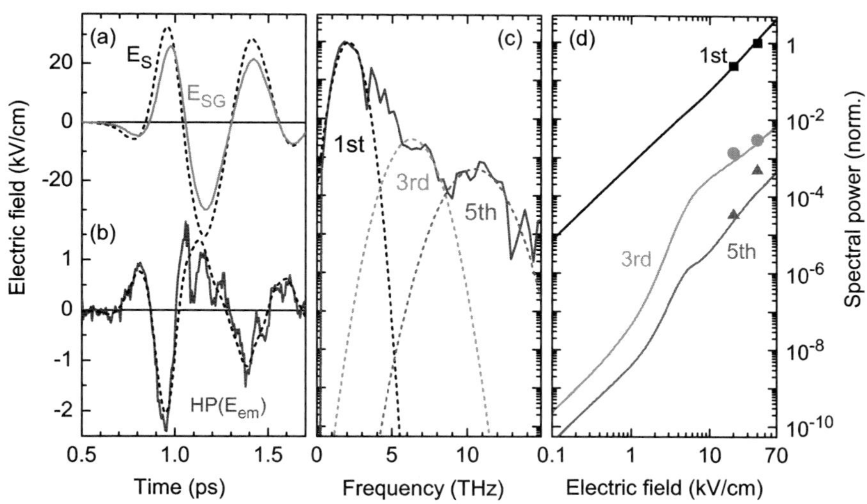

Fig. 8.5 (a) Electric field transients through the substrate (E_S, dashed line) and through the substrate and the graphene (E_{SG}, solid line) in the nonperturbative regime at $T = 50$ K. (b) Graphene response, the difference $E_{\rm em} = E_{SG} - E_S$, after high-pass filtering (HP) (solid line). The dashed line shows the result from Drude transmission with the same filtering. (c) Solid line: Spectral power $|E_{\rm em}(\omega)|^2$ from the data in (a) and (b). The dashed lines show Gaussian fits of the fundamental, the 3rd and the 5th harmonic. (d) Spectral power of the graphene response at the fundamental, the 3rd, and the 5th harmonics as a function of the incident field amplitude. The symbols are experimental results, the solid lines are calculations.

Next, single THz transients with field amplitudes in the nonperturbative regime were transmitted through the graphene sample to determine the anharmonicity of the intraband carrier motion. As in the linear interaction regime [Fig. 8.4(a)], $E_{SG}(t)$ and $E_S(t)$ in Fig. 8.5(a) are the transients through graphene and substrate and through the substrate alone. Both $E_{SG}(t)$ and $E_S(t)$ look similar with a reduced amplitude of $E_{SG}(t)$. This result already shows that the expected square wave behavior (Fig. 8.2) does not occur. To find out whether $E_{SG}(t)$ contains higher harmonics at all, we present the graphene response, i.e., the difference $E_{\text{em}}(t) = E_{SG}(t) - E_S(t)$, after high-pass filtering in Fig. 8.5(b). Without the strong fundamental, higher frequencies are borne out more clearly. The high-pass-filtered transient exhibits oscillations at frequencies higher than the fundamental THz carrier frequency with amplitudes considerably higher than the noise level. The latter is visible as fluctuations of the transient for zero field amplitude, e.g., at 0.6 ps. If we assume that the THz transmission at these field amplitudes is still dominated by the Drude response, we obtain, with the same filtering, the dashed line in Fig. 8.5(b). Thus, the observed high-frequency oscillations are not an artifact from the high-pass filtering but part of the intraband response of carriers in the strong THz field. The power spectrum $|E_{\text{em}}(\omega)|^2$ [Fig. 8.5(c)] clearly shows the first, third, and fifth harmonics with a ratio of $1 : 3 \times 10^{-3} : 5 \times 10^{-4}$. For a perfect square-wave shape of the intraband current, one expects a larger ratio of $1 : 1/3^2 : 1/5^2 = 1 : 1/9 : 1/25$ (Fig. 8.2) [Mikhailov (2007)]. This discrepancy has been discussed by [Paul *et al.* (2013)] who did not observe any higher harmonics, most probably because of the limited dynamic range of their experiments. Both decreasing the incident field amplitude and increasing the temperature leads to a decrease of the 3rd and 5th harmonics of the THz fundamental. In Fig. 8.5(d) we show the dependence of the fundamental and the 3rd and 5th harmonics of the graphene response on the incident field amplitude. The values are obtained from the experimental spectra by Gaussian fits with both the center frequencies and the widths fixed. The center frequency of the 3rd (5th) harmonic is 3 (5) times the center frequency of the fundamental and has $\sqrt{3}$ ($\sqrt{5}$) times the fundamental width, see dashed lines in Fig. 8.5(c).

8.4.3 *Two-dimensional THz spectroscopy*

In the 2D THz experiments, the electric fields of the two collinear phase-locked pulses B and A transmitted through the sample are measured as a

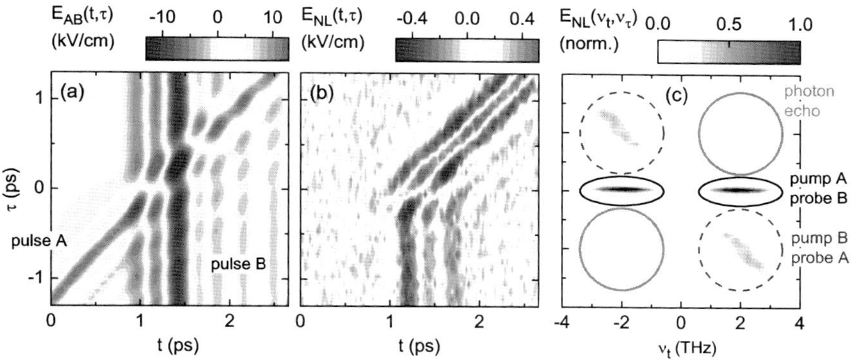

Fig. 8.6 Experimental results for the 40-layer graphene sample at a temperature of 20 K. (a) Electric field $E_{AB}(t, \tau)$ transmitted through the sample. (b) Nonlinear THz signal according to Eq. (8.6). (c) The 2D Fourier transform of the nonlinear signal allows for separating the various contributions. (Nonrephasing) pump-probe signals are present, photon echoes, i.e., rephasing signals, are missing. (For color reference, go to page 492.)

function of the delay τ between them and of the real time t. Pulse B is kept fixed in real time, whereas the position of pulse A is varied according to the delay τ (pulse A arrives earlier than pulse B for $\tau < 0$ and later for $\tau > 0$). The transients are measured for the three combinations of pulses, both pulses A and B incident on the sample, only pulse A, and only pulse B. The nonlinear signal is the difference between the electric field when both pulses are present minus the electric fields of the two individual pulses: $E_{\mathrm{NL}}(t, \tau) = E_{AB}(t, \tau) - E_A(t, \tau) - E_B(t)$. Experimental results obtained with a sample at 300 K are shown in Fig. 8.6. Figure 8.6(a) shows the transmitted field $E_{AB}(t, \tau)$, Fig. 8.6(b) the nonlinear signal $E_{\mathrm{NL}}(t, \tau)$ as a function of the real time t and the delay τ between the THz pulses A and B. The collinear geometry used in our setup does not allow for a separation of the various nonlinear signals based on the direction of their emission. However, the signal are readily separated in frequency space as has been discussed in detail in [Kuehn *et al.* (2009, 2011a); Woerner *et al.* (2013)]. A 2D Fourier transform of $E_{\mathrm{NL}}(t, \tau)$ yields the 2D spectrum plotted in Fig. 8.6(c) as a function of the two frequencies ν_t and ν_τ. The 2D spectrum displays two pump-probe signals in which either pulse A or pulse B interacts with the sample first. Photon-echo signals, which would occur within the solid circle, are absent although the data of Fig. 8.5 give evidence of the coherent character of the graphene response.

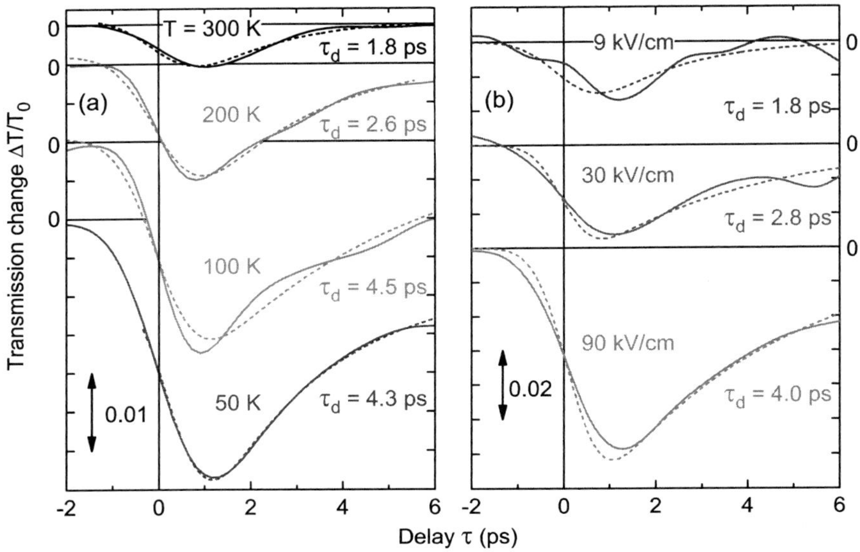

Fig. 8.7 Measured pump-probe signals (a) for different sample temperatures T (THz electric field amplitude 5 kV/cm) and (b) at room temperature for different electric field amplitudes. Solid lines: experimental results; dashed lines: single-exponential fits with decay times τ_d as indicated.

By Fourier filtering [Kuehn *et al.* (2011a); Woerner *et al.* (2013)] and back transform, we extract the THz-pump–THz-probe signals $E_{PP}(t,\tau)$ in the time domain. From the signal field $E_{PP}(t,\tau)$, the pump-induced differential change of transmitted intensity is derived by:

$$\frac{\Delta T}{T_0}(\tau) = \frac{2\int E_{\rm pr}(t,\tau)E_{PP}(t,\tau)\,dt}{\int E_{\rm pr}^2(t,\tau)\,dt}. \tag{8.7}$$

The probe field $E_{\rm pr}$ can be either E_A or E_B, depending on which signal was chosen in the frequency domain. Results for $\Delta T/T_0(\tau)$ from Eq. (8.7) are shown in Fig. 8.7. Figure 8.7(a) gives the temperature dependence for a THz amplitude of 5 kV/cm and Fig. 8.7(b) the dependence on THz field amplitude for 300 K. Excitation by the first THz pulse leads in all cases to a transmission decrease (induced absorption), independent of temperature and field amplitude. Single-exponential fits (dashed lines in Fig. 8.7) give recovery times τ_d between 1.8 and 4.5 ps, surprisingly short compared to the results in [Strait *et al.* (2011); Winnerl *et al.* (2011)]. The recovery times increase with the THz field amplitude and for decreasing temperature. Longer recovery times are correlated with a larger magnitude of differential transmission.

8.4.4 *Pump-probe measurements*

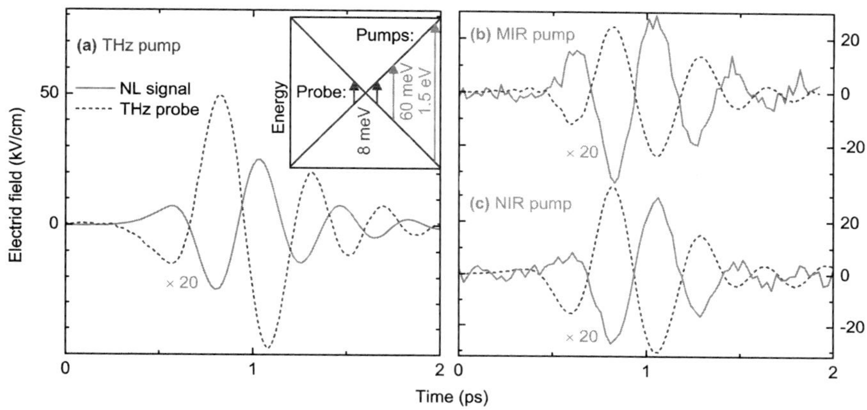

Fig. 8.8 Solid lines: Measured pump–THz-probe signals at room temperature at a delay $\tau = 0$ for different pump photon energies, (a) 8 meV, (b) 60 meV, (c) 1.5 eV. Dashed lines: THz probe pulses. The inset shows schematically the pump and probe transitions in graphene's band structure.

In addition to the THz-pump–THz-probe response extracted from the 2D data, we performed pump-probe measurements with mid- and near-infrared pump pulses (photon energies 60 meV and 1.5 eV) and THz probe pulses. The transmitted THz field is measured as a function of time with and without the pump pulse present. The difference of these transients is the nonlinear signal. Nonlinear signals for all three pump photon energies are shown in Fig. 8.8. In all cases, the nonlinear THz signal and the probe electric field are π phase shifted, corresponding to a transmission decrease. Thus, induced absorption occurs not only for THz pumping but for the other pump photon energies as well.

8.5 Discussion

We now discuss the following key results of the different THz experiments:

- THz transmission measurements reveal a coherent nonlinear response of the multilayer graphene sample, including the generation of the 3rd and 5th harmonic of the THz fundamental. The amplitudes of the harmonics are much smaller than predicted from an

elementary theoretical picture, even for finite Fermi energies, i.e., density of free carriers.
- THz photon echo signals are absent despite the coherent character of the nonlinear response.
- Pump–THz-probe measurements all show induced absorption, independent of pump photon energy.

At a first glance, the absence of photon-echo signals may be considered a manifestation of extremely fast scattering processes. Scattering processes such as electron-phonon, electron-impurity, and electron-electron scattering were shown to be responsible for the Drude width in linear absorption [Dawlaty *et al.* (2008); Nair *et al.* (2008); Stauber *et al.* (2008); Choi *et al.* (2009); Yan *et al.* (2011); Horng *et al.* (2011); Strait *et al.* (2011); George *et al.* (2008); Orlita *et al.* (2008); Neugebauer *et al.* (2009); Karch *et al.* (2011); Winnerl *et al.* (2011); Lao *et al.* (2013)] (see also Fig. 8.4) and for ultrafast thermalization and cooling [Breusing *et al.* (2011); Graham *et al.* (2013)]. In the nonperturbative regime of light-matter interaction explored in the present experiments, however, the coupling of carriers to the coherent THz field is much stronger than the interactions underlying the scattering processes. This fact is evident from a comparison of the Drude scattering rate $1/T_s = (40\text{ fs})^{-1} = 2.5 \times 10^{13}\text{ s}^{-1}$, which includes all scattering processes of electrons and is unchanged for strong THz fields [Malard *et al.* (2013)], with the Rabi frequency $\Omega = d\,F/\hbar = 6 \times 10^{14}\text{s}^{-1}$, which gives the interband transition rate (calculated for an amplitude of $F = 50\text{ kV/cm}$). The Rabi frequency is more than one order of magnitude larger than the scattering rate, a clear manifestation of the predominant coupling to the external THz field.

We describe the nonlinear response under nonperturbative conditions by a theoretical model neglecting any scattering. The key ingredients are the electronic band structure of graphene and the light-matter interaction. The electronic band structure has been calculated in the cone approximation [Ishikawa (2010)], a tight-binding approach [Castro Neto *et al.* (2009); Ishikawa (2013)], or a pseudopotential formalism [Cohen and Bergstresser (1966); van Haeringen and Junginger (1969); Mohan *et al.* (2012)]. In the neighborhood of the Dirac points all these calculations give essentially the same results but differ for larger distances q. For our calculations, we use the pseudopotential band structure, since it is easily combined with a gauge-invariant treatment of the light-matter interaction. A single Fourier coefficient of the potential, which can be determined from the energy difference between the K and the Γ points, is sufficient for a good description of the band structure. The light-matter interaction has to include both inter- and intraband transitions on the same footing, for high THz field amplitudes it is not possible to restrict oneself to one type of transition.

Radiative coupling [Stroucken *et al.* (1996); Shih *et al.* (2005)] needs to be included to describe the energy transfer from the multilayer graphene sample to the electromagnetic field appropriately. Even for multilayer graphene, the thickness of the entire stack is orders of magnitude smaller than the THz wavelength of typically 100 μm. Consequently, all layers experience the same driving field which is the field transmitted through the sample $E_{\text{tr}} = E_{\text{in}} + E_{\text{em}}$. It should be noted that without radiative coupling the driving field is the incident field. The field E_{em} emitted by the graphene sample is proportional to the sum of intra- and interband currents in all layers. For graphene layers on a SiC substrate, one also has to introduce the additional field $E_{\text{sub}}(t) = E_{\text{tr}}(t)(1 - n_{\text{sub}})/2$ emitted due to the refractive index jump at the interface [Sprinkle *et al.* (2009)].

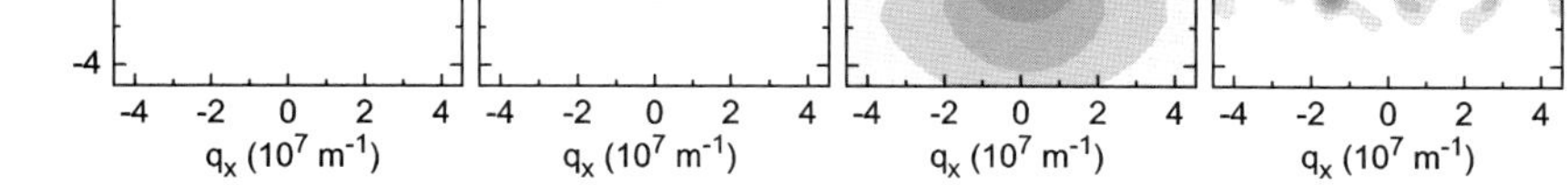

Fig. 8.9 (a), (b) Calculated electron populations in the conduction band $\rho_{22}(q_x, q_y)$, (c), (d) corresponding real part of the off-diagonal density matrix elements $2\,\text{Re}[\rho_{12}(q_x, q_y)]$, after interaction with a THz pulse polarized along x with an amplitude of (a), (c) 30 V/cm (perturbative regime) and (b), (d) 3 kV/cm (nonperturbative regime). (For color reference, go to page 492.)

The calculations show that the most important effect is the interplay between intra- and interband transitions. A good illustration for this behavior is the intensity of the THz harmonics. If there were only intraband transitions, one would get the results from Fig. 8.2, i.e., very strong odd harmonics. The simultaneous interband transitions [see Fig. 8.3(c)] increase the number of available free carriers, so that the amplitudes of third and fifth relative to the fundamental decrease. The same reason is behind the missing photon-echo signal. If one calculates the electron population in k-space for a weak THz field, one finds two maxima in k-space [Fig. 8.9(a)] where the transition frequency is equal to the THz frequency and where the interband matrix element [Fig. 8.3(b)] is large for the given field direction. At these low field amplitudes the extension of the trajectory is small, so that intraband motions do not smear out the spots. The corresponding polarization is shown in Fig. 8.9(c). At higher fields [Figs. 8.9(b) and (d)], both the population and polarization become chaotic, caused by the simultaneous interband transitions and the motion of the electrons together with their interband polarizations in k-space. In parallel, Rabi oscillations between valence and conduction bands occur at such high driving fields, with Rabi frequencies different at each point in k-space. This prevents coherent rephasing of emission from different transitions in k-space and results in a negligible photon-echo signal.

Interband excitation leads to an increase of the density of free carriers and a reduced interband absorption due to Pauli blocking of optically coupled states. In parallel, the higher carrier density enhances intraband absorption. The calculations show that the increase of intraband absorption dominates, explaining the experimental result that all pump–THz-probe measurements show induced absorption. In contrast to our sample with a low Fermi energy, in samples with large values of the Fermi energy reduced THz absorption was seen [Frenzel *et al.* (2013); Jnawali *et al.* (2013); Paul *et al.* (2013); Hwang *et al.* (2013); Hafez *et al.* (2014); Heyman *et al.* (2015)]. In this case, the increase of the density of free carriers is small compared to the density already present. Here, the reduced absorption is caused by heating of the electron distribution, which brings electrons into regions of k-space with higher effective masses and lower transition matrix elements.

The picosecond decay times of the pump-probe transients are strongly influenced by radiative coupling. It is illustrative to consider the following estimate. Considering interband transitions only in a fully coherent picture and assuming that all polarizations are parallel, the amplitude of the coherent interband current at a point in k-space is $e\,v_F$. Integrating this current over the part of the Brillouin zone with interband transition frequencies between ω_1 and ω_2 yields the total one-dimensional current density per layer of $J = e(\omega_2^2 - \omega_1^2)/(16\pi\, v_F)$. This current density corresponds to an emitted field of $E_{\rm em} = N\, Z_0\, J/(1 + n_{\rm sub})$ with $N = 40$ the number of layers, $Z_0 = 377\ \Omega$ the vacuum impedance, and $n_{\rm sub} = 3.1$ the refractive index of the substrate [Spitzer *et al.* (1959)]. The emitted field leads to an energy loss rate of $L = (1 + n_{\rm sub})E_{\rm em}^2/Z_0$. On the other hand, the energy content, i.e., $\hbar\omega/2$ integrated over the same part of the Brillouin zone is $W = N\hbar(\omega_2^3 - \omega_1^3)/(48\pi v_F^2)$. Thus, for a fully coherent emission the energy content has a radiative lifetime $T_{\rm rad} = W/L$. Using the frequency range relevant for our experiment ($\omega_1 = 2\pi \times 1$ THz, $\omega_2 = 2\pi \times 3$ THz), we find $T_{\rm rad} = 1$ ps, the same order of magnitude as the observed decay times of induced absorption in Fig. 8.7.

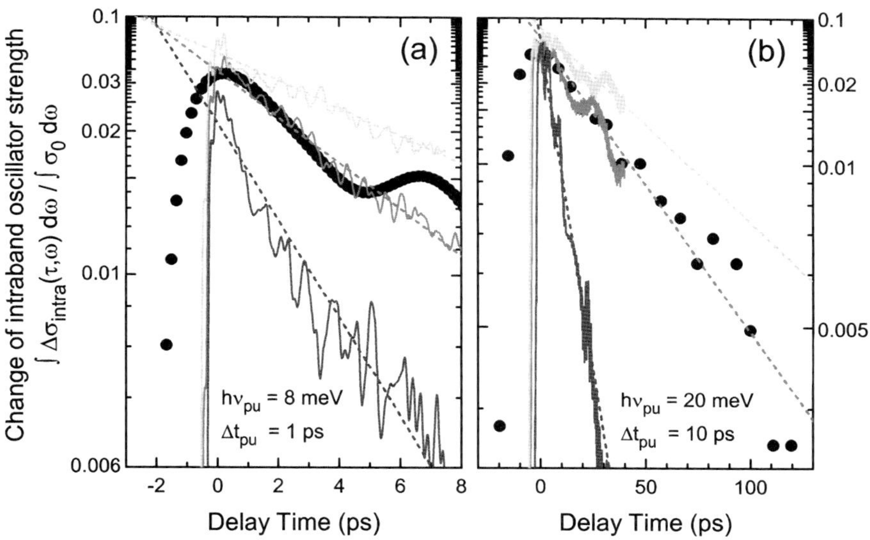

Fig. 8.10 (a), (b) Calculated electron-hole recombination transients due to radiative damping (solid lines) in comparison with various experimental situations (symbols). The transient change of the intraband oscillator strength is plotted on a reciprocal scale as a function of the delay for different THz pump pulses. In the reciprocal scale a signal decaying according to $1/[1 + a(t + b)]$ occurs as a straight line (cf. dashed lines). The red curves are calculated using the respective experimental parameters. The green curves are calculated with twice the electric field strength whereas the blue curves with half the field strength. Panel (a) corresponds to the $T = 50$ K transient shown in Fig. 8.7(a). The symbols in (b) are experimental results from [Winnerl *et al.* (2011)], i.e., the black circles in Fig. 3(b) of that reference. (For color reference, go to page 492.)

Finally, we discuss detailed calculations of the radiative electron-hole recombination in multilayer graphene. In Fig. 8.10 calculated electron-hole recombination transients (solid lines) are shown in comparison with two different experimental situations (symbols). Panel (a) corresponds to the pump-probe experiment, i.e., the $T = 50$ K transient shown in Fig. 8.7(a). In panel (b) the symbols represent experimental results from [Winnerl *et al.* (2011)]. The black circles in Fig. 3(b) of the latter article show a pump-probe transient measured using pump pulses at a photon energy of $h\nu_{\text{pu}} = 20$ meV, a pulse duration of $\Delta t_{\text{pu}} = 10$ ps, and a pump fluence of $\Phi = 0.4\ \mu\text{J cm}^{-2}$. As a measure of the electron-hole population, the transient change of the intraband oscillator strength is plotted on a reciprocal scale in Fig. 8.10. The calculated transients (solid lines) follow nicely a decay behavior according to $1/[1 + a(t + b)]$ [i.e., a solution of the differential equation $dn_{\text{eh}}/dt = -a\ n_{\text{eh}}^2$] which occurs as a straight line (cf. dashed lines) in the reciprocal scale of Fig. 8.10. The red curves in both panels were calculated using the respective experimental parameters. For comparison, calculations with twice the electric field strength (green curves) and half the field strength (blue curves) are shown as well. We would like to stress the fact that our theory based solely on light-matter interaction does not contain any fitting parameters. Nevertheless, it predicts the correct electron-hole recombination in multilayer graphene for two very different experimental situations.

These calculations clearly show that the decay rate decreases for higher field amplitudes. In this case the simplified assumption made above, namely that all interband transitions are parallel and in phase, is not true anymore. As seen in Figs. 8.9(b) and (d), a speckle-like pattern evolves. Here only the electrons within one speckle can emit coherently. Since the number of speckles increases for higher THz field amplitudes, one finds longer decay time for higher THz field amplitudes, as observed in the experiment [Fig. 8.7(b)]. The temperature dependence of the decay rate observed in Fig. 8.7(a) is primarily caused by a temperature-dependent amplitude of the driving field, which is reduced in multilayer graphene at elevated temperatures due to the intraband motion of thermally excited carriers.

We are aware of the fact that a theory based solely on light-matter interaction is not applicable for other situations, e.g., a single layer or a few layers of graphene at low THz fields where scattering processes dominate the relaxation scenario [Malic *et al.* (2016)]. Here, other recombination mechanisms like Auger recombination will dominate the population recovery. Scattering processes have also an influence on the radiative decay mechanism itself, since similar to the speckle-phenomenon discussed above scattering-induced decoherence can considerably reduce the radiative decay to the negligible spontaneous emission rate of THz photons.

8.6 Conclusions and Outlook

In conclusion, we have studied the nonlinear dynamics of electrons in multilayer epitaxial graphene by time-resolved THz spectroscopy in the regime of nonperturbative light-matter interaction. We demonstrate that the main features of the nonlinear response are reproduced by a theory including the electronic band structure and the light-matter interaction but neglecting scattering processes (for a detailed treatment of such scattering processes see the article by [Malic *et al.* (2016)] in this book). In contrast, a proper modeling of linear THz absorption requires the inclusion of scattering, here done on the level of Drude theory. The reason for this surprising behavior lies in the fact that light-matter interaction via the extremely large interband dipole in the THz range completely dominates over all other possible interactions. It is important to consider intra- and interband transitions simultaneously and to include radiative coupling to account for the complex nonlinear behavior.

Further experiments and theory are necessary to clarify the exact percentage of radiative vs. nonradiative decay in multilayer graphene. It is obvious that only a theory which includes both correct light-matter interaction and all scattering processes can explain the nonlinear THz response of multilayer graphene.

Acknowledgements

We would like to thank our former coworkers Pamela Bowlan and Elias Martinez-Moreno who performed the experiments discussed in this chapter. The authors thank the Deutsche Forschungsgemeinschaft for financial support (RE 806/9-1).

References

[1] Ashcroft, N. W. and Mermin, N. D. (1976). *Solid State Physics* (Saunders College, Philadelphia).

[2] Bartel, T., Gaal, P., Reimann, K., Woerner, M., and Elsaesser, T. (2005). Generation of single-cycle THz transients with high electric-field amplitudes, *Opt. Lett.* **30**, pp. 2805–2807, doi:10.1364/OL.30.002805.

[3] Berger, C., Song, Z., Li, T., Li, X., Ogbazghi, A. Y., Feng, R., Dai, Z., Marchenkov, A. N., Conrad, E. H., First, P. N., and de Heer, W. A. (2004). Ultrathin epitaxial graphite: 2D electron gas properties and a route toward graphene-based nanoelectronics, *J. Phys. Chem. B* **108**, pp. 19912–19916, doi:10.1021/jp040650f.

[4] Bowlan, P., Martinez-Moreno, E., Reimann, K., Elsaesser, T., and Woerner, M. (2014a). Terahertz radiative coupling and damping in multilayer graphene, *New J. Phys.* **16**, p. 013027, doi:10.1088/1367-2630/16/1/013027.

[5] Bowlan, P., Martinez-Moreno, E., Reimann, K., Elsaesser, T., and Woerner, M. (2014b). Ultrafast terahertz response of multi-layer graphene in the nonperturbative regime, *Phys. Rev. B* **89**, p. 041408(R), doi: 10.1103/PhysRevB.89.041408.

[6] Breusing, M., Kuehn, S., Winzer, T., Malić, E., Milde, F., Severin, N., Rabe, J. P., Ropers, C., Knorr, A., and Elsaesser, T. (2011). Ultrafast nonequilibrium carrier dynamics in a single graphene layer, *Phys. Rev. B* **83**, p. 153410, doi:10.1103/PhysRevB.83.153410.

[7] Castro Neto, A. H., Guinea, F., Peres, N. M. R., Novoselov, K. S., and Geim, A. K. (2009). The electronic properties of graphene, *Rev. Mod. Phys.* **81**, pp. 109–162, doi:10.1103/RevModPhys.81.109.

[8] Choi, H., Borondics, F., Siegel, D. A., Zhou, S. Y., Martin, M. C., Lanzara, A., and Kaindl, R. A. (2009). Broadband electromagnetic response and ultrafast dynamics of few-layer epitaxial graphene, *Appl. Phys. Lett.* **94**, p. 172102, doi:10.1063/1.3122348.

[9] Cohen, M. L. and Bergstresser, T. K. (1966). Band structures and pseudopotential form factors for fourteen semiconductors of the diamond and zinc-blende structures, *Phys. Rev.* **141**, pp. 789–796, doi: 10.1103/PhysRev.141.789.

[10] Dawlaty, J. M., Shivaraman, S., Strait, J., George, P., Chandrashekhar, M., Rana, F., Spencer, M. G., Veksler, D., and Chen, Y. (2008). Measurement of the optical absorption spectra of epitaxial graphene from terahertz to visible, *Appl. Phys. Lett.* **93**, p. 131905, doi:10.1063/1.2990753.

[11] Elsaesser, T., Reimann, K., and Woerner, M. (2015). Focus: Phase-resolved nonlinear terahertz spectroscopy—From charge dynamics in solids to molecular excitations in liquids, *J. Chem. Phys.* **142**, p. 212301, doi: 10.1063/1.4916522.

[12] Frenzel, A. J., Lui, C. H., Fang, W., Nair, N. L., Herring, P. K., Jarillo-Herrero, P., Kong, J., and Gedik, N. (2013). Observation of suppressed terahertz absorption in photoexcited graphene, *Appl. Phys. Lett.* **102**, p. 113111, doi:10.1063/1.4795858.

[13] George, P. A., Strait, J., Dawlaty, J., Shivaraman, S., Chandrashekhar, M., Rana, F., and Spencer, M. G. (2008). Ultrafast optical-pump terahertz-probe spectroscopy of the carrier relaxation and recombination dynamics in epitaxial graphene, *Nano Lett.* **8**, pp. 4248–4251, doi:10.1021/nl8019399.

[14] Graham, M. W., Shi, S.-F., Ralph, D. C., Park, J., and McEuen, P. L. (2013). Photocurrent measurements of supercollision cooling in graphene, *Nature Phys.* **9**, pp. 103–108, doi:10.1038/NPHYS2493.

[15] Grüneis, A., Saito, R., Samsonidze, G. G., Kimura, T., Pimenta, M. A., Jorio, A., Souza Filho, A. G., Dresselhaus, G., and Dresselhaus, M. S. (2003). Inhomogeneous optical absorption around the K point in graphite and carbon nanotubes, *Phys. Rev. B* **67**, p. 165402, doi: 10.1103/PhysRevB.67.165402.

[16] Hafez, H. A., Al-Naib, I., Oguri, K., Sekine, Y., Dignam, M. M., Ibrahim, A., Cooke, D. G., Tanaka, S., Komori, F., Hibino, H., and Ozaki, T. (2014). Nonlinear transmission of an intense terahertz field through monolayer graphene, *AIP Advances* **4**, p. 117118, doi:10.1063/1.4902096.

[17] Hass, J., Varchon, F., Millán-Otoya, J. E., Sprinkle, M., Sharma, N., de Heer, W. A., Berger, C., First, P. N., Magaud, L., and Conrad, E. H. (2008). Why multilayer graphene on $4H$-SiC $(000\bar{1})$ behaves like a single sheet of graphene, *Phys. Rev. Lett.* **100**, p. 125504, doi: 10.1103/PhysRevLett.100.125504.

[18] Heyman, J. N., Stein, J. D., Kaminski, Z. S., Banman, A. R., Massari, A. M., and Robinson, J. T. (2015). Carrier heating and negative photoconductivity in graphene, *J. Appl. Phys.* **117**, p. 015101, doi:10.1063/1.4905192.

[19] Horng, J., Chen, C.-F., Geng, B., Girit, C., Zhang, Y., Hao, Z., Bechtel, H. A., Martin, M., Zettl, A., Crommie, M. F., Shen, Y. R., and Wang, F. (2011). Drude conductivity of Dirac fermions in graphene, *Phys. Rev. B* **83**, p. 165113, doi:10.1103/PhysRevB.83.165113.

[20] Hwang, H. Y., Brandt, N. C., Farhat, H., Hsu, A. L., Kong, J., and Nelson, K. A. (2013). Nonlinear THz conductivity dynamics in p-type CVD-grown graphene, *J. Phys. Chem. B* **117**, p. doi:10.1021/jp407548a, doi: 10.1021/jp407548a.

[21] Ishikawa, K. L. (2010). Nonlinear optical response of graphene in time domain, *Phys. Rev. B* **82**, p. 201402(R), doi:10.1103/PhysRevB.82.201402.

[22] Ishikawa, K. L. (2013). Electronic response of graphene to an ultrashort intense terahertz radiation pulse, *New J. Phys.* **15**, p. 055021, doi: 10.1088/1367-2630/15/5/055021.

[23] Jnawali, G., Rao, Y., Yan, H., and Heinz, T. F. (2013). Observation of a transient decrease in terahertz conductivity of single-layer graphene induced by ultrafast optical excitation, *Nano Lett.* **13**, pp. 524–530, doi: 10.1021/nl303988q.

[24] Junginger, F., Mayer, B., Schmidt, C., Schubert, O., Mährlein, S., Leitenstorfer, A., Huber, R., and Pashkin, A. (2012). Nonperturbative interband response of a bulk InSb semiconductor driven off resonantly by terahertz electromagnetic few-cycle pulses, *Phys. Rev. Lett.* **109**, p. 147403, doi:10.1103/PhysRevLett.109.147403.

[25] Karch, J., Drexler, C., Olbrich, P., Fehrenbacher, M., Hirmer, M., Glazov, M. M., Tarasenko, S. A., Ivchenko, E. L., Birkner, B., Eroms, J., D.Weiss, Yakimova, R., Lara-Avila, S., Kubatkin, S., Ostler, M., Seyller, T., and Ganichev, S. D. (2011). Terahertz radiation driven chiral edge currents in graphene, *Phys. Rev. Lett.* **107**, p. 276601, doi: 10.1103/PhysRevLett.107.276601.

[26] Kuehn, W., Reimann, K., Woerner, M., and Elsaesser, T. (2009). Phase-resolved two-dimensional spectroscopy based on collinear n-wave mixing in the ultrafast time domain, *J. Chem. Phys.* **130**, p. 164503, doi: 10.1063/1.3120766.

[27] Kuehn, W., Reimann, K., Woerner, M., Elsaesser, T., and Hey, R. (2011a). Two-dimensional terahertz correlation spectra of electronic excitations in semiconductor quantum wells, *J. Phys. Chem. B* **115**, pp. 5448–5455, doi: 10.1021/jp1099046.

[28] Kuehn, W., Reimann, K., Woerner, M., Elsaesser, T., Hey, R., and Schade, U. (2011b). Strong correlation of electronic and lattice excitations in GaAs/AlGaAs semiconductor quantum wells revealed by two-dimensional terahertz spectroscopy, *Phys. Rev. Lett.* **107**, p. 067401, doi: 10.1103/PhysRevLett.107.067401.

[29] Kuzmenko, A. B., van Heumen, E., Carbone, F., and van der Marel, D. (2008). Universal optical conductance of graphite, *Phys. Rev. Lett.* **100**, p. 117401, doi:10.1103/PhysRevLett.100.117401.

[30] Lao, Y.-F., Perera, A. G. U., Shepperd, K., Wang, F., Conrad, E. H., and Williams, M. D. (2013). Temperature-dependent far-infrared response of epitaxial multilayer graphene, *Appl. Phys. Lett.* **102**, p. 231906, doi: 10.1063/1.4810757.

[31] Leitenstorfer, A., Hunsche, S., Shah, J., Nuss, M. C., and Knox, W. H. (1999). Detectors and sources for ultrabroadband electro-optic sampling: Experiment and theory, *Appl. Phys. Lett.* **74**, pp. 1516–1518, doi: 10.1063/1.123601.

[32] Mak, K. F., Sfeir, M. Y., Wu, Y., Lui, C. H., Misewich, J. A., and Heinz, T. F. (2008). Measurement of the optical conductivity of graphene, *Phys. Rev. Lett.* **101**, p. 196405, doi:10.1103/PhysRevLett.101.196405.

[33] Malard, L. M., Mak, K. F., Neto, A. H. C., Peres, N. M. R., and Heinz, T. F. (2013). Observation of intra- and inter-band transitions in the transient optical response of graphene, *New J. Phys.* **15**, p. 015009, doi:10.1088/1367-2630/15/1/015009.

[34] Malic, E., Winzer, T., Kadi, F., and Knorr, A. (2016). Microscopic view on the ultrafast carrier dynamics in graphene, in R. Binder (ed.), *Optical Properties of Graphene* (World Scientific, Singapore).

[35] Martin, J., Akerman, N., Ulbricht, G., Lohmann, T., Smet, J. H., von Klitzing, K., and Yacoby, A. (2007). Observation of electron–hole puddles in graphene using a scanning single-electron transistor, *Nature Phys.* **4**, pp. 144–148, doi:10.1038/nphys781.

[36] Mikhailov, S. A. (2007). Non-linear electromagnetic response of graphene, *Europhys. Lett.* **79**, p. 27002, doi:10.1209/0295-5075/79/27002.

[37] Mikhailov, S. A. and Ziegler, K. (2008). Nonlinear electromagnetic response of graphene: frequency multiplication and the self-consistent-field effects, *J. Phys.: Condens. Matter* **20**, p. 384204, doi:10.1088/0953-8984/20/38/384204.

[38] Mohan, B., Kumar, A., and Ahluwalia, P. K. (2012). A first principle study of interband transitions and electron energy loss in mono and bilayer graphene: Effect of external electric field, *Physica E* **44**, pp. 1670–1674, doi:10.1016/j.physe.2012.04.017.

[39] Mücke, O. D., Tritschler, T., Wegener, M., Morgner, U., and Kärtner, F. X. (2001). Signatures of carrier-wave Rabi flopping in GaAs, *Phys. Rev. Lett.* **87**, p. 057401, doi:10.1103/PhysRevLett.87.057401.

[40] Nair, R. R., Blake, P., Grigorenko, A. N., Novoselov, K. S., Booth, T. J., Stauber, T., Peres, N. M. R., and Geim, A. K. (2008). Fine structure constant defines visual transparency of graphene, *Science* **320**, p. 1308, doi: 10.1126/science.1156965.

[41] Neugebauer, P., Orlita, M., Faugeras, C., Barra, A.-L., and Potemski, M. (2009). How perfect can graphene be? *Phys. Rev. Lett.* **103**, p. 136403, doi:10.1103/PhysRevLett.103.136403.

[42] Novoselov, K. S., Geim, A. K., Morozov, S. V., Jiang, D., Katsnelson, M. I., Grigorieva, I. V., Dubonos, S. V., and Firsov, A. A. (2005). Two-dimensional gas of massless Dirac fermions in graphene, *Nature* **438**, pp. 197–200, doi: 10.1038/nature04233.

[43] Novoselov, K. S., Geim, A. K., Morozov, S. V., Jiang, D., Zhang, Y., Dubonos, S. V., Grigorieva, I. V., and Firsov, A. A. (2004). Electric field effect in atomically thin carbon films, *Science* **306**, pp. 666–669, doi: 10.1126/science.1102896.

[44] Orlita, M., Faugeras, C., Plochocka, P., Neugebauer, P., Martinez, G., Maude, D. K., Barra, A.-L., Sprinkle, M., Berger, C., de Heer, W. A., and Potemski, M. (2008). Approaching the Dirac point in high-mobility multilayer epitaxial graphene, *Phys. Rev. Lett.* **101**, p. 267601, doi: 10.1103/PhysRevLett.101.267601.

[45] Paul, M. J., Chang, Y. C., Thompson, Z. J., Stickel, A., Wardini, J., Choi, H., Minot, E. D., Norris, T. B., and Lee, Y.-S. (2013). High-field terahertz response of graphene, *New J. Phys.* **15**, p. 085019, doi:10.1088/1367-2630/15/8/085019.

[46] Reimann, K. (2007). Table-top sources of ultrashort THz pulses, *Rep. Prog. Phys.* **70**, pp. 1597–1632, doi:10.1088/0034-4885/70/10/R02.

[47] Shih, T., Reimann, K., Woerner, M., Elsaesser, T., Waldmüller, I., Knorr, A., Hey, R., and Ploog, K. H. (2005). Nonlinear response of radiatively coupled intersubband transitions of quasi–two-dimensional electrons, *Phys. Rev. B* **72**, p. 195338, doi:10.1103/PhysRevB.72.195338.

[48] Spitzer, W. G., Kleinman, D., and Walsh, D. (1959). Infrared properties of hexagonal silicon carbide, *Phys. Rev.* **113**, pp. 127–132, doi:10.1103/PhysRev.113.127.

[49] Sprinkle, M., Siegel, D., Hu, Y., Hicks, J., Tejeda, A., Taleb-Ibrahimi, A., Le Fèvre, P., Bertran, F., Vizzini, S., Enriquez, H., Chiang, S., Soukiassian, P., Berger, C., de Heer, W. A., Lanzara, A., and Conrad, E. H. (2009). First direct observation of a nearly ideal graphene band structure, *Phys. Rev. Lett.* **103**, p. 226803, doi:10.1103/PhysRevLett.103.226803.

[50] Stauber, T., Peres, N. M. R., and Geim, A. K. (2008). Optical conductivity of graphene in the visible region of the spectrum, *Phys. Rev. B* **78**, p. 085432, doi:10.1103/PhysRevB.78.085432.

[51] Strait, J. H., Wang, H., Shivaraman, S., Shields, V., Spencer, M., and Rana, F. (2011). Very slow cooling dynamics of photoexcited carriers in graphene observed by optical-pump terahertz-probe spectroscopy, *Nano Lett.* **11**, pp. 4902–4906, doi:10.1021/nl202800h.

[52] Stroucken, T., Knorr, A., Thomas, P., and Koch, S. W. (1996). Coherent dynamics of radiatively coupled quantum-well excitons, *Phys. Rev. B* **53**, pp. 2026–2033, doi:10.1103/PhysRevB.53.2026.

[53] Sun, D., Divin, C., Berger, C., de Heer, W. A., First, P. N., and Norris, T. B. (2010). Spectroscopic measurement of interlayer screening in multilayer epitaxial graphene, *Phys. Rev. Lett.* **104**, p. 136802, doi:10.1103/PhysRevLett.104.136802.

[54] Sun, D., Wu, Z.-K., Divin, C., Li, X., Berger, C., de Heer, W. A., First, P. N., and Norris, T. B. (2008). Ultrafast relaxation of excited Dirac fermions in epitaxial graphene using optical differential transmission spectroscopy, *Phys. Rev. Lett.* **101**, p. 157402, doi:10.1103/PhysRevLett.101.157402.

[55] Sun, D., Wu, Z.-K., Divin, C., Li, X., Berger, C., de Heer, W. A., First, P. N., and Norris, T. B. (2009). Ultrafast dynamics and interlayer thermal coupling of hot carriers in epitaxial graphene, *Phys. Status Solidi C* **6**, pp. 470–473, doi:10.1002/pssc.200880359.

[56] van Haeringen, W. and Junginger, H.-G. (1969). Pseudopotential approach to the energy band structure of graphite, *Solid State Commun.* **7**, pp. 1723–1725, doi:10.1016/0038-1098(69)90140-9.

[57] Varchon, F., Feng, R., Hass, J., Li, X., Nguyen, B. N., Naud, C., Mallet, P., Veuillen, J.-Y., Berger, C., Conrad, E. H., and Magaud, L. (2007). Electronic

structure of epitaxial graphene layers on SiC: Effect of the substrate, *Phys. Rev. Lett.* **99**, p. 126805, doi:10.1103/PhysRevLett.99.126805.

[58] Wegener, M. (2005). *Extreme Nonlinear Optics* (Springer, Berlin).

[59] Winnerl, S., Orlita, M., Plochocka, P., Kossacki, P., Potemski, M., Winzer, T., Malic, E., Knorr, A., Sprinkle, M., Berger, C., de Heer, W. A., Schneider, H., and Helm, M. (2011). Carrier relaxation in epitaxial graphene photoexcited near the Dirac point, *Phys. Rev. Lett.* **107**, p. 237401, doi: 10.1103/PhysRevLett.107.237401.

[60] Woerner, M., Kuehn, W., Bowlan, P., Reimann, K., and Elsaesser, T. (2013). Ultrafast two-dimensional terahertz spectroscopy of elementary excitations in solids, *New J. Phys.* **15**, p. 025039, doi:10.1088/1367-2630/15/2/025039.

[61] Wu, Q. and Zhang, X.-C. (1995). Free-space electro-optic sampling of terahertz beams, *Appl. Phys. Lett.* **67**, pp. 3523–3525, doi:10.1063/1.114909.

[62] Yan, H., Xia, F., Zhu, W., Freitag, M., Dimitrakopoulos, C., Bol, A. A., Tulevski, G., and Avouris, P. (2011). Infrared spectroscopy of wafer-scale graphene, *ACS Nano* **5**, pp. 9854–9860, doi:10.1021/nn203506n.

Chapter 9

Ultrafast Manipulation of Terahertz Waves using Graphene Metamaterials

Chihun In[1,2], Hyunyong Choi[1*]
[1]*School of Electrical and Electronic Engineering, Yonsei University, Seoul 120-749, South of Korea*
[2]*Center for Quantum-Beam-based Radiation Research Korea Atomic Energy Research Institute, Daejeon 305-353, South Korea*
[*] *hychoi@yonsei.ac.kr*

Engineering electromagnetic waves has been central to the development of modern optoelectronic devices. Global networks through the long-range communication, internet, and data transfer were enabled by the controlling the associated light. Although actively pursued electromagnetic wavelength for various optoelectronic applications is widely spanned from ultra-violet to far-infrared wavelength, it has been particularly difficult to manipulate the terahertz (THz) frequencies, which lie between microwave and infrared region. For the last two decades, worldwide researchers have found how to generate and detect THz waves, and after which optical technologies including THz spectroscopy, THz imaging, and THz sensing have been greatly advanced. Nonetheless, the small energy of THz photon (~ 4.14 meV at 1 THz) and the optical characteristics of long wavelength (300 μm at 1 THz) have still been imposing some technical limits on the THz applications: sensitive, and yet optically thin structure is required to compensate the energy loss of THz propagation. Recent discoveries of graphene have witnessed many potential for manipulating THz-related optical phenomena. This is especially because graphene provides both high sensitivity and atomically-thin dimension with a high THz absorption feature. Another technical advance of changing some characteristics of THz waves is to employ artificial structures, called metamaterials, which have also been extensively investigated to manage both absorption and transmission characteristics. By employing graphene and metamaterials, we have developed very effective methods in manipulating the THz spectra. Active modulation of THz waves was derived by layered composition of graphene metamaterials, where the electrons are migrated from graphene to metamaterials and vice versa. Emphasis should be made on the fact that the electrostatic and optical to the graphene is at the heart of the THz wave control. In this Chapter, two representative works of the THz manipulation in graphene metamaterials are presented. By changing the structure of metamaterials, distinguished functionalities of THz optics, refractive index change, and nonlinear wave mixing are demonstrated.

9.1. Introduction

Terahertz (THz) radiation is a type of electromagnetic radiation whose frequency ranges from the microwave to infrared wavelengths. Although we cannot see THz waves, thermal radiation from warmth allows us to feel them with far-infrared radiation. The free space propagation of THz radiation fills our lives; however, the engineering of THz radiation is the least explored aspect of the field of electromagnetic waves. We use the word "THz gap," because it was hard to develop THz wave sources and detectors [Lee 2009]. Although current THz science technology has allowed us to exploit advanced THz techniques of imaging [Mittleman *et al.* 1996; Dobroiu *et al.* 2006] and spectroscopy [Huber *et al.* 2001; Shan *et al.* 2003], the manipulation of THz waves remains an unexplored part of THz science. It has been found that most natural materials show fixed responses at THz frequencies. Metals and polymers have distinct but unchangeable THz responses unless their atomic or molecular compositions are modified [Yasuda and Hosako 2008; Wietzke *et al.* 2011]. In recent decades, the discovery of graphene has had a profound impact on a plethora of THz optics studies. Classified as a semi-metal, graphene has shown highly controllable spectra available at THz frequencies [Ren *et al.* 2012; Jnawali *et al.* 2013]. Due to the ultrathin atomic layer thickness, it is possible to change electron and hole densities with slight stimulation of the electrostatic field [Maeng *et al.* 2012; Li *et al.* 2015]. Not only are moderate conductivities provided by the high mobility of graphene carriers, but also the gated graphene structure has demonstrated controllable spectra at THz frequencies. The permanent doping of graphene is enabled by organic molecules applied to graphene, where the individual molecules effectively change the microscopic density of carriers [Wehling *et al.* 2008].

Recent studies on THz optics have noted great interest in the development of advanced THz devices that perform frequency-dependent THz manipulation. Because graphene exhibits rather flat transmission spectra at THz frequencies, researchers desired to engineer frequency-dependent constructions. Metamaterials, artificially designed structures, are also platforms to manipulate wave properties [Chen *et al.* 2006]. Composed of metal line arrangements on dielectric substrates, metamaterials have demonstrated exotic electromagnetic properties, such as clocking and imaging with unlimited resolution [Schurig *et al.* 2006; Zhu *et al.* 2011]. In particular, researchers have designed THz metamaterials that can interact with specific THz frequencies. Two contemporary but distinct branches of graphene and metamaterials are combined to perform THz wave manipulation in our method of design. The first demonstration of active THz control in graphene metamaterials was conducted by Lee *et al.* [2012], where electrostatic gating changed the THz response of metamaterials. The collaborated effects of graphene and metamaterials were enabled by charge interaction at the interface of the two materials. Further, Lee *et al.* investigated the manipulation of THz waves at the picosecond scale [Lee *et al.* 2013], where the transient control of THz waves was vividly illustrated by optical-pump THz-probe spectroscopy. Upon irradiation of the optical pulse onto graphene metamaterials, the free carriers in graphene effectively screened the frequency-dependent response of the metamaterials.

Recent studies on graphene have led us to more enhanced manipulation of THz waves. The attachment of silicon, organic molecules, ionic liquid, or two-dimensional structures on graphene represented a technological breakthrough in THz sciences [Han *et al.* 2012; Zhang *et al.* 2014; Wu *et al.* 2015; Li *et al.* 2015]. Here, we will focus on the THz effects in graphene metamaterials to show the frequency-dependent manipulation of THz waves. In the following, we will discuss how to generate and detect THz radiation (Section 9.2) followed by the experimental setup for optical-pump THz-probe spectroscopy (Section 9.3). To clarify the THz effect in graphene metamaterials, we will introduce a theoretical model for graphene (Section 9.4) and metamaterials (Section 9.5). In the last section, we will show the frequency-dependent manipulation of THz waves and THz nonlinear optics in graphene metamaterials (Section 9.6).

9.2. THz Spectroscopy: THz Generation and Detection

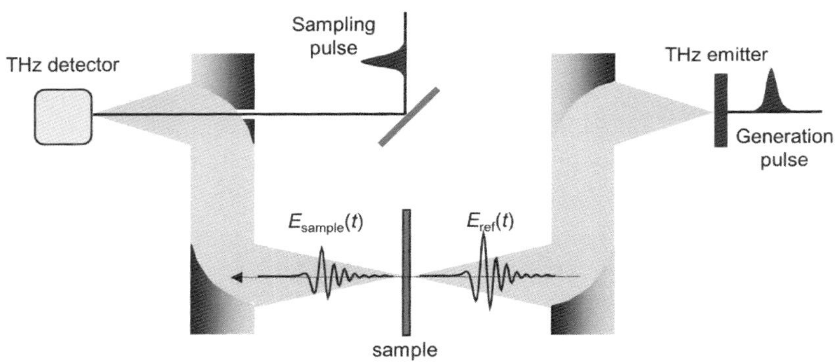

Fig. 9.1. Experimental setup for THz generation and detection.

In this section, we describe the experimental setup for THz spectroscopy. The purpose of THz spectroscopy is to obtain the Fourier-transformed transmission ratio through the sample $T(\omega)=E_{sample}(\omega)/E_{ref}(\omega)$, where $E_{sample}(\omega)$ and $E_{ref}(\omega)$ are the Fourier-transformed THz electric field with and without sample transmission, respectively. Because graphene metamaterials exhibit highly frequency-dependent spectra at THz frequencies, Fourier-transformed representation is more appropriate than time-scale representation. Figure 9.1 illustrates the typical experimental setup for THz spectroscopy. THz field $E_{ref}(t)$ is radiated from the THz emitter, and $E_{sample}(t)$ corresponds to the transmitted THz wave through the sample. Among the various methods to generate THz waves, optical nonlinearity is the most cost-efficient way to obtain pulsed THz radiation. Given that the bandwidth of the generation pulse is $\Delta\omega$, which lies between ω_{max} and ω_{min}, the difference-frequency mixing of the nonlinear crystal generates the THz frequency at $\omega_{THz}=\Delta\omega$, i.e. $\Delta\omega=\omega_{max}-\omega_{min}$. As shown in Fig. 9.2(a), the typically used THz emitter is a <110>-oriented ZnTe crystal, where the power of THz radiation is maximized when the generation pulse is polarized along 54.7 degrees with respect to the [001] crystal direction [Rice *et al.* 1994].

A commonly employed THz detection method is electro-optic (EO) sampling, where the rotation of sampling pulse polarization is used to detect the amplitude of the THz electric field. Figure 9.2(b) displays the illustration of electro-optic sampling using a <110>-oriented ZnTe crystal, where both the THz field and optical field are linearly polarized along the $[\bar{1}10]$ direction. The spatial and temporal overlap between the THz and optical fields leads to the rotation of sampling pulse polarization [Nahata *et al.* 1996]. One advantage of electro-optic sampling is that the full waveform of the transient THz field can be obtained by changing the time delay between the sampling pulse and the THz field.

Once $T(\omega)$ is obtained from electro-optic sampling, optical parameters such as conductivity $\sigma(\omega)$, permittivity $\varepsilon(\omega)$, and susceptibility $\chi(\omega)$ can be converted from $T(\omega)$. The numerical relation is

$$\sigma(\omega) = \frac{n+1}{Z_0}\left(\frac{E_{\text{ref}}}{E_{\text{sample}}} - 1\right) \tag{9.1}$$

given that the thickness d of the sample is optically thin ($\omega d/c \ll 1$) [Kaindl *et al.* 2002; Jnawali *et al.* 2013]. Here, Z_0=377 Ω is the impedance of free space, n is the refractive index of the substrate, and c is the speed of light. The physical meaning of $\sigma(\omega)$ is as follows: The positive real part of conductivity Re(σ) leads to a decrease in field amplitude, while the positive imaginary part of conductivity Im(σ) reveals the retarded phase of the THz field through sample transmission. Focusing on the graphene and graphene metamaterial studies, we will use $\sigma(\omega)$ to depict the frequency-dependent THz responses (Section 9.4 and Section 9.6), while $\varepsilon(\omega)$ is used to explain dipole oscillation in metamaterials (Section 9.5).

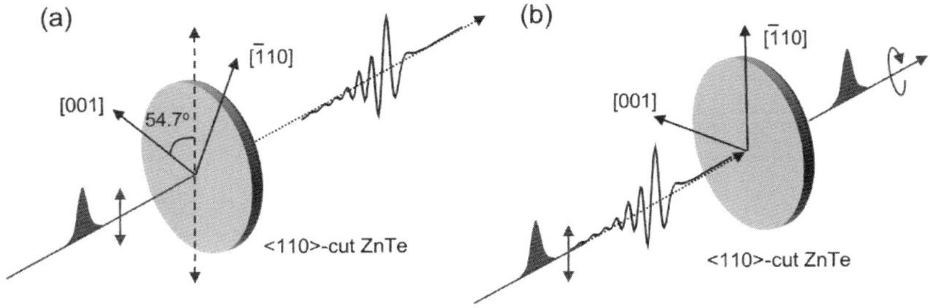

Fig. 9.2. (a) THz generation and (b) detection using <110>-oriented ZnTe crystal.

9.3. Optical-pump THz-probe Spectroscopy

Optical-pump THz-probe spectroscopy is an experimental technique involving two different light source frequencies. When the optical pump injects photocarriers, the THz wave simultaneously probes material conductivity. Consequently, we measure the photoexcited conductivity induced by the optical pump. We note that the photoexcitation occurs when the energy and momentum of photons are transferred to the electronic system. After the photoexcitation, the relaxation channel of the material itself restores the photoexcited state to the equilibrium. Most photoexcited carriers lose their energy via thermalization and electron-hole recombination. One advantage of optical-pump THz-probe spectroscopy is that it is possible to capture the transient response of photoexcited dynamics by changing the time delay between the optical pump and the THz probe. Given that the time duration of the optical pulse is tens of femtoseconds, a relaxation time ranging from picoseconds to nanoseconds is expected to follow [Kaindl *et al.* 2003; Suzuki and Shimano 2012]. We note that the fact that optical-pump THz-probe spectroscopy is contact free and non-invasive allows us to investigate nanostructured materials such as quantum dots, nanowires, and two-dimensional materials. Representative works on optical-pump

THz-probe spectroscopy have been published elsewhere [Ulbricht *et al.* 2011].

Figure 9.3 illustrates the experimental scheme for optical-pump THz-probe spectroscopy. An optical pulse is generated from a Ti:sapphire regenerative amplifier (Coherent RegA 9050) with a repetition rate of 250 kHz and a pulse energy of 6 μJ. The time width and wavelength of pulses are set to 50 fs and 800 nm, respectively. Parts of the pulse are used to generate and detect THz waves, while the remainder is used for the optical pump. There are two motion controllers: One changes the THz field delay t between the THz pulse and the sampling pulse, while the other is used to control the pump-probe delay Δt between the THz pulse and the pump pulse. The half-wave-plate ($\lambda/2$) and polarizer (PoL) regulate the polarization of each pulse. The waveform of the THz field is detected by an EO detector with a quarter-wave-plate ($\lambda/4$) and a Wollaston prism. We use a preamplifier and a lock-in amplifier to obtain a high signal-to-noise ratio with a chopper frequency of 10 kHz.

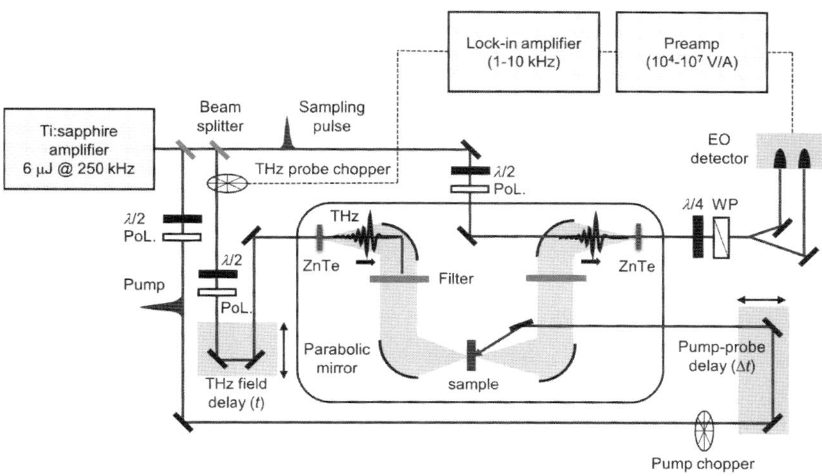

Fig. 9.3. Experimental setup for optical-pump THz-probe spectroscopy.

We use the term change in conductivity $\Delta\sigma(\omega)$ to depict the photoexcited response of graphene and graphene metamaterials. The following formula

$$\Delta\sigma(\omega) \approx -\frac{n+1}{Z_0}\frac{\Delta E_{\text{sample}}(\omega)}{E_{\text{sample}}(\omega)} \tag{9.2}$$

defines the conductivity change of optically thin materials ($\omega d/c << 1$), where $\Delta E_{\text{sample}}(\omega)$ is the photoexcited change in the transmitted THz field [Kaindl *et al.* 2009; Jnawali *et al.* 2013]. In the following discussions, the experimental setup depicted in Fig. 9.3 is used to measure the photoexcited THz response in graphene (Section 9.4) and graphene metamaterials (Section 9.6).

9.4. Basic Theories of THz Interaction with Graphene

Before showing the THz effects in graphene metamaterials, we use basic graphene theories to clarify the underlying mechanisms. The optical response of single-layer graphene is characterized by a quantum of conductance with a fixed value of $\sigma_0=2e^2/h$ [Nair *et al.* 2008]. The quantum of conductance is universal conductivity that is not affected by the details of the graphene band structure [Novoselov *et al.* 2005; Zhang *et al.* 2005]. Previous reports have found that graphene conductivity is quantized to σ_0 when the photon energy is larger than twice the graphene Fermi energy E_F [Wang *et al.* 2008]. On the other hand, it has been found that photon energy below $2|E_F|$ is largely absorbed by graphene-free carriers near the Fermi level [Mak *et al.* 2008]. Since the photon energy of a THz wave (E_{THz}=4.14 meV at 1 THz) is smaller than the typical $2|E_F|$ value of graphene, THz spectroscopy has been used to study the free-carrier response of graphene [Dani *et al.* 2012]. Within a few THz regions, previous researchers have shown that graphene conductivity at THz frequencies is a few times larger than that at visible frequencies [Sensale-Rodriguez *et al.* 2012; Ren *et al.* 2012].

Along with the strong THz response of graphene, the two-dimensional nature of graphene has attracted enormous interest in device physics. Within a simple gated structure, we were able to change the graphene conductivity via electrostatic gating [Maeng *et al.* 2012]. Figure 9.4(a) shows the typical gated graphene structure, where both p-doped Si and SiO_2 serve as the back gate. By applying a negative voltage to the back gate, the Coulomb attractive force induces the accumulation of positive charges on the graphene layer. The hole-doped graphene exhibits higher conductivity than the undoped graphene. The graphene band structure shown in Fig. 9.4(b) demonstrates the gate effect. While the slope of linear dispersion corresponds to the Fermi velocity of v_F=1.1×10^6 m/s, the horizontal dashed line indicates E_F at two different points. When E_F crosses the Dirac point, we call it a charge neutral point (CNP) where electron and hole density are balanced. When the graphene is hole-doped by the gate effect, E_F is shifted downward and free carriers support the current flow along the source-drain channel.

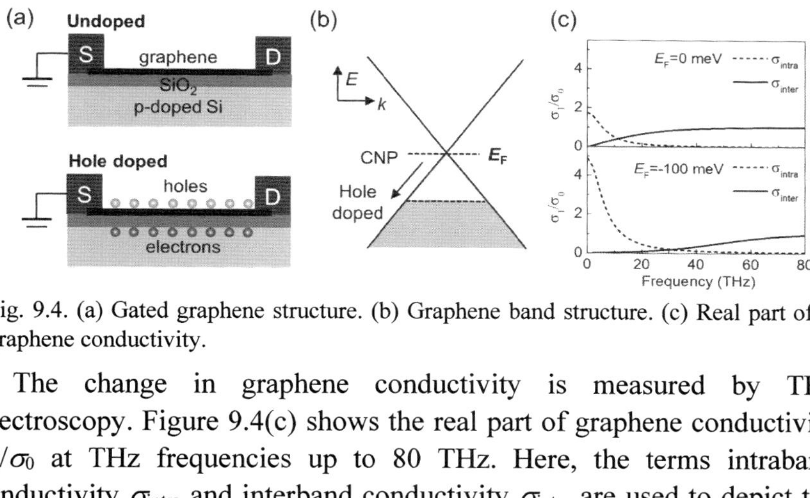

Fig. 9.4. (a) Gated graphene structure. (b) Graphene band structure. (c) Real part of graphene conductivity.

The change in graphene conductivity is measured by THz spectroscopy. Figure 9.4(c) shows the real part of graphene conductivity σ_1/σ_0 at THz frequencies up to 80 THz. Here, the terms intraband conductivity σ_{intra} and interband conductivity σ_{inter} are used to depict the light absorption below $2|E_F|$ and above $2|E_F|$, respectively. The plots are simulated from the Kubo formula:

$$\sigma_{\text{inter}}(\omega) = i\frac{e^2\omega}{\pi}\int_{\Delta}^{\infty} d\varepsilon \frac{(1+\Delta^2/\varepsilon^2)}{(2\varepsilon)^2-(\hbar\omega+i\Gamma)^2}\times[f(\varepsilon-E_F)-f(-\varepsilon-E_F)], \quad (9.3)$$

$$\sigma_{\text{intra}}(\omega) = i\frac{e^2/\pi\hbar}{\omega+i/\tau}\int_{\Delta}^{\infty} d\varepsilon(1+\Delta^2/\varepsilon^2)\times[f(\varepsilon-E_F)+f(\varepsilon+E_F)], \quad (9.4)$$

where $2\Delta=0$ is the bandgap of graphene, $\Gamma=10$ meV is the interband broadening constant, $\tau=25$ fs is the scattering time of free carriers, and $f(\varepsilon-E_F)$ is the Fermi distribution function with temperature $T=300$ K

[Dawlaty *et al.* 2008]. When E_F lies in 0 meV (CNP), σ_{inter} is nearly constant of σ_0, while σ_{intra} at 0 THz is about two times σ_0. When E_F is -100 meV, however, σ_{intra} at 0 THz is significantly increased above $4\sigma_0$, whereas the onset of σ_{inter} is shifted to a higher frequency. While the increase in σ_{intra} at E_F=-100 meV is assisted by free carriers, the shifted onset of σ_{inter} is explained by the increased value of $2|E_F|$. We note that the carrier density required to achieve E_F=-100 meV is $N=6.1\times10^{11}$ cm^{-2} from the relationship $E_F=\hbar v_F(N\pi)^{1/2}$ [Choi *et al.* 2009; Jnawali *et al.* 2013]. The electrostatic effect of the gated structure is determined by the relationship $N=C_g\Delta V/e$, where C_g is the gate capacitance and ΔV is the gate voltage away from CNP. Given that the literature value of C_g of 300-nm-thick SiO_2 is 16.9 nF/cm^2 [Maeng *et al.* 2012], ΔV=5.8 V induces a charge-accumulation density of $N=6.1\times10^{11}$ cm^{-2}. Recent developments in the ionic materials significantly increased the gate capacitance to 1 μF/cm^2 [Emani *et al.* 2012], where ΔV=0.1 V was sufficient to shift the E_F to -100 meV.

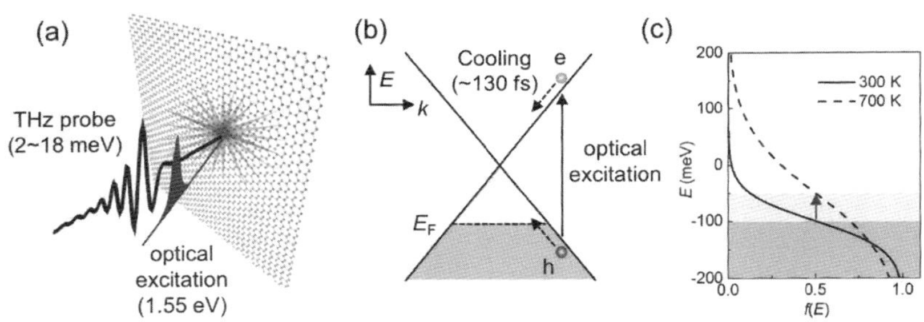

Fig. 9.5. (a) Scheme for optical-pump THz-probe spectroscopy in graphene. (b) Graphene band structure with optical excitation and cooling dynamics. (c) Fermi distribution at T=300 K and 700 K as a function of energy.

Having elucidated the electrostatic response of the gated graphene structure, we now show the optically controlled THz conductivity in the graphene layer. Figure 9.5(a) shows the scheme for optical-pump THz-probe spectroscopy on graphene. A pulse energy of 1.55 eV creates hot carriers in graphene, while a THz pulse energy ranging from 2 meV to 18 meV simultaneously probes the change in graphene conductivity. The energy of the optical pump is transferred to an electron-hole pair, as described in Figure 9.5(b). The hot carriers are cooled down within 130 fs due to the fast heat emission through the optical phonon [George *et al.*

2008; Gierz *et al.* 2013]. Because the time resolution of the THz probe is limited to 1 ps, we can detect the thermalized dynamics of graphene. Figure 9.5(c) describes the electron Fermi distribution at equilibrium temperature T=300 K and nonequilibrium temperature T=700 K near the Fermi energy [Frenzel *et al.* 2014]. The optical pump not only increases the electron temperature, but also shifts the chemical potential slightly upward at the moment of photoexcitation [Gierz *et al.* 2013]. Here, the increase in electron temperature increases σ_{intra}.

By employing the optical-pump THz-probe spectroscopy introduced in Section 1.3, we show the experimental result of optically controlled THz spectra in graphene [Kim *et al.* 2013]. Figure 9.6(a) displays the reference THz field $E_0(t)$ and the pump-induced change in the THz fields $\Delta E(t)$ at different Δt ranging from -1 ps to 5.2 ps. The opposite sign of $\Delta E(t)$ and $E_0(t)$ at t=0 ps implies that the amplitude of the THz probe is decreased through graphene transmission. Because the optical pump increases σ_{intra}, the drift of free carriers through the THz electric field enhances THz absorption. The well-known phenomenon in graphene is that graphene absorption saturates with increasing light intensities [Sun *et al.* 2012]. As shown in Fig. 9.6(b), $-\Delta E/E_0$ is not monotonically increased but saturated with increasing pump fluences. The nonlinear increase is explained by the temperature dependence of electron heat capacity: It is hard to increase electron temperature when electrons are heated by an intense pulse [Shi *et al.* 2014]. Figure 9.6(c) shows the time-resolved THz conductivity at Δt from -1 ps to 5.2 ps.

A stark contrast in $\Delta\sigma_1$ is observed between Δt=-1 ps and 0.2 ps attributed to the pump excitation. Due to the fast relaxation kinetics of graphene, the photoexcited conductivity returns to equilibrium within a few ps. The monotonic decrease in $\Delta\sigma_1$ as a function of frequency combined with positive $\Delta\sigma_2$ implies that the photoexcited THz response follows the Drude model [Horng *et al.* 2011]. The positive sign of $\Delta\sigma_1$ indeed shows that the electron temperature is increased by the optical pump, as depicted in Fig. 9.5(c). Recent studies on graphene have discovered that the sign of $\Delta\sigma_1$ becomes negative when the number of graphene residue carriers is increased [Jnawali *et al.* 2013]. While exfoliated graphene has demonstrated positive $\Delta\sigma_1$ dynamics [Strait *et al.* 2011], most chemical-vapor-deposition (CVD)-grown graphene tends to exhibit negative $\Delta\sigma_1$ dynamics [Frenzel *et al.* 2013; Jnawali *et al.* 2013]. Upon photoexcitation, a decrease in scattering time τ is prominent in CVD graphene, leading to negative photoexcited dynamics. A detailed analysis can be found elsewhere [Shi *et al.* 2014; Frenzel *et al.* 2014].

To summarize, we briefly reviewed the THz effects in graphene. In terms of conductivity, we showed that both electrostatic gating and the optical pump give rise to changes in graphene conductivity at THz frequencies. Both THz manipulation techniques have demonstrated a pronounced impact on THz optics. Recent studies have created more delicate and efficient devices to modulate THz waves. Graphene metamaterials, newly emerged THz devices, operate under the tunable THz characteristics of graphene.

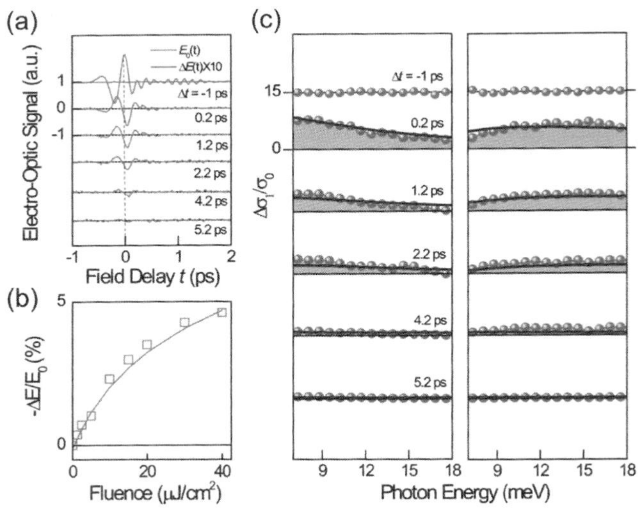

Fig. 9.6. (a) THz reference field $E_0(t)$ and pump-induced change in THz field $\Delta E(t)$. (b) $-\Delta E/E_0$ at t=0 ps as a function of fluence. (c) $\Delta\sigma_1/\sigma_0$ and $\Delta\sigma_2/\sigma_0$ at fluence of 40 μJ/cm^2. (Reprinted from [Kim *et al.* 2013])

9.5. THz Spectroscopy of Metamaterials

Metamaterials have served as a platform to manipulate electromagnetic waves in our method of design. The exotic electromagnetic properties of metamaterials enable us to access new functionalities such as invisibility [Schurig *et al.* 2006] and imaging with unlimited resolution

[Zhu *et al.* 2011]. The control of electromagnetism is derived from individual "meta-atoms" that exhibit frequency-dependent light–matter interaction [Liu and Giessen 2010]. Analogous to real atoms in solids, the shape of an individual meta-atom determines the optical property of the whole metamaterial structure. In this section, we study the basic electromagnetism of a meta-atom, which is characterized by resonant transmission spectra. This phenomenon, dipole oscillation, originates from the comparable length of the meta-atom and the incident light wavelength. We fabricated a two-dimensional array of meta-atoms to observe a single resonance in THz transmission [Jung *et al.* 2014].

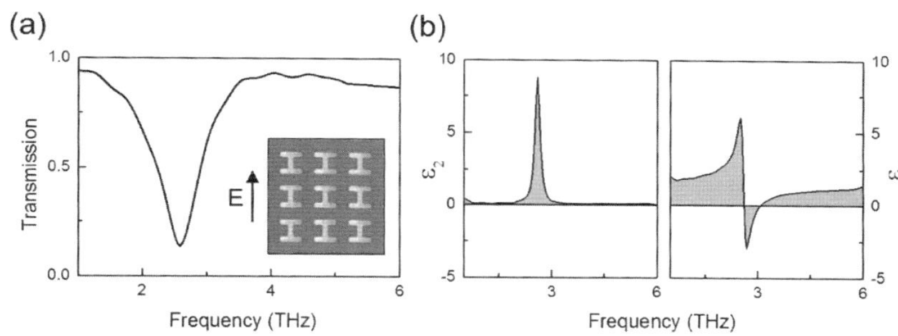

Fig. 9.7. (a) THz transmission through metamaterial. Inset depicts meta-atoms. (b) Real and imaginary part of permittivity of metamaterial.

The array of the fabricated meta-atoms is displayed in the inset of Fig. 9.7(a), where the THz electric field drifts charges along the metal line. In this structure, we observed a single resonance at 2.6 THz, which is attributed to the dipole oscillation. One of the characteristics of dipole oscillation is that optical constants are strongly changed near the resonant frequency. To show the effect, we obtain the real and imaginary parts of permittivity ε_1 and ε_2 using the THz spectroscopy depicted in Section 9.2. As shown in Fig. 9.7(b), ε_2 displays a sharp resonance, while ε_1 exhibits dispersive spectra at the resonant frequency of 2.6 THz. We understand that ε_2 stands for attenuation of the THz wave and ε_1 is related to electric displacement. It is worth noting that ε_1 becomes negative between 2.6 THz and 3.1 THz, while it is positive elsewhere. The following explanation clarifies the ε_2 sign: Electric displacement below the resonant frequency can follow the slow-varying THz electric field; however, the charge movement at frequencies slightly above the resonant frequency cannot follow the fast-varying THz electric field. The retarded movement of charges leads to the opposite direction between the electric field and electric displacement. When the frequency grows higher, the charges are barely drifted by the THz field and no light–matter interaction occurs.

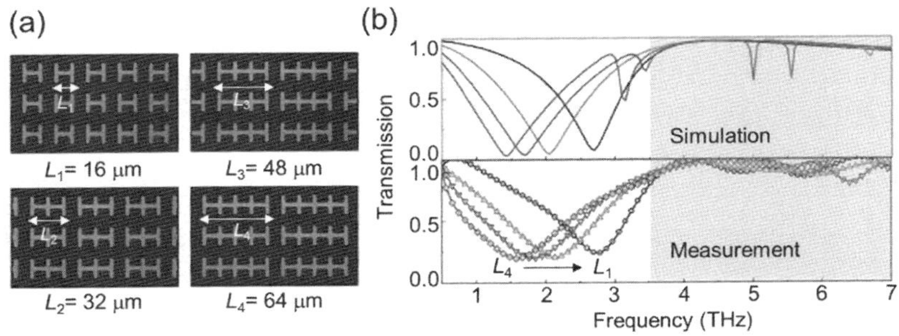

Fig. 9.8. (a) Different dimensions of four meta-atoms. Horizontal length ranges from L_1 to L_4. (b) THz transmission of the respective meta-atoms. (Reprinted from [Jung *et al.* 2014])

Having shown that the dielectric constants are strongly changed near the resonant frequency, we now present the movement of resonance that occurs when the meta-atom dimension is changed. Figure 9.8(a) illustrates the fabricated meta-atom structures, where the horizontal length increases by the unit length of 16 μm. As the dipole resonant frequency is inversely proportional to the length, it is assumed that the resonance is shifted to a lower frequency when the horizontal length increases [Jung *et al.* 2014]. Figure 9.8(b) shows both the simulated and measured THz transmission spectra, where the resonant frequency indeed shifts toward a lower frequency when the length is increased. The small dips in the shaded region reveal high harmonic resonances excited by the THz field. To summarize, we show the THz effects in metamaterials. Dipole oscillation in metamaterials exhibits distinct features to the graphene THz spectra. In the next section, we will present THz effects in graphene metamaterials. Our focus on graphene metamaterials is to manipulate THz waves using both the electrical and optical methods. The dipole resonance of metamaterials combined with the conductivity change in graphene constitutes the THz manipulation kinetics in graphene metamaterials.

9.6. Graphene Metamaterials

The hybridized effect of two different materials has sparked keen interest in the electromagnetic field. A preliminary example is the charge interaction between two-dimensional materials where the excitonic spectra are strongly modulated by the heterojunction structure [Hong *et al.* 2014; Huang *et al.* 2014]. Metamaterials exhibit distinct THz features from conventional materials due to their manipulated dimensions. Previous studies have tried to control the exotic THz transmission of metamaterials by employing the GaAs Schottky barrier [Chen *et al.* 2006] in which two-dimensional electron gas effectively screens the dipole oscillation. By controlling the Schottky barrier height through electrostatic gating, they demonstrate an amplitude modulation of 30% at THz frequencies. At the limit of atomic thickness, however, the metal/GaAs interface hardly provides the two-dimensional electron gas and no charge interaction is expected between the metamaterial and GaAs. Graphene, a single layer of carbon atoms, shows high mobility of free carriers and tunable absorption spectra available at THz frequencies. In particular, it is possible to synthesize flexible and ultrathin devices using atomically thin graphene that are otherwise difficult to construct with conventional bulk semiconductors. The large-area synthesis of graphene enables us to access the construction of layered graphene metamaterials. The direct transfer of graphene onto metamaterial is at the heart of THz control techniques based on graphene metamaterials. The first study of THz effects in graphene metamaterial was conducted by [Lee *et al.* 2012] where the amplitude modulation of THz waves reached 47% via electrostatic gating. Consecutive studies on optical effects in graphene metamaterials revealed ultrafast refractive index control [Lee *et al.* 2013] and nonlinear THz wave mixing [In *et al.* 2015]. In this section, we discuss the construction of graphene metamaterials and their THz effects. Strong modulation of THz waves is performed by electrically and optically controlled device operation.

9.6.1. *Refractive index manipulation in hexagonal meta-atoms*

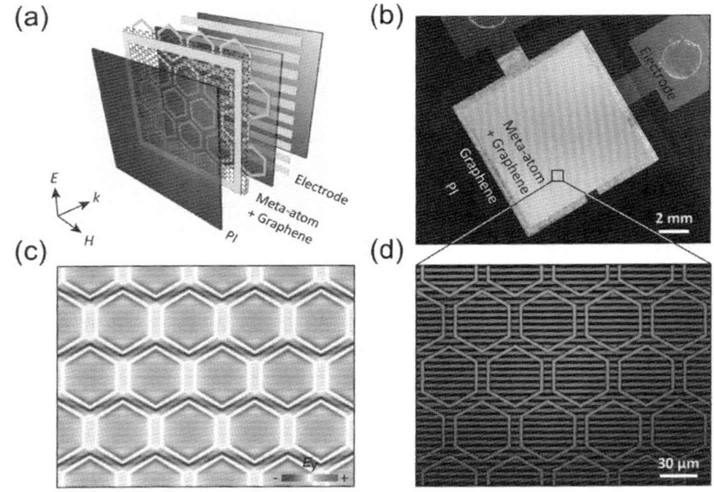

Fig. 9.9. (a) Scheme of graphene metamaterial. (b) Top image of graphene metamaterial. (c) Simulated THz electric field distributed on meta-atoms. (d) Microscopic image of meta-atoms. (Reprinted from [Lee *et al.* 2013]) (For color reference, go to page 493.)

The device construction of graphene metamaterial is not simple but contains several insulator layers. The composition can be divided into three parts: the bottom gate, metamaterial, and graphene. The bottom gate is composed of metal lines called the extraordinary transmission (EOT) gate. The attenuation and phase delay through the metal grids are trivial while they serve as gate electrodes. The EOT gate is sandwiched by polyimide (PI) layers that prevent charge leakage from the EOT. On top of the EOT gate, metamaterial is patterned by UV photolithography and metal lift-off. In Fig. 9.9(a), the hexagonal shapes of meta-atoms are fabricated in a honeycomb lattice. The inductance L of the metal lines and the capacitance C between meta-atoms give rise to LC resonance at THz frequencies. The CVD-grown graphene is transferred to metamaterials whose peripheries are enveloped by ground electrodes. To prevent the doping of graphene from ambient conditions, we covered the top of the device with a polyimide layer. The voltage difference between the EOT gate and the ground determines the density of charges in graphene. Figure 9.9(b) shows the image of the total device, where the consecutive layers of PI, graphene, graphene/meta-atoms, and electrodes are displayed. To depict the uniform alignment of meta-atoms, we show the simulated electric field (Figure 9.9(c)) and optical microscope image of the graphene metamaterial (Figure 9.9(d)). The highlighted field distribution in Fig. 9.9(c) is indicative of the capacitive response at the gap of the meta-atoms.

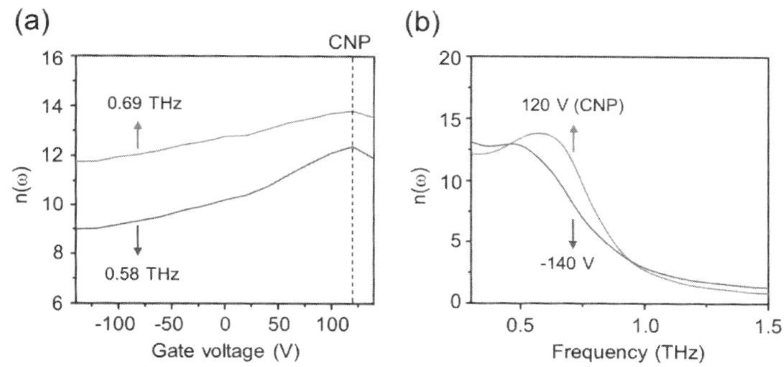

Fig. 9.10. (a) Refractive index $n(\omega)$ as a function of gate voltage at 0.58 THz and 0.69 THz. (b) $n(\omega)$ as a function of frequency at 120 V and -140 V. (Reprinted from [Lee *et al.* 2013])

Many metamaterial researchers are interested in achieving an unnaturally high refractive index [Choi *et al.* 2011; Campbell *et al.* 2013]. It is possible to efficiently control the propagation of the electromagnetic field if a high refractive index is provided. Provided that the screening effect of graphene regulates dipole oscillation, EOT gating can manage the refractive index of graphene metamaterial. The experimentally measured electrostatic change in the graphene metamaterial refractive index $n(\omega)$ is illustrated in Figure 9.10(a). We plot $n(\omega)$ at two different frequencies of 0.69 THz and 0.58 THz. When the graphene charge density is minimized at V=140 V (CNP), $n(\omega)$ at both frequencies reach their highest values. This indicates that the absence of graphene free-carrier screening leads to strengthened dipole oscillation. We note that $n(\omega)$ is higher at 0.69 THz than 0.58 THz, which corresponds to abnormal $n(\omega)$ dispersion. Most natural materials show a decrease in $n(\omega)$ when the frequency is increased [Ulbricht *et al.* 2011]. In contrast, the dipole oscillation in graphene metamaterial at THz frequencies gives rise to an increased $n(\omega)$ when the frequency is increased. The frequency-dependent plot of $n(\omega)$ shown in Figure 9.10(b) explicitly depicts the abnormal dispersion. While $n(\omega)$ rapidly drops near 0.7 THz, the frequency ranging from 0.1 THz to 0.6 THz at CNP corresponds to the abnormal dispersion. At V=-140 V, however, it is hard to determine the abnormal dispersion due to the relaxed strength of the dipole oscillation. The $n(\omega)$ spectra are changed by the electrostatic effect and reveal the charge interaction between graphene and metamaterial.

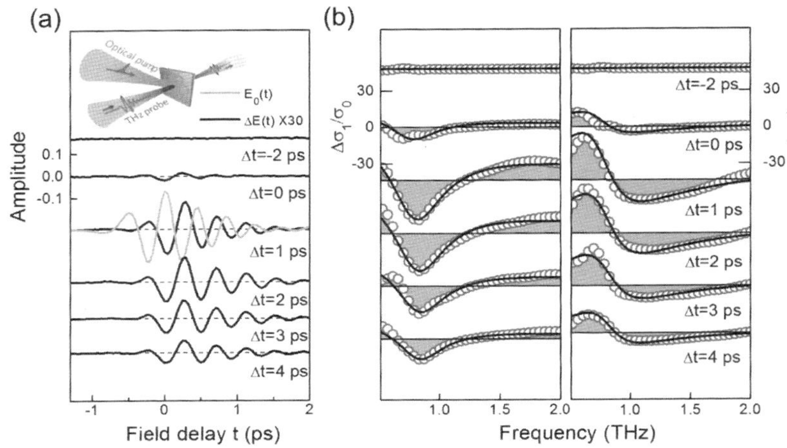

Fig. 9.11. (a) Reference THz field E_0(t) and pump-induced change in THz field $\Delta E(t)$ at Δt from -2 ps to 4 ps. (b) Fourier-transformed THz conductivity at Δt from -2 ps to 4 ps. (Reprinted from [Lee *et al.* 2013])

The photoexcited dynamics of graphene metamaterial exhibit resonant conductivity spectra provided by the dipole oscillation of meta-atoms. Figure 9.11(a) displays the reference THz field $E_0(t)$ and pump-induced change in THz field $\Delta E(t)$. The opposite sign of $E_0(t)$ and $\Delta E(t)$ indicates that the THz field is absorbed by graphene photocarriers. We understand the phase delay at Δt=1 ps is the result of photo-induced change in the refractive index. The Fourier-transformed representation of Fig. 9.11(b) unveils the real and imaginary parts of conductivity $\Delta\sigma_1/\sigma_0$ and $\Delta\sigma_2/\sigma_0$, which correspond to the field absorption and phase delay, respectively. The pronounced effect of graphene photocarriers is to screen the dipole oscillation of meta-atoms. Because the graphene resistance is effectively connected to the capacitance between meta-atoms in the parallel direction, the increased number of graphene photocarriers reduces the dipole strength as well as slightly shifts the resonance to the lower frequency. The dispersive relation at 0.7 THz in $\Delta\sigma_2/\sigma_0$ matches the $n(\omega)$ dispersion. We note that the relaxation speed of photoexcited conductivity in graphene metamaterial is similar to that in bare graphene.

9.6.2 *THz wave mixing in doubly-resonant structures*

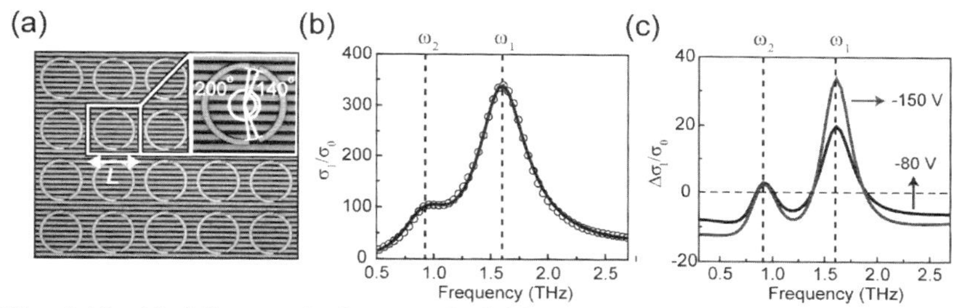

Fig. 9.12. (a) Microscopic image of aDSR meta-atom. (b) Real part of THz conductivity of aDSR meta-atom. (c) Change in real part of conductivity due to electrostatic gating. (Reprinted from [In *et al.* 2016])

While the hexagonal meta-atom structure yields single-dipole resonance at the THz frequency region, we introduce an asymmetric double split-ring-resonator (aDSR) in which two distinct resonances are created [In *et al.* 2015]. Figure 9.12(a) illustrates the meta-atom arrangement where a ring with two split-gaps corresponds to the unit meta-atom. The background grid lines represent the EOT gate. The resonant frequency here is determined by the angle of arcs, where a longer arc leads to a lower resonant frequency. As shown in Figure 9.12(c), there are two discrete resonances at $\omega_1/2\pi$=1.6 THz and $\omega_2/2\pi$ =0.9 THz. The fit line in Fig. 9.12(c) is the sum of Lorentz oscillators at $\omega_1/2\pi$ and $\omega_2/2\pi$, respectively. The hundreds of σ_0 in graphene metamaterial imply that the THz response here is dominated by the resonant effect. The electrostatic change in the real part of conductivity $\Delta\sigma_1/\sigma_0$ is described in Fig. 9.12(d) in which the negative background $\Delta\sigma_1/\sigma_0$ and increased oscillation strength at $\omega_1/2\pi$ and $\omega_2/2\pi$ are shown. The negative $\Delta\sigma_1/\sigma_0$ implies reduced conductivity in graphene due to the shifted Fermi level toward CNP. By reducing the free-carrier screening of dipole oscillation, the resonance strength at both $\omega_1/2\pi$ and $\omega_2/2\pi$ are increased. The enhancement at $\omega_1/2\pi$ is more substantial due to its larger intrinsic magnitude in equilibrium.

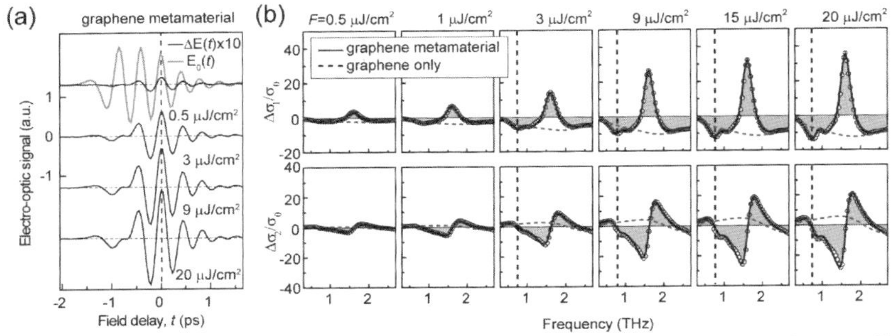

Fig. 9.13. (a) THz reference field $E_0(t)$ and pump-induced change in THz field $\Delta E(t)$ at pump fluence from 0.5 μJ/cm^2 to 20 μJ/cm^2. (b) Pump-induced change in THz conductivity in graphene (dashed line) and graphene metamaterial (circles). (Reprinted from [In *et al.* 2016])

We understand that the optical pump can increase the graphene conductivity when the electron is heated by intense pulse energy. One distinct feature of CVD-grown graphene is that it is more likely to possess defect sites that can assist with electron-phonon scattering. The increased scattering rate leads to the decreased conductivity of CVD graphene [Frenzel *et al.* 2013; Jnawali *et al.* 2013]. In Fig. 9.13(a), we plot $E_0(t)$ and $\Delta E(t)$ of graphene metamaterial with aDSR meta-atom as a function of pump fluence ranging from 0.5 μJ/cm^2 to 20 μJ/cm^2. When the optical pump excites graphene free carriers, the transmission of the THz field is increased at t=0. We interpret the positive sign of $\Delta E(t)$ here as being due to the increased scattering rate in graphene carriers. The Fourier-transformed conductivity changes $\Delta\sigma_1/\sigma_0$ and $\Delta\sigma_2/\sigma_0$ in graphene metamaterial and bare graphene are depicted in Fig. 9.13(b). We fit the curves with the sum of Lorentz oscillators and $\Delta\sigma_1/\sigma_0$ of bare graphene. While $\Delta\sigma_1/\sigma_0$ in graphene is negative with flat curves, $\Delta\sigma_1/\sigma_0$ in graphene metamaterial exhibits two resonances at $\omega_1/2\pi$ and $\omega_2/2\pi$. The optically controlled THz spectra in graphene metamaterial reveal similar features to the electrostatic gating effect (Fig. 9.12(c)).

One difference is the emergence of a new resonance at 0.7 THz indicated by the vertical dashed lines. The dispersive relation observed in

$\Delta\sigma_2/\sigma_0$ at 0.7 THz confirms the existence of the new resonant frequency. Given that the frequencies at $\omega_1/2\pi$ and $\omega_2/2\pi$ are located at 1.6 THz and 0.9 THz, the new resonant frequency at $\Delta\omega/2\pi$=0.7 THz is the manifestation of the second-order nonlinear response, i.e. $\Delta\omega/2\pi = \omega_1/2\pi - \omega_2/2\pi$. The difference-frequency mixing of the two dipole resonances at $\omega_1/2\pi$ and $\omega_2/2\pi$ is prominent above pump fluences of 3 μJ/cm^2. The negative sign of $\Delta\sigma_1/\sigma_0$ at $\Delta\omega/2\pi$ and $\omega_2/2\pi$ and the positive sign of $\Delta\sigma_1/\sigma_0$ at $\omega_1/2\pi$ match the energy conservation law in difference-frequency mixing [Boyd 2008].

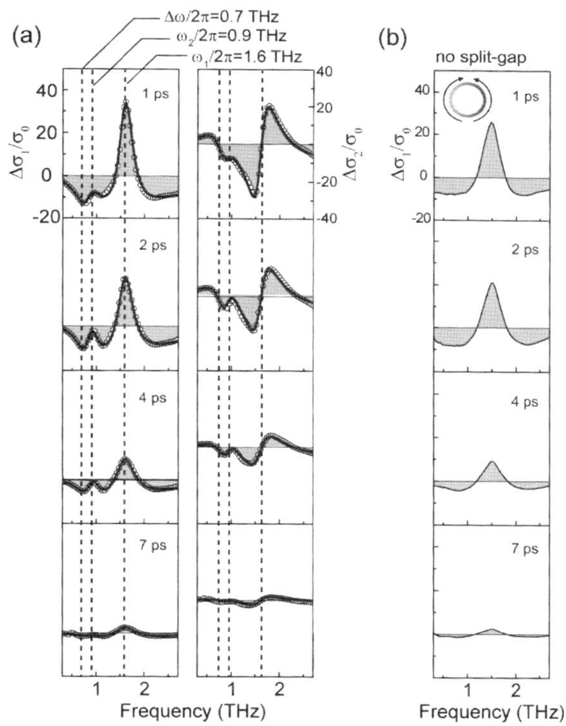

Fig. 9.14. (a) Pump-induced change in THz conductivity in aDSR graphene metamaterial at Δt from 1 ps to 7 ps. (b) Pump-induced change in THz conductivity in graphene metamaterial with no split-gap at Δt from 1 ps to 7 ps. (Reprinted from [In *et al.* 2016])

Previous researchers have found that the second-order nonlinear response in graphene is prohibited due to the lack of inversion symmetry [Hendry *et al.* 2010; Cheng *et al.* 2014]. Although the third-order wave mixing at the near-infrared region has been demonstrated in bare graphene, it has been hard to detect the second-order nonlinearity. Here, we found that the difference-frequency mixing in graphene metamaterial is the collaborated effect of graphene and metamaterial. Upon photoexcitation, the split-gap of a meta-atom exhibits nonlinearity due to the transient conductivity change in graphene. The absence of difference-frequency mixing upon electrostatic gating in Fig. 9.12(c) indicates that the negative conductivity change is not sufficient to break the harmonic response in graphene metamaterial. Along with the negative $\Delta\sigma_1/\sigma_0$, the transient response of graphene $\Delta\sigma_1/\sigma_0$ is required to show the effect. When the optical pump excites graphene free carriers, the split-gap of the aDSR meta-atom serves as nonlinear capacitance through which the second-order dipole oscillation is provided. The experimental results are analogous to the previous demonstration of nonlinear wave mixing at microwave frequencies where a varactor diode was inserted in a split-ring-resonator [Rose *et al.* 2011; Lapine *et al.* 2014]. The nonlinear THz effects in graphene metamaterial differ from those of previous studies in that i) the transient photoresponse of graphene is employed to develop optical nonlinearity at THz frequencies and ii) ultrathin and compact devices are available to generate THz frequencies.

The theoretical derivation of the *LC* circuit model is provided elsewhere [In *et al.* 2015]. A circuit analysis containing graphene resistance and the *LC* circuit of a meta-atom reveals that the split-gap exhibits second-order nonlinearity upon photoexcitation. Here, we present time-resolved THz wave mixing in graphene metamaterials to supplement the nonlinear kinetics. Figure 9.14(a) shows $\Delta\sigma_1/\sigma_0$ in graphene metamaterials with Δt ranging from 1 ps to 7 ps. The fit lines are obtained from Lorentz oscillators and graphene $\Delta\sigma_1/\sigma_0$. The resonance at $\Delta\omega/2\pi$ rapidly disappears as graphene free carriers are relaxed. Because the location of resonant frequencies at $\omega_1/2\pi$ and $\omega_2/2\pi$ is well defined by the aDSR meta-atom, the resonant frequency at $\Delta\omega/2\pi$ is fixed regardless of Δt. The dispersive relations in $\Delta\sigma_2/\sigma_0$ at $\omega_1/2\pi$, $\omega_2/2\pi$, and $\Delta\omega/2\pi$ ascertain the resonant state of graphene metamaterial. In Fig. 9.14(b), we show the time-resolved THz conductivity in graphene metamaterial with no split-gap. Because the nonlinearity is provided by the split-gap, the absence of a split-gap leads to the lack of distinct features of difference-frequency mixing. Here, the THz response is dominated by the single-dipole oscillation of the meta-atom, as depicted in the inset of Fig. 9.14(b).

To summarize, we reviewed the previous works on THz effects in graphene metamaterials. The exotic THz spectra are manipulated by screened and unscreened dipole oscillation of graphene metamaterials. The electrostatic effect induces charge accumulation onto graphene, while optical excitation results in transient THz dynamics. Electrically and optically engineered THz spectra are at the heart of device performance in graphene metamaterials. Flexible and compact device construction is enabled by atomically thin graphene combined with the two-dimensional arrangement of meta-atoms. We confirmed that the high refractive index at THz frequencies is inherent in the dipole oscillation of metamaterials. The ultrafast kinetics of graphene not only allow picosecond modulation of THz spectra but also produce nonlinear wave mixing through the transient dynamics of graphene conductivity. The frequency-dependent manipulation of THz waves developed by graphene metamaterials is expected to contribute to novel optoelectronic operation in future THz devices

Acknowledgements

C. In and H. Choi were supported by the National Research Foundation of Korea (NRF) through the government of Korea (MSIP) (Grant Nos 2016R1A4A1012929, NRF-2015R1A2A1A10052520), Global Frontier Program (2014M3A6B3063709), the Yonsei University Yonsei-SNU Collaborative Research Fund of 2014, and the Yonsei University Future-leading Research Initiative of 2014. C. In was also supported by NRF through MSIP (Grant No. NRF-2015H1A2A1034809).

Bibliography

[1] Boyd, R. W. (2008) *Nonlinear Optics*, 3rd Ed. (Academic Press, USA).

[2] Campbell, T., Hibbins, A. P., Sambles, J. R., and Hooper, I. R. (2013). Broadband and low loss high refractive index metamaterials in the microwave regime, *Appl. Phys. Lett.*, 102, pp. 091108.

[3] Chen, H., Padilla, W. J., Zide, J. M. O., Gossard, A. C., Taylor, A. J., and Averitt, R. D. (2006). Active terahertz metamaterial devices, *Nature*, 444, pp. 597–600.

[4] Cheng, J. L., Vermeulen, N., and Sipe, J. E. (2014). Third order optical nonlinearity of graphene, *New J. Phys.*, 16, pp. 053014.

[5] Choi, H., Borondics, F., Siegel, D. A., Zhou, S. Y., Martin, M. C., Lanzara, A., and Kaindl, R. A. (2009). Broadband electromagnetic response and ultrafast dynamics of few-layer epitaxial graphene, *Appl. Phys. Lett.*, 94, pp. 172102.

[6] Choi, M., Lee, S. H., Kim, Y., Kang, S. B., Shin, J., Kwak, M. H., Kang, K., Lee, Y., Park, N., and Min, B. (2011). A terahertz metamaterial with unnaturally high refractive index, *Nature*, 470, pp. 369–373.

[7] Dani, K. M., Lee, J., Sharma, R., Mohite, A. D., Galande, C. M., Ajayan, P. M., Dattelbaum, A. M., Htoon, H., Taylor, A. J., and Prasankumar, R. P. (2012). Intraband conductivity response in graphene observed using ultrafast infrared-pump visible-probe spectroscopy, *Phys. Rev. B*, 86, pp. 125403.

[8] Dawlaty, J. M., Shivaraman, S., Strait, J., George, P., Chandrashekhar, M., Rana, F., Spencer, M. G., Veksler, D., and Chen, Y. (2008). Measurement of the optical absorption spectra of epitaxial graphene from terahertz to visible, *Appl. Phys. Lett.*, 93, pp. 131905.

[9] Dobroiu, A., Otani, C., and Kawase, K. (2006). Terahertz-wave sources and imaging applications, *Meas. Sci. Technol.*, 17, pp. R161–R174.

[10] Emani, N. K., Chung, T.-F., Ni, X., Kildishev, A. V, Chen, Y. P., and Boltasseva, A. (2012). Electrically tunable damping of plasmonic resonances with graphene, *Nano Lett.*, 12, pp. 5202–5206.

[11] Frenzel, A. J., Lui, C. H., Fang, W., Nair, N. L., Herring, P. K., Jarillo-Herrero, P., Kong, J., and Gedik, N. (2013). Observation of suppressed terahertz absorption in photoexcited graphene, *Appl. Phys. Lett.*, 102, pp. 113111.

[12] Frenzel, A. J., Lui, C. H., Shin, Y. C., Kong, J., and Gedik, N. (2014). Semiconducting-to-metallic photoconductivity crossover and temperature-dependent Drude weight in graphene, *Phys. Rev. Lett.*, 113, pp. 056602.

[13] George, P. A., Strait, J., Dawlaty, J., Shivaraman, S., Chandrashekhar, M., Rana, F., and Spencer, M. G. (2008). Ultrafast optical-pump terahertz-probe spectroscopy of the carrier relaxation and recombination dynamics in epitaxial graphene, *Nano Lett.*, 8, pp. 4248–4251.

[14] Gierz, I., Petersen, J. C., Mitrano, M., Cacho, C., Turcu, I. C. E., Springate, E., Stöhr, A., Köhler, A., Starke, U., and Cavalleri, A. (2013). Snapshots of non-equilibrium Dirac carrier distributions in graphene, *Nat. Mater.*, 12, pp. 1119–1124.

[15] Han, T.-H., Lee, Y., Choi, M.-R., Woo, S.-H., Bae, S.-H., Hong, B. H., Ahn, J.-H., and Lee, T.-W. (2012). Extremely efficient flexible organic light-emitting diodes with modified graphene anode, *Nat. Photonics*, 6, pp. 105–110.

[16] Hendry, E., Hale, P. J., Moger, J., Savchenko, A. K., and Mikhailov, S. A. (2010). Coherent nonlinear optical response of graphene, *Phys. Rev. Lett.*, 105, pp. 097401.

[17] Hong, X., Kim, J., Shi, S.-F., Zhang, Y., Jin, C., Sun, Y., Tongay, S., Wu, J., Zhang, Y., and Wang, F. (2014). Ultrafast charge transfer in atomically thin MoS_2/WS_2 heterostructures, *Nat. Nanotechnol.*, 9, pp. 682–686.

[18] Horng, J., Chen, C., Geng, B., Girit, C., Zhang, Y., Hao, Z., Bechtel, H. A., Martin, M., Zettl, A., Crommie, M. F., Shen, Y. R., and Wang, F. (2011). Drude conductivity of Dirac fermions in graphene, *Phys. Rev. B*, 83, pp. 165113.

[19] Huang, C., Wu, S., Sanchez, A. M., Peters, J. J. P., Beanland, R., Ross, J. S., Rivera, P., Yao, W., Cobden, D. H., and Xu, X. (2014). Lateral heterojunctions within monolayer $MoSe_2$–WSe_2 semiconductors, *Nat. Mater.*, 13, pp. 1096–1101.

[20] Huber, R., Tauser, F., Brodschelm, A., Bichler, M., Abstreiter, G., and Leitenstorfer, A. (2001). How many-particle interactions develop after ultrafast excitation of an electron-hole plasma., *Nature*, 414, pp. 286–289.

[21] In, C., Kim, H., Min, B., and Choi, H. (2015). Photoinduced nonlinear mixing of terahertz dipole resonances in graphene metadevices, *Adv. Mater.*, 28, pp. 1495–1500.

[22] Jnawali, G., Rao, Y., Yan, H., and Heinz, T. F. (2013). Observation of a transient decrease in terahertz conductivity of single-layer graphene induced by ultrafast optical excitation, *Nano Lett.*, 13, pp. 524–530.

[23] Jung, H., In, C., Choi, H., and Lee, H. (2014). Anisotropy modeling of terahertz metamaterials: polarization dependent resonance manipulation by meta-atom cluster., *Sci. Rep.*, 4, pp. 5217.

[24] Kaindl, R. A., Carnahan, M. A., Hägele, D., Lövenich, R., and Chemla, D. S. (2003). Ultrafast terahertz probes of transient conducting and insulating phases in an electron-hole gas, *Nature*, 423, pp. 734–738.

[25] Kaindl, R. A., Carnahan, M. A., Orenstein, J., Chemla, D. S., Christen, H. M., Zhai, H. Y., Paranthaman, M., and Lowndes, D. H. (2002). Far-infrared optical

conductivity gap in superconducting MgB_2 films, *Phys. Rev. Lett.*, 88, pp. 270031–270034.

[26] Kaindl, R. A., Hägele, D., Carnahan, M. A., and Chemla, D. S. (2009). Transient terahertz spectroscopy of excitons and unbound carriers in quasi-two-dimensional electron-hole gases, *Phys. Rev. B - Condens. Matter Mater. Phys.*, 79.

[27] Kim, J., Lim, S. C., Chae, S. J., Maeng, I., Choi, Y., Cha, S., Lee, Y. H., and Choi, H. (2013). Ultrafast zero balance of the oscillator-strength sum rule in graphene., *Sci. Rep.*, 3, pp. 2663.

[28] Lapine, M., Shadrivov, I. V., and Kivshar, Y. S. (2014). *Colloquium*: Nonlinear metamaterials, *Rev. Mod. Phys.*, 86, pp. 1093–1123.

[29] Lee, S. H., Choi, J., Kim, H.-D., Choi, H., and Min, B. (2013). Ultrafast refractive index control of a terahertz graphene metamaterial., *Sci. Rep.*, 3, pp. 2135.

[30] Lee, S. H., Choi, M., Kim, T.-T., Lee, S., Liu, M., Yin, X., Choi, H. K., Lee, S. S., Choi, C.-G., Choi, S.-Y., Zhang, X., and Min, B. (2012). Switching terahertz waves with gate-controlled active graphene metamaterials, *Nat. Mater.*, 11, pp. 936–941.

[31] Lee, Y.-S. (2009) *Principles of Terahertz Science and Technology*, (Springer, Germany).

[32] Li, Q., Tian, Z., Zhang, X., Singh, R., Du, L., Gu, J., Han, J., and Zhang, W. (2015). Active graphene–silicon hybrid diode for terahertz waves, *Nat. Commun.*, 6, pp. 7082.

[33] Liu, N., and Giessen, H. (2010). Coupling effects in optical metamaterials, *Angew. Chem. Int. Ed. Engl.*, 49, pp. 9838–9852.

[34] Maeng, I., Lim, S., Chae, S. J., Lee, Y. H., Choi, H., and Son, J. (2012). Gate-controlled nonlinear conductivity of Dirac fermion in graphene field-effect transistors measured by terahertz time-domain spectroscopy, *Nano Lett.*, 12, pp. 551–555.

[35] Mak, K. F., Sfeir, M. Y., Wu, Y., Lui, C. H., Misewich, J. A., and Heinz, T. F. (2008). Measurement of the optical conductivity of graphene, *Phys. Rev. Lett.*, 101, pp. 196405.

[36] Mittleman, D. M., Jacobsen, R. H., and Nuss, M. C. (1996). T-ray imaging, *IEEE J. Sel. Top. Quantum Electron.*, 2, pp. 679–692.

[37] Nahata, A., Weling, A. S., and Heinz, T. F. (1996). A wideband coherent terahertz spectroscopy system using optical rectification and electro-optic sampling, *Appl. Phys. Lett.*, 69, pp. 2321.

[38] Nair, R. R., Blake, P., Grigorenko, A. N., Novoselov, K. S., Booth, T. J., Stauber, T., Peres, N. M. R., and Geim, A. K. (2008). Fine structure constant defines visual transparency of graphene, *Science*, 320, pp. 1308.

[39] Novoselov, K. S., Geim, A. K., Morozov, S. V., Jiang, D., Katsnelson, M. I., Grigorieva, I. V., Dubonos, S. V., and Firsov, A. A. (2005). Two-dimensional gas of massless Dirac fermions in graphene, *Nature*, 438, pp. 197–200.

[40] Ren, L., Zhang, Q., Yao, J., Sun, Z., Kaneko, R., Yan, Z., Nanot, S., Jin, Z., Kawayama, I., Tonouchi, M., Tour, J. M., and Kono, J. (2012). Terahertz and infrared spectroscopy of gated large-area graphene, *Nano Lett.*, 12, pp. 3711–3715.

[41] Rice, A., Jin, Y., Ma, X. F., Zhang, X.-C., Bliss, D., Larkin, J., and Alexander, M. (1994). Terahertz optical rectification from ⟨110⟩ zinc-blende crystals, *Appl. Phys. Lett.*, 64, pp. 1324.

[42] Rose, A., Huang, D., and Smith, D. R. (2011). Controlling the second harmonic in a phase-matched negative-index metamaterial, *Phys. Rev. Lett.*, 107, pp. 063902.

[43] Schurig, D., Mock, J. J., Justice, B. J., Cummer, S. A., Pendry, J. B., Starr, A. F., and Smith, D. R. (2006). Metamaterial electromagnetic cloak at microwave frequencies, *Science*, 314, pp. 977–980.

[44] Sensale-Rodriguez, B., Yan, R., Kelly, M. M., Fang, T., Tahy, K., Hwang, W. S., Jena, D., Liu, L., and Xing, H. G. (2012). Broadband graphene terahertz modulators enabled by intraband transitions, *Nat. Commun.*, 3, pp. 780.

[45] Shan, J., Wang, F., Knoesel, E., Bonn, M., and Heinz, T. F. (2003). Measurement of the frequency-dependent conductivity in sapphire., *Phys. Rev. Lett.*, 90, pp. 247401.

[46] Shi, S.-F., Tang, T.-T., Zeng, B., Ju, L., Zettl, A., and Wang, F. (2014). Controlling graphene ultrafast hot carrier response from metal-like to semiconductor-like by electrostatic gating, *Nano Lett.*, 14, pp. 1578–1582.

[47] Strait, J. H., Wang, H., Shivaraman, S., Shields, V., Spencer, M., and Rana, F. (2011). Very slow cooling dynamics of photoexcited carriers in graphene observed by optical-pump terahertz-probe spectroscopy, *Nano Lett.*, 11, pp. 4902–4906.

[48] Sun, D., Aivazian, G., Jones, A. M., Ross, J. S., Yao, W., Cobden, D., and Xu, X. (2012). Ultrafast hot-carrier-dominated photocurrent in graphene, *Nat. Nanotechnol.*, 7, pp. 114–118.

[49] Suzuki, T., and Shimano, R. (2012). Exciton Mott Transition in Si Revealed by Terahertz Spectroscopy, *Phys. Rev. Lett.*, 109, pp. 046402.

[50] Ulbricht, R., Hendry, E., Shan, J., Heinz, T. F., and Bonn, M. (2011). Carrier dynamics in semiconductors studied with time-resolved terahertz spectroscopy, *Rev. Mod. Phys.*, 83, pp. 543–586.

[51] Wang, F., Zhang, Y., Tian, C., Girit, C., Zettl, A., Crommie, M., and Shen, Y. R. (2008). Gate-variable optical transitions in graphene, *Science*, 320, pp. 206–209.

[52] Wehling, T. O., Novoselov, K. S., Morozov, S. V., Vdovin, E. E., Katsnelson, M. I., Geim, A. K., and Lichtenstein, A. I. (2008). Molecular doping of graphene, *Nano Lett.*, 8, pp. 173–177.

[53] Wietzke, S., Jansen, C., Reuter, M., Jung, T., Kraft, D., Chatterjee, S., Fischer, B. M., and Koch, M. (2011). Terahertz spectroscopy on polymers: A review of morphological studies, *J. Mol. Struct.*, 1006, pp. 41–51.

[54] Wu, Y., La-O-Vorakiat, C., Qiu, X., Liu, J., Deorani, P., Banerjee, K., Son, J., Chen, Y., Chia, E. E. M., and Yang, H. (2015). Graphene terahertz modulators by ionic liquid gating, *Adv. Mater.*, 27, pp. 1874–1879.

[55] Yasuda, H., and Hosako, I. (2008). Measurement of terahertz refractive index of metal with terahertz time-domain spectroscopy, *Jpn. J. Appl. Phys.*, 47, pp. 1632–1634.

[56] Zhang, W., Chuu, C.-P., Huang, J.-K., Chen, C.-H., Tsai, M.-L., Chang, Y.-H., Liang, C.-T., Chen, Y.-Z., Chueh, Y.-L., He, J.-H., Chou, M.-Y., and Li, L.-J. (2014). Ultrahigh-gain photodetectors based on atomically thin graphene-MoS_2 heterostructures., *Sci. Rep.*, 4, pp. 3826.

[57] Zhang, Y., Tan, Y.-W., Stormer, H. L., and Kim, P. (2005). Experimental observation of the quantum Hall effect and Berry's phase in graphene., *Nature*, 438, pp. 201–204.

[58] Zhu, J., Christensen, J., Jung, J., Martin-Moreno, L., Yin, X., Fok, L., Zhang, X., and Garcia-Vidal, F. J. (2011). A holey-structured metamaterial for acoustic deep-subwavelength imaging, *Nat. Phys.*, 7, pp. 52–55.

Part 4

Saddle Point Spectroscopy

Chapter 10

Spectroscopy of Graphene at the Saddle Point

Daniela Wolf[1], Dong-Hun Chae[2], Tobias Utikal[3], Patrick Herlinger[4], Jurgen Smet[4,**], Harald Giessen[5,†], and Markus Lippitz[1,‡]

[1] *Experimental Physics III, University of Bayreuth, Universitaetsstrasse 30, 95440 Bayreuth, Germany*

[2] *Center for Electricity & Magnetism, Korean Metrology Institute of Standards & Science, Daejeon 305-340, Republic of Korea*

[3] *Max Planck Institute for the Science of Light, Guenther-Scharowsky-Strasse 1, 91058 Erlangen, Germany*

[4] *Max Planck Institute for Solid State Research, Heisenbergstrasse 1, 70569 Stuttgart, Germany*

[5] *4th Physics Institute and Research Center SCoPE, University of Stuttgart, Pfaffenwaldring 57, 70569 Stuttgart, Germany*

[*] *J.Smet@fkf.mpg.de*, [†] *Giessen@physik.uni-stuttgart.de*, [‡] *Markus.Lippitz@uni-bayreuth.de*

In contrast to the previous chapters of this book, in the following sections the optical properties of graphene far away from the Dirac points will be treated. As has been shown earlier, at frequencies above the far-infrared region, the optical response of graphene is dominated by interband transitions between the valence and conduction band. The band structure of graphene exhibits a saddle point between two neighboring Dirac cones, so that the optical absorption will deviate from its constant value of $\pi\alpha$ for photon energies in the visible and ultraviolet regime.

This chapter covers the linear optical response of graphene near the saddle point. After a theoretical treatment of saddle point phenomena like the van Hove singularity (Sec. 10.1), a general introduction to optical spectroscopy of graphene in the ultraviolet and visible wavelength regime will be given in Section 10.2. We will then show that the optical properties of graphene near the saddle point can only be explained when many-body effects like electron-hole interactions are taken into account (Sec. 10.3). A simple model based on the classical theory by Fano is presented in Section 10.4 which is able to describe the response down to near-infrared energies. After examining the dependence of the excitonic resonance in graphene on influences like doping or strain (Sec. 10.5), an overview over similar phenomena in graphene-related materials is given is Section 10.6. Finally, we show how the optical constants of graphene can be extracted from a fit to the Fano model in combination with the Kramers-Kronig relations.

10.1 Critical Points and Van Hove Singularities in the Band Structure

In a solid state system, the electronic states form bands, described by a dispersion relation $E(\vec{k})$ which relates the energy of the state with the electron wavevector $\vec{k}$. Many different wavevectors can result in the same energy. The density of states ρ gives the number of states that fall within an energy range E to $E + dE$

$$\rho(E) = \frac{1}{4\pi^3} \int \frac{dS_{\vec{k}}}{|\nabla_{\vec{k}} E(\vec{k})|}, \tag{10.1}$$

where $dS_{\vec{k}}$ is a surface of constant energy E. Obviously, the density of states diverges at points where the dispersion relation $E(\vec{k})$ exhibits a horizontal tangent, i.e., when $\nabla_{\vec{k}} E = 0$, as pointed out by van Hove [van Hove (1953)]. The wavevectors at which van Hove singularities occur are the critical points of the Brillouin zone.

Optical spectra are related to transitions between electronic states. The transition probability W_{if} of transitions excited by a photon with energy $E = \hbar\omega$ are described by Fermi's golden rule as

$$W_{if} \cong \frac{2\pi}{\hbar} |\langle f|\hat{H}'|i\rangle|^2 \rho_j(E). \tag{10.2}$$

$|i\rangle$ and $|f\rangle$ represent the initial and final states of the transition. $\hat{H}'$ is the perturbation operator, which for optical transitions is proportional to the dot product of the vector potential $\vec{A}$ and the dipole operator $\hat{p}$ [Yu and Cardona (2005)]. $\rho_j(E)$ is the joint density of states [Fox (2001)]. It is calculated according to Eq. 10.1, replacing the *electronic* band $E(\vec{k})$ with

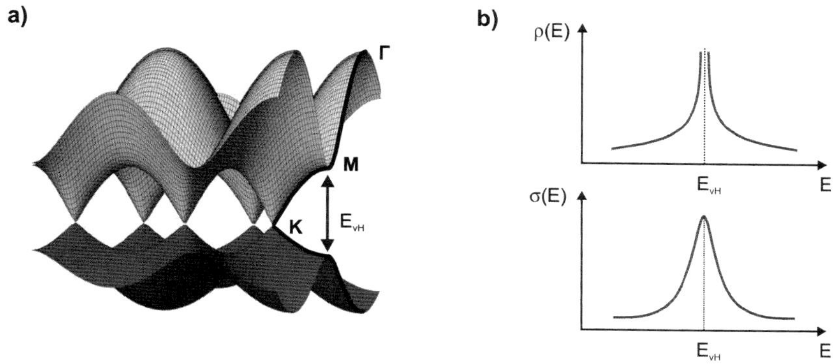

Fig. 10.1 a) Band structure of graphene with Dirac point K, center of the Brillouin zone Γ, and saddle point M. E_{vH} is the energy gap at the saddle point. b) Schematic dependence of the density of states and the optical conductivity on the energy in the vicinity of the van Hove singularity E_{vH} at a saddle point in a two dimensional system.

the *optical* band $E_{cv}(\vec{k}) = E_c(\vec{k}) - E_v(\vec{k})$. E_c and E_v are the dispersion relations of conduction and valence band, respectively [Wooten (1972)]. So-called *optical* critical points occur when $\nabla_{\vec{k}} E_{cv} = \nabla_{\vec{k}} E_c - \nabla_{\vec{k}} E_v = 0$ [Wooten (1972)].

Van Hove singularities are classified by the curvature of the bands when moving away from the critical points. In two and three dimensions, a critical point can represent a minimum or a maximum in the band separation or else a saddle point, whereas in the one dimensional case only a minimum or a maximum is possible. The behavior of the density of states in the vicinity of a critical point depends on the type of the van Hove singularity as well as the dimensionality of the system [Yu and Cardona (2005)].

Looking at the band structure of graphene shown in Fig. 10.1a, we can identify a saddle point at the M point in the Brillouin zone. Here, both electronic bands and hence also the optical band E_{cv} have a horizontal tangent. Consequently, an electronic and an optical critical point coincide at the M point of graphene. At a saddle point with energy gap E_{vH} in a two dimensional system, the joint density of states is given by

$$\rho(E) \propto -\ln|E - E_{\mathrm{vH}}|. \tag{10.3}$$

Since the density of states is symmetric around the singularity, from Fermi's golden rule one would expect the optical spectrum to be also symmetric around E_{vH} as schematically shown in Fig.10.1b. Towards lower energies, the linear dispersion at the Dirac cones shapes the optical response. Here, the universal value of the optical conductivity of $\sigma(E) = \pi e^2/2h$ should be recovered, corresponding to a constant absorbance of $\pi\alpha \approx 2.3\%$ for monolayer graphene [Kuzmenko *et al.* (2008); Nair *et al.* (2008)].

10.2 Spectroscopy of Graphene at Visible and Ultraviolet Wavelengths

To investigate the optical properties of graphene in the visible and ultraviolet wavelength regime, different sample and measurement geometries are possible. In the following, we will describe in more detail a transmission measurement of free-standing graphene which has been published in [Chae *et al.* (2011)].

Free-standing graphene samples can be fabricated by transferring graphene from a substrate to an aperture of polymer resist. This geometry is advantageous as it eliminates potential influences of the substrate, so that the response is purely determined by the optical properties of the graphene sheet. The graphene flakes are first exfoliated from natural graphite on an oxidized silicon substrate. The number of layers can be determined by the optical contrast in a microscope image and by Raman spectroscopy. The sample is coated with a polymer resist and apertures with a diameter of a few μm are written on top of the flakes using electron beam lithography. After development, the structured polymer layer together with the graphene flake are removed from the substrate by etching the silicon dioxide in an alkaline solution. A typical free-standing graphene sample is shown in Fig. 10.2a. Due to the 2.3% absorption in the visible, we can identify the graphene sheets that extend over the holes in the polymer layer.

The transmission of the samples is measured in a confocal microscope (see Fig. 10.2b). The light source combines a tungsten halogen bulb and a deuterium lamp, providing unpolarized white light with photon energies from 1.5 to 5.5 eV. The light is focused onto the sample by an all-reflective Cassegrain objective and the transmitted light is collected with a UV microscope objective. The spatial resolution on the sample is better than 1.5 μm, ensuring that only the free-standing part of the graphene flake is measured. The transmitted light is analyzed in a spectrometer consisting of a monochromator and a CCD camera as detector. The spectrum is combined from exposures with two different gratings.

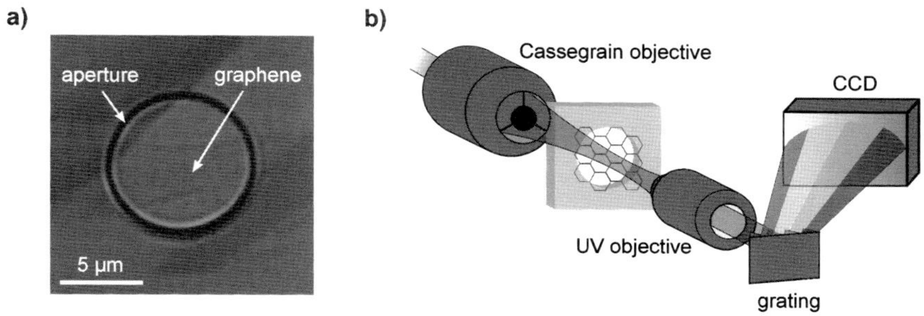

Fig. 10.2 a) Optical microscope image of a free-standing graphene layer, taken in transmission mode. b) Schematic view of an experimental setup for transmission spectroscopy. Adapted with permission from [Chae *et al.* (2011)]. Copyright (2011) American Chemical Society. (For color reference, go to page 493.)

The transmittance T is defined as

$$T = \frac{T_g - T_r}{T_r}, \qquad (10.4)$$

where T_g is the transmission through the graphene layer and T_r the transmission through an empty reference aperture. This normalization eliminates all spectral dependencies of the setup such as the lamp spectrum, possible dispersion of the optical elements etc. The weak absorption of the thin graphene sheets implies a negligible reflectance, hence the absorbance A follows directly from the transmittance as $A = 1 - T$.

The measured absorption spectra for monolayer and bilayer graphene are shown in Fig. 10.3a. At low photon energies below 2 eV, the absorption spectra are almost flat and approach the predicted value of $\pi\alpha$ and $2\pi\alpha$ for the monolayer and the bilayer sample, respectively. Towards higher energies, the absorbance increases monotonically until it reaches a maximum value at about 4.6 eV. Beyond the peak, it steeply decreases, thus giving rise to an asymmetric line shape. The absorbance of graphene in the UV reaches a maximum of about 11%, a strikingly high value for a single-atom thick system.

Our results are in accordance with those obtained from reflectivity measurements on graphene on a silicon dioxide substrate [Mak *et al.* (2011)] as well as the absorption and transmission curves from ellipsometric data [Kravets *et al.* (2010)]. Both find an asymmetric line shape and a redshift from the van Hove singularity. For graphene supported on a transparent

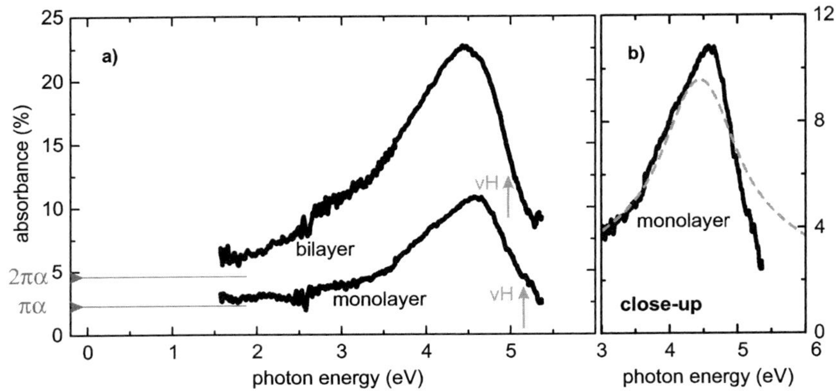

Fig. 10.3 a) Measured absorbance $A = 1 - T$ spectrum of free-standing mono- and bilayer graphene (black lines). The curves approach the universal value of $\pi\alpha$ and $2\pi\alpha$ at lower energies, but the peaks are red-shifted compared to the energy of the van Hove singularity (vH). b) Close-up showing that the asymmetric line shape of the monolayer spectrum (black line) cannot be brought into agreement with a model neglecting electron–hole interactions (grey dashed line). Adapted with permission from [Chae *et al.* (2011)]. Copyright (2011) American Chemical Society.

substrate with negligible absorption and refractive index n_s, absorbance, transmittance, and reflectance are related via [Hecht (1998); Mak *et al.* (2011)]

$$\sigma(E) \propto A = \frac{n_s - 1}{2} T = \frac{n_s^2 - 1}{4} R. \tag{10.5}$$

The reflectance contrast R is defined and measured in analogy to the transmittance T (see Eqn. 10.4). Due to these simple relations, the data obtained with the different measurement methods can easily be compared.

To model the measured data, we first draw on the relations developed in the previous section. We have seen that the spectral shape of the interband transitions near the saddle point in a two dimensional system is proportional to the joint density of states. To take inhomogeneous broadening due to carrier relaxation into account, the resulting line shape can be convoluted with a Gaussian function of width γ. This simple single-particle model based on band-to-band transitions thus predicts a symmetric absorption spectrum of the shape

$$A(E) = B + C \left[\exp\left(-\frac{(E - E_{\mathrm{vH}})^2}{\gamma^2} \right) \otimes [-\ln|E - E_{\mathrm{vH}}|] \right]. \tag{10.6}$$

E_{vH} is given by the transition energy at the van Hove singularity, which is about 5.15 eV for monolayer and 5.0 eV for bilayer graphene [Yang *et al.* (2009)]. The constant absorption at low energies from the Dirac cones is modeled as an offset B, so that the constant scaling factor C and the width γ are the only free parameters.

However, this model utterly fails to describe the measured absorbance spectrum as is shown in the close-up in Fig. 10.3b. The measured absorbance peak is red-shifted compared to the position of the van Hove singularity to about 4.6 eV. However, even when the van Hove singularity is shifted to this energy as in Fig. 10.3b, the pronounced asymmetry of the line shape disagrees with the predicted symmetric response. Obviously, the absorption spectrum of free-standing graphene in the visible and UV regime cannot be described considering band-to-band transitions only. In this approximation, the charge carriers are uncorrelated and interactions are not taken into account. We therefore conclude that many-body effects cannot be neglected here and we will address this issue in the next sections.

10.3 Saddle Point Excitons and Many-body Effects

As we have seen in the previous section, the single non-interacting particle picture fails to describe the absorption spectrum of graphene in the visible and UV regime. A similar effect has been observed in the optical spectra of bulk semiconductors such as CdTe or ZnTe [Pétroff and Balkanski (1971)], where the discrepancies between experiment and theory based on band-to-band transitions have been explained by Coulomb effects between electrons and holes at the critical points [Kane (1969)]. When an electron is excited from the valence to the conduction band, it will interact with the remaining electrons in the valence band, which can also be described in terms of the interaction with a hole. Due to the attractive Coulomb interaction between electron and hole, their motion will be correlated, resulting in a bound electron–hole pair known as an exciton. Assuming that excitonic states exist in the vicinity of the saddle point, the optical spectra of CdTe and ZnTe have been explained successfully. Since the constant energy surfaces near a saddle point have hyperbolic shape, such excitonic states are called hyperbolic or saddle point excitons [Yu and Cardona (2005)]. In analogy to the semiconductor example, we will in the following assume an excitonic state near the saddle point in graphene.

Depending on the screening of the Coulomb interaction by the valence electrons, excitons can be weakly to tightly bound. In semiconductors, where the binding energy of an exciton can be described in analogy to the hydrogenic atom, the screening scales inversely with the dielectric constant of the material and is usually on the order of a few meV [Yu and Cardona (2005)]. In three dimensional bulk metals, the screening is large and thus usually prevents correlated electron–hole pairs [Ashcroft and Mermin (1976)]. However, when the dimensionality is reduced such as in one-dimensional metallic carbon nanotubes, the screening is weakened so that excitonic correlations can occur [Wang *et al.* (2007)].

The screening depends on the density of states near the Fermi energy. Graphene is a semi-metal, so that the density of states vanishes at the Dirac point and then grows linearly. In bilayer graphene, however, the density of states is constant and nonzero near the Dirac point [Castro Neto *et al.* (2009)]. For Fermi energies near the Dirac point, the screening ability of electrons in monolayer graphene is therefore reduced compared to the bilayer. Hence, we expect that the exciton binding energy, corresponding to the energy difference between the saddle point and the discrete excitonic state, is larger in monolayer than in bilayer graphene.

The influence of excitonic effects on the optical response of graphene has first been investigated theoretically by Yang et al. [Yang *et al.* (2009)]. Using first-principles calculations based on a many-body Green's function, this approach indicates remarkable effects of Coulomb interactions on the optical response of graphene. According to these calculations, the attractive interactions between electrons and holes dominates over the repulsive interaction between the electrons. Overall, a red-shift of the absorption peak to about 4.5 eV with an asymmetric line shape is predicted, which is in very good agreement with the effects observed in the measurements.

Consequently and in accordance with the interpretation of other groups [Kravets *et al.* (2010); Mak *et al.* (2011)], it is a clear signature of saddle point excitons that is observed in the optical spectrum of graphene in the visible and ultraviolet wavelength regime. Due to the attractive interaction between electron and hole, these excitons form a discrete state at an energy slightly below the saddle point as indicated in Fig. 10.4. In the next section, we introduce a simple model to calculate the optical response of graphene including excitonic effects.

10.4 Interpretation with a Classical Fano Model

To explain the influence of electron–hole interactions on the optical spectra of semiconductors, Phillips drew on the classical theory developed by Fano

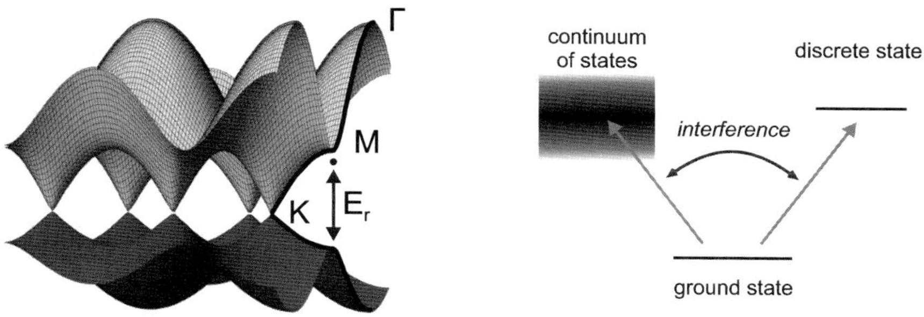

Fig. 10.4 Band structure of graphene with the discrete excitonic state (marked as a dot) near the saddle point M with resonance energy E_{r}. Note that this sketch merges one- and two-particle pictures for a more intuitive visualization. As schematically shown on the right, a Fano resonance can be interpreted as interference between transitions to a discrete state and a continuum of states.

[Phillips (1966); Fano (1965)]. In this picture, a resonance occurs when a discrete state couples to a continuum of states, giving rise to an asymmetric line shape. Fano interpreted this in a general approach by configuration interaction between two quantum mechanical channels, belonging to transitions to the discrete state and the continuum, respectively. In graphene, the discrete excitonic state near the M point in the band structure overlaps with the continuum of states descending from the saddle point down to the Dirac point, the so-called Dirac continuum, as shown in Fig. 10.4.

With the Fano model, the resulting absorption spectrum has the form

$$A_{\mathrm{Fano}}(E) = C\,\frac{(s+q)^2}{1+s^2} = C\left(1 + \frac{q^2-1}{1+s^2} + \frac{2qs}{1+s^2}\right), \tag{10.7}$$

with

$$s = \frac{E - E_{\mathrm{r}}}{\gamma/2}. \tag{10.8}$$

The Fano resonance can be interpreted as interference between the two channels, i.e., between transitions into the continuum and to the discrete state (see Fig. 10.4). Here, the discrete exciton state is modeled as a Lorentzian resonance of width γ and energy E_{r}. These quantities determine the parameter s. The unit-free Fano parameter q is given by the ratio of the transition probabilities to the discrete state and to the continuum. C is an overall scaling factor.

The three terms in the parentheses of Eq. 10.7 correspond to the spectrally constant absorption of the Dirac continuum, the discrete state, and the interference, respectively. The interference term leads to the asymmetric line shape, which can be ascribed to destructive interference between the channels. Alternatively, it can be understood in terms of the phase change by π of the Lorentzian in the vicinity of the resonance. For a negative Fano parameter q, the interference term is positive below the resonance ($s < 0$) and negative above ($s > 0$). Coming from low energies, this leads to the slow increase of the absorbance toward the resonance and the steep decrease beyond. In the limit of vanishing transitions into the continuum, the Fano parameter q becomes infinite and Eq. 10.7 reduces to a Lorentzian, as expected. Far away from the resonance E_r, the transitions to the continuum dominate the response.

As can be seen in Fig. 10.5, a Fano model using Eq. 10.7 excellently describes the measured data over the entire spectral range for both monolayer and bilayer graphene. The asymmetric peak in the ultraviolet is well reproduced in line shape as well as position and there is a very good agreement also down to low photon energies. Table 10.1 summarizes the resulting values of the fit parameters for the two measured samples as well as two other mono- and bilayer samples. The low-energy limit $A(0)$ of the absorption

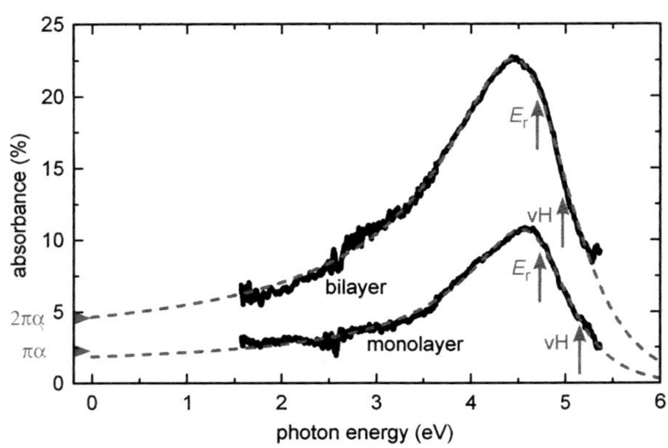

Fig. 10.5 The measured absorbance $A = 1 - T$ spectrum of free-standing mono- and bilayer graphene (solid lines) are well described by a Fano model (dashed lines). The difference between the resonance energy E_r of the exciton and the van Hove singularity (vH) determines the exciton binding energy. Adapted with permission from [Chae *et al.* (2011)]. Copyright (2011) American Chemical Society.

Table 10.1 Summary of the Fano fitting parameters and deduced values for different samples. The data sets marked with (*) are shown in Fig. 10.3 and Fig. 10.5. $A(0)$ is the absorbance at zero energy in units of $\pi\alpha$. E_b is the exciton binding energy calculated from the difference between E_r and the van Hove singularity E_{vH}. Reprinted with permission from [Chae *et al.* (2011)]. Copyright (2011) American Chemical Society.

data set	E_r (eV)	γ (eV)	q	C (%)	$A(0)/(\pi\alpha)$	E_b (meV)
monolayer*	4.73	1.30	−3.3	0.9	0.82	420
monolayer	4.78	1.58	−3.6	0.7	0.75	370
bilayer*	4.70	1.63	−3.2	2.0	2.0	270
bilayer	4.73	1.39	−3.3	1.8	1.7	240

obtained from the fit parameters agrees well with the universal value of $n\pi\alpha$ ($n = 1, 2$). The relevance of the other fit parameters is discussed in the next section in the context of the intrinsic properties of the excitonic state.

Note that no signature of the direct interband transition at the energy of the van Hove singularity appears in the measured and fitted spectra. This can be explained in terms of a redistribution of oscillator strength by the resonant excitons, in accordance with the theory presented above [Yang *et al.* (2009)].

Finally, we remark that the optical response of graphene is fully determined by the topology of the band structure, whereas its detailed shape seems to be less relevant. To illustrate this idea further, imagine the band dispersion of graphene as a rubber sheet which is pinned down at all points corresponding to the Dirac points in the band structure. If now the sheet is pulled up at the center of each Brillouin zone, automatically a saddle point as well as a continuum of states evolves. Following this topological reasoning, we suggest that the system can always be described with a single resonance, irrespective of the details of the band structure such as the slope of the Dirac cones. Hence, it is not surprising that we find an excellent agreement between our measured data and a Fano model fit, even when ignoring details of the band structure.

10.5 Properties of the Excitonic Resonance in Graphene

The fit parameters from the Fano interference model as listed in Table 10.1 reveal several properties of the excitonic resonance.

The binding energy of the exciton is calculated as the energy difference from the resonance to the van Hove singularity as $E_{\mathrm{vH}} - E_{\mathrm{r}}$ (see Section 10.3). We find values of about 400 meV for monolayer and 250 meV for bilayer graphene. This confirms the reasoning given in Section 10.3 that the binding energy is larger in monolayer graphene due to reduced screening at the Fermi energy. The obtained values of the exciton binding energy are surprisingly large, as compared to a few tens of meV in metallic carbon nanotubes [Wang *et al.* (2007)].

Whereas there is a systematic variation of the excitonic binding energy between monolayer and bilayer, the other fit parameters γ and q are seemingly only subject to sample-to-sample variation. The linewidth γ corresponds to a lifetime $\tau = \hbar/\gamma$ of the excitonic state. The large values of γ emerging from the Fano fit indicate very short exciton lifetimes on the order of 0.5 fs. Moreover, the linewidth exceeds the exciton binding energy by about a factor of three. Hence, the exciton is strongly modified due to the coherent superposition with the continuum, or, in other words, the exciton quickly decays into the continuum of states. Consequently, one speaks of an excitonic resonance rather than of truly bound excitons in graphene. In this context, a resonant state describes a bound state whose energy level overlaps with the band states of its host [Yu and Cardona (2005)].

The influence of doping on the excitonic resonance has been studied both experimentally and theoretically by Mak et al. [Mak *et al.* (2014)]. The measured and calculated optical conductivity spectra for different levels of

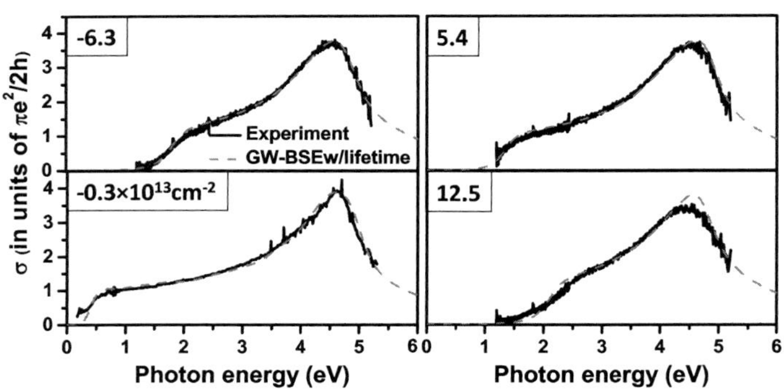

Fig. 10.6 The calculated conductivity spectra $\sigma(E)$ with quasiparticle lifetime effects included (dashed lines) are compared to the experimental spectra (solid lines) for four different doping densities, given by the carrier density n. Adapted with permission from [Mak *et al.* (2014)]. Copyright (2014) by the American Physical Society.

doping are shown in Fig. 10.6. In the experiments, a graphene field-effect transistor is prepared on a transparent quartz substrate and contacted with a transparent polymer electrolyte top gate. That way, high carrier concentrations on the order of $n \approx 10^{14}\text{cm}^{-2}$ of both holes and electrons can be obtained, corresponding to a Fermi energy of up to 1.2 eV. At low photon energies, a progressive suppression of the absorption is observed with increasing doping, as expected due to Pauli blocking of interband transitions [Wang *et al.* (2008)]. The absorption peak in the UV shows a redshift and becomes broader as well as more symmetric for both electron and hole doping, whereas the absolute peak height is not affected. All effects can be explained by the increased dielectric screening which leads to a modification of the Coulomb interactions. The enhanced screening weakens the electron–hole interactions, hence giving rise to a more symmetric absorption peak. This agrees with the fact that the symmetric line shape emerging from the band-to-band transitions should be recovered in the limit of vanishing excitonic effects. The redshift of the resonance is attributed to the repulsive electron–electron interaction which with increasing doping dominates over the attractive electron–hole interaction. The increased width of the peak corresponds to a decreased lifetime of the excitonic state. This is the result of an increased density of states at the Fermi energy due to the doping and hence an increased number of available decay channels.

Also strain has intriguing effects on the properties of graphene as it introduces profound modifications in the electronic band structure. Regarding the optical properties of graphene, mainly the effect of uniaxial strain has been investigated, i.e., when the graphene sheet is stretched along a certain direction. The symmetry breaking in the hexagonal lattice goes hand in hand with a symmetry breaking in the Brillouin zone as shown schematically in Fig. 10.7. Hence, the dependence of the optical spectrum will be strongly anisotropic. Liang et al. [Liang *et al.* (2012)] studied the influence of strain from a first-principles approach under consideration of electron–hole interactions. A summary of the optical spectra for the different cases is shown in Fig. 10.7. With increasing amount of strain, the absorption peak is predicted to split into two peaks. This is due to the lifting of the degeneracy between two neighboring saddle points M and M' in the band structure. The positions are related to the transition energies at the M and M' points, corrected by electron–hole interactions. The amplitudes of the two peaks as well as the exact peak positions depend on the polarization of the probing light and the orientation of the flake, respectively. Quali-

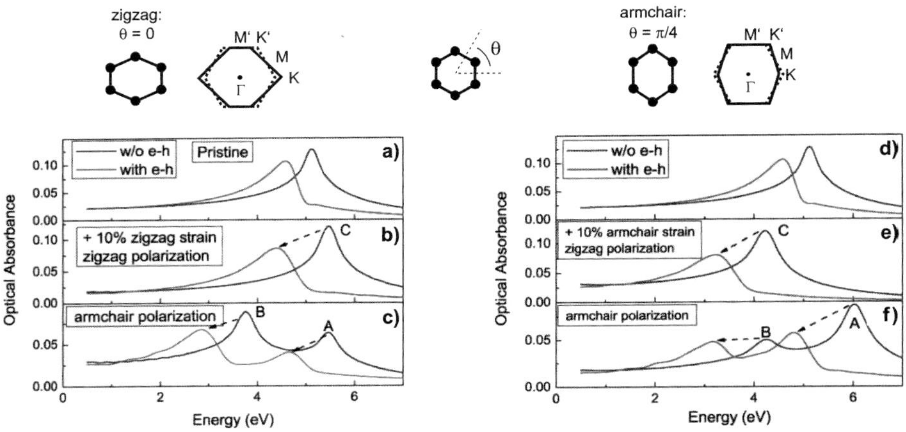

Fig. 10.7 Optical absorption spectra of pristine graphene (a,d) compared to graphene under zigzag (b,c) and armchair (e,f) strain for light polarization along the zigzag and armchair directions. The curves calculated including excitonic effects (light grey) are always redshifted with respect to those not taking many-body effects into account (dark grey). Adapted with permission from [Liang *et al.* (2012)] Copyright (2012) by Cambridge University Press.

tatively similar results have been obtained by other groups [Pereira *et al.* (2010); Pellegrino *et al.* (2010)], however, without taking many-body effects into account. To our knowledge, the effects described here have so far not been observed experimentally.

10.6 Excitons in Graphene-related Materials

In this section, we will discuss the occurrence of excitonic states in the optical response of graphene-related systems such as multi-layer or nano-structured graphene.

First, we address the dependence of the excitonic state on the number of graphene layers. As shown in Sections 10.4 and 10.5, bilayer graphene exhibits a slightly redshifted excitonic resonance, so that the binding energy of the exciton is reduced compared to monolayer graphene. With increasing layer number $N = 1 - 5$, interlayer interactions modify the band structure, also near the saddle point [Grüneis *et al.* (2008); Klintenberg *et al.* (2009)]. Due to the symmetry of the bands, however, only optical transitions with energies comparable to that of the monolayer are allowed. On the other hand, screening increases with the thickness of the sample due to the increasing density of states at the Fermi energy [Castro Neto *et al.* (2009)]. The peak position is a result of the competing contributions of attractive electron–hole and repulsive electron–electron interaction, as discussed above. While the attractive interaction dominates for monolayer graphene, both contributions are increasingly screened and cancel for $N > 3$ [Yang *et al.* (2009)]. This is underlined by experiments showing that the optical conductivity increases almost linearly with N and that the peak position exhibits only a negligible redshift [Mak *et al.* (2011)] which quickly converges towards the value of bulk graphite [Taft and Philipp (1965); Yang *et al.* (2009)].

So far, our discussions have focused on graphene sheets with the most common stacking, so-called Bernal stacking, where adjacent layers are shifted by one atomic spacing. When the layers are twisted compared to the Bernal stacking configuration, interlayer interactions perturb the band structure and additional van Hove singularities emerge, as has been observed for example in bilayer graphene with scanning tunneling spectroscopy [Li *et al.* (2009)]. These features are very interesting as they can be tuned via the rotation angle. The rotation-induced van Hove singularities give rise to additional peaks in the optical spectra of twisted bilayer graphene. In addition to the redshift of the peaks due to excitonic effects, coherent interactions between multiple van Hove singularities seem to cause a redistribution of oscillator strength [Havener *et al.* (2014)].

When the dimensionality of graphene is further reduced by structuring it for example into nanoribbons, excitonic effects will dominate the spectrum as a consequence of enhanced Coulomb interactions. However, the band structure is completely altered compared to graphene so that the excitons observed in graphene nanoribbons are of different nature. They are not related to a saddle point but to van Hove singularities at the band gap or represent so-called dark edge excitons [Yang *et al.* (2008); Prezzi *et al.* (2008)]. The emerging excitonic peaks in the optical spectrum depend on the structure and width of the ribbons and are in the visible energy range for few nanometer wide ribbons [Denk *et al.* (2014)]. The behavior is very similar to that of one dimensional carbon nanotubes [Spataru *et al.* (2004); Wang *et al.* (2005); Dresselhaus *et al.* (2007)]. This does not come as a surprise, since graphene nanoribbons essentially are unzipped carbon nanotubes.

Besides graphene, a lot of attention has been paid recently to other two dimensional crystals that exhibit a hexagonal arrangement of atoms, the so-called graphene analogues. Extensively studied systems are monolayers of hexagonal boron nitride as well as transition metal dichalcogenides (XY_2, with X=Mo,W and Y=S,Se,Te). These systems offer an interesting alternative to graphene especially regarding optoelectronic applications as they possess sizable band gaps of a few hundred meV up to a few eV, hence allowing to build switchable devices. Excitonic states have also been observed in the optical response of these materials in the visible and ultraviolet wavelength regime [Watanabe *et al.* (2004); Arnaud *et al.* (2006); Wang *et al.* (2012); Mak *et al.* (2013)]. As for nanoribbons, these excitonic states are related to the band gaps at the K point of the band structure.

10.7 Kramers Kronig Fit of the Optical Constants from the Fano Model

In the last section of this chapter, we would like to address an additional benefit of the Fano model. In combination with the Kramers-Kronig relations, the optical constants can be recovered from measured optical spectra. This represents an interesting alternative to ellipsometric measurements that are often used to determine optical constants.

In general, the complex refractive index and the complex dielectric function are linked via

$$\widetilde{n} = n + i\kappa = \sqrt{\widetilde{\epsilon}} = \sqrt{\epsilon_1 + i\epsilon_2}. \tag{10.9}$$

The real part n of the complex refractive index describes the dispersion. The imaginary part κ is called the extinction coefficient and captures the absorption of the material.

Kramers-Kronig relations describe the fundamental connection between the real and imaginary parts of a system's response to a sinusoidal perturbation. An important application in optical spectroscopy is the calculation of the real refractive index from the extinction coefficient typically measured in transmission [Wooten (1972)]. When both quantities are measured, the Kramers-Kronig relations are used to check the consistency of the data.

In the formulation by Kramers [Kramers (1927)], the real and imaginary part of the dielectric function are connected by integral relationships as follows:

$$\epsilon_1(\omega) - 1 = \frac{2}{\pi}\mathcal{P}\int_0^\infty \frac{\epsilon_2(\omega')\omega'}{\omega'^2 - \omega^2}d\omega' \tag{10.10}$$

$$\epsilon_2(\omega) = -\frac{2\omega}{\pi}\mathcal{P}\int_0^\infty \frac{\epsilon_1(\omega') - 1}{\omega'^2 - \omega^2}d\omega' + \frac{4\pi\sigma_{DC}}{\omega}, \tag{10.11}$$

where $\mathcal{P}$ denotes the Cauchy principal value. Note that the DC conductivity, σ_{DC}, vanishes for many materials, so that the last term in Eqn. 10.11 is often neglected. To do this calculation correctly, in principle a measurement over the whole frequency range from zero up to infinity is required. Unavoidably, the finite range of any measurement leads to inaccuracies. Several methods have been suggested to improve the accuracy of the transformation and its convergence [Lucarini *et al.* (2005)]. However, even a finite spectral range leads to good results when the main features of the function are captured and proper extrapolations are made. An important condition is that the considered functions decay faster than $1/\omega$ so that $\epsilon_1 \to 1$ for $\omega \to \infty$. This behavior corresponds to a driven harmonic oscillator that at large frequencies cannot follow the excitation anymore.

The accuracy of the optical constants derived with the Kramers-Kronig relation is usually comparable to those derived from ellipsometric measurements which critically depend on the choice of the physical model. The Kramers-Kronig relations are therefore extensively used in many areas of spectroscopy, ranging from photoemission and electron energy loss spectroscopy up to high energy particle physics and acoustics.

In our approach, the optical constants of graphene are calculated from the Fano fit to the measured absorbance spectrum shown in Sec. 10.4. That way, the spectral range is extended from the measured range of $1.5-5.5$ eV to the range $0.5-12$ eV in which the Fano is considered to be valid, hence improving the accuracy of the deduced optical constants. For the calculation, the infinite integral in Eqn. 10.10 is transformed into a sum over a finite number of N energy increments:

$$\epsilon_1(E) - 1 = \frac{2}{\pi}\Delta E \sum_{k=1}^{N} \frac{\epsilon_2(k)E(k)}{E(k)^2 - E(j)^2}, \tag{10.12}$$

where the summand for $k = j$ is zero to exclude the pole from the integration. The absorption (or transmission/reflection) curve resulting from a Kramers-Kronig consistent set of optical constants ϵ_1, ϵ_2 is calculated with a transfer matrix method[Born and Wolf (1999)]. Based on the Fresnel

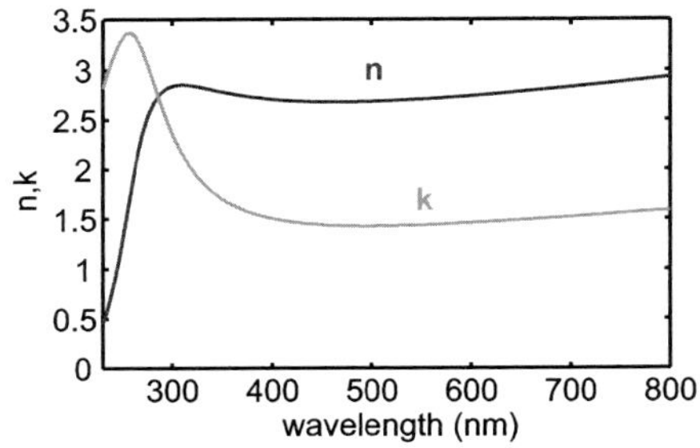

Fig. 10.8 The Kramers-Kronig consistent optical constants n and κ of graphene derived from the Fano model in the visible and UV wavelength regime.

coefficients, this method describes the propagation of light through a multilayer system as a function of the thicknesses and the optical constants of the individual layers. It has been used already in the early days of graphene to study its visibility on different substrates [Blake *et al.* (2007)]. In our fitting routine, the optical constants are varied until the absorption spectrum calculated with the transfer matrix method fits the measured curve best. The optical constants derived from the Fano model are shown in Fig. 10.8. Within the measured wavelength regime, there is a good qualitative agreement with the optical constants of graphene [Kravets *et al.* (2010); Weber *et al.* (2010)] as well as of graphite [Taft and Philipp (1965); Greenaway *et al.* (1969)].

Having the optical constants at hand, the response of graphene on an arbitrary substrate can be calculated with the transfer matrix formalism

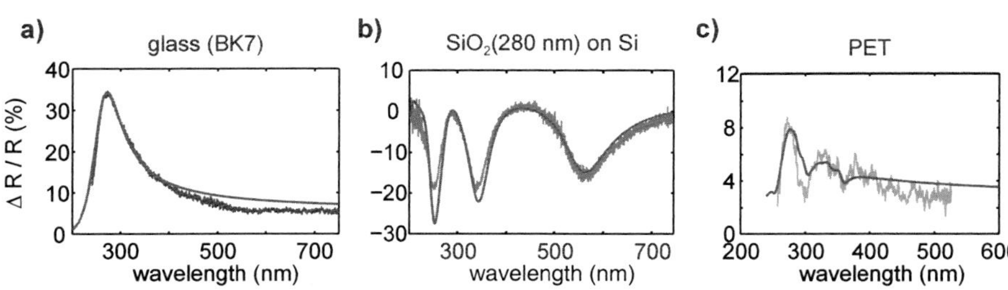

Fig. 10.9 The reflectivity of graphene on different substrates: a) standard cover glass (BK7), b) oxidized silicon (SiO_2 thickness 280 nm), and c) PET. The measured spectra are compared to the response calculated with a transfer matrix method from the optical constants derived above and a graphene thickness of 0.34 nm. The response on PET is weakly damped, as PET itself also shows absorption with $\kappa \neq 0$.

[Blake *et al.* (2007)]. In this case, usually reflectivity contrast measurements are performed as described in Sec. 10.2. The calculations are a simple method to predict the value and spectral dependence of the optical response. That way, they enable an estimation of the visibility of the graphene flakes on the substrate. To support this claim, Fig. 10.9 compares the measured reflectivity contrast of graphene on three different substrates with simulated curves. On standard cover glass, the visibility peaks in the UV but is still at about 10% in the visible regime. On the commonly used oxidized silicon substrate, interference fringes due to the SiO_2 thin film are observed. Tuning the thickness of the SiO_2 layer, the visibility at different wavelengths can be optimized with the help of the calculations [Blake *et al.* (2007)]. PET, a transparent plastic, serves as an example for a substrate with non-negligible absorption. In this case, the reflectivity contrast is strongly reduced.

10.8 Summary

In this chapter, the linear optical properties of graphene near the saddle point have been described. We have shown that electron–hole interactions define the response in this regime, leading to an asymmetric lineshape and a redshift of the peak in the UV. The coupling between the excitonic state near the saddle point and the continuum of states down to the Dirac point can be modeled with a simple Fano model. It reproduces very well the observed features in the spectrum and even the constant value of the absorption in the low-energy limit. From the model, the binding energy of the excitonic state is estimated to be on the order of a few hundred meV. The short lifetime below 1 fs indicates a quick decay of the exciton into the continuum of states.

Besides the intrinsic properties, we presented an overview of possible influences on the excitonic resonance in graphene and related systems. When the carrier density is varied by doping, the Coulomb interactions are modified due to increased screening, leading to a redshift of the resonance and a decreased lifetime of the excitonic state. Also strain is predicted to strongly influence the spectrum of graphene near the saddle point. As it breaks the symmetry of the lattice, the degeneracy between neighboring saddle points is lifted, which should lead to a splitting of the peak in the UV and an anisotropic response. When more than one graphene layer is considered, the excitonic effects are weakened with increasing layer number and the response quickly converges towards that of bulk graphite. Twisted bilayer graphene constitutes an interesting system, as additional van Hove singularities emerge which can be tuned via the rotation angle. Excitonic effects are also found in nanostructured graphene and other low-dimensional materials. However, the response in these systems is not comparable to graphene since it is related to excitonic effects at the band gap.

Finally, we have shown that the optical constants of graphene can be derived from a measured spectrum using the Kramers-Kronig relations. This allows to calculate the optical spectrum of graphene in arbitrary multilayer systems with the transfer matrix formalism.

Bibliography

[1] Arnaud, B., Lebègue, S., Rabiller, P., and Alouani, M. (2006). Huge Excitonic Effects in Layered Hexagonal Boron Nitride, *Physical Review Letters* **96**, p. 026402.

[2] Ashcroft, N. and Mermin, N. (1976). *Solid State Physics* (Thomas Learning, Toronto).

[3] Blake, P., Hill, E., Castro Neto, A., Novoselov, K., Jiang, D., Yang, R., Booth, T., and Geim, A. (2007). Making graphene visible, *Applied Physics Letters* **91**, p. 063124.

[4] Born, M. and Wolf, E. (1999). *Principles of Optics*, 7th edn. (Pergamon Press).

[5] Castro Neto, A., Guinea, F., Peres, N., Novoselov, K., and Geim, A. (2009). The electronic properties of graphene, *Review of Modern Physics* **81**, pp. 109–162.

[6] Chae, D.-H., Utikal, T., Weisenburger, S., Giessen, H., von Klitzing, K., Lippitz, M., and Smet, J. (2011). Excitonic Fano Resonance in Free-Standing Graphene, *Nano Letters* **11**, pp. 1379–1382.

[7] Denk, R., Hohage, M., Zeppenfeld, P., Cai, J., Pignedoli, C., Söde, H., Fasel, R., Feng, X., Müllen, K., Wang, S., Prezzi, D., Ferretti, A., Ruini, A., Molinari, E., and Ruffieux, P. (2014). Exciton-dominated optical response of ultra-narrow graphene nanoribbons, *Nature Communications* **5**, p. 4253.

[8] Dresselhaus, M., Dresselhaus, G., Saito, R., and Jorio, A. (2007). Exciton Photophysics of Carbon Nanotubes, *Annual Reviews of Physical Chemistry* **58**, pp. 719–747.

[9] Fano, U. (1965). Effects of Configuration Interaction on Intensities and Phase Shifts, *Physical Review* **124**, 6, pp. 1866–1878.

[10] Fox, M. A. (2001). *Optical Properties of Solids* (Oxford University Press).

[11] Greenaway, D., Harbeke, G., Bassani, F., and Tosatti, E. (1969). Anisotropy of the Optical Constants and the Band Structure of Graphite, *Physical Review* **178**, 3, pp. 1340–1348.

[12] Grüneis, A., Attaccalite, C., Wirtz, L., Shiozawa, H., Saito, R., Pichler, T., and Rubio, A. (2008). Tight-binding description of the quasiparticle

dispersion of graphite and few-layer graphene, *Physical Review B* **78**, p. 205425.

[13] Havener, R., Liang, Y., Brown, L., Yang, L., and Park, J. (2014). Van Hove Singularities and Excitonic Effects in the Optical Conductivity of Twisted Bilayer Graphene, *Nano Letters* **14**, pp. 3353–3357.

[14] Hecht, E. (1998). *Optics* (Addison-Wesley, Reading, MA).

[15] Kane, E. (1969). Coulomb Effects at Saddle-Type Critical Points, *Physical Review* **180**, pp. 852–858.

[16] Klintenberg, M., Lebègue, S., Sanyal, B., Fransson, J., and Eriksson, O. (2009). Evolving properties of two-dimensional materials: from graphene to graphite, *Journal of Physics: Condensed Matter* **21**, p. 335502.

[17] Kramers, H. (1927). La diffusion de la lumière par les atomes, in *Atti del Congresso Internazionale dei Fisici* (Zanichelli, Bologna), pp. 545–557.

[18] Kravets, V., Grigorenko, A. N., Nair, R. R., Blake, P., Anissimova, S., Novoselov, K. S., and Geim, A. K. (2010). Spectroscopic ellipsometry of graphene and an exciton-shifted van Hove peak in absorption, *Physical Review B* **81**, p. 155413.

[19] Kuzmenko, A., van Heumen, E., Carbone, F., and van der Marel, D. (2008). Universal Optical Conductance of Graphite, *Physical Review Letters* **100**, p. 117401.

[20] Li, G., Luican, A., Lopes dos Santos, J., Castro Neto, A., Reina, A., Kong, J., and Andrei, E. (2009). Observation of Van Hove singularities in twisted graphene layers, *Nature Physics* **6**, pp. 109–113.

[21] Liang, Y., Huang, S., and Yang, L. (2012). Many-electron effects on optical absorption spectra of strained graphene, *Journal of Materials Research* **27**, pp. 403–409.

[22] Lucarini, V., Saarinen, J., Peiponen, K.-E., and Vartiainen, E. (2005). *Kramers-Kronig Relations in Optical Materials Research* (Springer).

[23] Mak, K., da Jornada, F., He, K., Deslippe, J., Petrone, N., Hone, J., Shan, J., Louie, S., and Heinz, T. (2014). Tuning Many-Body Interactions in Graphene: The Effects of Doping on Excitons and Carrier Lifetimes, *Physical Review Letters* **112**, p. 207401.

[24] Mak, K., He, K., Lee, C., Lee, G., Hone, J., Heinz, T., and Shan, J. (2013). Tightly bound trions in monolayer MoS_2, *Nature Materials* **12**, pp. 207–211.

[25] Mak, K., Shan, J., and Heinz, T. F. (2011). Seeing Many-Body Effects in Single- and Few-Layer Graphene: Observation of Two-Dimensional Saddle-Point Excitons, *Physical Review Letters* **106**, p. 046401.

[26] Nair, R., Blake, P., Grigorenko, A., Novoselov, K., Booth, T., Stauber, T., Peres, N., and Geim, A. (2008). Fine Structure Constant Defines Visual Transparency of Graphene, *Science* **320**, p. 1308.

[27] Pellegrino, F., Angilella, G., and Pucci, R. (2010). Strain effect on the optical conductivity of graphene, *Physical Review B* **81**, p. 035411.

[28] Pereira, V., Ribeiro, R., Peres, N., and Castro Neto, A. (2010). Optical Properties of Strained Graphene, *Europhysics Letters* **92**, p. 67001.

[29] Pétroff, Y. and Balkanski, M. (1971). Coulomb Effects at Saddle-Type Critical Points in CdTe, ZnTe, ZnSe, and HgTe, *Physical Review B* **3**, 10, pp. 3299–3302.

[30] Phillips, J. (1966). Excitons, in J. Tauc (ed.), *Optical properties of Solids* (Academic Press, London).

[31] Prezzi, D., Varsano, D., Ruini, A., Marini, A., and Molinari, E. (2008). Optical properties of graphene nanoribbons: The role of many-body effects, *Physical Review B* **77**, p. 041404(R).

[32] Spataru, C., Ismail-Beigi, S., Benedict, L., and Louie, S. (2004). Excitonic effects and optical spectra of single-walled carbon nanotubes, *Physical Review Letters* **92**, p. 077402.

[33] Taft, E. and Philipp, H. (1965). Optical Properties of Graphite, *Physical Review* **138**, 1A, pp. 197–202.

[34] van Hove, L. (1953). The Occurence of Singularities in the Elastic Frequency Distribution of a Crystal, *Physical Review* **56**, 6, pp. 1189–1193.

[35] Wang, F., Cho, D., Kessler, B., Deslippe, J., Schuck, P., Louie, S., Zettl, A., Heinz, T., and Shen, Y. (2007). Observation of Excitons in One-Dimensional Metallic Single-Walled Carbon Nanotubes, *Physical Review Letters* **99**, p. 227401.

[36] Wang, F., Dukovic, G., Brus, L., and Heinz, T. (2005). The optical resonances in carbon nanotubes arise from excitons, *Science* **308**, pp. 838–841.

[37] Wang, F., Zhang, Y., Tian, C., Girit, C., Zettl, A., Crommie, M., and Shen, Y. (2008). Gate-Variable Optical Transitions in Graphene, *Science* **320**, pp. 206–209.

[38] Wang, Q., Kalantar-Zadeh, K., Kis, A., Coleman, J., and Strano, M. (2012). Electronics and optoelectronics of two-dimensional transition metal dichalcogenides, *Nature Nanotechnology* **7**, pp. 699–712.

[39] Watanabe, K., Taniguchi, T., and Kanda, H. (2004). Direct-bandgap properties and evidence for ultraviolet lasing of hexagonal boron nitride single crystal, *Nature Materials* **3**, pp. 404–409.

[40] Weber, J., Calado, V., and van de Sanden, M. (2010). Optical constants of graphene measured by spectroscopic ellipsometry, *Applied Physics Letters* **97**, p. 091904.

[41] Wooten, F. (1972). *Optical Properties of Solids* (Academic Press).

[42] Yang, L., Cohen, M., and Louie, S. (2008). Excitonic Effects in the Optical Spectra of Graphene Nanoribbons, *Nano Letters* **7**, 10, pp. 3312–3115.

[43] Yang, L., Deslippe, J., Park, C.-H., Cohen, M. L., and Louie, S. G. (2009). Excitonic Effects on the Optical Response of Graphene and Bilayer Graphene, *Physical Review Letters* **103**, p. 186802.

[44] Yu, P. Y. and Cardona, M. (2005). *Fundamentals of Semiconductors*, 3rd edn. (Springer).

Chapter 11

Nonlinear Saddle Point Spectroscopy and Electron-Phonon Interaction in Graphene

Rolf Binder[1,*], Adam T. Roberts[1,2], Nai-Hang Kwong[1], Arvinder Sandhu[3] and Henry O. Everitt[2,4]

[1] *College of Optical Sciences, University of Arizona, Tucson, Arizona 85721, USA*

[2] *U.S. Army Aviation and Missile Research, Development, and Engineering Center, Redstone Arsenal, Alabama 35898, USA*

[3] *Department of Physics, University of Arizona, Tucson, Arizona 85721, USA*

[4] *Department of Physics, Duke University, Durham, North Carolina 27708, USA*

[*] *binder@optics.arizona.edu*

This chapter focuses on ultrafast nonlinear spectroscopy at the ultraviolet M- or saddle-point in the electronic bandstructure of graphene where the position and dynamical evolution of its absorption peak are especially sensitive to electron-phonon interactions. Specifically, we explore how these absorption peak changes are caused by optically-induced modifications of the phonon temperature by way of several electron-phonon scattering processes. We present a detailed theoretical model for electron-phonon interactions based on the concept of deformation potentials. We also include a discussion of the phonon dispersion obtained from dynamical matrices. We derive the electronic self-energy to lowest order in the electron-phonon interaction Hamiltonian, then use it to calculate the interband susceptibility and the differential transmission spectrum. Using literature values for deformation potentials, we find good agreement between theory and experiment, indicating that this formalism provides a good understanding of the microscopic electron-phonon coupling processes that renormalize the electronic transitions close to the M-point and produce the observed differential transmission spectra.

11.1 Introduction to Nonlinear Saddle-point Spectroscopy

One of the important benefits of using optical spectroscopy to study the intrinsic physics of graphene is that optical experiments can relatively easily be applied to any spectral region and hence to any optical transition in the graphene bandstructure. Therefore, optical studies are not restricted to transitions close to the Dirac point (or K-point), but they also include higher frequencies such as those close to the saddle point (M-point). Those frequencies are in the near ultraviolet (UV) and therefore well within the range of existing optical spectroscopic techniques.

The density of states (DOS) approaches infinity at the M-point, and this leads to a peak structure for optical transitions between valence and conduction band states near that point. In reality, an infinite DOS does not result in infinite absorption. The absorption line shape, shown schematically in Fig. 11.1, is determined by electron-scattering or dephasing processes that involve electron-phonon and possibly electron-impurity scattering, and also by the excitonic electron-hole Coulomb interaction. Since the bands look like a saddle, excitonic effects are quite different from those in conventional III-V semiconductors such as GaAs, where the bands have local minima separated by an energy gap E_g. In GaAs for example, the joint density of states is zero for energies below E_g and non-zero above E_g, so without exciton effects there is no maximum in the density of states. (In an ideal 2-dimensional situation, it is just a step function, as discussed in the Introductory Tutorial, chapter 1.) In those systems, no discrete exciton resonances appear in the spectrum unless one takes into account Coulomb interactions binding the electron to the hole. With excitonic effects included, more or less spectrally narrow resonances appear. Exciton resonances are especially useful for the study of processes that result from external modifications. The peaks can shift in frequency, they can broaden, they can get smaller (saturate), or they can generally change their shape if the crystal is subject external modifications. Such modifications include strong light fields that probe the system's nonlinear optical response, temperature changes, mechanical deformations inducing strain in the system, or applied magnetic fields. Quite generally, the narrower the excitonic resonances are, the easier it is to detect those external modifications, as even a small shift can induce large changes in the optical absorption or transmission spectrum. On the other hand, without any discrete resonances, it is more difficult to detect modifications. In graphene, we have such a structureless absorption spectrum at infrared and visible frequencies, and we do not have sharp excitonic resonances in the visible. But the saddle point absorption does at least give a peak in the absorption, albeit a very broad peak with a width of order eV, much larger than those of excitonic peaks in GaAs quantum wells that, at low temperature, are typically on the order of meV. Nevertheless, the fact that we have an absorption peak in the ultraviolet frequency regime near 4.5eV means that we can look for the same signatures – shifts, broadening, and saturation of the peak position – as we usually do with excitonic peaks in semiconductors like GaAs. This is schematically shown in Fig. 11.1. The relatively sensitive detection of saddle point peak modifications can then be used to draw conclusions about the microscopic processes that give rise to those modifications. It can also help establish values for basic parameters entering the theory, such as the coupling strength in the electron-phonon interaction which is often parameterized by deformation potentials. In particular, values for so-called effective deformation potentials describing the coupling of electrons with acoustic phonons, which is important not only in an optical context, but also in the context of electric transport, vary substantially. The experimental estimates obtained from electrical transport measurements [1–3] range from $16 - 50$ eV, while theoretical predictions indicate acoustic deformation potentials in the range of $\sim 2.8 - 7$ eV [4–7]. Given that many observables are proportional to the square of the effective deformation potential, this is a substantial uncertainty.

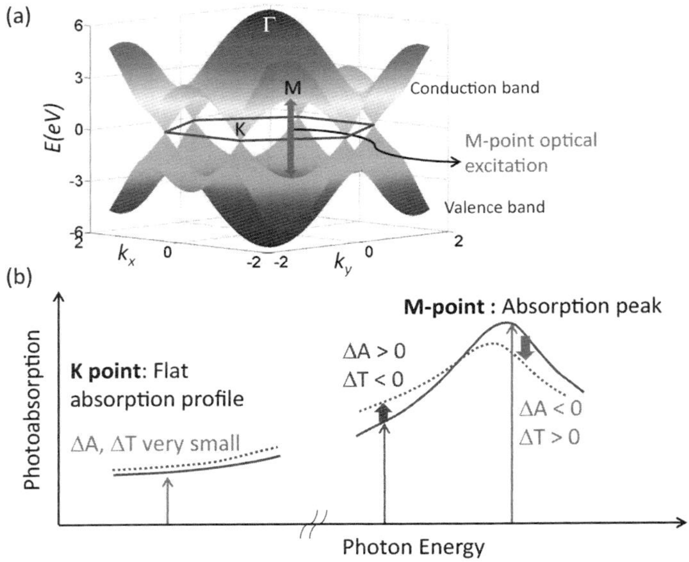

Fig. 11.1 (a) Electronic bandstructure of graphene. (b) Schematic variation of absorption with photon energy (solid curve). The M-point absorption peak corresponds to the saddle point in the bandstructure. We measure the pump-induced change in probe absorption (transmission) ΔA ($\Delta T = 1 - \Delta A$) which corresponds to the modification of the absorption profile (dashed curve) due to electron-phonon renormalization. Around the M-point, the energy-resolved differential transmission has a large magnitude and exhibits a sign change, enabling easy identification of many-body effects. From Ref. [8].

In the following, we present results of pump and probe experiments using a variety of frequencies and delay times. Generally, the pump pulse is applied first, followed by a weak probe pulse, Fig. 11.2. We distinguish two kinds of pump–probe experiments: degenerate pump-probe, where both pulses have the same center-frequency, and non-degenerate, which allows us to scan a certain frequency range of the probe frequency while leaving the pump frequency fixed. Depending on the frequency and on the delay time, we can see a large variety of nonlinear optical phenomena and study the underlying microscopic processes. In this chapter, we will mainly focus on electron-phonon interactions observed through non-degenerate pump-probe experiments with delay times ranging from several picoseconds to several hundred picoseconds. First, however, we will describe our sample and present results from degenerate pump-probe experiments that cover very different pump frequencies and a large interval of pump-probe delay times.

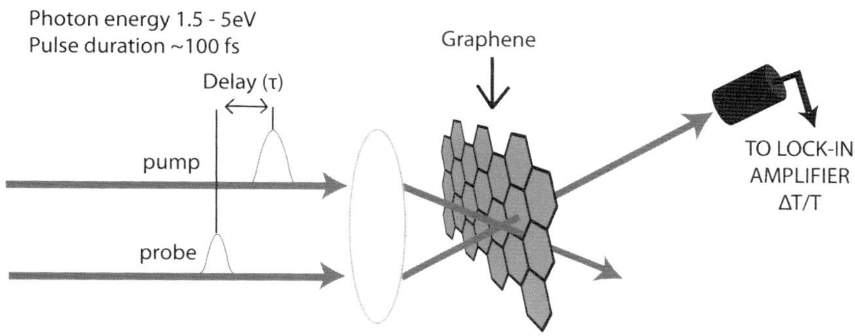

Fig. 11.2 Sketch of pump-probe experiment.

The graphene used in our study was grown through chemical vapor deposition [9] on a copper foil. When the growth was complete, the sample was coated with poly(methyl methacrylate), PMMA, and then the copper was dissolved away so that we were left with a film of PMMA with graphene on one side. This floats on top of water, and the PMMA keeps the graphene from crumpling. To deposit the graphene onto the 200 μm thick sapphire substrate, we lifted the graphene/PMMA out of a water bath with the substrate. After drying and baking, the PMMA was dissolved with acetone, leaving just the graphene on the substrate. We repeated the process for each of the layers.

In order to enhance the signal level in our measurements, up to ten individually grown graphene layers were stacked on top of each other. Raman spectroscopy, reviewed in chapter 3 of this book, verifies that monolayer behavior remains. In Fig. 11.3 we show a Lorentzian fit to the so-called 2D Raman line, and such a fit has been shown to indicate single-layer behavior [10]. Another indication is the observed ratio of the G to 2D peak Raman intensities and areas, which is about 1:4.8 [11]. Prior linear absorption measurements [12] also indicated that there is minimal interlayer coupling in CVD graphene stacks compared to exfoliated samples or multilayer films on SiC. Additionally, we used Raman spectra to determine the doping of our sample, which shows that our sample is slightly hole doped with the Fermi level at approximately -100 meV.

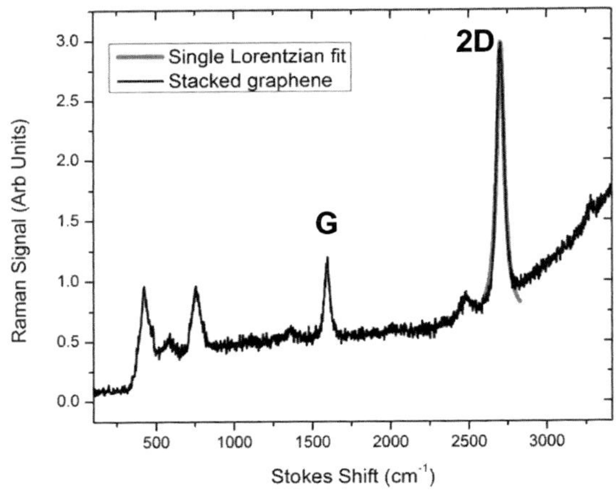

Fig. 11.3 Raman spectrum of a 10 layer sample. A Lorentzian fit of the 2D line is hardly recognizable since it traces the experimental data almost perfectly.

In our pump-probe experiments, we used a 1kHz Ti:Sapphire laser amplifier and an optical parametric amplifier. The pump and probe pulses, with an approximate duration of 100fs, were focused slightly non-collinearly on the sample. The pump beam was chopped, and lock-in amplification was used to detect changes in the transmission of the time delayed probe over a wide range of photon energies on either side of the 4.6 eV M-point resonance.

We first look at time-resolved, degenerate ($\hbar\omega_{pu} = \hbar\omega_{pr}$) pump-probe data for differential transmission, measuring the relative change in transmittance caused by the pump, $\Delta T/T$. Figure 11.4 shows degenerate differential transmission signals for three different frequencies, red (1.6eV), blue (3.2eV), and UV (4.8eV). Figures 11.4(a) and (b) show a positive transient lasting a few hundred femtoseconds for UV and blue photons, respectively. This can be attributed to Pauli blocking of the probe absorption by pump-induced electrons (holes) in the conduction (valence) band [13]. The carriers relax through carrier-carrier and carrier-phonon scattering mechanisms on a picosecond time scale [13–21]. After 2 ps, the plots in Figures 11.4(a) and (b) exhibit a slow relaxation to the zero baseline. We note that the sign of $\Delta T/T$ at long timescales (> 2 ps) is opposite in these two cases. For the 4.8 eV case, we observe a strong positive differential signal, $\Delta T/T > 0$, while, for 3.2 eV case, it is negative. The negative signal indicates the generation of additional absorption channels, which cannot be explained in the independent particle picture. Recently, increasing emphasis has been put on the understanding of differential optical transmission spectra that are negative, i.e. on situations where optical pumping leads to an increase of absorption [8,22–25]. Such negative differential transmission is usually related to complex many-body effects, in contrast to optically induced phase-space filling that is conceptually more simple and leads to positive differential transmission. We note that, in Fig. 11.4, the $\Delta T/T$ signal lasts for the same duration, approximately 300ps in the cases of frequencies at 3.2 and 4.8eV. This suggests a common origin despite opposite signs. Since 300ps is much longer than carrier relaxation and recombination times, the signals must be dominated by optically-induced electron-phonon interactions.

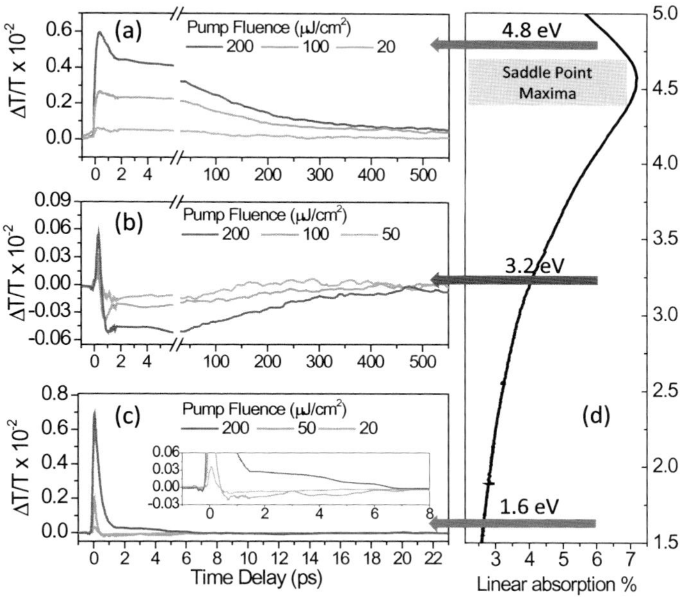

Fig. 11.4 $\Delta T/T$ of the probe pulse at pump-probe energies of (a) 4.8 eV (b) 3.2 eV and (c) 1.6 eV. In each case different curves correspond to different pump fluence levels. (d) The linear absorption profile indicating the probing energies relative to the saddle point absorption maxima. From. Ref. [8]. (For color reference, go to page 494.)

It is instructive to compare the signal obtained with 3.2 and 4.8 eV pumping with the more frequently studied case of $\hbar\omega_{pu} = \hbar\omega_{pr} = 1.6$ eV. As can be seen from Fig. 11.4(c), our 1.6eV data exhibits a positive transient followed by a decay within 1ps and a very small differential signal beyond a time delay of $2ps$. These observations are consistent with

Pauli-blocking and carrier relaxation mechanisms discussed extensively in the literature [13, 16–18] and indicate that substantial phonon excitation and carrier recombination occurs within the first 2 ps. The transient signals beyond 2ps are thus expected to be dominated by phonon-induced renormalization. As we can see from the zoomed-in view shown in the inset of Fig. 11.4(c), the positive and negative transients in the 1.6eV data are rather weak beyond 2ps. Hence, the M-point spectroscopy may indeed give more sensitive signals under certain excitation conditions.

We can summarize the basic physical processes in our experiment as follows: The optical pump pulse creates charge carriers: electrons in the conduction band and holes in the valence band. The carriers relax through the excitation of phonons [14]. After the pump-produced carriers have substantially relaxed through phonon emission, the interaction of probe-excited electrons with the phonon population leads to a renormalization-induced change of the bandstructure that modifies the probe absorption spectrum (Fig. 11.5). Depending on the photon energy, the probe can thus experience a decrease or increase in the absorption, and the change in the sign of differential transmission provides unambiguous evidence of band renormalization due to electron-phonon interactions.

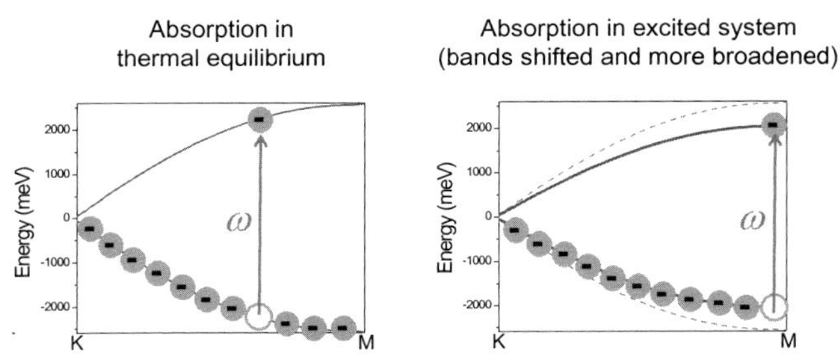

Fig. 11.5 Sketch of bandstructure renormalization and corresponding change of absorption channels.

In the following, we concentrate on the regime of long pump-probe delay times (≥4ps). As already mentioned, in that regime the differential transmission signal must be dominated by long-lived modifications to the phonons and to the electron-phonon interaction. In order to prove this, we have used non-degenerate pump-probe spectroscopy that allows us to obtain frequency resolved differential spectra in the vicinity of the absorption maximum at 4.6eV as a function of pump-probe delay time. Before presenting those experimental data, we will first develop a theory that predicts the expected change in the absorption due to an increase in the phonon temperature. As a prerequisite for this we first need to set up a theory that describes electron phonon coupling in graphene. This will be done in the next section, followed by a discussion of the absorption and transmission changes.

11.2 Electron-phonon Interaction

11.2.1 *Many-electron tight binding model*

We have seen in the Introductory Tutorial, chapter 1), that the electronic bandstructure of graphene can be described within a tight-binding model. In the present chapter, we use the tight-binding model to describe the many-electron system, but in contrast to the Introductory Tutorial, where we used a Bloch representation, we now use a representation that is particularly intuitive and similar to many conventional tight-binding theories, and also easy to motivate.

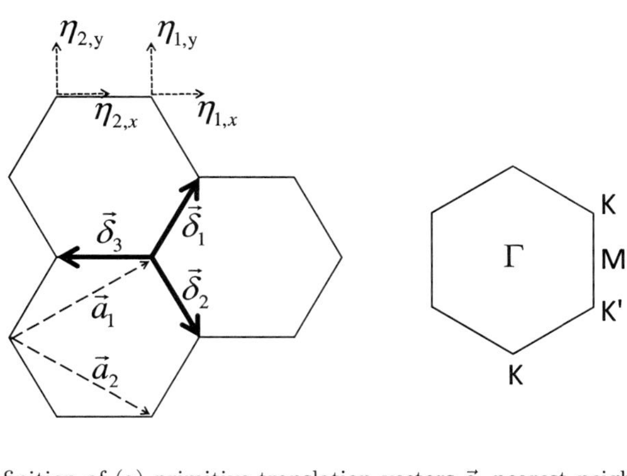

Fig. 11.6 Definition of (a) primitive translation vectors $\vec{a}$, nearest neighbor vectors $\vec{\delta}$, phonon mode vectors $\vec{\eta}$ defined through Eq. (11.12), and (b) Brillouin zone used in this chapter. From Ref. [26].

Our starting point is a phenomenological Hamiltonian in the basis of localized electrons. The parameters that enter this Hamiltonian are the on-site energies and nearest-neighbor hopping terms:

$$H = \sum_{jn\sigma} \varepsilon_{2p} c^{\dagger}_{jn\sigma} c_{jn\sigma} - \sum_{\langle jn, j'n' \rangle \sigma} t c^{\dagger}_{jn\sigma} c_{j'n'\sigma}, \tag{11.1}$$

where ε_p is the on-site energy of the 2p electrons in the graphene lattice (not the 2p energy of the individual carbon atoms), t is the hopping (or transfer) integral, and $\langle jn, j'n' \rangle$ denotes summation over nearest neighbor pairs. The creation (annihilation) operators $c^{\dagger}_{jn\sigma}$ ($c_{jn\sigma}$) are labeled by the unit cell index j, the basis atom index $n = 1, 2$, and the spin σ. Note that the spin quantum number had been suppressed in chapter 1. The on-site energy ε_{2p} of an electron at site n, j can be thought of as having two contributions: a kinetic energy contribution at site j, n and an electron-ion interaction contribution of the form

$$W_{el-ion}(\mathbf{r})|_{\mathbf{r}=\mathbf{R}_{jn}} = -\sum_{j'n'} W(\mathbf{r} - \mathbf{R}_{j'n'})|_{\mathbf{r}=\mathbf{R}_{jn}} = -\sum_{j'n'} W(\mathbf{R}_{jn} - \mathbf{R}_{j'n'}), \tag{11.2}$$

where $\mathbf{R}_{jn}$ is the position vector of the ion at the site n, j. The condition $\mathbf{r} = \mathbf{R}_{jn}$ means the electron orbital at site n, j moves rigidly with the ion at that site.

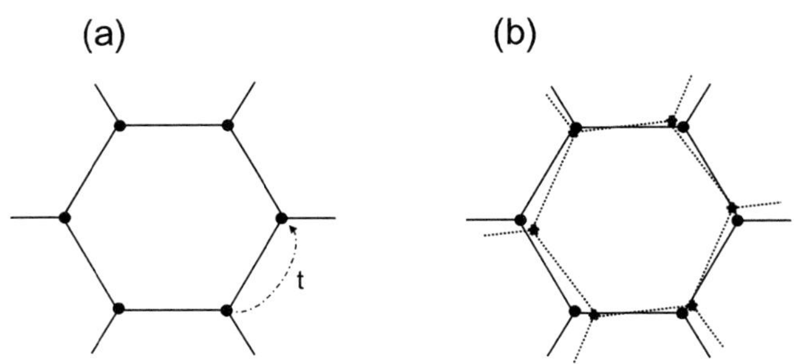

Fig. 11.7 Sketch of the graphene structure with (a) nearest-neighbor hopping indicated and (b) an arbitrary deformation of the lattice.

A deviation of the graphene lattice from its ideal lattice position will, in general, result in a modification of the electronic bandstructure. An arbitrary example is shown in Fig. 11.7. The electronic bandstructure of the ideal lattice, when the atoms are not displaced from the ideal positions, has been discussed in the Introductory Tutorial, chapter 1. There, the on-site energy ε_p and the hopping parameter (or hopping integral) t was introduced. The values of these parameters will change if the lattice undergoes a distortion. In the present chapter, we will denote by $\varepsilon_p^{(0)}$ and t_0 the on-site energy and hopping parameter, respectively, for the ideal lattice. We assume the hopping parameter t to be a function of the nearest-neighbor distance, i.e. $t = t(|\mathbf{R}_{jn} - \mathbf{R}_{j'n'}|)$, where it is implied that the site indices are restricted to nearest neighbor pairs. To distinguish between the ideal lattice positions and the actual position in the deformed lattice, we denote the ion positions as $\mathbf{R}_{jn} = \mathbf{R}_{jn}^{(0)} + \vec{\xi}_{jn}$, where $\mathbf{R}_{jn}^{(0)}$ are the ideal lattice position vectors ($\mathbf{R}_{j1}^{(0)} = \mathbf{R}_{j}^{(0)}$, $\mathbf{R}_{j2}^{(0)} = \mathbf{R}_{j}^{(0)} + \vec{\delta}_1$, $\mathbf{R}_{j}^{(0)} = j_1\mathbf{a}_1 + j_2\mathbf{a}_2$), and $\vec{\xi}_{jn}$ are the displacements due to lattice vibrations. Our notational convention for defining various lattice vectors in real and reciprocal spaces is shown in Fig. 11.6.

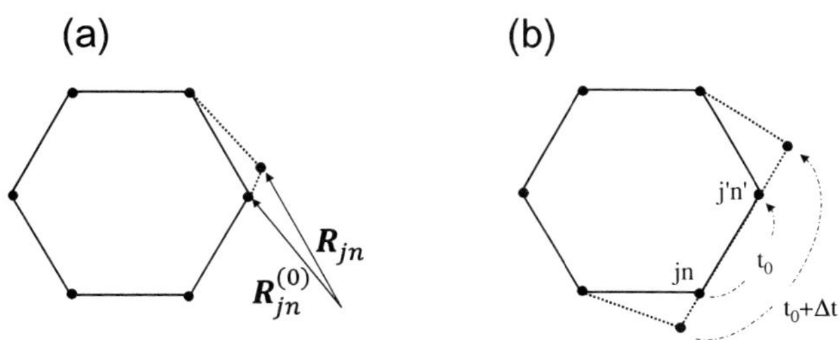

Fig. 11.8 (a) Sketch of the displacement of an atom leading to a change of the on-site energy. (b) Sketch of the modification of the hopping integral.

To obtain the electron-phonon interaction appropriate for graphene we follow Ref. [27]. In the deformed lattice, we replace the on-site energy in Eq. (11.1) in the following way:

$$\varepsilon_{2p} \to \varepsilon_{2p}^0 - \sum_{j'n'} [W(\mathbf{R}_{jn} - \mathbf{R}_{j'n'}) - W(\mathbf{R}_{jn}^{(0)} - \mathbf{R}_{j'n'}^{(0)})], \tag{11.3}$$

where $\varepsilon_{2p}^0 = \varepsilon_{2p}(\mathbf{R}_{jn} = \mathbf{R}_{jn}^{(0)})$. Similarly, we replace the hopping energy:

$$t \to t_0 + (t - t_0), \tag{11.4}$$

where $t_0 = t(|\mathbf{R}_{jn}^{(0)} - \mathbf{R}_{j'n'}^{(0)}|)$ and $t = t(|\mathbf{R}_{jn} - \mathbf{R}_{j'n'}|)$. This is schematically shown in Fig. 11.8. In the following, we label as DP the electron-phonon interaction that results from the change of the on-site energy by the deformation potential coupling, sometimes called conventional deformation potential coupling. We label the electron-phonon interaction that results from the change of the hopping energy by MH, for modulated hopping.

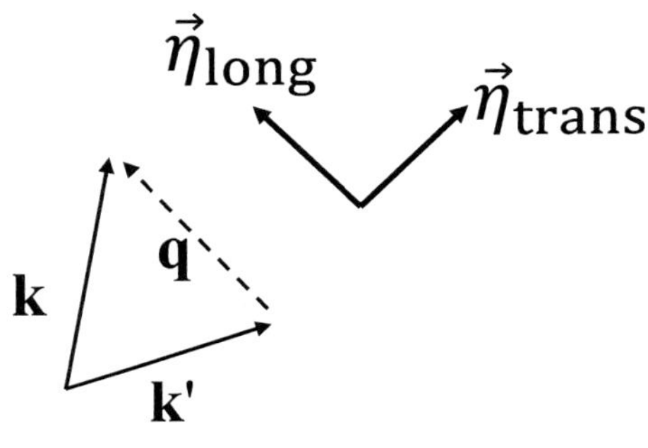

Fig. 11.9 Sketch of a momentum-conserving scattering process. Also shown are longitudinal and transverse phonon displacement vectors.

To first order in the displacements, the Hamiltonian for the DP term, i.e. the difference between ε_{2p} and ε_{2p}^0 in Eq. 11.3, is

$$H_{DP} = -\frac{1}{A} \sum_{jn\sigma j'n'\mathbf{q}} i\mathbf{q} \cdot (\vec{\xi}_{jn} - \vec{\xi}_{j'n'}) e^{i\mathbf{q}\cdot(\mathbf{R}_{jn}^{(0)} - \mathbf{R}_{j'n'}^{(0)})} W(\mathbf{q}) c_{jn\sigma}^{\dagger} c_{jn\sigma} \quad (11.5)$$

This expression follows after writing the potential energy W in terms of its 2-dimensional Fourier transform, $W(\mathbf{r}) = (1/A)\sum_{\mathbf{q}} e^{i\mathbf{q}\cdot\mathbf{r}} W(\mathbf{q}))$, where A is the area of the system. It is customary to take only the second term with $\vec{\xi}_{j'n'}$ in Eq. (11.5) for the deformation potential approximation called Nordheim's rigid ion model (*e.g.* [28–30]). The argument for this is that in the general electron-ion interaction, which is of the form $W_{el-ion}(\mathbf{r}) = -\sum_{j'n'} W(\mathbf{r} - \mathbf{R}_{j'n'})$, only the ion positions $\mathbf{R}_{j'n'}$ should be displaced, not the electron positions $\mathbf{r}$. But even if the first term would be taken into account, the strength of the electron-phonon interaction would not change much, as we have shown in Ref. [26].

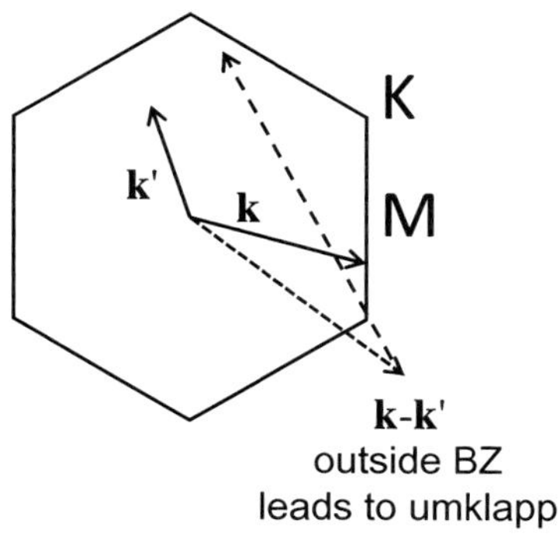

Fig. 11.10 Sketch of an umklapp process.

For the modulated hopping (MH) Hamiltonian we find from the first-order expansion in $\vec{\xi}_{jn}$ in Eq. 11.4

$$H_{MH} = -\left.\frac{\partial t}{\partial r}\right|_0 \frac{1}{a} \sum_{\langle jn,j'n'\rangle\sigma} [\mathbf{R}_{jn}^{(0)} - \mathbf{R}_{j'n'}^{(0)}] \cdot [\vec{\xi}_{jn} - \vec{\xi}_{j'n'}] c_{jn\sigma}^{\dagger} c_{j'n'\sigma}, \quad (11.6)$$

where the spatial gradient of the hopping parameter, $\vec{\nabla} t(r) = (\partial t/\partial r)\mathbf{r}/r$, $\left.\frac{\partial t}{\partial r}\right|_0 = \left.\frac{\partial t}{\partial r}\right|_{r=|\mathbf{R}_{jn}^{(0)} - \mathbf{R}_{j'n'}^{(0)}|}$ and $|\mathbf{R}_{jn}^{(0)} - \mathbf{R}_{j'n'}^{(0)}| = a$, has been used. We now define the deformation potentials as follows. For the conventional deformation potential, we define

$$D_{DP} = \frac{2}{A_0} W(\mathbf{0}) \quad (11.7)$$

$$D_{DP}(\mathbf{q}) = D_{DP} \frac{W(\mathbf{q})}{W(\mathbf{0})} = \frac{2}{A_0} W(\mathbf{q}), \quad (11.8)$$

and for the modulated hopping we use the definition

$$D_{MH} = \frac{3}{4} \left.\frac{\partial t}{\partial r}\right|_0 a. \quad (11.9)$$

An estimate for D_{MH} could be based on the bondlength derivative of the bond energy, which in Ref. [31] is found to be $-6.4\text{eV}/\text{Å}$. Using $a = 1.42\text{Å}$, we have $D_{MH} \simeq -6.82\text{eV}$, which is, apart from the sign, slightly larger than the value of 2.5eV used in [32].

In conventional deformation potential theories, only the long-wavelength limit is important. Hence, the deformation potential is one number, usually taken as $D_{DP}(\mathbf{0})$. This assumes that the interaction $W(\mathbf{q})$ does not exhibit a singularity at $\mathbf{q} = \mathbf{0}$. This is true in metals, where plasma screening changes the long-range Coulomb potential into a short-range potential, and in wave vector space, short range potentials have a finite Fourier component at zero wave vector. In the context of deformation potentials, wavevector-dependent deformation potentials have been used, for example, by Woods and Mahan [33].

With the preceding definitions, the electron-phonon interactions read

$$H_{DP} = -i\frac{1}{N} \sum_{jn\sigma j'n'\mathbf{q}} D_{DP}(\mathbf{q})\mathbf{q} \cdot (\vec{\xi}_{jn} - \vec{\xi}_{j'n'}) e^{i\mathbf{q}\cdot(\mathbf{R}_{jn}^{(0)} - \mathbf{R}_{j'n'}^{(0)})} c_{jn\sigma}^{\dagger} c_{jn\sigma} \tag{11.10}$$

and

$$H_{MH} = -\frac{4}{3a^2} D_{MH} \sum_{\langle jn,j'n' \rangle \sigma} [\mathbf{R}_{jn}^{(0)} - \mathbf{R}_{j'n'}^{(0)}] \cdot [\vec{\xi}_{jn} - \vec{\xi}_{j'n'}] c_{jn\sigma}^{\dagger} c_{j'n'\sigma}, \tag{11.11}$$

where N is the number of carbon atoms, with $N/2 = A/A_0$ being the number of unit cells. We stress that the $\mathbf{q}$ sum in Eq. (11.10) is not restricted to a Brillouin zone because in principle the interaction can vary in real (configuration) space on a scale smaller that the crystal's unit cell. Therefore it has Fourier components beyond the BZ.

We next convert H_{DP} and H_{MH} in Eqs. (11.10) and (11.11) into a sum over the free phonon and Bloch electron eigenmodes. We expand the ionic displacements $\vec{\xi}_{jn}$ and electron operators $c_{jn\sigma}$ ($c_{jn\sigma}^{\dagger}$) in their respective eigenmodes. For $\vec{\xi}_{jn}$, standard phonon quantization theory [34] yields

$$\vec{\xi}_{jn} = \frac{i}{\sqrt{A}} \sum_{\mu,\mathbf{Q}\in BZ} e^{i\mathbf{Q}\cdot\mathbf{R}_{jn}^{(0)}} \vec{\eta}_{\mu n}(\mathbf{Q}) \sqrt{\frac{\hbar A_0}{2\omega_{\mathbf{Q}}^{\mu} M_c}} (d_{-\mathbf{Q}\mu}^{\dagger} + d_{\mathbf{Q}\mu}), \tag{11.12}$$

where $\omega_{\mathbf{Q}}^{\mu}$ are the phonon mode frequencies of branch μ at wavevector $\mathbf{Q}$, M_c is the mass of the ion, $d_{\mathbf{Q}\mu}$ ($d_{\mathbf{Q}\mu}^{\dagger}$) are the phonon annihilation and creation operators, and $\vec{\eta}_{\mu n}(\mathbf{Q})$ are the 2-dimensional phonon mode vectors. Below, we restrict ourselves to the in-plane transverse and longitudinal acoustic and optic modes, μ=TA, LA, LO, TO. We will discuss a simple and specific model for the phonon frequencies and mode vectors below in Sec. 11.2.2. Quite generally, the mode vectors are normalized according to

$$\sum_{n=1}^{2} |\vec{\eta}_{\mu n}(\mathbf{Q})|^2 = 1,$$

Furthermore, they have the following symmetries within our convention for $\vec{\xi}_{jn}$ containing a factor of i. The fact that the operators $\vec{\xi}_{jn}$ are Hermitian requires that

$$\vec{\eta}_{\mu n}(\mathbf{Q}) = -\vec{\eta}^{\,*}_{\mu n}(-\mathbf{Q}).$$

Given the lattice periodicity of the classical phonon Hamiltonian, the phonon mode vectors $\vec{\eta}_{\mu n}(\mathbf{Q})$ satisfy the following relations under translations in reciprocal space (see Sec. 11.2.2):

$$\vec{\eta}_{\mu 1}(\mathbf{Q}+\mathbf{G}) = \vec{\eta}_{\mu 1}(\mathbf{Q}) \tag{11.13}$$

$$e^{i\mathbf{G}\cdot\vec{\delta}_1}\vec{\eta}_{\mu 2}(\mathbf{Q}+\mathbf{G}) = \vec{\eta}_{\mu 2}(\mathbf{Q}), \tag{11.14}$$

where $\mathbf{G}$ is a reciprocal lattice vector. In practice, we choose to make one component of $\vec{\eta}_{\mu 1}(\mathbf{Q})$ purely real for all $\mathbf{Q}$.

Finally, we transform the localized electron operators to Bloch states

$$c_{jn\sigma} = \sqrt{\frac{2}{N}} \sum_{\mathbf{k}\in BZ,\nu=\pm 1} e^{i\mathbf{k}\cdot\left(\mathbf{R}_j^{(0)}+\vec{\delta}_1\delta_{n,2}\right)} u_{\mathbf{k}\nu n} c_{\mathbf{k}\nu\sigma}$$

labeled by the 2-dimensional wavevector $\mathbf{k}$, where $c_{\mathbf{k}\nu\sigma}$ denotes the electron annihilation operator for a Bloch electron in band ν with two-dimensional crystal momentum $\hbar\mathbf{k}$, $\delta_{n,2}$ is the Kronecker delta, and $u_{\mathbf{k}\nu n}$ is the lattice-periodic part of the Bloch wave function.

Our approach here is similar to one used throughout much of the graphene literature, see for example [35–37]. Within this approach, we have two electronic bands, called π and π^*. In our formalism, we denote the valence (conduction) band by $\nu = -1$ ($\nu = 1$). It is important to note that the quality of this two-band approximation decreases with increasing photon energy. It is most valid close to the Dirac and saddle points and least valid close to the Γ point. Nevertheless, we believe it is useful and instructional to use this simple model for quantitative studies close to the Dirac and saddle point and to elucidate basic trends beyond those points.

11.2.2 *Phonons in graphene*

Before we proceed with the discussion of the electron-phonon interaction, let us briefly insert a relatively simple model for the phonons in graphene. There are many textbook treatments of phonons in a one-dimensional chain with harmonic nearest-neighbor interactions, see for example [34]. In those treatments, harmomic nearest-neighbor interactions are related to potential energies that are quadratic functions of the atomic displacements. Assuming the displacements form traveling waves, and inserting a traveling wave ansatz in the Newton's equations of motion for the displacements, one finds easily the relation between frequency and wave vector, i.e. the phonon dispersion. We now follow the same strategy, but the task is more complicated as we have a two-dimensional crystal and we have, in principle, three directions in which each atom can oscillate (x,y, and z). Still, one can find an expression for the potential energy that is quadratic in the displacements and from there obtain equations of motion that contain forces linear in the displacements. For the specific case of graphene, how to do this was shown in Refs. [33, 38].

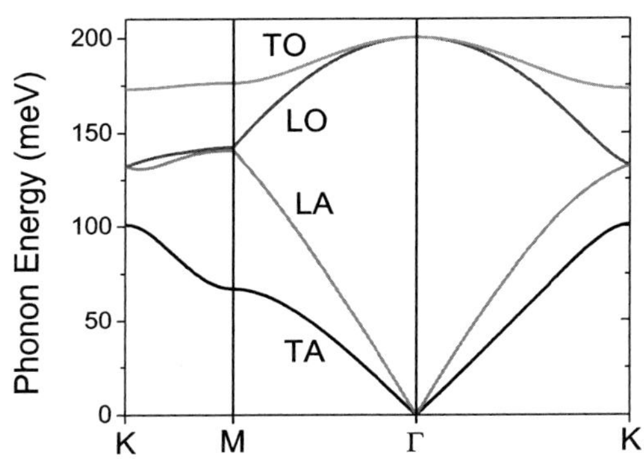

Fig. 11.11 The phonon dispersion diagram for the in-plane branches. From Ref. [26].

The classical Hamilton function for the ionic motion is written as

$$H_{ph} = \sum_{jn} \frac{P_{jn}^2}{2M} + V_{bs} + V_{bb}, \tag{11.15}$$

where the first term represents the kinetic energy of the ions, and the potential energies account for bond stretching (bs) and bond bending (bb). The nearest neighbor bond stretching potential energy is written as

$$V_{bs} = \frac{\alpha}{2} \sum_{\langle jn,j'n' \rangle} (|\mathbf{R}_{jn} - \mathbf{R}_{j'n'}| - |\mathbf{R}_{jn}^{(0)} - \mathbf{R}_{j'n'}^{(0)}|)^2 \tag{11.16}$$

with α being a free parameter. For small displacements, the distances can be expanded to first order and written as

$$|\mathbf{R}_{jn} - \mathbf{R}_{j'n'}| \approx |\mathbf{R}_{jn}^{(0)} - \mathbf{R}_{j'n'}^{(0)}| + (\xi_{jn} - \xi_{j'n'}) \cdot \frac{(\mathbf{R}_{jn}^{(0)} - \mathbf{R}_{j'n'}^{(0)})}{|\mathbf{R}_{jn}^{(0)} - \mathbf{R}_{j'n'}^{(0)}|} + ... \tag{11.17}$$

so that the bond stretching potential energy becomes quadratic in the displacements,

$$V_{bs} = \frac{\alpha}{2} \sum_{\langle jn,j'n' \rangle} \left((\xi_{jn} - \xi_{j'n'}) \cdot \frac{(\mathbf{R}_{jn}^{(0)} - \mathbf{R}_{j'n'}^{(0)})}{|\mathbf{R}_{jn}^{(0)} - \mathbf{R}_{j'n'}^{(0)}|} \right)^2 . \tag{11.18}$$

Similarly, the nearest neighbor bond bending potential energy is written as

$$V_{bb} = \frac{\beta a^2}{2} \sum_{\langle jn:j'n'j''n'' \rangle} [\cos\theta_{jnj'n'j''n''} - \cos\theta_0]^2, \tag{11.19}$$

where β is another free parameter, $\theta_0 = 2\pi/3$ is the bond angle of the ideal lattice, and $\theta_{jnj'n'j''n''}$ is the angle between two adjacent bonds, i.e. between the vectors $\mathbf{R}_{j'n'} - \mathbf{R}_{jn}$ and $\mathbf{R}_{j''n''} - \mathbf{R}_{jn}$:

$$\cos\theta_{jnj'n'j''n''} = \frac{(\mathbf{R}_{j'n'} - \mathbf{R}_{jn}) \cdot (\mathbf{R}_{j''n''} - \mathbf{R}_{jn})}{|\mathbf{R}_{j'n'} - \mathbf{R}_{jn}||\mathbf{R}_{j''n''} - \mathbf{R}_{jn}|}. \tag{11.20}$$

We can now linearize (11.20) and obtain a potential energy quadratic in the displacements. The cumbersome but straightforward algebra is omitted here.

The classical equations of motion are then given by

$$M\frac{d^2\xi_{jn}}{dt^2} = -\frac{\partial H_{ph}}{\partial \xi_{jn}} = -\frac{\partial V_{bs}}{\partial \xi_{jn}} - \frac{\partial V_{bb}}{\partial \xi_{jn}}. \tag{11.21}$$

In a typical textbook treatment of a one-dimensional chain, the dispersion relation is obtained from the equation of motion when a plane wave ansatz is used as the condition for non-trivial, non-zero solutions. In the case of graphene, the displacements of the two basis atoms in three dimensions are coupled, and the condition for non-trival solutions becomes an eigenvalue equation of the dimension 2×3. If we restrict ourselves to the in-plane motions of the atoms and omit the motion perpendicular to the graphene layer, the dimension is 2×2. The displacements of each of the two basis atoms are defined in Fig. 11.6. The four displacements $\eta_{1x}, \eta_{1y}, \eta_{2x}, \eta_{2y}$ are the four components of the eigenvalue problem that yield the phonon frequencies as eigenvalues and the phonon displacement vectors as corresponding eigenvectors. We use the label μ for the 4 solutions of the eigenvalue problem. The derivation of the eigenvalue problem starting from (11.21) is quite lengthy and will be omitted here. The result can be written as

$$(\hbar\omega^\mu)^2 \begin{bmatrix} \eta_{1x}^\mu \\ \eta_{1y}^\mu \\ \eta_{2x}^\mu \\ \eta_{2y}^\mu \end{bmatrix} = (\widehat{D}_{bs} + \widehat{D}_{bb}) \begin{bmatrix} \eta_{1x}^\mu \\ \eta_{1y}^\mu \\ \eta_{2x}^\mu \\ \eta_{2y}^\mu \end{bmatrix}, \tag{11.22}$$

where the bond stretching dynamical matrix is

$$\widehat{D}_{bs} = \bar{K}_1 \begin{pmatrix} \frac{3}{2} & 0 & h_3 & -h_2 \\ 0 & \frac{3}{2} & -h_2 & h_1 \\ h_3^* & -h_2^* & \frac{3}{2} & 0 \\ -h_2^* & h_1^* & 0 & \frac{3}{2} \end{pmatrix}, \tag{11.23}$$

and the bond-bending dynamical matrix reads

$$\widehat{D}_{bb} = \bar{K}_2 \begin{pmatrix} h_6 & -h_5 & 6h_1 & 6h_2 \\ -h_5^* & h_4 & 6h_2 & 6h_3 \\ 6h_1^* & 6h_2^* & h_6 & -h_5^* \\ 6h_2^* & 6h_3^* & -h_5 & h_4 \end{pmatrix}. \tag{11.24}$$

The elements entering the above two matrices are

$$h_1(\mathbf{q}) = -\frac{3}{2}e^{iq_x a_L/(2\sqrt{3})}\cos\left(\frac{q_y a_L}{2}\right) \tag{11.25}$$

$$h_2(\mathbf{q}) = i\frac{\sqrt{3}}{2}e^{iq_x a_L/(2\sqrt{3})}\sin\left(\frac{q_y a_L}{2}\right) \tag{11.26}$$

$$h_3(\mathbf{q}) = -e^{-iq_x a_L/\sqrt{3}} - \frac{1}{2}e^{iq_x a_L/(2\sqrt{3})}\cos\left(\frac{q_y a_L}{2}\right) \tag{11.27}$$

$$h_4(\mathbf{q}) = 7 + \sin^2\left(\frac{q_y a_L}{2}\right) + 2\cos\left(\frac{q_y a_L}{2}\right)\cos\left(\frac{\sqrt{3}}{2}q_x a_L\right) \tag{11.28}$$

$$h_5(\mathbf{q}) = \sqrt{3}ie^{i\sqrt{3}q_x a_L/2}\sin\left(\frac{q_y a_L}{2}\right) - i\frac{\sqrt{3}}{2}\sin(q_y a_L) \tag{11.29}$$

$$h_6(\mathbf{q}) = 9 - 3\sin^2\left(\frac{q_y a_L}{2}\right). \tag{11.30}$$

The eigenvectors of the phonon modes are normalized according to: $|\eta^{\mu}_{1x}|^2 + |\eta^{\mu}_{1y}|^2 + |\eta^{\mu}_{2x}|^2 + |\eta^{\mu}_{2y}|^2 = 1$. In our calculations, we choose the y-component of $\vec{\eta}_{\mu 1}$ to be purely real for all $\mathbf{q}$. We use the parameter values $\bar{K}_1 = 0.01$ $(\text{eV})^2$, and $\bar{K}_2 = 0.0565\bar{K}_1$ [38]. Figure 11.11 shows the phonon dispersion obtained from this model, and Fig. 11.12 shows the eigenvectors across the entire Brillouin zone. Again, we have omitted the out-of-plane phonons, so that we have only 2 acoustic and 2 optical branches. They are called longitudinal and transverse corresponding to the displacement vectors close

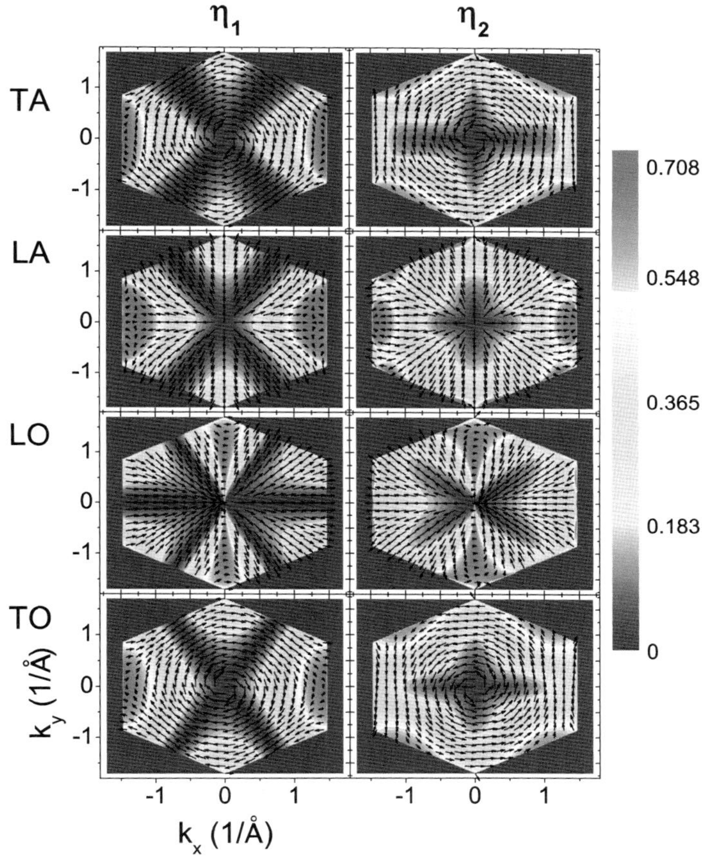

Fig. 11.12 The eigenvector contour plots of the complex-valued two-dimensional displacement vectors η_1 and η_2 corresponding to the phonon branches in Fig. 11.11. The real part of the eigenvectors are shown as arrows, and the magnitude of the imaginary part as color contour. From Ref. [26]. (For color reference, go to page 495.)

to the Γ point. From Fig. 11.12, we can clearly see the longitudinal and transverse behavior of these modes close to the zone center. In the limit of small wave vectors, the acoustic branches are, to linear order in $\mathbf{q}$ [38],

$$\hbar\omega_{\mathbf{q}}^{TA} = qa_L\sqrt{\frac{3}{2}\frac{\bar{K}_1\bar{K}_2}{\bar{K}_1 + 6\bar{K}_2}} \tag{11.31}$$

$$\hbar\omega_{\mathbf{q}}^{LA} = qa_L\sqrt{\frac{3}{2}\frac{\bar{K}_1\bar{K}_2}{\bar{K}_1 + 6\bar{K}_2} + \frac{1}{8}\bar{K}_1}, \tag{11.32}$$

Furthermore, in this limit the two-dimensional eigenvectors of atoms 1 and 2 become identical in the case of acoustic phonons,

$$\vec{\eta}_1^{TA} = \vec{\eta}_2^{TA} = -\frac{1}{\sqrt{2}q}\begin{pmatrix} -q_y \\ q_x \end{pmatrix} \tag{11.33}$$

$$\vec{\eta}_1^{LA} = \vec{\eta}_2^{LA} = \frac{1}{\sqrt{2}q}\begin{pmatrix} q_x \\ q_y \end{pmatrix}, \tag{11.34}$$

and anti-parallel, but otherwise equal, in the case optical phonons,

$$\vec{\eta}_1^{TO} = -\vec{\eta}_2^{TO} = -\frac{1}{\sqrt{2}q}\begin{pmatrix} -q_y \\ q_x \end{pmatrix} \tag{11.35}$$

$$\vec{\eta}_1^{LO} = -\vec{\eta}_2^{LO} = \frac{1}{\sqrt{2}q}\begin{pmatrix} q_x \\ q_y \end{pmatrix}. \tag{11.36}$$

We can see from Fig. 11.12 that away from the zone center the two eigenvectors of atom 1 and 2 are not identical (with directions parallel and antiparallel for acoustic and optic phonons, respectively) anymore. Also, away from the zone center they are no longer real-valued, purely transverse, or purely longitudinal.

11.2.3 *Hamiltonian and self-energy*

In this section we first write the Hamiltonian for the electron-phonon interaction and then the electronic self-energies that result from that Hamiltonian.

The deformation potential Hamiltonian Eq. (11.10) reads, in terms of the free phonon and Bloch electron modes,

$$H_{DP} = -\frac{1}{\sqrt{A}} \sum_{\substack{\mathbf{k}\mathbf{k}' \epsilon BZ \\ \nu\nu'\sigma\mu}} \left(\frac{\hbar A_0}{2M_c a^2 \omega^{\mu}_{\mathbf{k}-\mathbf{k}'}} \right)^{1/2} D_{DP} W_{\mu}^{DP}(\mathbf{k}\nu, \mathbf{k}'\nu')$$
$$\times \left[d^{\dagger}_{-\mathbf{k}+\mathbf{k}',\mu} + d_{\mathbf{k}-\mathbf{k}',\mu} \right] c^{\dagger}_{\mathbf{k}\nu\sigma} c_{\mathbf{k}'\nu'\sigma} \tag{11.37}$$

which describes scattering of electrons from state $\mathbf{k}$ to $\mathbf{k}'$ through a phonon with wave vector $\mathbf{q}$, see Fig. 11.9. The matrix element entering (11.37) is

$$W_{\mu}^{DP}(\mathbf{k}\nu, \mathbf{k}'\nu') = \sum_{\mathbf{G} \in \mathfrak{R}} \frac{W\left(\mathbf{k} - \mathbf{k}' - \mathbf{G}\right)}{W\left(\mathbf{0}\right)}$$

$$\times a\left(\mathbf{k}-\mathbf{k}'-\mathbf{G}\right)\cdot\left[\vec{\eta}_{\mu 1}\left(\mathbf{q}\right)+\vec{\eta}_{\mu 2}\left(\mathbf{q}\right)e^{-i(\mathbf{G}_u-\mathbf{G})\cdot\vec{\delta}_1}\right]$$
$$\times\frac{1}{2}\left(1+\nu\nu' e^{i\left(\phi(\mathbf{k})-\phi\left(\mathbf{k}'\right)-\mathbf{G}\cdot\vec{\delta}_1\right)}\right), \tag{11.38}$$

where $\mathbf{q}=\mathbf{k}-\mathbf{k}'-\mathbf{G}_u$, $\mathfrak{R}$ denotes the set of reciprocal vector $\mathbf{G}_u$ is chosen such that $\mathbf{q}$ is reduced to the first BZ. If $\mathbf{G}_u$ is not zero, we call the scattering process an umklapp process, which is schematically shown in Fig. 11.10.

The last line in Eq. (11.38) is a generalization of the well-known Bloch factor $\langle u_{\mathbf{k}\nu}|u_{\mathbf{k}'\nu'}\rangle=\frac{1}{2}\left(1+\nu\nu' e^{i\left(\phi(\mathbf{k})-\phi\left(\mathbf{k}'\right)\right)}\right)$. Our generalization includes contributions from phonons in Brillouin zones other than the first. These processes are determined by the range of $W\left(\mathbf{k}-\mathbf{k}'-\mathbf{G}\right)$ in reciprocal space.

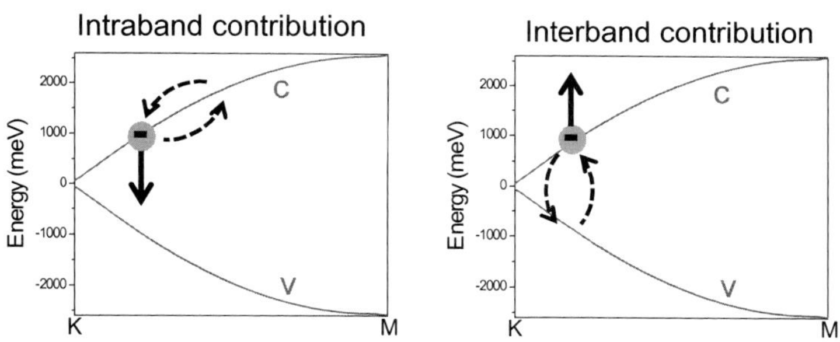

Fig. 11.13 Schematics indicating intraband and interband contributions to the self-energy.

The modulated hopping Hamiltonian is obtained in a similar way. From Eq. (11.11) we find

$$H_{MH}=-\frac{1}{\sqrt{A}}\sum_{\substack{\mathbf{k}\mathbf{k}'\epsilon BZ\\ \nu\nu'\sigma\mu}}\left(\frac{\hbar A_0}{2M_c a^2\omega^{\mu}_{\mathbf{k}-\mathbf{k}'}}\right)^{1/2}\frac{4}{3}D_{MH}W^{MH}_{\mu}\left(\mathbf{k}s,\mathbf{k}'s'\right)$$
$$\times\left[d^{\dagger}_{-\mathbf{k}+\mathbf{k}',\mu}+d_{\mathbf{k}-\mathbf{k}',\mu}\right]c^{\dagger}_{\mathbf{k}\nu\sigma}c_{\mathbf{k}'\nu'\sigma}, \tag{11.39}$$

where

$$W^{MH}_{\mu}(\mathbf{k}\nu,\mathbf{k}'\nu')=\frac{i}{2}\left[\nu e^{i\phi(\mathbf{k})}\left(\mathbf{F}(-\mathbf{k})\cdot\vec{\eta}_{\mu 1}(\mathbf{q})-\mathbf{F}(-\mathbf{k}')\cdot\vec{\eta}_{\mu 2}(\mathbf{q})e^{-i\mathbf{G}_u\cdot\vec{\delta}_1}\right)\right.$$
$$\left.+\ \nu' e^{-i\phi(\mathbf{k}')}\left(\mathbf{F}(\mathbf{k}')\cdot\vec{\eta}_{\mu 1}(\mathbf{q})-\mathbf{F}(\mathbf{k})\cdot\vec{\eta}_{\mu 2}(\mathbf{q})e^{-i\mathbf{G}_u\cdot\vec{\delta}_1}\right)\right] \tag{11.40}$$

with $\mathbf{q}$ and $\mathbf{G}_u$ the same as defined above. The function $\mathbf{F}(\mathbf{k})$ is defined by

$$\mathbf{F}(\mathbf{k}) = \frac{1}{a}\sum_{j=1}^{3} \vec{\delta}_j e^{i\mathbf{k}\cdot\vec{\delta}_j}, \tag{11.41}$$

and for our choice of coordinate system, this function reads as

$$\mathbf{F}(\mathbf{k}) = \left[e^{i\frac{1}{2}k_x a}\cos(\sqrt{3}k_y a/2) - e^{-ik_x a}\right]\hat{x} + \left[i\sqrt{3}e^{i\frac{1}{2}k_x a}\sin\left(\sqrt{3}k_y a/2\right)\right]\hat{y}.$$

We can now write the total electron-phonon Hamiltonian that includes both the conventional deformation potential and modulated hopping contributions:

$$\begin{aligned} H_{el-phon} &= H_{MH} + H_{DP} \\ &= \frac{1}{\sqrt{A}} \sum_{\substack{\mathbf{k}\mathbf{k}'\epsilon BZ \\ \nu\nu'\sigma\mu}} \left(\frac{\hbar A_0}{2M_c a^2 \omega^{\mu}_{\mathbf{k}-\mathbf{k}'}}\right)^{1/2} M_\mu\left(\mathbf{k}\nu, \mathbf{k}'\nu'\right) \\ &\quad \times \left[d^{\dagger}_{-\mathbf{k}+\mathbf{k}',\mu} + d_{\mathbf{k}-\mathbf{k}',\mu}\right] c^{\dagger}_{\mathbf{k}\nu\sigma} c_{\mathbf{k}'\nu'\sigma} \end{aligned}$$

The transition matrix elements in the Hamiltonian are

$$M_\mu\left(\mathbf{k}\nu, \mathbf{k}'\nu'\right) = -\frac{4}{3}D_{MH}W^{MH}_\mu\left(\mathbf{k}\nu, \mathbf{k}'\nu'\right) - D_{DP}W^{DP}_\mu(\mathbf{k}\nu, \mathbf{k}'\nu'). \tag{11.42}$$

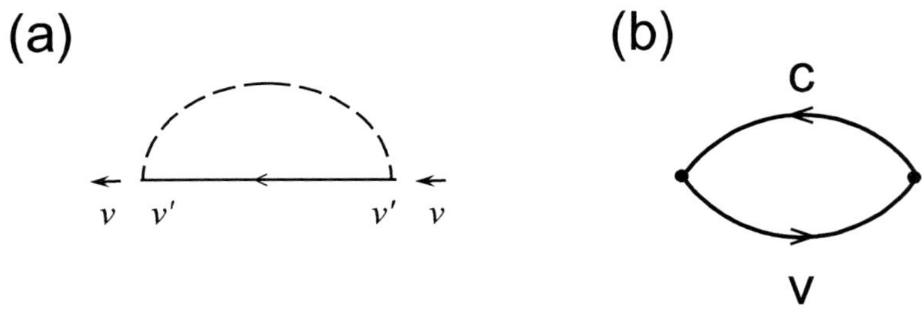

Fig. 11.14 Feynman diagrams for (a) the electron self-energy due to electron-phonon interaction and (b) for the interband susceptibility. The solid lines represent electron Green's functions, and the dashed line represents the phonon Green's function.

We will now use $H_{el-phon}$ and evaluate the electron self-energy in lowest order of $H_{el-phon}$, using a Green's function or Feynman diagram approach. The self-energy $\Sigma_\nu(\mathbf{k})$ gives a renormalization to the unperturbed electron energies from the unperturbed bandstructure. More specifically, the real part of $\Sigma_\nu(\mathbf{k})$ gives the shift of the energies, and the imaginary part is a measure of the lifetime of the electron states. For example, if the electron initially is in the band ν at wavevector $\mathbf{k}$, then due to electron-phonon scattering it will leave that state after a typical time of the order of $\hbar/\mathrm{Im}\Sigma_\nu(\mathbf{k})$. The basic formula for the self-energy can be found, for example as Eq. (3.5.10) in [39], and is shown in Fig. 11.14a (see also, e.g. [22,40–43]). We use an extension of the formula in [39], since we include both intra and interband contributions, and furthermore the squared matrix element depends on the electron wavevectors $\mathbf{k}$ and $\mathbf{k}'$ separately, not just on the difference $\mathbf{k} = \mathbf{k}-\mathbf{k}'$. Applying the quasi-particle approximation $\Sigma_\nu(\mathbf{k}) = \Sigma_\nu(\mathbf{k}, \hbar\omega = E_\nu^{(0)}(\mathbf{k}))$, the self-energy reads

$$\Sigma_\nu(\mathbf{k}) = \frac{1}{A} \sum_{\mathbf{k}' \in BZ, \mu\nu'} C^{\mu}_{s,s'}(\mathbf{k}, \mathbf{k}') \times \left[\frac{1 - f_{\nu'}(\mathbf{k}') + n^{ph}_\mu(\mathbf{k} - \mathbf{k}')}{E_\nu^{(0)}(\mathbf{k}) - E_{\nu'}^{(0)}(\mathbf{k}') - \hbar\omega^\mu_{\mathbf{k}-\mathbf{k}'} + i\zeta} + \frac{f_{\nu'}(\mathbf{k}') + n^{ph}_\mu(\mathbf{k}'-\mathbf{k})}{E_\nu^{(0)}(\mathbf{k}) - E_{\nu'}^{(0)}(\mathbf{k}') + \hbar\omega^\mu_{\mathbf{k}'-\mathbf{k}} + i\zeta} \right], \tag{11.43}$$

with $(1/A)\sum_{\mathbf{k}' \in BZ} = \int_{BZ} d^2k'/(2\pi)^2$, and ζ is a small positive constant ensuring the retarded nature of the self-energy. The coupling coefficients, which are the squared transition matrix elements, are

$$C^{\mu}_{\nu,\nu'}(\mathbf{k}, \mathbf{k}') = \frac{\hbar}{2M_c a^2} \frac{A_o}{\omega^\mu_{\mathbf{k}-\mathbf{k}'}} \left| \frac{4}{3} D_{MH} W^{MH}_\mu(\mathbf{k}\nu, \mathbf{k}'\nu') + D_{DP} W^{DP}_\mu(\mathbf{k}\nu, \mathbf{k}'\nu') \right|^2 . \tag{11.44}$$

As already mentioned, the self-energy contains intraband ($\nu = \nu'$) and interband contributions ($\nu \neq \nu'$), which are schematically shown in Fig. 11.13. The electronic distribution functions are denoted by $f_\nu(\mathbf{k})$, and $n^{ph}_\mu(\mathbf{k}) = 1/(e^{\hbar\omega^\mu_{\mathbf{k}}/k_B T} - 1)$ is the Bose function for the phonons. If the electrons are in thermal equilibrium, the distributions $f_\nu(\mathbf{k})$ are Fermi functions. At zero temperature, the valence band is completely full, $f_v(\mathbf{k}) = 1$, and the conduction band completely empty, so $f_c(\mathbf{k}) = 0$. At non-zero temperature, the chemical potential determines the number and distribution of electrons thermally excited from the valence to the conduction band. In the numerical results for differential transmission spectra shown below, the change of the thermal excitation of electrons is negligible.

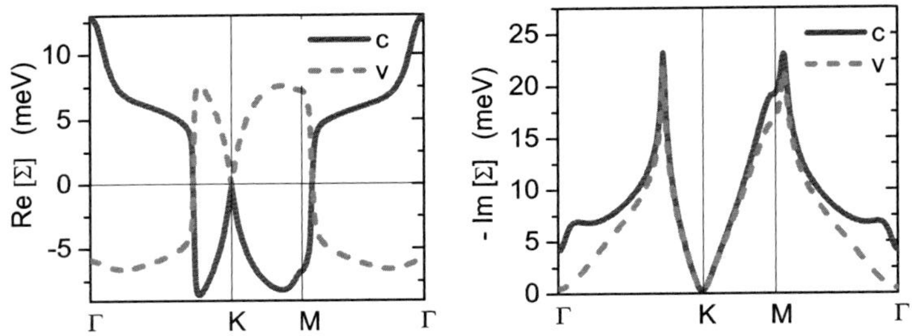

Fig. 11.15 Self-energy (left: real part, right: imaginary part times -1) at temperature T=300K including both deformation potential (DP) and modulated hopping (MH) contributions for the conduction (c) band, $\nu = +1$, and valence (v) band, $\nu = -1$. From Ref. [26].

As pointed out in Ref. [44], the real part of the self-energy is structurally very similar to the energy shift that one would obtain from second-order perturbation theory. Second-order energy corrections are of the form

$$E_i^{(2)} = \sum_{j \neq i} |V_{i,j}|^2/(E_i^{(0)} - E_j^{(0)}), \tag{11.45}$$

where $V_{i,j}$ are perturbation matrix elements between states i and j. We see that neglecting the small phonon energy in the denominator of Eq. (11.43) makes the self-energy structure similar to calculations from conventional perturbation theory. Considering first the conduction (c) band ($\nu = +1$) in Eq. (11.43), we have a subset of energies with lower and upper bound, and the expression for $E_i^{(2)}$ shows that the shift of the lower (upper) bound is negative (positive). We see from Fig. 11.15 that indeed the shift from the real part of Σ is positive at the highest c-band energy (i.e. at Γ), and negative at the smallest c-band energy (i.e. at K). The shifts naively expected from second-order perturbation theory, if states are restricted to one band are indicated schematically in Fig. 11.16. However, it is clear that the conical intersection is only preserved if, approaching the Dirac point, the bands either do not shift at all, or they shift parallel by the same amount, which is indicated schematically in Fig. 11.17.

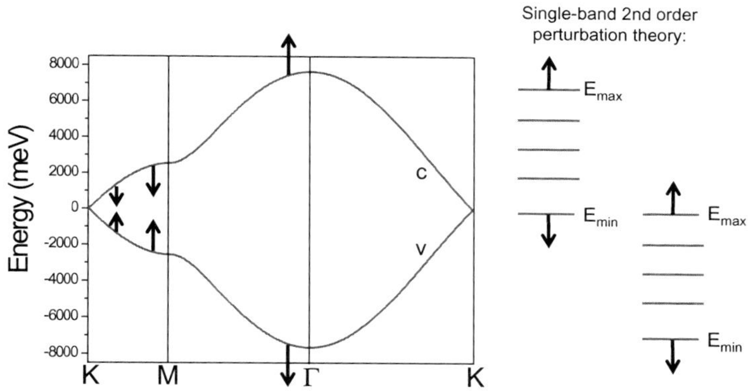

Fig. 11.16 Sketch of analogy with second-order perturbation theory.

We see from Fig. 11.15 that the real part of the c- and v-band self-energy becomes indeed very small at the Dirac point. To understand the origin of this behavior, we show in Fig. 11.18 the intra (cc) and inter (cv) band contributions. For semiconductors like GaAs which have a gap, it is common practice to restrict calculations of the electron self-energy due to electron-phonon coupling to just intraband contributions. This is based on the fact that the interband contributions are small due to the appearance of the gap in the energy denominator. For graphene, however, such an approximation would not be appropriate. Figure 11.18 shows a strong cancellation between intra- and interband contributions (Fig. 11.13) that becomes gradually more perfect as we approach the K-point. If the characteristic conical intersection of c- and v-band is to be preserved by the electron-phonon interaction, then either both c- and v-self-energies must vanish at K, or both must be equal. In other words, they cannot have opposite signs. The former is the case for T=0K, and the latter has been found to be true in Ref. [40]. We also note that the imaginary part of the interband self-energy is very small except very close to K, as the transition energies from v to c are much larger than the phonon energies.

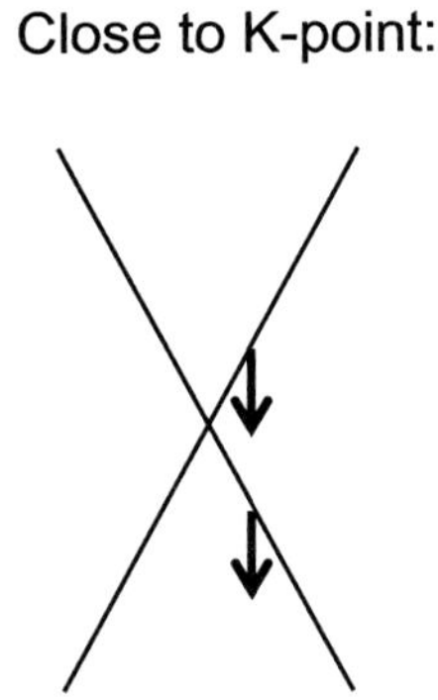

Fig. 11.17 Sketch of the renormalization close to the Dirac point required to preserve the conical intersection.

Figure 11.15 shows that the real part of Σ changes sign slightly above the M-point. According to the Kramers-Kronig relations there will be a peak in the imaginary part of Σ close to where the real part changes sign, as seen in Fig. 11.15. A similar discussion holds for the valence (v-band, $\nu = -1$) band, except that the highest energy is at the K-point and the lowest at Γ.

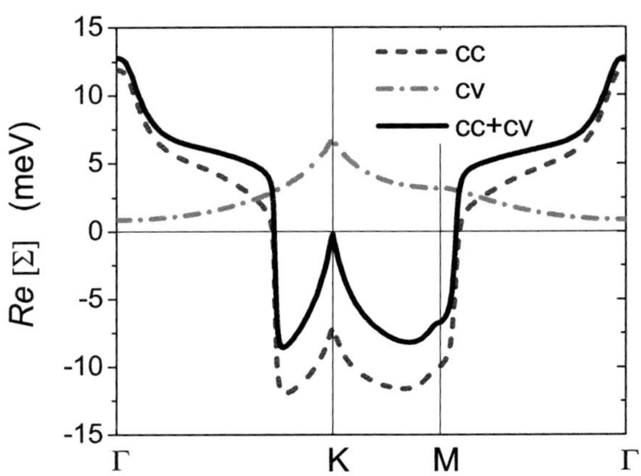

Fig. 11.18 Conduction band self-energy and its intra and interband contributions. From Ref. [26].

For LA phonons, our intraband coupling coefficient close to the Dirac point can be found by approximating the electron wavevector to be in the vicinity of the Dirac point, as outlined in chapter 1. We find

$$C_{1,1}^{LA}(\mathbf{k},\mathbf{k}') = \frac{\hbar}{4M_c}\frac{A_o}{\omega_{\mathbf{q}}^{LA}}q^2 \left| D_{MH}\sin\frac{1}{2}\left(\theta_{\mathbf{p}} + \theta_{\mathbf{p}'} + 4\theta_{\mathbf{q}}\right) \right.$$
$$\left. + D_{DP}\cos\frac{1}{2}\left(\theta_{\mathbf{p}} - \theta_{\mathbf{p}'}\right)\right|^2, \tag{11.46}$$

which can be compared directly to the corresponding expression given in Ref. [32]. We see that the DP term depends only on the difference in the directions of the electron wave vectors before and after the scattering, $(\theta_{\mathbf{p}} - \theta_{\mathbf{p}'})$, which means that the scattering process behaves as if the medium is isotropic. This is expected for scattering in the long-wavelength limit, as in that limit the crystal structure becomes unimportant. However, we see that the MH contribution is not isotropic, even in the long-wavelength limit. This may not be surprising as modulated hopping always involves certain bonds. For example, the electron-phonon interaction might result from a stretching of the horizontal bond in Fig. 11.7. In the long-wavelength limit, this stretching would be the same for all horizontal bonds. However, it is clear that this deformation is not isotropic.

We also note that sometimes it is useful to introduce a so-called effective deformation potential. Here, the contributions from modulated hopping (with longitudinal and transverse phonons) are absorbed in the conventional (longitudinal phonon) deformation potential terms. This means that the angular dependence of the electron-phonon coupling is assumed to be that of the conventional deformation potential, whereas the numerical value of the conventional deformation potential is adjusted in a way that it approximately includes the modulated hopping contribution. This concept has been used for scattering involving acoustical and optical phonons. For example, Ref. [32] introduces an effective deformation potential $\Xi \simeq 5$eV for electron-acoustic phonon scattering, using the simple isotropic angular dependence of the DP term in Eq. 11.46, rather than using the more complex, anisotropic angular dependence of the MH term in that equation.

In our recent study, Ref. [8], we adopted the effective deformation potential concept and found good agreement with our experimental pump-probe measurements near the M-point. Similarly, an effective optical deformation potential with a value of 11eV/Å was used in Ref. [6], using an angular dependence of the DP term. In Ref. [8], we adopted the concept of an effective optical deformation potential with the angular dependence of the conventional longitudinal acoustical deformation potential coupling and a value of $D^{op}_{eff} = 11\text{eV}/\text{Å}$. The results of that study will be discussed below, in Sec. 11.3.

The electron-phonon self-energy $\Sigma_\nu(\mathbf{k})$ critically contributes to many physical observables such as electrical conductivity and optical spectra. Electrical conductivity is largely determined by scattering processes close to the Dirac point. However, as already mentioned, optical techniques can access basically any point in the Brillouin zone, so knowledge of $\Sigma_\nu(\mathbf{k})$ over the entire Brillouin zone becomes more important. Furthermore, the signal in a differential transmission study does not depend on the self-energy itself, but only on the optically-induced changes of the self-energy. Clearly, without any such change, the differential transmission signal must be zero, independent of the form of the self-energy. If the changes are small enough, the signal will be linear in the change of the self-energy, which we denote by $\Delta\Sigma_s(\mathbf{k})$. A small change of the self-energy, in turn, is linearly related to the change of the electronic and phonon distribution functions:

$$\Delta\Sigma_\nu(\mathbf{k}) = \frac{1}{A} \sum_{\mathbf{k}' \in BZ, \mu s'} C^{\mu}_{\nu,\nu'}(\mathbf{k},\mathbf{k}') \times \left[\frac{-\Delta f_{\nu'}(\mathbf{k}') + \Delta n^{ph}_{\mu}(\mathbf{k}-\mathbf{k}')}{E^{(0)}_\nu(\mathbf{k}) - E^{(0)}_{\nu'}(\mathbf{k}') - \hbar\omega^{\mu}_{\mathbf{k}-\mathbf{k}'} + i\zeta} + \frac{\Delta f_{\nu'}(\mathbf{k}') + \Delta n^{ph}_{\mu}(\mathbf{k}'-\mathbf{k})}{E^{(0)}_\nu(\mathbf{k}) - E^{(0)}_{\nu'}(\mathbf{k}') + \hbar\omega^{\mu}_{\mathbf{k}'-\mathbf{k}} + i\zeta} \right], \tag{11.47}$$

where $\Delta f_\nu(\mathbf{k})$ and $\Delta n^{ph}_{\mu}(\mathbf{k})$ denote the optically-induced changes to the electronic and phonon distribution functions, respectively. Here, we do not calculate the self-energy self-consistently, but rather we use unrenormalized bandenergies $E^{(0)}_\nu(\mathbf{k})$ in the denominators. A self-consistent calculation would lead to higher-order, and therefore even smaller, corrections to the self-energy.

The basic functional form of the differential self-energy is quite similar to that of the self-energy, but of course the values are different and depend sensitively on the condition under which the changes are observed. In our optical experiments the differential transmission signal probes the differential absorption via observation of transmission changes. If the delay time between pump and probe is larger than the electron-hole recombination time and larger than the time it takes for the phonons to come to a quasi-thermal equilibrium – which is described by an elevated phonon temperature entering the phonon distribution $n^{\text{ph}}_{\mu}(\mathbf{k})$ – then the observed change depends only on the initial temperature T_{equ} and the actual temperature T. This in turn depends on the delay time and approaches T_{equ} in the limit of infinite delay times.

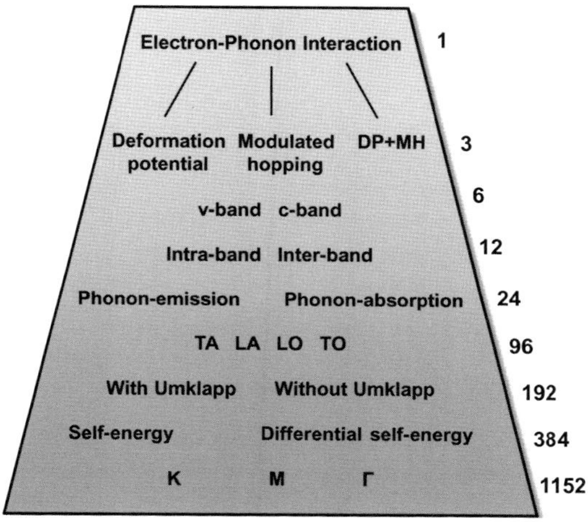

Fig. 11.19 Sketch showing that obtaining a comprehensive understanding of electron-phonon interaction in graphene is challenging. For the basic parameters entering the theory, the number of combinations (indicated on the right) grows rapidly.

We would like to finish this section with a general comment. In principle, we should be able to use the theory developed so far to obtain a comprehensive understanding of the electron-phonon process that occur in graphene. However, we have also seen that there are a number of distinct physical processes involved, and the observation can be either a direct observation of the self-energy or of the differential self-energy. We would also like to have that information for various regions in the k-space, for example close to the K, M, or Γ point. As we see in Fig. 11.19, obtaining a comprehensive understanding is challenging simply because of the relatively large number of processes and modes of observation.

11.3 Differential Transmission Spectra

As mentioned above, we have performed differential transmission spectroscopy in the UV regime, using a pump pulse with fixed frequency and a probe that scans the frequencies between 3.5 and 5 eV. To compute differential transmission spectra theoretically, we use the susceptibility and transmission formula found in the Introductory Tutorial, chapter 1. More specifically, we calculate the transmission twice, first for the sample at room temperature, $T_{equ} = 300\text{K}$, and then for an elevated phonon temperature. At the elevated temperature, the band energies are renormalized by the differential self-energies:

$$\chi(\omega) = -\frac{2e^2}{m^2\omega^2} \int_{BZ} \frac{d^2k}{(2\pi)^2} \langle [\mathbf{p}(\mathbf{k}) \cdot \hat{\mathbf{e}}_0]^2 \rangle_\theta \frac{f_{-1}(\mathbf{k}) - f_1(\mathbf{k})}{\hbar\omega - \Delta E(\mathbf{k}) + i\gamma(\omega, \mathbf{k})} L_F(\mathbf{k}), \tag{11.48}$$

where

$$\Delta E(\mathbf{k}) = E_1(\mathbf{k}) - E_{-1}(\mathbf{k}) + \text{Re}\Delta\Sigma_1(\mathbf{k}) - \text{Re}\Delta\Sigma_{-1}(\mathbf{k}) \tag{11.49}$$

and

$$\gamma(\omega, \mathbf{k}) = \gamma_0(\omega) - \text{Im}\Delta\Sigma_1(\mathbf{k}) - \text{Im}\Delta\Sigma_{-1}(\mathbf{k}). \tag{11.50}$$

We use $\gamma_0(\omega) = \gamma_0 \tanh(0.4\omega/\omega_m)$ with $\gamma_0 = 0.8$ eV and $\hbar\omega_m = 4.6$ eV, and a phenomenological Fano-like lineshape factor after Ref. [45], $L_F(\mathbf{k}) = (-1 + x)^2/(1 + x^2)$, $x = 2(\Delta E(\mathbf{k}) - \hbar\omega_F)/\gamma_F$ with $\hbar\omega_F = 5.12$ eV and $\gamma_F = 0.95$ eV. The single-particle energies $E_\nu(\mathbf{k})$ entering Eq. (11.48) differ from $E_s^{(0)}(\mathbf{k})$ in that we take the hopping parameter to be $\tilde{t} = 2.38$ eV in order to shift the absorbance maximum to coincide with that of the experimental spectrum. In the calculation of self-energies we use the value of $t = 2.55$ eV [36] for the hopping parameter.

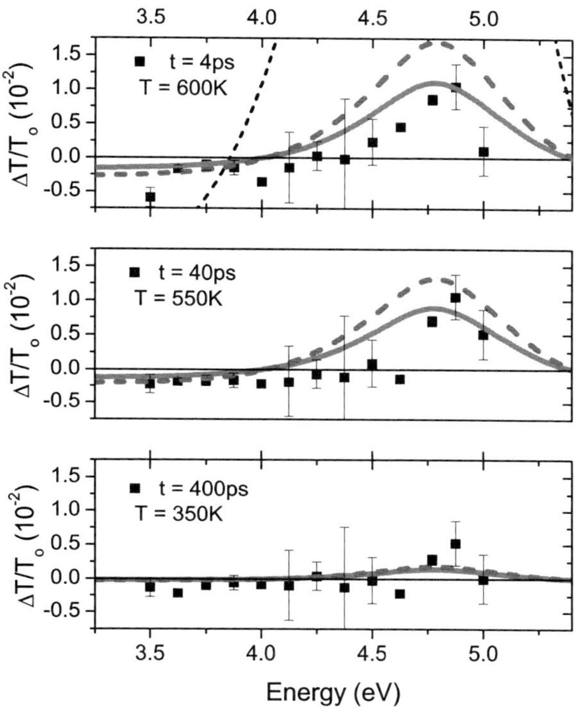

Fig. 11.20 $\Delta T/T$ as a function of probe photon energy at three time delays (4, 40, and 400 ps). The pump pulse fluence is 400 μJ/cm^2 with fixed excitation energy of 4.8 eV. Each panel is labeled with a specific phonon temperature value. The calculated results for coupling to acoustic phonons only with $D^{\rm ac}_{\rm eff} = 5.3$ eV are shown as solid lines, and acoustic+optical phonons with $D^{\rm op}_{\rm eff} = 11$ eV/Å are shown as long dashed lines. The short dashed line (top panel) shows acoustic-only coupling with $D^{\rm ac}_{\rm eff} = 20$ eV. From Ref. [8].

Since the light beam irradiates many different flakes of graphene with random in-plane orientation in each layer, we average $\mathbf{p}(\mathbf{k})\cdot\hat{\mathbf{e}}_0$ over the angle θ, where $\mathbf{p}(\mathbf{k})$ is the momentum matrix element, and $\hat{\mathbf{e}}_0 = (\cos\theta, \sin\theta)$ is the polarization unit vector of the light field. The relationship between the momentum matrix element and the k-gradient dipole matrix element was discussed in detail in the Introductory Tutorial, chapter 1, where it was shown that $e\hbar\mathbf{p}(\mathbf{k}) = im_0[E^{(0)}_{-1}(\mathbf{k}) - E^{(0)}_{1}(\mathbf{k})]\mathbf{d}(\mathbf{k})$ with $\mathbf{d}(\mathbf{k}) = e\frac{1}{2}\vec{\nabla}_{\mathbf{k}}\phi(\mathbf{k})$, $(-e)$ is the electron charge, and m_0 is the free electron mass in vacuum.

More details regarding the relation between momentum and dipole matrix elements in crystals can be found in Ref. [46].

To compute the transmitted and reflected light field amplitudes of the multi-layer system (in our case 10 layers), we use a conventional transfer matrix method with transfer matrices for the two-dimensional graphene sheets. Details of the transfer matrix method can be found in the Introductory Tutorial, chapter 1.

Figures 11.20 show experimental differential spectra for delay times 4, 40 and 400 ps, and Figures 11.22 overlay the 4 ps experimental differential spectra with calculated spectra for an elevated temperature of 600K. For the delay time of 4ps, we estimate the initial temperature $\mathcal{T}_0$ by assuming that within a short time $\tau < 2$ ps [13,14,16–18,21], the photoexcited carriers relax and the supplied laser pulse energy E_L is transferred to the phonons with internal energy given by

$$U_{ph}(\mathcal{T}_0) = \sum_{\mathbf{q}\mu} \hbar\omega^{\mu}_{\mathbf{q}}[n^{\text{ph}}_{\mu}(\mathbf{q}, \mathcal{T}_0) + 1/2]. \tag{11.51}$$

By calculating the internal energy of phonons as a function of temperature, the initial temperature is found using $U_{ph}(\mathcal{T}_0) = U_{ph}(300\text{K}) + E_L$, as illustrated in Fig. 11.21. For the highest incident fluence of 400 μJ/cm^2, the estimated initial temperature of the phonon system is $\mathcal{T}_0 = 610$ K. The temperature will decrease as the phonons decay and transfer energy to the substrate. We find that in the electron-phonon coupling dominated region ($\tau > 2$ ps), the $\Delta T/T$ at 4.8 eV is linearly related to the temperature, so the temperature decay profile may be estimated by fitting an exponential decay curve to $\Delta T/T$ at 4.8eV in figure 11.20(a). Using this method, the phonon temperatures as a function of delay time, $\mathcal{T}(\tau)$, are estimated to be approximately 600, 550, and 350 K at time delays of $\tau = 4$, 40, and 400 ps, respectively.

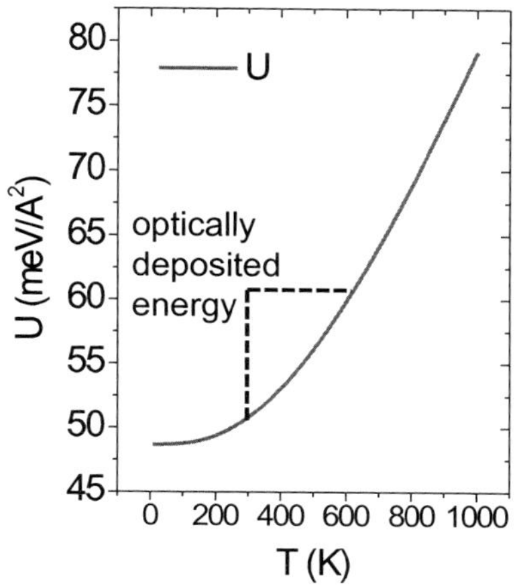

Fig. 11.21 Internal energy of phonon system. Also indicated is the increase of the phonon temperature due to the estimated energy deposited by the pump beam.

The agreement between the experiment and the theory based on effective deformation potentials shown in Fig. 11.20 is relatively good, which in turn helps to establish the approximate values of the effective deformation potentials. We can also obtain results based on the more complete theory that contains both DP and DM contributions and both longitudinal and transverse acoustic and optical phonon branches, then compare them to experimental data. Figure 11.22 shows that comparison, where we use values for the deformation potentials consistent with Ref. [32] but with negative signs: $D_{MH} = -2.5\text{eV}$, $D_{DP} = -2.8\text{eV}$. First we notice that these theoretical results agree even better than the ones shown in Fig. 11.20. When we take all phonons (optical and acoustic) into account, we see that the MH contribution dominates over the DP contribution to the signal. This is consistent with similar findings in, for example, Ref. [47]. We also see that ingoring umklapp process reduces the signal to approximately half of its value. That is not surprising, as the signal is dominated by transitions at the BZ boundary (the M-point), and for that point the umklapp region is approximately half of the BZ. (See Ref. [26] for more details.) On the other hand, when we take both MH and DP contributions into account, we see that the acoustic phonon contribution dominates that from the optical phonons. This insight is one of the advantages of optical differential transmission spectroscopy. In many other experiments, optical phonon contributions dominate over those from acoustic phonons. We are particularly interested in acoustic phonons, and from Eq. (11.47) we see that the phonon emission contribution described by the "1" in the first fraction in the self-energy (11.43) is not present in the differential self-energy. Hence, both fractions in Eq. (11.47) require populations to be nonzero. Since the differential electron populations are negligible in our case, and the fractions entering the differential self-energy are dominated by the differential phonon occupations. Generally acoustic phonons have larger occupations than optical phonons, so their differential phonon occupation also turns out to be larger than that of the optical phonons. Even though the coupling coefficients $C'^{\mu}_{\nu,\nu'}(\mathbf{k},\mathbf{k}')$ are generally smaller for acoustic than optical phonons, the overall contribution of the acoustic phonons can still be substantial and even larger than that from the optical phonons. This is clearly what we see from Fig. 11.22.

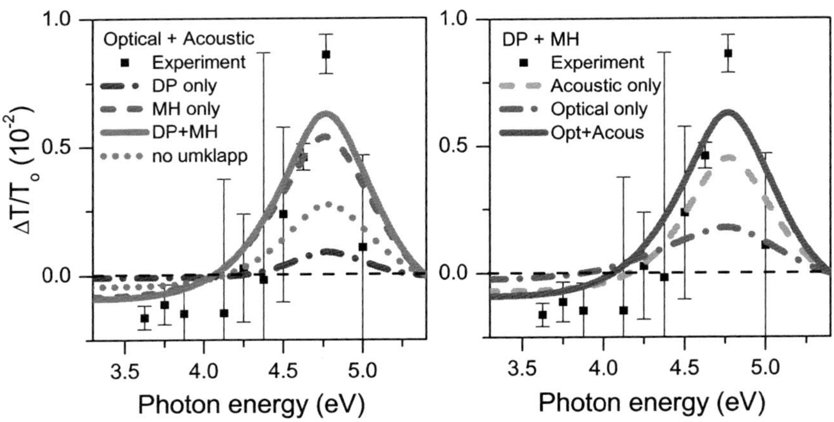

Fig. 11.22 Calculated differential transmission spectra (solid lines) and experimental data (squares, after Ref. [8]), for delay time 4ps. The equilibrium temperature is 300K, and the elevated phonon temperature is 600K. (a) The contributions from DP and MH for both optical and acoustic phonons. (b) The contribution from optical and acoustic phonons for DP and MH. From Ref. [26].

Bibliography

[1] K. I. Bolotin, K. J. Sikes, J. Hone, H. L. Stormer and P. Kim, Temperature-dependent transport in suspended graphene, *Phys. Rev. Lett.* **101**, p. 096802 (2008).

[2] J. H. Chen, C. Jang, S. D. Xiao, M. Ishigami and M. S. Fuhrer, Intrinsic and extrinsic performance limits of graphene devices on SiO2, *Nature Nanotechnology* **3**, pp. 206–209 (2008).

[3] C. R. Dean, A. F. Young, I. Meric, C. Lee, L. Wang, S. Sorgenfrei, K. Watanabe, T. Taniguchi, P. Kim, K. L. Shepard and J. Hone, Boron nitride substrates for high-quality graphene electronics, *Nature Nanotechnology* **5**, 10, pp. 722–726 (2010).

[4] K. Kaasbjerg, K. S. Thygesen and K. W. Jacobsen, Unraveling the acoustic electron-phonon interaction in graphene, *Phys. Rev. B* **85**, 16, p. 165440 (2012).

[5] K. M. Borysenko, J. T. Mullen, E. A. Barry, S. Paul, Y. G. Semenov, J. M. Zavada, M. B. Nardelli and K. W. Kim, First-principles analysis of electron-phonon interactions in graphene, *Phys. Rev. B* **81**, 12, p. 121412 (2010).

[6] T. Low, V. Perebeinos, R. Kim, M. Freitag and P. Avouris, Cooling of photoexcited carriers in graphene by internal and substrate phonons, *Phys. Rev. B* **86**, p. 045413 (2012).

[7] V. Perebeinos and P. Avouris, Inelastic scattering and current saturation in graphene, *Phys. Rev. B* **81**, 19, p. 195442 (2010).

[8] A. T. Roberts, R. Binder, N. H. Kwong, D. Golla, D. Cormode, B. J. LeRoy, H. O. Everitt and A. Sandhu, Optical characterization of electron-phonon interactions at the saddle point in graphene, *Phys. Rev. Lett.* **112**, 18, p. 187401 (2014).

[9] X. S. Li, W. W. Cai, J. H. An, S. Kim, J. Nah, D. X. Yang, R. Piner, A. Velamakanni, I. Jung, E. Tutuc, S. K. Banerjee, L. Colombo and R. S. Ruoff, Large-area synthesis of high-quality and uniform graphene films on copper foils, *Science* **324**, 5932, pp. 1312–1314 (2009).

[10] L. Malard, M. Pimenta and G. Dreselhaus, Raman spectroscopy in graphene, *Phys. Rep.* **473**, pp. 51 – 87 (2009).

[11] D. Graf, F. Molitor, K. Ensslin, C. Stampfer, A. Jungen, C. Hierold and

L. Wirtz, Spatially resolved Raman spectroscopy of single- and few-layer graphene, *Nano Letters* **7**, 2, pp. 238–242 (2007).
[12] T. Yu, J. Z. Shang, J. Y. Lin and G. G. Gurzadyan, Ultrafast Electron-Optical Phonon Scattering and Quasiparticle Lifetime in CVD-Grown Graphene, *ACS Nano* **5**, 4, pp. 3278–3283 (2011).
[13] M. Breusing, S. Kuehn, T. Winzer, E. Malic, F. Milde, N. Severin, J. P. Rabe, C. Ropers, A. Knorr and T. Elsaesser, Ultrafast nonequilibrium carrier dynamics in a single graphene layer, *Phys. Rev. B* **83**, 15, p. 153410 (2011).
[14] E. Malic, T. Winzer, E. Bobkin and A. Knorr, Microscopic theory of absorption and ultrafast many-particle kinetics in graphene, *Phys. Rev. B* **84**, 20, p. 205406 (2011).
[15] S. Winnerl, M. Orlita, P. Plochocka, P. Kossacki, M. Potemski, T. Winzer, E. Malic, A. Knorr, M. Sprinkle, C. Berger, W. A. de Heer, H. Schneider and M. Helm, Carrier relaxation in epitaxial graphene photoexcited near the dirac point, *Phys. Rev. Lett.* **107**, 23, p. 237401 (2011).
[16] D. Sun, C. Divin, C. Berger, W. A. de Heer, P. N. First and T. B. Norris, Hot carrier cooling by acoustic phonons in epitaxial graphene by ultrafast pump-probe spectroscopy, *Physica Status Solidi C: Current Topics in Solid State Physics* **8**, 4, pp. 1194 – 1197 (2011).
[17] R. W. Newson, J. Dean, B. Schmidt and H. M. van Driel, Ultrafast carrier kinetics in exfoliated graphene and thin graphite films, *Opt. Exp.* **17**, pp. 2326–2333 (2009).
[18] J. M. Dawlaty, S. Shivaraman, M. Chandrashekhar, F. Rana and M. G. Spencer, Measurement of ultrafast carrier dynamics in epitaxial graphene, *Appl. Phys. Lett.* **92**, 4, p. 042116 (2008).
[19] D. Sun, Z. K. Wu, C. Divin, X. B. Li, C. Berger, W. A. de Heer, P. N. First and T. B. Norris, Ultrafast Relaxation of Excited Dirac Fermions in Epitaxial Graphene Using Optical Differential Transmission Spectroscopy, *Phys. Rev. Lett.* **101**, p. 157402 (2008).
[20] J. H. Strait, H. N. Wang, S. Shivaraman, V. Shields, M. Spencer and F. Rana, Very slow cooling dynamics of photoexcited carriers in graphene observed by optical-pump terahertz-probe spectroscopy, *Nano Lett.* **11**, 11, pp. 4902–4906 (2011).
[21] J. Z. Shang, Z. Q. Luo, C. X. Cong, J. Y. Lin, T. Yu and G. G. Gurzadyan, Femtosecond UV-pump/visible-probe measurements of carrier dynamics in stacked graphene films, *Appl. Phys. Lett.* **97**, 16, p. 163103 (2010).
[22] B. Sun and M. Wu, Negative differential transmission in graphene, *Phys. Rev. B* **88**, pp. 235422 –(10) (2013a).
[23] B. Sun and M. Wu, Microscopic theory of ultrafast dynamics of carriers photoexcited by THz and near-infrared linearly polarized laser pulses in graphene, *New Journal of Physics* **15**, pp. 083038 –(30) (2013b).
[24] L. Malard, K.F.Mal, A. C. Neto, N. Peres and T. Heinz, Observation of intra- and inter-band transitions in the transient optical response of graphene, *New Journal of Physics* **15**, pp. 015009 –(14) (2013).
[25] F. Kadi, T. Winzer, E. Malix, A. Knorr, F. Gottfert, M. Mittendorff, S. Winnerl and M. Helm, Microscopic description of intraband absorption

in graphene: The occurrence of transient negative differential transmission, *Phys. Rev. Lett.* **113**, pp. 035502 –(5) (2014).

[26] R. Binder, A. Roberts, N. Kwong, A. Sandhu and H. Everitt, Global k-space analysis of electron-phonon interaction in graphene and application to M-point spectroscopy, *Phys. Rev. B* **93**, pp. 085414 – (18) (2016).

[27] L. Pietronero, S. Strössler and H. Zeller, Electrical conductivity of a graphite layer, *Phys. Rev. B* **22**, pp. 904 – 910 (1980).

[28] L. Nordheim, Zur Elektronentheorie der Metalle. I, *Ann. d. Physik* **9**, p. 607 (1931).

[29] G. D. Whitfield, Theory of electron-phonon interactions, *Phys. Rev.* **121**, p. 720 (1961).

[30] O. Madelung, *Introduction of Solid-State Theory*, 1st edn. Springer-Verlag, Heidelberg (1978).

[31] V. M. Pereira, A. H. C. Neto and N. M. R. Peres, Tight-binding approach to uniaxial strain in graphene, *Phys. Rev. B* **80**, 4, p. 045401 (2009).

[32] K. Kaasbjerg, K. Thygesen and K. Jacobsen, Unraveling the acoustic electron-phonon interaction in graphene, *Phys. Rev. B* **85**, pp. 165440 – (8) (2012).

[33] L. Woods and G. Mahan, Electron-phonon effects in graphene and armchair (10,10) single-wall carbon nanotubes, *Phys. Rev. B* **61**, pp. 10651 – 10663 (2000).

[34] N. W. Ashcroft and N. D. Mermin, *Solid State Physics*. Holt, Rinehart and Winston, Dumfries, NC (1976).

[35] R. Saito, G. Dresselhaus and M. Dresselhaus, *Physical Properties of Carbon Nanotubes*, 1st edn. World Scientific Publishing Co., London (1998).

[36] G. Samsonidze, E. Barros, R. Saito, J. Jiang, G. Dresselhaus and M. Dresselhaus, Electron-phonon coupling mechanism in two-dimensional graphite and single-wall carbon nanotubes, *Phys. Rev. B* **75**, pp. 155420 – (8) (2007).

[37] A. H. Castro-Neto, F. Guinea, N. Peres, K. Novoselov and A. Geim, The electronic properties of graphene, *Rev. Mod. Phys.* **81**, pp. 110 – 155 (2009).

[38] H. Suzuura and T. Ando, Phonons and electron-phonon scattering in carbon nanotubes, *Phys. Rev. B* **65**, p. 235412 (2002).

[39] G. D. Mahan, *Many-Particle Physics*. Plenum Press, New York (1981).

[40] C.-H. Park, F. Giustino, M. L. Cohen and S. G. Louie, Velocity renormalization and carrier lifetime in graphene from the electron-phonon interaction, *Phys. Rev. Lett.* **99**, pp. 086804 –(4) (2007).

[41] M. Calandra and F. Mauri, Electron-phonon coupling and electron self-energy in electron-doped graphene: Calculation of angular-resolved photoemission spectra, *Phys. Rev. B* **76**, 20, pp. 205411 – (9) (2007).

[42] Z. Li and J. Carbotte, Phonon structure in dispersion curves and density of states of massive dirac fermions, *Phys. Rev. B* **88**, p. 045417 (2013).

[43] Z. Li and J. Carbotte, Electron-phonon correlations on spin texture of gapped helical dirac fermions, *Eur. Phys. J. B* **88**, p. 87 (2015).

[44] G. D. Mahan, *Many-Particle Physics*. Plenum Press, New York (1986).

[45] K. Mak, J. Shan and T. Heinz, Seeing many-body effects in single- and few-layer graphene: Observation of two-dimensional saddle-point excitons, *Phys. Rev. Lett.* **106**, pp. 046401 – (4) (2011).

[46] B. Gu, N. Kwong and R. Binder, Relation between the interband dipole and momentum matrix elements in semiconductors, *Phys. Rev. B* **87**, pp. 125301 – (17) (2013).
[47] C. H. Park, N. Bonini, T. Sohier, G. Samsonidze, B. Kozinsky, M. Calandra, F. Mauri and N. Marzari, Electron-phonon interactions and the intrinsic electrical resistivity of graphene, *Nano Letters* **14**, 3, pp. 1113–1119 (2014).

Part 5

Device Applications

Chapter 12

Femtosecond Pulse Generation with Voltage-Controlled Graphene Saturable Absorbers

Işınsu Baylam[1], Sarper Özharar[2], Nurbek Kakenov[3], Coşkun Kocabaş[3], and Alphan Sennaroğlu[1*]

[1]*Laser Research Laboratory, Departments of Physics and Electrical-Electronics Engineering, Koç University, Istanbul 34450, Turkey*

[2]*Department of Electrical-Electronics Engineering, Bahçeşehir University, Istanbul 34353, Turkey*

[3]*Department of Physics, Bilkent University, Ankara 06800, Turkey*

* *asennar@ku.edu.tr*

We describe experiments where voltage controlled graphene supercapacitor devices are used as saturable absorbers to initiate femtosecond pulses from tunable solid-state lasers. The device architectures allow the control of the Fermi level and hence the overall insertion loss of these devices at low operating voltages on the order of a few volts. Hence, they can be readily incorporated into the resonator of low-gain solid-state lasers. After the background sections on laser fundamentals and ultrafast pulse generation, the second part of the

Chapter focuses on the spectroscopic characteristics of graphene and use of graphene supercapacitor devices as saturable absorbers to initiate femtosecond pulses. We present comparative, experimental data taken with a multi-pass cavity, Cr^{4+}:forsterite laser, where two different graphene-based supercapacitor device architectures were used to generate sub-100-fs pulses near 1250 nm.

12.1. Introduction and Overview of the Chapter

This Chapter focuses on the use of graphene based devices, in particular, graphene supercapacitors, as saturable absorbers for the initiation of femtosecond pulse generation from solid-state lasers. Since single-layer graphene already works well as a saturable absorber over a broad wavelength range, one naturally asks the question "why should we go to more complicated graphene device structures if monolayer graphene already functions well as a versatile saturable absorber?." To answer this question, we note that when monolayer graphene is introduced into the resonator of tunable solid-state lasers to initiate pulsed operation, its small-signal insertion loss may cause a considerable reduction in the laser power efficiency. Hence, it is important to develop techniques to control the absorption level and also, the insertion loss of monolayer graphene.

In this Chapter, we describe an approach where the absorption level of single layer graphene can be lowered so that it can be incorporated into the resonator of low-gain solid-state lasers. In particular, we focus on graphene based supercapacitors of two different design architectures. Our experimental results clearly demonstrate that by application of moderate bias voltages on the order of a few volts, the supercapacitor structures can be introduced into the laser resonators with low loss. At these bias voltages, the devices further have sufficient saturation characteristics to effectively initiate femtosecond pulse generation.

The background sections of the Chapter introduce the reader to tunable solid-state lasers, methods of femtosecond pulse generation, effect of Kerr nonlinearities and dispersion on ultrashort pulses, power analysis of solid-state lasers, and design considerations of laser resonators and multi-pass cavities. The remaining sections focus on the spectroscopic characteristics of graphene and use of graphene supercapacitor devices as saturable absorbers to initiate femtosecond pulses. We further present experimental data taken with a multi-pass cavity, Cr^{4+}:forsterite laser, where two different graphene-based supercapacitor devices were successfully utilized to generate sub-100-fs pulses near 1250 nm.

12.2. Tunable Solid-State Lasers and Ultrashort Pulse Generation

Lasers are oscillators in which amplification at optical frequencies is combined with positive feedback provided by highly reflecting cavity mirrors to generate coherent, directional light [1-3]. Here, amplification is based on stimulated emission and may occur in gaseous, liquid, or solid media, provided that the energy level structure allows for the possibility to obtain population inversion (in other words, a larger number of atoms/molecules in the excited state than in the lower state) under a suitable pumping scheme. Oscillation results once the gain exceeds the losses over one round trip inside the laser cavity. In this Chapter, we will focus primarily on the use of graphene supercapacitor structures as saturable absorbers in pulse generation from tunable solid-state lasers. In this class of lasers, the gain medium used inside the resonator is typically an insulating crystal, glass, or ceramic doped with laser-active ions. Laser-active ions mainly come from the rare-earth (such as thulium, ytterbium, erbium, and so on) or transition metal (such as chromium, titanium, cobalt, and so on) series of the periodic table [4, 5]. In solid-state lasers, optical pumping with another laser or flashlamp is typically employed to obtain population inversion. The center operation wavelength of a tunable solid-state laser depends on the ion-host combination used. Electron-phonon coupling or splitting of the degenerate angular momentum states enables tunability of the output wavelength with wavelength-selective elements such as prisms or birefringent tuning plates placed inside the cavity [6, 7]. For example, titanium-doped sapphire (Ti^{3+}:Al_2O_3) gain medium can generate coherent radiation in the 660-1180 nm wavelength range when pumped in the green near 532 nm [8]. In this Chapter, all of the pulse generation experiments utilizing voltage controlled graphene devices were conducted inside a Cr^{4+}:forsterite tunable solid-state laser, which produces tunable output in the 1130-1367 nm wavelength range when pumped at 1064 nm [9]. The available broad gain bandwidth of tunable solid-state lasers also enables the generation of femtosecond pulses as we will discuss soon.

Coming back to the laser cavity, the boundary conditions imposed by the cavity architecture further determine the spatial intensity distribution (beam profile) as well as the resonant frequencies of the electromagnetic modes supported by the laser cavity. Typically, there are infinitely many resonant electromagnetic modes of a laser cavity, each described by three integers (m,n,q). Conventionally, the integers m and n give information about the transverse intensity distribution of the confined beam (in the plane which is perpendicular to the axis of the resonator along which the beam propagates) that satisfies the boundary conditions, whereas the integer q describes the longitudinal variation of the field along the axis of the resonator. Different complete orthogonal sets of functions may be used to describe the field distribution of the laser modes and, in general, resonant mode frequencies depend on all three integers (m,n,q). Furthermore, the output of the laser may in general be a superposition of several of these orthogonal modes. Here, in the discussion of mode locking, we will limit ourselves to the case of transverse electromagnetic modes which are denoted as $TEM_{m,n,q}$. In particular, we will consider the case where only the fundamental transverse mode $TEM_{0,0,q}$, which has a Gaussian transverse profile, is present with no other higher-order transverse modes. In this specific case, there are still infinitely many modes, each with the same transverse beam profile, but different longitudinal field distributions.

For a standard two-mirror cavity producing a Gaussian output beam, it can be easily shown that the frequencies of the modes with different longitudinal variation are equally spaced in the frequency domain. The frequency separation between two adjacent modes is referred to as the free spectral range (ν_{FSR}) and is given by the inverse of the round-trip time of the cavity:

$$\nu_{FSR} = \frac{c}{2L} = \frac{1}{\tau_c}, \tag{12.1}$$

where, c is the speed of light, L is the cavity length, and τ_c is the cavity round-trip time. Here, we assumed that the cavity is nearly empty (neglecting the length of the gain medium in relation to the total length of the cavity) and the refractive index of the medium between the end mirrors is nearly unity. In a typical laser, only a few of these longitudinal modes with frequencies close to the peak of the amplification band will oscillate with no fixed phase relation among themselves. As a result, the output spectrum of the laser will typically have a narrow linewidth and the superposition of the randomly oscillating modes will result in an approximately constant output power as a function of time. In this case, we say that the laser is operating in the continuous-wave (cw) mode.

12.2.1 *What is mode locking?*

There are different regimes of operation for a laser. In the earlier section, we already touched upon the cw regime of operation where the output of the laser stays nearly constant as a function of time. In addition to cw operation, a laser can be operated in pulsed mode by using different schemes such as Q switching, gain switching, or mode locking. Here, we will focus primarily on mode locking which is the method commonly employed to produce pulses with picosecond and femtosecond duration. Also, in the experiments described in later sections, graphene supercapacitor devices were employed to initiate mode-locked operation of Cr^{4+}:forsterite lasers.

Mode locking is simply a method to distribute the output energy of a laser among multiple longitudinal modes and locking their relative phase, resulting in the production of a periodic train of short pulses with broad spectral width [10]. To delineate the role of mode locking, we refer to Fig. 12.1, which shows the temporal profile of the laser output for different mode spectra and relative phase between modes.

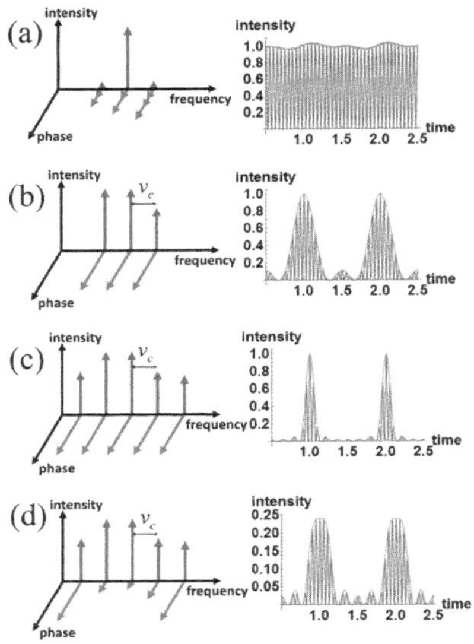

Figure 12.1. Temporal profile of the laser output for different mode spectra and relative phase, showing the basic characteristics of mode locking: (a) cw operation with one major lasing mode and randomly varying relative phase between neighboring modes. (b) Mode-locked operation with multiple modes oscillating with the same phase. (c) Mode-locked operation with a larger number of oscillating modes, leading to shorter output pulses (d) Mode locking in the presence of dispersion: the spectral bandwidth is the same as in (c) but, the relative phase varies with frequency, giving rise to temporally broader pulses.

As seen in Fig. 12.1(a), a few modes oscillating with random phases lead to a cw output with some random fluctuation of power over time. If the laser is mode-locked, then the spectral components are in phase, and as the number of locked modes increases, pulses formed in the time domain get shorter (Figures 12.1(b) and 12.1(c)). When all the modes oscillate with the same phase, the resulting pulse is said to be transform limited and has the shortest possible temporal duration for a given spectral bandwidth. For example, the pulses shown in Figs. 12.1(b) and 12.1(c) are transform limited. For a transform limited pulse, the pulse duration and spectral bandwidth are related through

$$\tau_p = \frac{K}{M \cdot v_{FSR}} = \frac{K}{\Delta v}. \tag{12.2}$$

In this equation τ_p is the full-width-at-half-maximum (FWHM) pulse duration (also referred to as the pulse width), M is the number of locked modes within the spectral bandwidth (FWHM) Δv, v_{FSR} is the free spectral range, and K is a constant on the order of unity, with a value that depends on the specific pulse shape (for example, K=0.441 and 0.315 for Gaussian and sech^2 shaped pulses, respectively) [11]. For example, a laser with a 100 MHz free spectral range and 50,000 locked modes (in other words, a bandwidth of 5 THz=5×10^{12}Hz) can generate a sech^2 pulse with a duration of 63 fs (63×10^{-15} seconds).

Note that, as can be seen in Fig. 12.1(d), the relative phase relation between the modes also affects the output pulse duration during mode locking. If the oscillating modes have a varying spectral phase, then the pulse width gets longer than the transform limited width. Such a relative phase difference among the modes may arise under different circumstances such as pulse propagation in a dispersive medium or self phase modulation in a Kerr medium. For example, in a dispersive medium, varying phase arises because the refractive index of the medium changes with wavelength. In this case, different wavelengths travel at different speeds, resulting in the broadening of the optical pulse.

The spectral phase of the pulse can be expressed as a Taylor series of the angular frequency ω around the center frequency ω_0 of the pulse as follows:

$$\Phi(\omega) = \Phi_0 + \frac{\partial \Phi}{\partial \omega}(\omega - \omega_0) + \frac{\partial^2 \Phi}{\partial \omega^2}\frac{(\omega - \omega_0)^2}{2} + \frac{\partial^3 \Phi}{\partial \omega^3}\frac{(\omega - \omega_0)^3}{6} + \ldots \tag{12.3}$$

In this expansion, the first derivative term ($\frac{\partial \Phi}{\partial \omega}$) is called the group delay and determines the transit time of the pulse in a dispersive medium. The second derivative term $\left(\frac{\partial^2 \Phi}{\partial \omega^2}\right)$ is called the group delay dispersion (GDD). While an optical pulse is traveling through a medium with a positive GDD, the low frequency (or long wavelength) components will travel faster than the high frequency (or short wavelength) components. The opposite is true for a medium with a negative GDD. Hence, irrespective of the sign of GDD, pulse broadening will occur. If an initially unchirped (i.e. transform limited) Gaussian pulse, having a width (FWHM) of τ_0, undergoes a group delay dispersion of GDD, pulse broadening occurs and the output pulse width τ becomes

$$\tau = \tau_0 \sqrt{1 + \left(\ln(16) \frac{GDD}{\tau_0^2} \right)^2} \, . \tag{12.4}$$

The third derivative term $\left(\frac{\partial^3 \Phi}{\partial \omega^3} \right)$ is called third order dispersion (TOD) and causes distortion of the pulse profile. For pulses with broad optical spectra, TOD and higher order terms (which are not shown in Eq. (12.3)) must be taken into account to fully describe the evolution of the pulse.

12.2.2 *Methods of mode locking*

Active or passive mode locking techniques can be employed to force the modes of a laser to oscillate with a fixed relative phase. In either case, the locking mechanism produces more loss (or less gain) for the cw regime than for the pulsed mode of operation. In this section, we will first provide a qualitative description of active mode locking based on amplitude modulation. Next, we will discuss Kerr lens mode locking and saturable absorber mode locking as examples of passive mode locking.

12.2.2.1 *Active mode locking*

In active mode locking, an intensity or a phase modulator is introduced into the cavity and is driven by an external RF source. The depth of modulation can be controlled externally and does not depend on the pulsewidth. Modulation can be done at different frequencies. For example, in fundamental mode locking, the loss is modulated at the frequency ν_m which is equal to the cavity fundamental frequency

($\nu_m = \nu_c = \nu_{FSR}$) [12]. Or, in harmonic mode locking of the nth order, the modulator is driven at an integer multiple of the cavity fundamental frequency ($\nu_m = n \cdot \nu_c = n\nu_{FSR}$) [13, 14]. Other modulation frequencies are also possible [15, 16]. Here, we will only describe amplitude modulation (AM) fundamental mode locking.

When an optical signal at the frequency ν_0 is amplitude modulated at the frequency of ν_m, two sidebands are generated at frequencies $\nu_0 + \nu_m$ and $\nu_0 - \nu_m$ that are in phase (see Fig. 12.2(a)). In other words, a portion of the mode energy at the frequency ν_0 is distributed to spectral lines at $\nu_0 + \nu_m$ and $\nu_0 - \nu_m$. If these sidebands are also supported by the laser cavity (in other words, if $\nu_m = \nu_c$), then they will not be filtered out by the cavity and reach the modulator after one round trip. As a result, these sidebands will generate their own sidebands. Hence, after each round trip, the total number of sidebands will increase and the bandwidth will get broader until some bandwidth limiting effect (such as dispersion or intracavity filtering) starts to dominate and balance the spectral bandwidth. This process is schematically shown in Fig. 12.2.

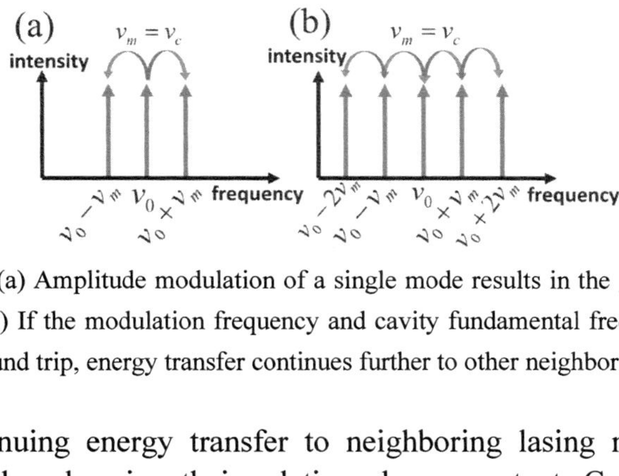

Figure 12.2. (a) Amplitude modulation of a single mode results in the generation of two sidebands. (b) If the modulation frequency and cavity fundamental frequency are equal, after each round trip, energy transfer continues further to other neighboring modes.

This continuing energy transfer to neighboring lasing modes "locks" them together, keeping their relative phase constant. Correspondingly, this process results in the generation of a periodic train of optical pulses in the time domain.

It can be shown that when dispersion and self phase modulation effects are neglected, the steady-state pulse duration (FWHM) during AM fundamental mode locking can be obtained from the Kuizenga-Siegman formula [10, 12, 17]:

$$\tau = \frac{\sqrt{2\ln 2}}{\pi} \cdot \left(\frac{g}{M}\right)^{1/4} \cdot \sqrt{\frac{1}{v_m \Delta v_g}}. \tag{12.5}$$

In this equation, g is the gain at the center frequency, M is the modulation strength, v_m is the modulation frequency, and Δv_g is the gain bandwidth of the laser medium.

12.2.2.2 *Passive mode locking*

Unlike active mode locking, no externally driven modulator is employed in passive mode locking. Instead, an intensity-dependent saturable loss mechanism initiates the generation of mode-locked pulses [10]. A commonly employed scheme is based on the self focusing of the intracavity laser beam due to the Kerr effect (typically inside the gain medium) and is known as Kerr lens mode locking (KLM). The refractive index of a Kerr medium depends on the light intensity and can be expressed as

$$n(r,t) = n_0 + n_2 I(r,t), \tag{12.6}$$

where n_0 is the low-intensity, wavelength-dependent refractive index of the Kerr medium, n_2 is the nonlinear refractive index coefficient, and $I(r,t)$ is the intensity of the light. Since, the intensity of an optical pulse depends both on position and time, the refractive index of the Kerr medium also becomes a function of position and time.

Let us first explore the spatial variation of the refractive index to describe the KLM action. In the spatial domain, the refractive index of the Kerr medium varies due to the position dependence of the beam intensity. For example, if the beam intensity has a Gaussian distribution, the center of the pulse at high intensity will experience a high refractive index (for $n_2>0$), whereas the parts further away from the optical axis will experience a lower refractive index (see Fig. 12.3(a)). As a result, the Kerr medium will behave as a lens and for $n_2>0$, a beam with a sufficiently high intensity will undergo self focusing. This is referred to as Kerr induced self focusing or Kerr lensing (see Fig. 12.3(b)). To initiate mode locking with KLM action, the pumping volume inside the gain medium of a laser is typically adjusted to be smaller than the lasing volume. Then, focusing around the gain medium is optimized to initiate self focusing of the laser beam. If sufficient self focusing can be initiated so that the lasing volume is reduced to overlap better with the pump volume, the laser beam experiences higher gain. Because self focusing becomes more effective at higher peak powers, Kerr lensing induces mode-locked operation since the resulting high-peak pulses have higher gain in comparison with cw regime of operation. One advantage of the KLM technique is that, most of the solid-state gain media also serve as the Kerr medium, requiring no additional components in the medium. On the other hand, critical cavity alignment is often necessary to initiate KLM action.

The focusing strength of Kerr lensing increases with the power of the pulse. The power at which the Kerr induced lensing balances the natural divergence of a Gaussian beam, is called the critical power (P_{cr}) and is given by

$$P_{cr} = \frac{\alpha\lambda^2}{8\pi n_0 n_2}, \tag{12.7}$$

where, α is a dimensionless correction factor between 3.77 and 6.4, λ is the center wavelength of the pulse, and n_0 and n_2 are as defined above [18]. Once the optical power is comparable to the critical power, effects due to Kerr induced focusing begin to play a major role in the shaping of the beam. Above the critical power, self focusing may cause instabilities in the mode-locked pulse train or even catastrophic damage inside the Kerr medium.

Now let us explore the Kerr effect in the temporal domain. In this case, the refractive index changes as a function of time across the temporal extend of the pulse and leads to a time varying phase. This effect is referred to as self-phase-modulation (SPM), since its strength depends on the intensity of the pulse (see Fig. 12.3 (c)). As can be seen in Fig. 12.3 (c), the phase change is maximum at the center of the pulse and decreases toward the wings. Also note that the instantaneous frequency of the signal which is given by the time derivative of its phase, varies as shown in Fig. 12.3(d). In other words, the pulse acquires a chirp similar to what it would experience in the presence of positive GDD.

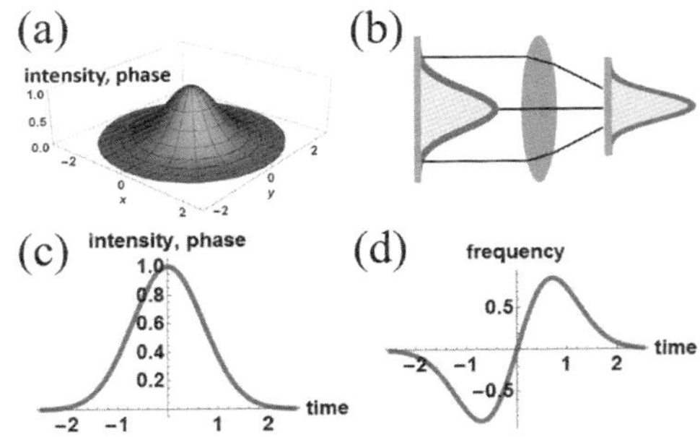

Figure 12.3. (a) Spatial intensity profile and the accompanying Kerr phase of a Gaussian beam. (b) Kerr induced self focusing. (c) Temporal intensity profile and the resulting Kerr phase of a Gaussian pulse. (d) Variation of the carrier frequency of a pulse as a function of time due to self phase modulation.

Another passive mode locking technique utilizes saturable absorbers to initiate a pulse train. In fact, in the experiments that are described later, graphene based supercapacitors are utilized as saturable absorbers to generate mode-locked pulses. For a saturable absorber, the absorption saturates and decreases with increasing intensity of the incident light. Hence, a low-intensity signal will experience a high loss, whereas a higher intensity signal will experience lower loss after passing through a saturable absorber. As a result, an optical pulse entering a saturable absorber will get shorter since the wings will experience more loss than the pulse center. Saturable absorbers can be classified in two groups as fast and slow saturable absorbers, depending on their recovery time compared to the duration of the pulse they are interacting with. Both can be used for initiating mode-locked operation [10, 19, 20]. If the saturable absorber recovery time is shorter than the pulse width, then it acts as a fast saturable absorber and its loss (Q) is determined by the instantaneous intensity of the pulse [10]:

$$Q = \frac{Q_0}{1 + \dfrac{I}{I_{sat}}}. \tag{12.8}$$

In this equation, Q_0 is the small signal loss, I is the intensity and I_{sat} is the saturation intensity at which the loss value becomes half of the small signal loss. For a slow saturable absorber, its recovery dynamics cannot be neglected and the loss can be modeled as

$$\frac{dQ}{dt} = -\frac{Q(t) - Q_0}{\tau_A} - \frac{Q(t)P(t)}{E_A}, \tag{12.9}$$

where τ_A is the decay time of the absorber, *P(t)* is the instantaneous power of the laser as a function time, and E_A is the saturation energy of the absorber [10, 20]

In tunable solid-state lasers, both types of passive mode locking (KLM and saturable absorber mode locking) are routinely employed to generate pulses in the picosecond and femtosecond regime. The balance between self-phase modulation and dispersion play an important role in the shaping of the steady-state pulse. In particular, since the effect of SPM is similar to positive GDD, a sufficient amount of negative GDD added to the laser cavity can cancel pulse distortions due to SPM. When SPM and negative GDD are balanced, solitary pulses can be generated from the laser resonator. The initiation scheme may depend on KLM, saturable absorber action or other intensity-dependent loss mechanism. In the solitary pulse regime, the pulses have sech^2 pulse profile and satisfy the well-known soliton area theorem [10]:

$$W\tau = 1.76\frac{|GDD| \cdot \lambda \cdot A_{eff}}{\pi \cdot n_2 \cdot L_k}, \tag{12.10}$$

where W is the intracavity pulse energy, τ is the FWHM pulse duration, L_k is the length of the Kerr medium where SPM occurs (in most cases it is the length of the gain crystal), A_{eff} is the effective cross-sectional area of the optical pulse, and GDD is the net negative dispersion of the laser cavity in one round trip. Note that, for a given Kerr medium and cavity design, A_{eff}, λ, n_2, and L_k are fixed, although A_{eff} may vary somewhat depending on where, in the stability region of the cavity, KLM is initiated. In such a case, for a fixed amount of GDD, a larger intracavity energy (W) results in the production of a shorter pulsewidth, provided that the reflectivity bandwidth of the cavity mirrors, gain bandwidth of the medium, and higher-order dispersion do not limit the pulsewidth.

12.3. Graphene Supercapacitors

12.3.1 *Overview of bulk graphene*

Graphene is a nano-structured material consisting of a mono-atomic layer of carbon atoms in a two-dimensional hexagonal arrangement. Because of its high electron mobility, good electrical conductivity, and unique optical properties [21], graphene has stimulated intense research in the fields of electronics and optics [22]. Monolayer graphene further has zero band gap, with the massless charge carriers obeying a linear dispersion relation [23]:

$$E = \hbar v_F |\mathbf{k}|. \tag{12.11}$$

Above, **k** is the two-dimensional (2D) wave-vector, v_F is the Fermi velocity (1.1×10^6 m/s), and $\hbar$ is Planck's constant. Even a mono-atomic layer of graphene interacts strongly with light and has a small-signal absorption (A) of about 2.3%. In addition, A is nearly constant over a very broad wavelength range and can be expressed in terms of fundamental constants as [24]

$$A = \pi\alpha \approx 2.3\%. \tag{12.12}$$

Here, α is the fine structure constant (1/137).

These novel physical properties of graphene, along with the availability of simple and low-cost growth techniques, make it an important candidate for the exploration of cutting-edge technological applications in electronics and photonics [21, 22]. Since this Chapter primarily focuses on the application of graphene based capacitor devices for the generation of ultrashort optical pulses, we provide here a brief overview of recent studies about the ultrafast optical properties of graphene. Apart from the ultrabroad absorption described above, the excited carriers of monolayer graphene have fast decay times [25] of the order of picoseconds. Furthermore, saturation of absorption can be readily obtained in graphene with saturation fluence levels on the order of 20 μJ/cm^2 [26, 27] These favorable nonlinear optical properties of graphene make it a versatile saturable absorber for ultrashort pulse generation as discussed in Section 12.2.2.2.

Ultrashort pulse generation has been demonstrated in different solid-state or fiber lasers [28], operating in different wavelength ranges from near to mid infrared, by utilizing graphene as a saturable absorber [27-34]. For example, at the short wavelength limit, a graphene mode-locked Ti:sapphire laser was shown to generate 63 fs pulses near 800 nm [32]. In another study, 226-fs pulses were generated at the wavelength of 2500 nm by using a graphene mode-locked Cr:ZnSe laser [34]. Experiments also show that graphene saturable absorbers can be used without damage up to fluence levels of 3×10^{12} W/cm^2 [35]. These results obtained from different lasers clearly establish graphene as a robust, low-cost saturable absorber that can operate over a very broad spectral range, much broader than what has been achieved with other saturable absorbers such as semiconductor saturable absorber mirrors [20] or carbon nanotubes [28].

12.3.2 *Why graphene based supercapacitors?*

Based on the foregoing discussion, one naturally raises the question mentioned in the Introduction: "why should we go to more complicated graphene device structures if monolayer graphene already functions well as a versatile saturable absorber?." The major reason is that when placed inside a laser resonator, even a mono-atomic layer of graphene results in 2.3% of small-signal absorption per pass, giving a round-trip loss of about 5%, in addition to the other sources of losses. Such a level of loss may not be tolerable in low-gain lasers where the net optical amplification is comparable to this loss level. This would also be an issue in lasers operated at low power output by using low-power, low-cost diode pump lasers [36, 37].

This limitation calls for approaches to control the absorption level of monolayer graphene, while maintaining its saturation characteristics. Obviously, reducing the thickness is not an option here!

Our studies clearly show that graphene based supercapacitor structures can offer a solution for controlling the loss level in graphene. To explain the operation of the device, we refer to Fig. 12.4 which shows the energy band structure of graphene for two different positions of the Fermi level. The energy-band diagram for monolayer graphene, where the Fermi level (E_F) lies at the Dirac point (neglecting the unintentional doping due to the substrate) is shown in Fig. 12.4(a). As can be seen from Fig. 12.4(a), absorption occurs readily for all photon energies (E_P) since the band gap energy of graphene is zero.

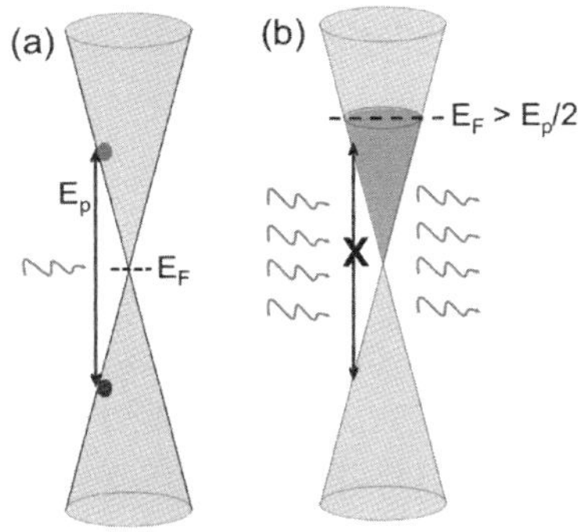

Figure 12.4. Sketch of the energy band diagram for graphene for the case when (a) the Fermi level lies at the Dirac point and (b) the Fermi level is shifted up.

If the Fermi level is shifted up from the Dirac point, no absorption occurs for photon energies less than twice the Fermi level shift, due to Pauli blocking. In other words, graphene becomes transparent for $E_p < 2\Delta E_F$ or equivalently, for $\lambda_p > hc/(2\Delta E_F)$, where E_p is the photon energy, ΔE_F is the shift in the Fermi level position, λ_p is the wavelength of light, and $hc = 1240\ eV - nm$. At room temperature, due to the nature of the Fermi-Dirac distribution, the transition of graphene from the absorbing to transparent state does not occur sharply at λ_p. Hence, for wavelengths near λ_p, it is possible to control the absorption level of graphene and therefore, its loss level inside a low-gain laser resonator by controlling the Fermi level position. As we will discuss in the coming sections, graphene based supercapacitor structures will be the preferred choice since they can be used to readily shift the Fermi level position with low applied bias voltages of the order of few volts.

12.3.3 *Chemically doped graphene and solid-state graphene capacitors*

Different methods have been proposed and demonstrated for the control of the Fermi level position and hence, the amount of absorption in graphene. In the case of chemical doping, a fixed shift of the Fermi level position can be achieved. To date, n-doped (electron) and p-doped (hole) graphene structures with shifted Fermi level have been reported by several groups [23, 38-41]. However, chemical doping is not suitable for controlling the loss level of a laser resonator because it leads to an irreversible change in the band structure and the resulting fixed Fermi level shift cannot be dynamically controlled. Another alternative involves electrostatic gating of the graphene by using solid-state capacitor structures. Here, graphene based solid-state capacitors have been demonstrated to obtain dynamic control of the Fermi level position, with important potential applications in both electronics and photonics [29, 42-46]. However, as far as laser mode locking applications are concerned, solid-state graphene capacitor architectures only operate over a limited wavelength range. This is because high bias voltages are typically needed to obtain sufficient Fermi level shift for the modulation of visible or near infrared absorption and this usually leads to dielectric breakdown. It is precisely at this point that we see the advantage of the graphene supercapacitor structures: the graphene based supercapacitor structures can be used effectively to control the absorption from the near infrared to the visible region at low operating voltages with a minimum risk of dielectric breakdown.

12.3.4 *Operational characteristics of graphene supercapacitors*

A capacitor, often used as an energy storage device, consists of two conducting plates (also referred to as electrodes in our discussion) separated with an insulating medium. Usually, a solid insulating material is sandwiched between these plates. In a supercapacitor, the conducting plates are separated with a high-dielectric-constant ionic electrolyte which can be organic or inorganic. In this case, formation of ultrathin electrical double layers (EDL) enhances the energy storage capability of the capacitor. A simplified explanation for EDL formation is offered in Fig. 12.5. More detailed analysis of the mechanism can be found in Ref. [47]. As shown in Fig. 12.5, application of a bias voltage between the electrodes of the supercapacitor polarizes the high dielectric constant electrolyte and leads to the formation of few-nanometer-thick electrical double layers. In particular, positively charged ions of the electrolyte are attracted by the negatively charged plate and these positive ions further attract the negative ions of the electrolyte. The same process also continues near the other plate, resulting in the formation of several EDLs inside the electrolyte. These EDLs act as separate supercapacitors connected in series and since the thickness of each EDL is determined by the ionic radii, which are of the order of few nanometers, large electric fields are produced within the supercapacitor at low operating voltages.

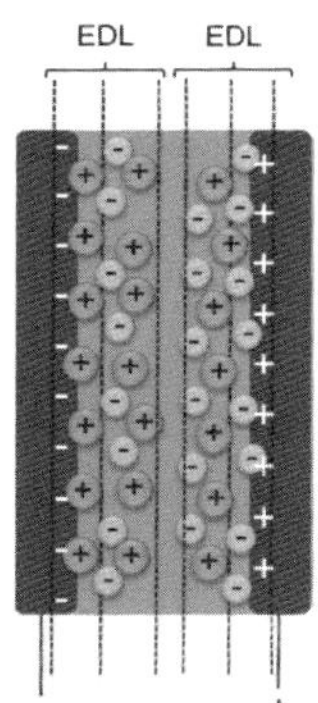

Figure 12.5. Formation of electrical double layers (EDLs) inside a supercapacitor. (For color reference, go to page 496.)

In the case of graphene-based supercapacitors, the capacitor structure is formed with graphene-graphene or graphene-metal electrodes. It has been shown that the supercapacitor architecture can electrostatically dope the graphene electrodes [48]. This is similar to the operation of solid-state graphene-based capacitors. However, due to the EDL formation, low bias voltages of the order of few volts (safe enough to avoid the dielectric breakdown of the electrolyte) are sufficient to produce a large enough shift of the Fermi level to modulate the absorption up to the visible wavelengths. Furthermore, as opposed to chemically doped graphene, electrostatic doping of the graphene layers by using supercapacitor architectures allows dynamical control of the Fermi level position and hence, the level of loss.

Graphene-based supercapacitor devices have been demonstrated by several groups for electronic and photonic applications [49-53]. Their application as fast saturable absorbers with controllable optical insertion loss for ultrashort pulse generation was first demonstrated by our groups [48, 51, 52]. In the following sections, two specific graphene-based supercapacitor structures (graphene-graphene and graphene-gold supercapacitors) will be described and experimental results obtained in femtosecond pulse generation will be presented.

12.3.5 *Voltage controlled absorption spectra of graphene supercapacitors*

Figure 12.6 shows one of the voltage controlled graphene supercapacitor saturable absorber (VCG-SA) architectures used for femtosecond pulse generation. The supercapacitor here consists of two graphene electrodes. The fabrication method of the graphene electrodes and the detailed production steps of the overall device can be found in Ref [51]. The electrolyte diethymethyl (2-methoxyethyl) ammonium bis (trifluro-methylsulfony) imide [DEME][TFSI] with a relative dielectric constant of 14 was used between the electrodes [54].

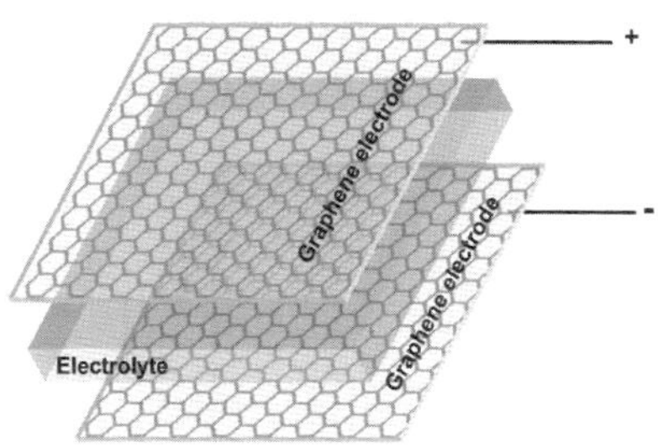

Figure 12.6. Schematic of the voltage controlled graphene supercapacitor saturable absorber (VCG-SA).

As described in the previous section, when a voltage bias is applied to the graphene electrodes, the electrolyte is polarized and electrical double layers (EDLs) are formed. These EDLs electrostatically dope the graphene electrodes, resulting in a shift of the Fermi level position.

To explore the voltage dependent insertion loss of the VCG-SA, steady-state optical transmission spectra of the device were measured at normal incidence for different bias voltages. Figure 12.7 shows the measured fractional change in the transmission as a function of applied voltage at six selected wavelengths. Full range of voltage-dependent steady-state transmission spectra may be found in Ref [51].

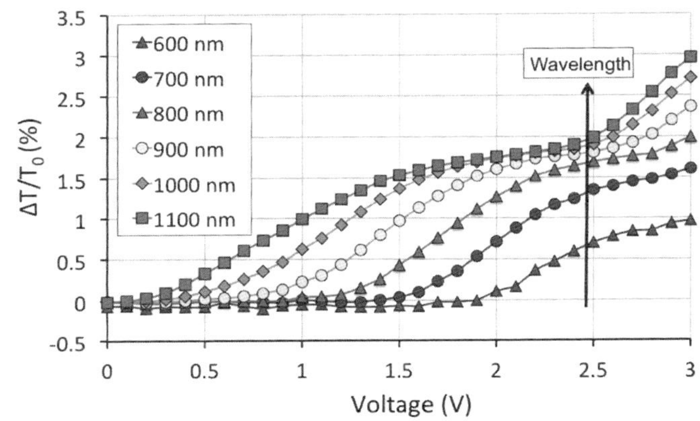

Figure 12.7. Measured variation of the fractional change in the steady-state transmission of the VCG-SA as a function of the applied voltage at different wavelengths.

To explain the behavior observed in Fig. 12.7, we first note that according to measurements, the resulting shift in the Fermi level varies nearly linearly with the applied bias voltage [48, 51]. As was discussed in Section 12.3.2 and as can be seen from Fig. 12.7, the transmission of the VCG-SA can be increased by applying a bias voltage. For example, the change in the transmission starts at around 2V of applied bias for 600 nm (corresponding to approximately 2 eV of photon energy), whereas, for 800 nm (nearly 1.5 eV of photon energy), transmission change begins at voltages close to 1.25 V. This effect clearly demonstrates the Pauli blocking principle for the VCG-SA described in Section 12.3.2. As expected, when the wavelength increases, the photon energy decreases and a lower bias voltage is required to cause the same amount of fractional change in the transmission. Note that, the two-step behavior in the voltage-dependent transmission curve of Fig. 12.7 is mainly due to the use of two graphene electrodes. As the voltage bias is increased at a given photon energy, one of the graphene electrodes reaches transparency before the other graphene electrode and at high enough voltages, the device eventually becomes completely transparent. The mechanism of this two-step change in the fractional transmission, which is possibly due to the asymmetry in the radii of the ions forming the EDLs, requires further investigation.

In our ultrashort pulse generation experiments, we developed another supercapacitor architecture consisting of one graphene electrode and one gold electrode (voltage controlled graphene-gold saturable absorber, VCG-gold-SA), as shown in Fig. 12.8 [52]. Use of only one graphene electrode further reduces the optical insertion loss of this novel design. The spacing between the two electrodes was filled with the same electrolyte. A hole with a diameter of 5mm was patterned on the gold electrode so that the laser beam can pass through the supercapacitor. The fabrication steps of the VCG-gold-SA are similar to those of the previous device and may be found in Ref. [54].

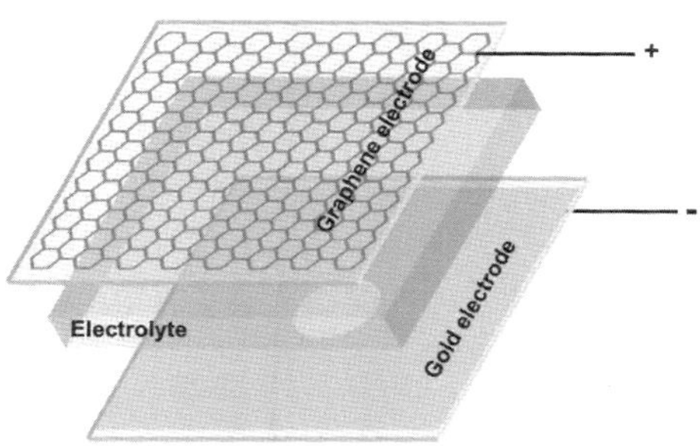

Figure 12.8. Schematic of the voltage controlled graphene gold supercapacitor saturable absorber (VCG-gold-SA).

The operation principle of the VCG-gold-SA is similar to the previous device and is based on Pauli blocking. However, it had two different features when compared with the supercapacitor having two graphene electrodes. First, even at zero bias, the Fermi level for VCG-gold-SA is already 0.5 eV above the Dirac point due to the mismatch of the work functions for graphene and gold electrodes [52]. Second, the optical transmission of the VCG-gold-SA did not exhibit a two-step variation with applied bias since only one graphene electrode was used. Figure 12.9 shows the measured fractional change in the optical transmission of the VCG-gold-SA as a function of the applied voltage. Similar bias voltage dependence is observed in this case. At the wavelength of 650 nm, the change in the transmission starts at around 2V and for higher wavelengths, the required voltage bias to obtain the same fractional transmission change decreases. Full range of voltage dependent optical transmission data for the VCG-gold-SA can be found in Ref. [52].

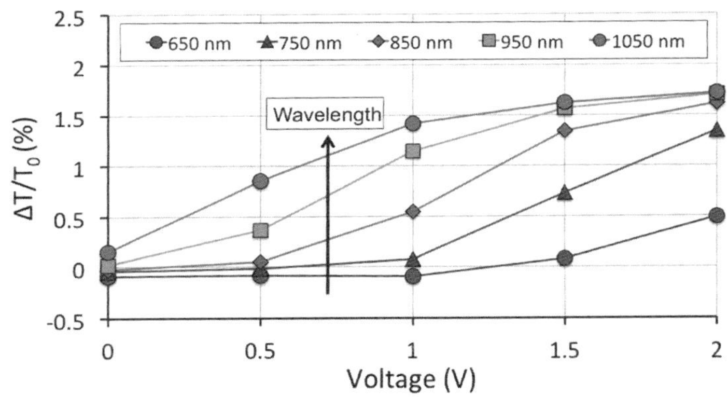

Figure 12.9. Measured variation of the fractional change in the optical transmission of the VCG-gold-SA as a function of the applied voltage.

Since only the optical insertion loss due to the graphene electrode(s) was modulated by using the supercapacitor design, the losses originating from the substrates and the electrolyte are not modified by the bias voltage. Hence, the electrolyte and the substrates for the electrodes should be carefully chosen to minimize the insertion losses at a given wavelength. In our experiments, a Cr^{4+}:forsterite tunable solid-state laser, operating around 1.25 μm, was used to test the fast saturable absorber performance of our devices. Therefore, graphene and/or gold electrodes were synthesized on an infrared transparent substrate (infrasil or quartz). Figure 12.10 further shows the optical transmission spectrum of the electrolyte Diethymethyl (2-methoxyethyl) ammonium bis (trifluro- methylsulfony) imide [DEME][TFSI], that was used during the device fabrication [52]. Note that absorption losses near 1.25 μm are minimal for this electrolyte.

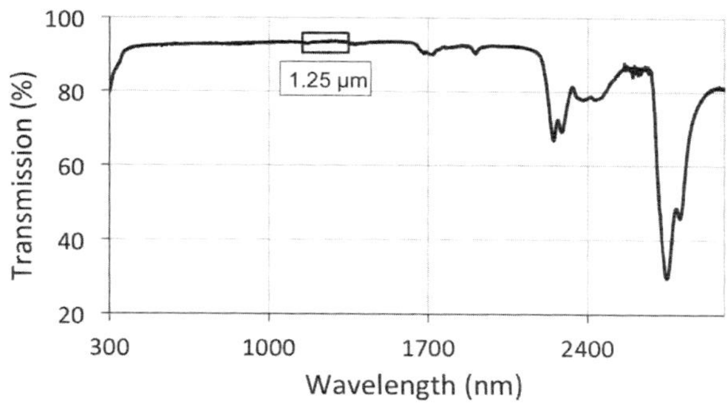

Figure 12.10. Measured transmission spectrum of the electrolyte Diethymethyl (2-methoxyethyl) ammonium bis (trifluro- methylsulfony) imide [DEME][TFSI].

12.4. Design Considerations for Mode-Locked Oscillators

12.4.1 *Output power analysis of a laser oscillator*

Here, we provide a simplified overview of how a laser operates and identify the important parameters that affect the power performance. The interested reader may consult other references that provide a detailed analysis of laser operation [1, 55]).

A laser is an oscillator, consisting of 4 main components as sketched in Fig. 12.11: 1) a gain medium to provide amplification, 2) a pumping mechanism to achieve amplification by stimulated emission, 3) a resonator consisting of highly reflecting mirrors (HR in Fig. 12.11) to provide positive feedback, and 4) a partially transmitting end mirror, commonly referred to as an output coupler (OC), where the output beam exits the oscillator. Pumping can be done in different ways. In the tunable solid-state lasers considered here, we typically employ optical end pumping by using another laser.

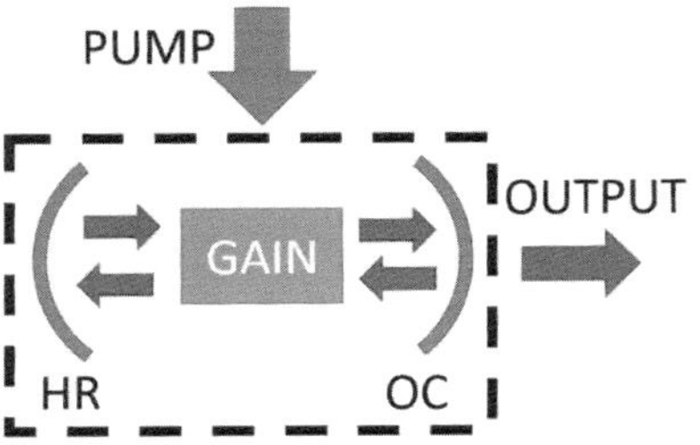

Figure 12.11. A simplified block diagram of a laser (HR: highly reflecting end mirror, OC: output coupler).

Figure 12.12 shows how the cw output power (P_{out}) of a laser oscillator changes with the pump power (P_p). As the pump power is increased, the optical gain due to stimulated emission will increase inside the cavity. Once the gain is sufficient to overcome all of the intracavity losses, lasing action begins. The pump power where lasing begins is known as the threshold pump power P_{th} and for an end pumped solid-state laser, is given by

$$P_{th} = \frac{\pi}{4} \frac{h\nu_P}{\tau_f \sigma_e} \frac{\left(w_L^2 + w_P^2\right)}{\eta_a} (L+T) \qquad (12.13)$$

Here, $h\nu_P$ is the pump photon energy, $\eta_a = \left(1 - \mathrm{Exp}(-\alpha_P \ell)\right)$ is the total small-signal pump absorption (α_p=small-signal differential pump absorption coefficient, ℓ=crystal length), w_L and w_p are the respective laser and pump spot sizes inside the gain medium, σ_e is the stimulated emission cross section of the gain medium, τ_f is the fluorescence lifetime of the upper laser level, L is the total round-trip loss of the resonator, and T is the transmission of the output coupler. Note that P_{th} varies linearly with (L+T). By using output couplers with different transmission values, one can use this fact to estimate the loss L of the resonator from pump threshold measurements.

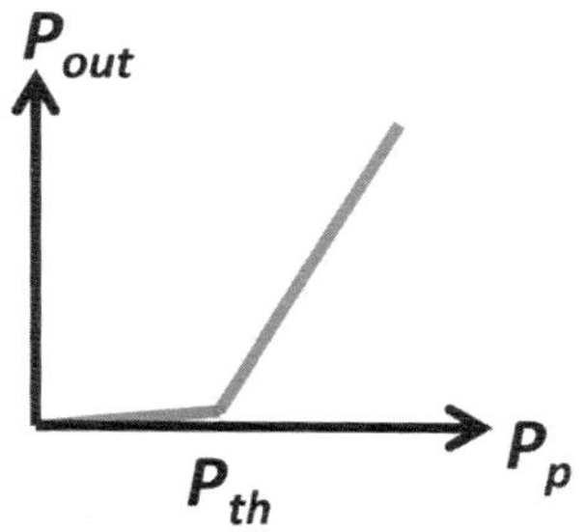

Figure 12.12. Typical variation of the output power (P_{out}) of a laser as a function of the pump power (P_p). The lasing begins when P_p is equal to the threshold power P_{th}.

For pump powers below P_{th}, the output power of a laser is typically negligible in comparison with what can be obtained above lasing threshold. For $P_p > P_{th}$, the laser produces a finite output power P_{out}, which can be expressed in terms of the pump power as

$$P_{out} = \eta \, (P_p - P_{th}). \tag{12.14}$$

Here, η is the laser power slope efficiency with respect to the input power and can be calculated from

$$\eta = \frac{\lambda_p}{\lambda_L} \cdot \frac{T}{T+L} \cdot \eta_a, \tag{12.15}$$

where λ_p is the pump wavelength and λ_L is the laser wavelength [56]. Other parameters appearing in Eq. (12.15) are as defined in Eq. (12.13). If we look at the output power as a function of the absorbed pump power, λ_p / λ_L corresponds to the maximum power slope efficiency that can be obtained in the limit of negligible resonator losses. This corresponds to the case where every absorbed pump photon leads to the generation of one laser photon from the resonator and gives the ideal quantum efficiency of the laser.

12.4.2 *Design and analysis of the laser cavity with ABCD matrices*

In this Section, we briefly describe how the beam parameters of the TEM_{00} mode (such as spotsize and radius of curvature) supported by a laser resonator are determined. We limit our discussion to the case of paraxial optics where the rays/optical beams remain close to the optical axis. Looking first at the ray model of light propagation, we can use 2×2 matrices known as the ABCD matrices to represent an optical system such as mirrors, lenses, or interfaces. In this approach, an optical ray is described by its height (r) and the angle (θ) made with respect to the optical axis. The effect of the optical system can then be calculated by using the matrix equation

$$\begin{bmatrix} r_{out} \\ \theta_{out} \end{bmatrix} = \begin{bmatrix} A & B \\ C & D \end{bmatrix} \cdot \begin{bmatrix} r_{in} \\ \theta_{in} \end{bmatrix}, \tag{12.16}$$

where r_{in} and θ_{in} are the input ray height and input ray angle before the optical element, and r_{out} and θ_{out} are the output ray height and angle after the optical element. Examples of ABCD matrices for common optical elements are shown in Table 12.1.

Table 12.1. ABCD Matrices for common optical elements

Name	ABCD Matrix	Remarks
Translation	$\begin{bmatrix} 1 & d \\ 0 & 1 \end{bmatrix}$	d is the distance traveled along the optical axis.
Thin lens / Curved mirror	$\begin{bmatrix} 1 & 0 \\ -\frac{1}{f} & 1 \end{bmatrix}$	f is the focal length and is positive for a converging lens/mirror.
Curved optical interface	$\begin{bmatrix} 1 & 0 \\ \frac{1}{R}\left(\frac{n_1}{n_2}-1\right) & \frac{n_1}{n_2} \end{bmatrix}$	n_1 and n_2 are the refractive indices of the first and second medium, respectively. R is the radius of curvature of the surface, and is positive for a convex surface.

ABCD matrix formalism can also be used to determine the parameters of the TEM_{00} Gaussian beam (the spotsize function $w(z)$ and the radius of curvature function $R(z)$) supported by a laser resonator. One round trip through the optical resonator will typically involve several optical elements such as the gain medium, curved mirrors around the gain medium, free space propagation over various distances, and so on. In this case, the resultant ABCD matrix that describes one round trip will be given by the product of the respective ABCD matrices of the individual optical elements. Let us assume that the elements of this round-trip matrix are A_T, B_T, C_T, and D_T. It can be shown that if the laser resonator is stable, in other words, if it supports a confined Gaussian beam, then the following condition is satisfied [56]:

$$-1 \le \frac{A_T + D_T}{2} \le 1. \tag{12.17}$$

For a Gaussian beam, the spotsize function $w(z)$ and the radius of curvature function $R(z)$ can be calculated from the q parameter which is given by [56]:

$$\frac{1}{q(z)} = \frac{1}{R(z)} - i\frac{\lambda}{n_0 \pi w^2(z)}. \quad (12.18)$$

Here, n_0 is the refractive index of the medium and λ is the wavelength. One can use the ABCD matrix of an optical system to determine its effect on a Gaussian beam as well by using the equation [56]:

$$q_{out} = \frac{Aq_{in} + B}{Cq_{in} + D}. \quad (12.19)$$

Above, q_{in} and q_{out} are the q parameters of the input and output beams, respectively. Hence, the spotsize and radius of curvature of the self consistent Gaussian beam supported by an optical resonator are determined by requiring that

$$q = \frac{A_T q + B_T}{C_T q + D_T}, \quad (12.20)$$

at each position inside the cavity. The values of A_T, B_T, C_T, and D_T will vary with position within the cavity but $A_T + D_T$ remains invariant. We used this approach to design the specific laser resonators used in the mode locking experiments with graphene supercapacitors. Refs [1-3, 55] provide a more detailed analysis of optical resonators.

12.4.3 *Cavity extension with q-preserving multi-pass cavities*

As discussed in the previous sections, the physical length of the cavity determines the free spectral range (v_{FSR}), and hence, the pulse repetition rate during mode-locked operation. In other words, the cavity length determines how many pulses will be generated by the mode-locked laser in one second. As the cavity length is increased, the repetition rate decreases and the laser generates fewer pulses per second. This means that the energy per pulse will increase, provided that the average power of the laser remains the same. In the experiments described in Section 12.5, we used this approach to scale up the output energy of the graphene mode-locked oscillators by introducing an additional optical path into the laser cavity.

A practical method to increase the optical path length with a compact mirror configuration is based on the multi-pass cavity (MPC) concept. In a MPC, the light beam reflects multiple times, typically between two mirrors, so that the effective optical path is increased while keeping the physical distance between the mirrors constant. In the MPC configurations used here, we employed two highly reflecting mirrors, each with a notch, to inject and extract the beam as shown in Fig. 12.13. Once inside the MPC, the beam completes a certain number of round trips before exit. For example, if the distance between the two MPC mirrors is *d*, and if the bouncing beam completes 14 round trips inside the MPC, the effective optical path length is extended by 14×d.

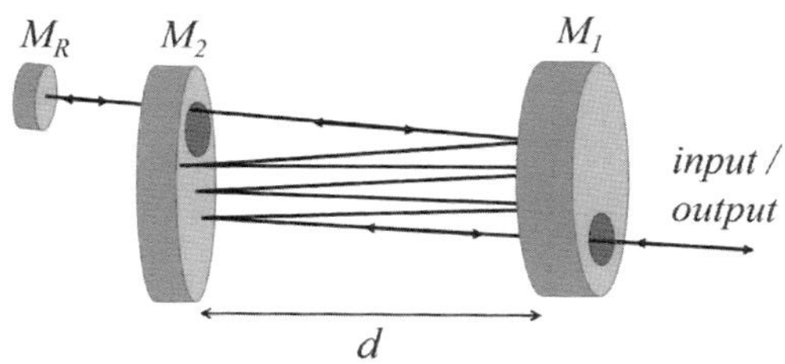

Figure 12.13. A simple layout of a multi-pass cavity (MPC), consisting of two notched mirrors.

In a practical MPC design, we also pay attention to the fact that the MPC is q preserving. The q parameter of a Gaussian beam was introduced in the previous section. If the MPC is q preserving, the input as well as the returning beam has the same q parameter, which means that the original spotsize distribution and stability condition of the short cavity remains the same. The ABCD matrix formalism introduced above can be used to determine the specific mirror separations for constructing a q-preserved MPC and the resulting extension in the optical path length. For a detailed analysis of q preserving MPCs, we refer the reader to Refs. [56, 57].

12.5. Femtosecond Pulse Generation Experiments with VCG-SA and VCG-gold-SA

12.5.1 *Experimental setup of the MPC Cr^{4+}:forsterite laser*

In order to test the fast saturable absorber performance of the voltage controlled graphene and graphene-gold supercapacitor devices, a multipass-cavity (MPC) Cr^{4+}:forsterite solid-state laser was used. The Cr^{4+}:forsterite lasers emit tunable radiation around 1.3 μm when optically excited near 1 μm. The first lasing action in Cr^{4+}:forsterite was reported in 1988 [58] and the first KLM operation was demonstrated in 1991 [59]. By utilizing the broad gain bandwidth of Cr^{4+}:forsterite, pulses as short as 14 fs have been generated [60]. Since the operation wavelength of the Cr^{4+}:forsterite laser is long enough to minimize Rayleigh scattering losses and also below the water absorption window, it is a suitable candidate for biomedical applications requiring deep tissue penetration, including multi-photon microscopy [61, 62] and optical coherence tomography [63, 64]. Therefore, it is important to develop alternative methods of mode locking for femtosecond pulse generation from Cr^{4+}:forsterite lasers.

The experimental setup of the MPC Cr^{4+}:forsterite laser with voltage controlled graphene-based saturable absorber and dispersion compensation optics is shown in Fig. 12.14(a).

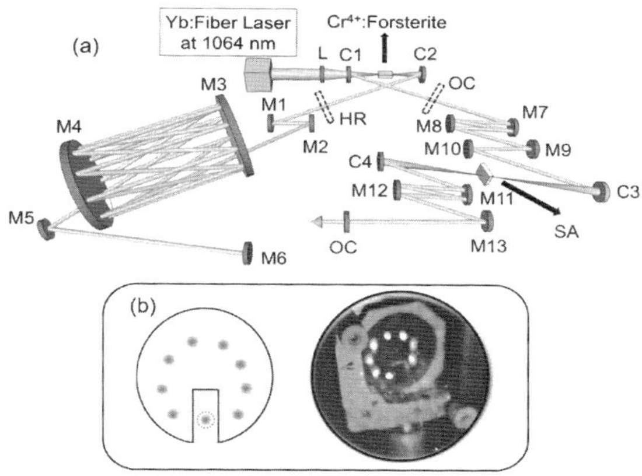

Figure 12.14. (a) The multipass-cavity (MPC) Cr^{4+}:forsterite laser that was used during the mode locking experiments with VCG-SA and VCG-gold-SA. (b) A sketch of the spot distribution on the flat MPC mirror and the infrared image of the actual spot pattern. (For color reference, go to page 496.)

Initially, the short x-cavity was constructed with two curved mirrors (C_1 and C_2 in Fig. 12.14(a)), a high reflector (HR, dashed), and an output coupler (OC, dashed). A 20-mm-long, Brewster-cut Cr^{4+}:forsterite crystal with 70 % absorption at the pump wavelength of 1.064 μm was pumped by a commercial Yb:fiber laser and kept at a fixed temperature of 20 ^{0}C by water cooling. The pump beam was focused inside the crystal with the help of the input lens (L) which had a focal length of 20 cm. By taking into account the cavity parameters such as mirror separations, mirror radii of curvature, and the crystal length, the laser beamwaist inside the crystal was estimated to be 30 μm by using the ABCD matrix method described in Section 12.4.2.

In order to reduce the pulse repetition rate of the resonator and to scale up the pulse energy at low average output powers, the high reflector (HR) was removed and the short cavity was extended with the help of a q-preserving multi-pass cavity (MPC). Two flat high reflectors (M1 and M2) in the optical path of the MPC were used to adjust the inclination angle and the offset of the beam entering the MPC. The MPC consisted of a curved (M3, radius of curvature=4 m) and a flat (M4) high reflector with notches to allow beam injection and extraction. Details of the design rules used in the construction of the q-preserving MPC may be found in [27, 51, 52, 65]. In this MPC, the injected laser beam completed 9 round-trips between the mirrors before exit. In Fig. 12.14(b), one sees 8 spots on the flat MPC mirror, because the last spot leaves the MPC through the notch and is retro-reflected by using another curved mirror (M6, radius of curvature=2 m) at a distance of 165 cm from the flat MPC mirror M4. Use of a retroreflecting mirror with a radius of curvature of 2 m makes the MPC q-preserving as proven in Ref. [57].

As mentioned in Section 12.2.2.2, to generate solitary optical pulses from the mode-locked laser, self phase modulation due to the Kerr effect inside the gain crystal should be compensated with net negative group delay dispersion (GDD) in one round trip. At the wavelength of operation of the Cr^{4+}:forsterite laser, GDD due to the crystal and air are positive. For GDD compensation, flat Gires-Tournouis interferometer (GTI) mirrors (M7-M8 and M11-M12 in Fig. 12.14(a)) with high reflection and negative dispersion near 1250 nm were added to the cavity on the output coupler side. Flat and highly reflecting dispersion control mirrors (DCMs, C1, C2, M1, M2, M5, M9, M10, M13) with negative dispersion were also introduced to the laser resonator. Each GTI and DCM had a GDD value of -250 fs^2 and -150 fs^2, respectively. The net GDD over one round trip of the MPC Cr^{4+}:forsterite laser was estimated to be -4440 fs^2. The output coupler arm of the resonator also contained a pair of concave high reflectors (C3 and C4), each with a radius of curvature of 50 cm. These curved mirrors were used to create a beamwaist and to obtain a sufficiently high intensity to saturate the absorption of the VCG-SA or VCG-gold-SA. After the addition of all the optics (MPC, dispersion compensation optics, C3, and C4), the pulse repetition rate of the laser cavity was reduced from 144 MHz to 4.51 MHz. Refs [27, 51, 52, 65] provide further information about the detailed design of the MPC Cr^{4+}:forsterite laser used in the mode locking experiments with different carbon-based saturable absorbers.

12.5.2 *Power efficiency and mode locking data*

The power efficiency of the Cr^{4+}:forsterite laser with different configurations involving multi-pass cavity (MPC) and dispersion compensation optics was determined by measuring the continuous-wave output power of the laser as a function of the input pump power. For example, the slope efficiency (η) of the short x-cavity laser was around 17.5% with the 4.7% transmitting output coupler and decreased to 13 % after the addition of the MPC and dispersion compensation optics. The insertion loss of the various optics added into the laser resonator was estimated by using the methods discussed in Section 12.4.1 and further information can be found in Ref. [27]. As discussed in Ref. [27], it was not possible to operate the Cr^{4+}:forsterite laser with a total intracavity loss that exceeded 10%. Since the small-signal insertion loss of the voltage controlled graphene saturable absorber (VCG-SA) and voltage controlled graphene-gold (VCG-gold-SA) saturable absorber were above 5%, the 4.7% output coupler was replaced with a 2.4% transmitting output coupler before adding the saturable absorber into the laser resonator. The VCG-SA/VCG-gold-SA was positioned at Brewster incidence between the two curved high reflectors as shown in Fig. 12.14. Keeping the saturable absorber at Brewster incidence minimizes reflection losses and prevents etalon effects which are unwanted during mode-locked operation. Due to the small-signal insertion loss of the devices at zero applied bias, the output power of the laser was reduced upon the addition of the saturable absorber. However, as the bias voltage was increased, the output power of the laser at the wavelength of 1.25 μm started to increase due to the onset of Pauli blocking. Fig. 12.15 shows the measured output power of the Cr^{4+}:forsterite laser which exhibits this behavior. The input pump power was 7.1W in this case.

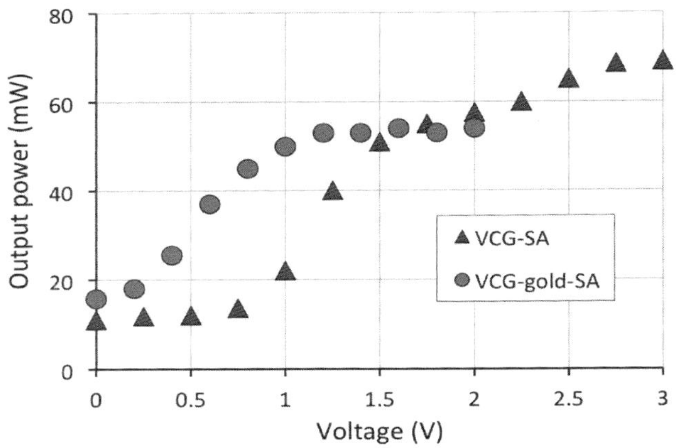

Figure 12.15. Continuous-wave output power of the Cr^{4+}:forsterite laser as a function of the applied bias at a fixed input power of 7.1W. The resonator contains VCG-SA or VCG-gold-SA in addition to the MPC and dispersion control optics.

For VCG-SA, the output power at 0V bias was around 11 mW and increased up to 69 mW when the bias voltage was increased to 3V [51]. The continuous-wave output power curve for VCG-SA shows a two-step behavior similar to the fractional transmission change spectra that were shown previously in Fig. 12.7. This two-step behavior is mainly because of the use of two graphene electrodes. For this supercapacitor, Pauli blocking happens for both of the graphene electrodes as explained earlier. In the case of VCG-gold-SA, since only one graphene electrode was used, two-step behavior was not observed. For VCG-gold-SA, the output power starts from nearly 16 mW at 0V bias and increases up to 54 mW for 2V of applied bias [52]. At bias voltages below 1.5 V, the output power performance with the VCG-gold-SA is superior in comparison with that for VCG-SA, due to the use of only one graphene electrode.

To further show the advantage of the using one graphene electrode, small-signal insertion losses of both devices were estimated by using the threshold pump power data as explained in Section 12.4.1. Figure 12.16 shows the estimated double-pass losses of the two different graphene supercapacitors as a function of applied bias. Since the growth process of graphene and use of different substrates affect the loss levels, it is not possible to do a direct comparison of loss for the two architectures. However, Fig. 12.16 clearly shows that the double-pass insertion loss of the VCG-gold-SA containing one graphene electrode is lower for the operating voltage range (0-1 V) that was utilized at the wavelength of 1.25 μm (corresponding to approximately 1eV of photon energy). As can be seen from Fig. 12.16, at sufficiently large applied voltages, absorption due to graphene is completely blocked and only residual losses coming from the substrate and the electrolyte remain for both devices. Hence, the slightly lower loss obtained with VCG-SA at voltages near 3V is possibly due to the fact that substrates with better optical quality were used.

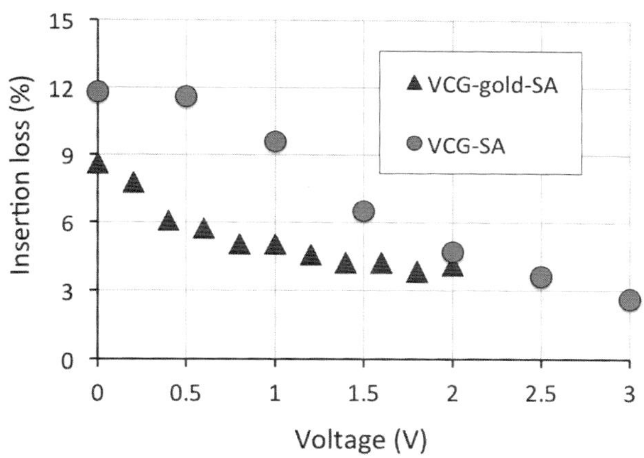

Figure 12.16. Estimated double-pass insertion loss for VCG-SA and VCG-gold-SA as a function of the applied bias voltage.

After the insertion of the supercapacitor saturable absorbers into the laser resonator, mode-locked operation was initiated by translating the output coupler. Figure 12.17 shows typical ultrashort pulse characterization data with (a) the optical spectrum, (b) radio frequency (RF) spectrum, and (c) autocorrelation trace for VCG-gold-SA mode-locked Cr^{4+}:forsterite laser.

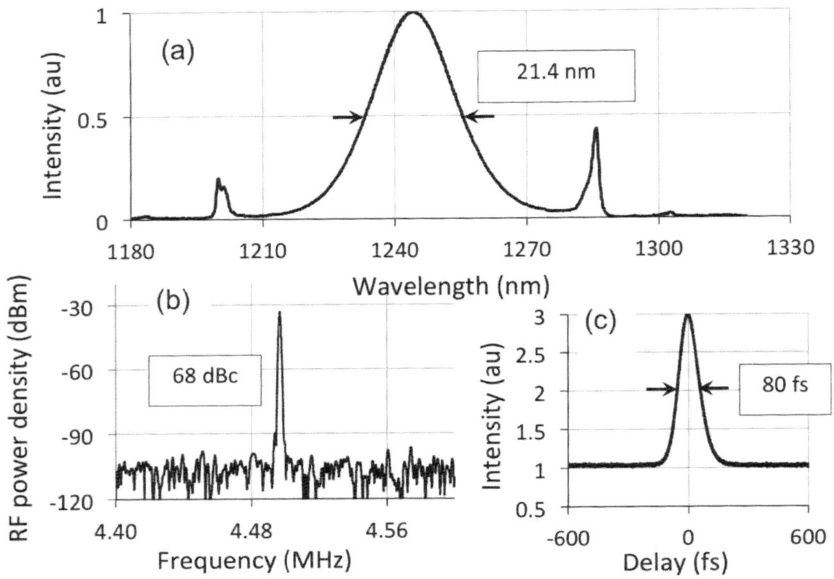

Figure 12.17. (a) Optical spectrum, (b) RF spectrum, and (c) autocorrelation trace of the femtosecond pulses generated with VCG-gold-SA mode-locked Cr^{4+}:forsterite laser at the bias voltage of 0.8V.

The Cr^{4+}:forsterite laser resonator mode-locked with VCG-SA or VCG-gold-SA produced femtosecond pulses with similar characteristics as shown in Table 5.1. VCG-SA or VCG-gold-SA acts as the initiator of the pulse train and the ultimate pulse parameters such as temporal duration, peak power, and time-bandwidth product depend on the balance between the Kerr nonlinearity strength and overall dispersion inside the resonator. In particular, pulse durations were in the 80-90 fs range. In addition, the time-bandwidth products of the generated pulses were in the 0.33-0.37 range, close to the ideal value of 0.315 for solitary pulses with $sech^2$ profile. The RF spectrum of the generated pulses (Fig. 12.17(b)) clearly shows that the sideband noise of the photodetected pulse train was at least 68 dB below the fundamental tone of the carrier at the pulse repetition frequency of 4.51 MHz. The incident fluence on the saturable absorbers (shown in Table 12.2) was determined by using the beamwaist size at the saturable absorber and the intracavity pulse energy. The beamwaists were estimated by using the ABCD matrix method described before. The incident fluence levels were well above the saturation fluence of the graphene saturable absorber (approximately in the range of 290-424 μJ/cm^2 [51, 52]) and guaranteed that the graphene-based supercapacitors were sufficiently saturated during mode-locked operation. Furthermore, up to these fluence levels (maximum value of 424 μJ/cm^2 in Table 12.2), no optical damage was observed in these devices.

Table 12.2. Characteristics of the femtosecond pulse train generated with the Cr^{4+}:forsterite laser mode-locked with VCG-SA and VCG-gold-SA at different bias voltages.

Supercapacitor type	**Voltage (V)**	**Spectral width (nm)**	**Pulse duration (fs)**	**Time-bandwidth product**	**Peak power of the pulse (kW)**	**Incident fluence (μJ/cm^2)**
VCG-SA	0.5	21	89	0.35	30	364
VCG-SA	1	19.6	84	0.37	37	424
VCG-gold-SA	0.4	20.6	89	0.35	32	290
VCG-gold-SA	0.8	21.4	80	0.33	42	335

In the mode locking experiments, output peak powers of as high as 42 kW could be generated at low output power levels (about 15 mW) due to the use of the MPC extension that lowers the pulse repetition rate. Since the photon energy of the laser emission wavelength of 1.25 μm is nearly 1 eV, Pauli blocking takes place for bias voltages greater than 1V and our supercapacitor based saturable absorbers become nearly transparent without any intensity-dependent saturation left. Hence, no mode locking could be obtained for bias voltages beyond 0.8V for VCG-gold-SA and 1V for VCG-SA at the laser operation wavelength. Furthermore, due to the lower insertion loss of VCG-gold-SA (see Fig. 5.3 above), the mode locking threshold pump power was also reduced in the experiments involving the VCG-gold-SA with one graphene electrode and mode locking could also be achieved for voltages below

0.5V. Further information about the mode locking results obtained with each of the devices can be found in Refs. [51, 52].

Further increase in average output power could be obtained by increasing the input pump power. However, this also resulted in pulse instabilities such as multi-pulse operation and/or mode-locked operation with cw background. This is mainly attributed to the nonlinearities arising from self focusing inside the gain medium. To investigate this limit further, nonlinear refractive index (n_2) of the Cr^{4+}:forsterite medium was estimated from the mode locking data obtained with different saturable absorbers and the critical power was determined for each of the cases. The soliton area theorem (Eq. (12.10)) was used in the determination of n_2. The critical power (P_{cr}) was determined by using Eq. (12.7). Table 12.3 summarizes the estimated results obtained for n_2 and P_{cr} from the mode locking data of Cr^{4+}:forsterite by using single-walled carbon nanotube [65], single layer graphene [27], VCG-gold [52] saturable absorbers. Details of the calculations may be found in Refs. [27, 65]. Since the maximum obtainable intracavity peak powers were comparable to the critical power for self focusing, it may be concluded from this analysis that self focusing in the Cr^{4+}:forsterite gain medium appears as the limiting factor for further power scaling as was noted in other studies [66, 67].

Table 12.3. Values of the nonlinear refractive index and critical power obtained from mode-locking data of the Cr^{4+}:forsterite laser with different saturable absorbers.

Saturable absorber type	**Nonlinear refractive index n_2 (m^2/W)**	**Critical power P_{cr} (MW)**	**Maximum intracavity power (MW)**
Single-walled carbon nanotube	9.36×10^{-20}	1.5	0.9
Single layer graphene	11.83×10^{-20}	1.21	1.13
VCG-gold-SA	11×10^{-20}	1.3	1.5

12.6. Conclusions

In conclusion, our experimental results show that graphene-based supercapacitors can be used in an effective way as saturable absorbers to initiate femtosecond pulses from tunable solid-state lasers. Furthermore, the loss limitation posed by a monolayer graphene saturable absorber can be overcome by adjusting the absorption level of these devices with low bias voltages on the order of a few volts. This enables their use in low-gain tunable solid-state lasers, where even a monolayer graphene can cause considerable reduction in power performance. Our experimental results further demonstrated that graphene-gold supercapacitor structure is more favorable since it has only one graphene electrode and therefore introduces less loss into the laser resonator, in comparison with the supercapacitor structure involving two graphene electrodes. Due to their advantages such as low-cost fabrication and broadband operation, we foresee that voltage-controlled graphene supercapacitor devices will have widespread use in the generation of femtosecond pulses from different types of lasers operating over a broad wavelength range.

Acknowledgments

The authors thank Abdullah Muti for his assistance in the preparation of the manuscript. Sarper Özharar further acknowledges the support of the Science Academy through the BAGEP Young Scientist Award Program. The experiments were partially funded by TUBITAK (The Scientific and Technological Research Council of Turkey) under Project No. 113F278.

References

[1] A. Sennaroglu, *Photonics and laser engineering : principles, devices, and applications* (McGraw-Hill, New York, 2010).

[2] W. T. Silfvast, *Laser Fundamentals* (Cambridge University Press, 2004).

[3] O. Svelto, *Principles of Lasers* (Plenum Press, New York, 1989).

[4] A. A. Kaminskii, *Laser Crystals* (Springer, Berlin, 1990).

[5] W. Koechner, *Solid-State Laser Engineering,* 4 ed., Springer Series in Optical Sciences (Springer, New York, 1996), Vol. 1.

[6] P. F. Moulton, "Tunable solid-state lasers," Proceedings of the IEEE **80**, 348-364 (1992).

[7] A. Sennaroglu, "Broadly tunable Cr^{4+} -doped solid-state lasers in the near infrared and visible," Progress in Quantum Electronics **26**, 287-352 (2002).

[8] P. F. Moulton, "Spectroscopic and Laser Characteristics of Ti-Al2O3," Journal of the Optical Society of America B-Optical Physics **3**, 125-133 (1986).

[9] V. G. Baryshevskii, M. V. Korzhik, A. E. Kimaev, M. G. Livshits, V. B. Pavlenko, M. L. Meil'man, and B. I. Minkov, "Tunable chromium forsterite laser in the near IR region," Journal of Applied Spectroscopy (USSR) **53**, 675-676 (1990).

[10] H. A. Haus, "Mode-locking of lasers," IEEE J. Sel. Top. Quantum Electron. **6**, 1173-1185 (2000).

[11] J. C. Diels and W. Rudolph, *Ultrashort Laser Pulse Phenomena: Fundamentals, Techniques, and Applications on a Femtosecond Time Scale* (Academic Press, 2006).

[12] D. J. Kuizenga and A. E. Siegman, "FM and AM mode locking of the homogeneous laser-Part I: Theory," IEE Journal of Quantum Electronics **QE-6**, 694-708 (1970).

[13] M. Becker, D. Kuizenga, and A. Siegman, "Harmonic mode locking of the Nd:YAG laser," IEEE Journal of Quantum Electronics **8**, 687-693 (1972).

[14] S. Longhi, P. Laporta, S. Taccheo, and O. Svelto, "Third-order-harmonic mode locking of a bulk erbium:ytterbium:glass laser at a 2.5-GHz repetition rate," Opt Lett **19**, 1985-1987 (1994).

[15] S. Ozharar, S. Gee, F. Quinlan, S. Lee, and P. J. Delfyett, "Pulse-amplitude equalization by negative impulse modulation for rational harmonic mode locking," Optics Letters **31**, 2924-2926 (2006).

[16] C. M. Wu and N. K. Dutta, "High-repetition-rate optical pulse generation using a rational harmonic mode-locked fiber laser," IEEE Journal of Quantum Electronics **36**, 145-150 (2000).

[17] D. Kuizenga and A. Siegman, "FM and AM mode locking of the homogeneous laser - Part II: Experimental results in a Nd:YAG laser with internal FM modulation," IEEE Journal of Quantum Electronics **6**, 709-715 (1970).

[18] D. Huang, M. Ulman, L. H. Acioli, H. A. Haus, and J. G. Fujimoto, "Self-Focusing-Induced Saturable Loss for Laser Mode-Locking," Optics Letters **17**, 511-513 (1992).

[19] F. X. Kartner, I. D. Jung, and U. Keller, "Soliton mode-locking with saturable absorbers," IEEE J. Sel. Top. Quantum Electron. **2**, 540-556 (1996).

[20] U. Keller, K. J. Weingarten, F. X. Kartner, D. Kopf, B. Braun, I. D. Jung, R. Fluck, C. Hönninger, N. Matuschek, and J. A. d. Au, "Semiconductor saturable absorber mirrors (SESAM's) for femtosecond to nanosecond pulse generation in solid-state lasers," IEEE J. Sel. Top. Quantum Electron. **2**, 435-453 (1996).

[21] K. S. Novoselov, V. I. Falko, L. Colombo, P. R. Gellert, M. G. Schwab, and K. Kim, "A roadmap for graphene," Nature **490**, 192-200 (2012).

[22] F. Bonaccorso, Z. Sun, T. Hasan, and A. C. Ferrari, "Graphene photonics and optoelectronics," Nature Photonics **4**, 611-622 (2010).

[23] A. H. Castro Neto, F. Guinea, N. M. R. Peres, K. S. Novoselov, and A. K. Geim, "The electronic properties of graphene," Reviews of Modern Physics **81**, 109-162 (2009).

[24] A. B. Kuzmenko, E. van Heumen, F. Carbone, and D. van der Marel, "Universal Optical Conductance of Graphite," Physical Review Letters **100**, 117401 (2008).

[25] J. M. Dawlaty, S. Shivaraman, M. Chandrashekhar, F. Rana, and M. G. Spencer, "Measurement of ultrafast carrier dynamics in epitaxial graphene," Appl Phys Lett **92**, 042116 (2008).

[26] W. B. Cho, J. W. Kim, H. W. Lee, S. Bae, B. H. Hong, S. Y. Choi, I. H. Baek, K. Kim, D.-I. Yeom, and F. Rotermund, "High-quality, large-area monolayer graphene for efficient bulk laser mode-locking near 1.25 um," Optics Letters **36**, 4089-4091 (2011).

[27] S. Ozharar, I. Baylam, M. N. Cizmeciyan, O. Balci, E. Pince, C. Kocabas, and A. Sennaroglu, "Graphene mode-locked multipass-cavity femtosecond Cr4+: forsterite laser," J. Opt. Soc. Am. B **30**, 1270-1275 (2013).

[28] Z. Sun, T. Hasan, and A. C. Ferrari, "Ultrafast lasers mode-locked by nanotubes and graphene," Physica E: Low-dimensional Systems and Nanostructures **44**, 1082-1091 (2012).

[29] C.C. Lee, T. R. Schibli, G. Acosta, and J. S. Bunch, "Ultra-short optical pulse generation with single-layer graphene," Journal of Nonlinear Optical Physics & Materials **19**, 767-771 (2010).

[30] W. D. Tan, C. Y. Su, R. J. Knize, G. Q. Xie, L. J. Li, and D. Y. Tang, "Mode locking of ceramic Nd:yttrium aluminum garnet with graphene as a saturable absorber," Appl Phys Lett **96**, 031106-031103 (2010).

[31] Q. Bao, H. Zhang, Z. Ni, Y. Wang, L. Polavarapu, Z. Shen, Q.H. Xu, D. Tang, and K. Loh, "Monolayer graphene as a saturable absorber in a mode-locked laser," Nano Research **4**, 297-307 (2011).

[32] I. H. Baek, H. W. Lee, S. Bae, B. H. Hong, Y. H. Ahn, D.-I. Yeom, and F. Rotermund, "Efficient Mode-Locking of Sub-70-fs Ti:Sapphire Laser by Graphene Saturable Absorber," Applied Physics Express **5**, 032701 (2012).

[33] J. Ma, G. Q. Xie, P. Lv, W. L. Gao, P. Yuan, L. J. Qian, H. H. Yu, H. J. Zhang, J. Y. Wang, and D. Y. Tang, "Graphene mode-locked femtosecond laser at 2 um wavelength," Optics Letters **37**, 2085-2087 (2012).

[34] M. N. Cizmeciyan, J. W. Kim, S. Bae, B. H. Hong, F. Rotermund, and A. Sennaroglu, "Graphene mode-locked femtosecond Cr:ZnSe laser at 2500nm," Optics Letters **38**, 341-343 (2013).

[35] A. Roberts, D. Cormode, C. Reynolds, T. Newhouse-Illige, B. J. Leroy, and A. S. Sandhu, "Response of graphene to femtosecond high-intensity laser irradiation," Appl Phys Lett **99**(2011).

[36] F. Canbaz, E. Beyatli, L. J. Chen, A. Sennaroglu, F. X. Kartner, and U. Demirbas, "Highly efficient and robust operation of Kerr-lens mode-locked Cr:LiSAF lasers using gain-matched output couplers," Optics Letters **39**, 327-330 (2014).

[37] C. Cihan, E. Beyatli, F. Canbaz, L. J. Chen, B. Sumpf, G. Erbert, A. Leitenstorfer, F. X. Kartner, A. Sennaroglu, and U. Demirbas, "Gain-Matched Output Couplers for Efficient Kerr-Lens Mode-Locking of Low-Cost and High-Peak Power Cr:LiSAF Lasers," IEEE J. Sel. Top. Quantum Electron. **21**(2015).

[38] D. C. Wei, Y. Q. Liu, Y. Wang, H. L. Zhang, L. P. Huang, and G. Yu, "Synthesis of N-Doped Graphene by Chemical Vapor Deposition and Its Electrical Properties," Nano Lett **9**, 1752-1758 (2009).

[39] H. T. Liu, Y. Q. Liu, and D. B. Zhu, "Chemical doping of graphene," J Mater Chem **21**, 3335-3345 (2011).

[40] T. Ohta, A. Bostwick, T. Seyller, K. Horn, and E. Rotenberg, "Controlling the electronic structure of bilayer graphene," Science **313**, 951-954 (2006).

[41] S. Y. Zhou, G. H. Gweon, A. V. Fedorov, P. N. First, W. A. De Heer, D. H. Lee, F. Guinea, A. H. Castro Neto, and A. Lanzara, "Substrate-induced bandgap opening in epitaxial graphene," Nat Mater **6**, 770-775 (2007).

[42] C. C. Lee, C. Mohr, J. Bethge, S. Suzuki, M. E. Fermann, I. Hartl, and T. R. Schibli, "Frequency comb stabilization with bandwidth beyond the limit of gain lifetime by an intracavity graphene electro-optic modulator," Optics Letters **37**, 3084-3086 (2012).

[43] C. C. Lee, S. Suzuki, W. Xie, and T. R. Schibli, "Broadband graphene electro-optic modulators with sub-wavelength thickness," Optics Express **20**, 5264-5269 (2012).

[44] M. Liu, X. B. Yin, E. Ulin-Avila, B. S. Geng, T. Zentgraf, L. Ju, F. Wang, and X. Zhang, "A graphene-based broadband optical modulator," Nature **474**, 64-67 (2011).

[45] S. H. Lee, M. Choi, T. T. Kim, S. Lee, M. Liu, X. Yin, H. K. Choi, S. S. Lee, C. G. Choi, S. Y. Choi, X. Zhang, and B. Min, "Switching terahertz waves with gate-controlled active graphene metamaterials," Nat Mater **11**, 936-941 (2012).

[46] X. T. Gan, R. J. Shiue, Y. D. Gao, K. F. Mak, X. W. Yao, L. Z. Li, A. Szep, D. Walker, J. Hone, T. F. Heinz, and D. Englund, "High-Contrast Electrooptic Modulation of a Photonic Crystal Nanocavity by Electrical Gating of Graphene," Nano Lett **13**, 691-696 (2013).

[47] R. M. Lynden-Bell, A. I. Frolov, and M. V. Fedorov, "Electrode screening by ionic liquids," Physical Chemistry Chemical Physics **14**, 2693-2701 (2012).
[48] E. O. Polat and C. Kocabas, "Broadband Optical Modulators Based on Graphene Supercapacitors," Nano Lett **13**, 5851-5857 (2013).
[49] J. J. Yoo, K. Balakrishnan, J. S. Huang, V. Meunier, B. G. Sumpter, A. Srivastava, M. Conway, A. L. M. Reddy, J. Yu, R. Vajtai, and P. M. Ajayan, "Ultrathin Planar Graphene Supercapacitors," Nano Lett **11**, 1423-1427 (2011).
[50] Y. W. Zhu, S. Murali, M. D. Stoller, K. J. Ganesh, W. W. Cai, P. J. Ferreira, A. Pirkle, R. M. Wallace, K. A. Cychosz, M. Thommes, D. Su, E. A. Stach, and R. S. Ruoff, "Carbon-Based Supercapacitors Produced by Activation of Graphene," Science **332**, 1537-1541 (2011).
[51] I. Baylam, M. N. Cizmeciyan, S. Ozharar, E. O. Polat, C. Kocabas, and A. Sennaroglu, "Femtosecond pulse generation with voltage-controlled graphene saturable absorber," Optics Letters **39**, 5180-5183 (2014).
[52] I. Baylam, O. Balci, N. Kakenov, C. Kocabas, and A. Sennaroglu, "Graphene-gold supercapacitor as a voltage controlled saturable absorber for femtosecond pulse generation," Optics Letters **41**(2016).
[53] E. J. Lee, S. Y. Choi, H. Jeong, N. H. Park, W. Yim, M. H. Kim, J. K. Park, S. Son, S. Bae, S. J. Kim, K. Lee, Y. H. Ahn, K. J. Ahn, B. H. Hong, J. Y. Park, F. Rotermund, and D. I. Yeom, "Active control of all-fibre graphene devices with electrical gating," Nat Commun **6**(2015).
[54] M. M. Huang, Y. P. Jiang, P. Sasisanker, G. W. Driver, and H. Weingartner, "Static Relative Dielectric Permittivities of Ionic Liquids at 25 degrees C," Journal of Chemical and Engineering Data **56**, 1494-1499 (2011).
[55] A. E. Siegman, *Lasers* (University Science Books, Sausalito, California, 1986), p. 1283.
[56] A. Sennaroglu, *Solid-State Lasers and Applications* (Taylor & Francis, 2010).
[57] A. Sennaroglu, A. M. Kowalevicz, E. P. Ippen, and J. G. Fujimoto, "Compact femtosecond lasers based on novel multipass cavities," IEEE Journal of Quantum Electronics **40**, 519-528 (2004).

[58] V. Petricevic, S. K. Gayen, R. R. Alfano, K. Yamagishi, H. Anzai, and Y. Yamaguchi, "Laser action in chromium-doped forsterite," Appl Phys Lett **52**, 1040-1042 (1988).
[59] A. Seas, Petricevic, V., Alfano, R. R., "Continuous-wave mode-locked operation of a chromium-doped forsterite laser," Optics Letters **16**(1991).
[60] C. Chudoba, J. G. Fujimoto, E. P. Ippen, H. A. Haus, U. Morgner, F. X. Kartner, V. Scheuer, G. Angelow, and T. Tschudi, "All-solid-state Cr : forsterite laser generating 14-fs pulses at 1.3 mu m," Optics Letters **26**, 292-294 (2001).
[61] V. V. Yakovlev, A. Ivanov, and V. Shcheslavskiy, "High-energy femtosecond Cr4+:forsterite oscillators and their applications in biomedical and material sciences," Applied Physics B **74**, s145-s152 (2002).
[62] C.-K. Sun, C.-C. Chen, S.-W. Chu, T.-H. Tsai, Y.-C. Chen, and B.-L. Lin, "Multiharmonic-generation biopsy of skin," Optics Letters **28**, 2488-2490 (2003).
[63] B. E. Bouma, G. J. Tearney, I. P. Bilinsky, B. Golubovic, and J. G. Fujimoto, "Self-phase-modulated Kerr-lens mode-locked Cr:forsterite laser source for optical coherence tomography," Optics Letters **21**, 1839-1841 (1996).
[64] G. J. Tearney, M. E. Brezinski, B. E. Bouma, S. A. Boppart, C. Pitris, J. F. Southern, and J. G. Fujimoto, "In vivo endoscopic optical biopsy with optical coherence tomography," Science **276**, 2037-2039 (1997).
[65] I. Baylam, S. Ozharar, H. Cankaya, S. Y. Choi, K. Kim, F. Rotermund, U. Griebner, V. Petrov, and A. Sennaroglu, "Energy scaling of a carbon nanotube saturable absorber mode-locked femtosecond bulk laser," Optics Letters **37**, 3555-3557 (2012).
[66] B. Chassagne, A. Ivanov, J. Oberle, G. Jonusauskas, and C. Rulliere, "Experimental determination of the nonlinear refractive index in an operating Cr:forsterite femtosecond laser," Optics Communications **141**, 69-74 (1997).
[67] B. Chassagne, G. Jonusauskas, J. Oberle, and C. Rulliere, "Multipulse operation regime in a self-mode-locked Cr4+: forsterite femtosecond laser," Optics Communications **150**, 355-362 (1998).

Chapter 13

Graphene-Based Optical Modulators

Sinan Balci[1], and Coskun Kocabas[2]

[1]*Department of Astronautical Engineering, University of Turkish Aeronautical Association, 06790 Ankara, Turkey*

[2]*Department of Physics, Bilkent University, 06800 Ankara, Turkey*

In this chapter, we summarize the recent progress on graphene based optical modulators. Ability to control density of high mobility electrons on large area graphene surface enables realization of new type of electro-optical modulators in optoelectronics. Due to the low electronic density of states, accumulation of charges on graphene significantly shifts the Fermi energy up to 1 eV giving rise to profound optical effects in the infrared and visible spectra. On the other hand, graphene operates as a tunable Drude metal in long wavelengths such as THz and microwave. This unique broadband activity of graphene has stimulated a great deal of interest in graphene community due to its potential use in new optoelectronic devices. After discussing the electrically tunable optical properties of graphene, we highlight the key achievements in the field.

1. Introduction

The two-dimensional crystal of carbon, graphene, yields a new platform to control light-matter interaction in a very broad spectrum [1, 2]. Atomic thickness together with the tunable high mobility charge carriers on graphene provides gate-tunable optical properties, which have attracted great interest in the graphene community in order to realize novel optoelectronic devices [3]. Although graphene is one-atom thick, it can yield profound optical absorption due to interband and intraband electronic transitions. Figure 13.1(a) and 1(b) show free standing graphene sheet and its linear band structure. Since the band structure of graphene does not have any energy specific transitions, the optical response of graphene is flat over a very broad spectrum. In the visible and near infrared wavelengths, graphene yields 2.3% optical absorption thanks to the interband transitions between valence and conduction bands [4]. These vertical transitions yield a universal constant optical conductivity of $\sigma_0 = \pi e^2/2h$, which yields 2.3% optical absorption in the free standing graphene monolayer. However, on a dielectric substrate, the optical absorption is reduced due the dielectric properties of the environment.

Electrostatic gating makes graphene transparent due to Pauli blocking. Doped graphene has a gap in the optical absorption when the photon energy is less than $2E_F$ (E_F refers to the shift in Fermi energy from the Dirac point). Due to unintentional doping and charge puddle formations, pristine graphene has Fermi energy around 0.2 eV which makes graphene transparent for energies less than 0.4 eV. (see Fig. 1.1(c))

For longer wavelengths, far-IR, terahertz, and microwave graphene shows metallic like optical response, which can be modeled with Drude like optical conductivity as

$$\sigma(\omega) = \frac{e^2}{\pi\hbar}\frac{iE_F}{\omega+i\tau^{-1}} \tag{13.1}$$

where, E_F is the Fermi energy, τ is the transport scattering time. Figure 13.1(c) shows the calculated optical absorption of graphene with varying doping levels. As the charge density is varied from 10^{12} to 10^{14} cm^{-2} the Fermi energy increases from 0.2 to 1 eV resulting in optical transparency in the visible and near IR whereas the absorption of graphene increases up to 50% in THz and microwave regions. These profound activity yields a new platform to control light in a very broad spectrum. This intriguing response originates from the high mobility free electrons on graphene surface. Since the thickness of graphene (~0.3 nm) is much smaller than the wavelength of the incident light (400 nm to 1 m), the optical activity solely originates from the tunable electrons. The lack of bulk material eliminates the

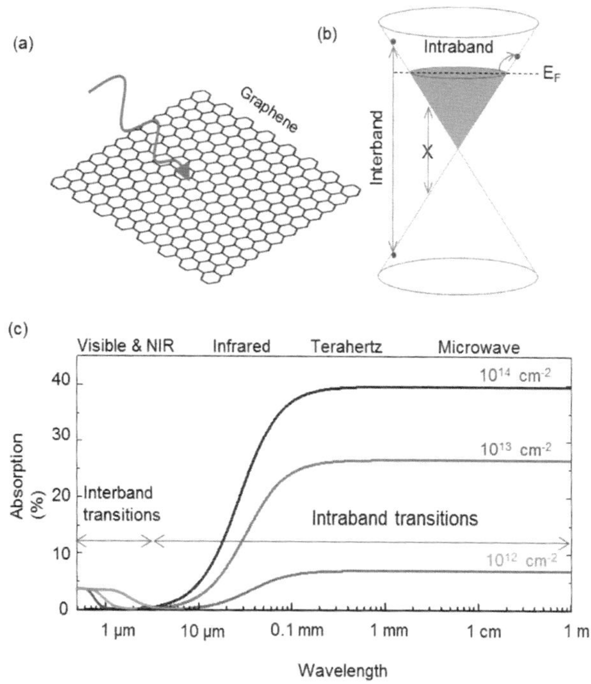

Fig. 13.1. (a,b) Schematic showing of a free standing graphene sheet and its band structure. The interband and intraband electronic transitions are shown on the band structure. Due to Pauli blocking, doped graphene has a gap in the optical absorption. Graphene is transparent for photon energies less than $2E_F$. (c) Calculated optical absorption of graphene at different doping levels. (τ=200 fs) Graphene is electrically tunable in a very broad spectrum ranging from visible to microwave wavelengths. The activity in the visible is due to Pauli blocking principle. Shifting the Fermi energy blocks the interband transitions and makes graphene transparent for photon energies less than $2E_F$. For long wavelengths, however, graphene operates as a tunable Drude metal whose response is defined by charge density and transport scattering time.

unnecessary absorption, which gives rise to large unmodulated insertion losses.

Various electrostatic gating schemes have been developed to control electron density on graphene. Figure 13.2 shows the schematic drawing of the electrostatic gating methods. The most common approach is the back-gating scheme that uses a doped silicon as a back-gate electrode and a thin oxide layer as the gate dielectric. Graphene layer is transferred onto the oxide surface by using polymer assisted mechanical exfoliation transfer process. In this device configuration, charge densities of 10^{12} cm^{-2} and Fermi energies of 0.3 eV can be easily achieved with large gate voltages (>50 V). However, dielectric breakdown at high gate voltages and common gate electrode are the limiting factors of the back-gating scheme.

Top-gating scheme (Fig. 13.2(b)) is developed to address individual devices on a single chip. A thin dielectric layer is deposited on graphene and followed by a metallization process to define the local gate electrodes. The main advantage of the top-gating scheme is that each device on a single chip can be individually addressable. This is a key requirement for multipixel devices. However, due to hydrophobic nature of graphene surface and incompatible nature of graphene with plasma assisted processes the quality of dielectric films deposited on graphene is very poor. The third approach is the electrolyte gating scheme that uses mobile ions in an electrolyte to gate graphene (Fig. 13.2(c)). Application of a bias voltage between graphene and electrolyte polarizes the ions and forms an electrical double layer (EDL) on the graphene-electrolyte interface. Since the thickness of the EDL is only a few nanometers, the voltage drops rapidly across the EDL and hence leading to a significant electric field that is not possible to obtain by using dielectric based gating schemes (Fig. 13.2(d)). By using this approach, charge density of 10^{14} and Fermi energy of 1 eV can be easily achieved. The electrochemical window of the electrolyte is the limiting factor of the achievable charge densities on graphene. Ionic liquids are the most promising electrolyte with large electrochemical window. Alternatively, liquid crystal based proton conducting gel electrolytes have been recently introduced as an electrolyte in graphene optical modulators and graphene Fermi energies of around 1.5 eV has been successfully reached [5].

Based on these gating schemes, rich variety of graphene based optical modulators have been demonstrated [2]. We can classify these modulators into four main categories: (1) waveguide-based modulators, (2) photonic crystal based modulators, (3) supercapacitor-based modulators, and (4) plasmonic based modulators. Although the operation principle of these modulators are the same -Pauli blocking principle- they differ in the way incident light interacts with graphene layer that defines the interaction path.

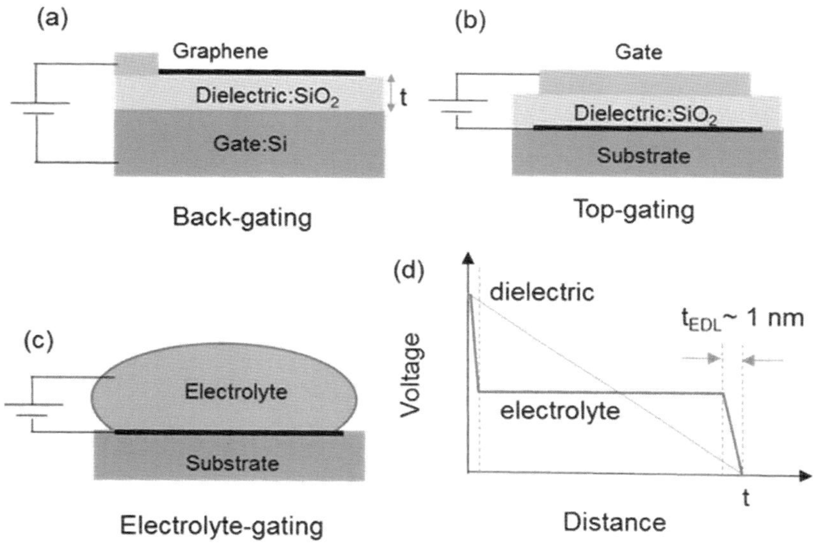

Fig. 13.2. (a-c) Common gating schemes used for controlling electronic and optical properties of graphene. (d) Schematic showing of the voltage drop across the gate dielectric and electrolyte based devices. For solid dielectrics, voltage drops linearly between the graphene and gate electrode; however, for liquid electrolytes, voltage drops very rapidly on the interface due the thin electrical double layers which yield very large electric field on graphene.

13.2 Waveguide-based optical modulators

The first class of graphene based electro-optical modulators use a waveguide structure integrated with gated-graphene layer [6-8]. In this configuration, high refractive index waveguide –usually Si- operates as the gate electrode and a thin oxide layer coated on the waveguide serves as the gate dielectric. Graphene is transferred onto the waveguide structure. Figure 12.3(a) shows the first demonstration of the graphene optical modulators operating at the telecommunication wavelengths [7]. The graphene layer on the slot waveguide introduces additional losses due to interband transitions. The evanescent field of the waveguide mode overlaps with the graphene layer. Application of a bias voltage between the waveguide and the graphene layers shifts the Fermi energy around 0.3 eV, and blocks the

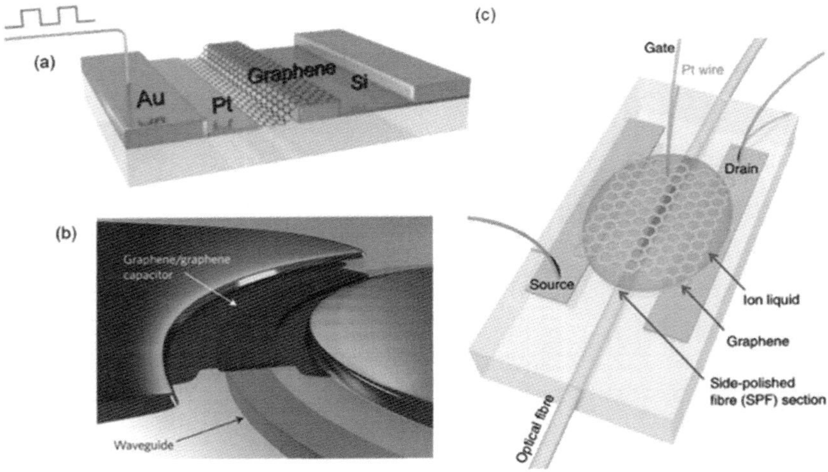

Fig. 13.3. Waveguide based optical modulators. (a) Graphene integrated on a Si slot waveguide. (b) Graphene integrated on a ring resonator. (c) Graphene integrated on a fiber optic waveguide. Figure adapted with permission from: a, ref. 7, b, Ref. [8], and c, Ref. [6], Nature Publishing Group. (For color reference, go to page 497.)

interband transitions at 1550 nm wavelength. The advantage of waveguide configuration is that, the interaction path can be increased with the length of the waveguide. They were able to demonstrate a strong electroabsorption modulation of 0.1 dBμm^{-1} over a broad range of wavelengths ranging from 1.35 μm to 1.6 μm [7]. Small footprint of the waveguide structure, 25 μm^2, allows switching speed up to 1.2 GHz at 3 dB. Later, different designs including double layer graphene structure have been demonstrated [9]. Very recently Phare et al. implemented the same principle on a ring-shaped optical waveguide with switching speed up to 30 GHz [8]. Figure 12.3(b) shows the schematic drawing of the proposed device structure. Although the switching speed of these devices are promising, the operation wavelength of these modulators are limited to near-IR wavelengths due to limited tunability of the Fermi energy with the back-gating. To overcome this problem, Lee et al. introduced electrolyte gating, which yields Fermi energy shifts up to 0.5 eV [6].

In their optic modulator design, they used a side-polished optical fiber and placed graphene onto the polished side of the optical fiber such that the evanescent field of the fiber can overlap with the graphene layer, Fig. 12.3(c). To achieve high level of doping, they used electrolyte gating. Thanks to the centimeter-long optical path, they were able to show absorption tunability as large as 70%. For the waveguide based optical modulators, there is a tradeoff between switching speed and operation wavelengths. Dielectric based gating provides small footprint with fast switching at IR frequencies (low frequency), whereas electrolyte gating provides slow operation in the visible spectrum (high frequency).

13.2.2 *Photonic crystal based optical modulators*

The second class of graphene based optical modulators are based on photonic crystals integrated with active graphene layers. The presence of graphene in the photonic crystal cavity introduces additional optical loses that can be controlled via electrical means. Gan et al. demonstrated high-contrast electro-optic modulation of a photonic crystal nanocavity integrated with an electrically gated monolayer graphene [10]. They used ionic gel electrolyte to control Fermi energy of graphene up to 0.8 eV, which yields significant modulation in the quality factor and the resonance wavelength of the cavity. Independently, Majumdar et al. demonstrated a very similar photonic crystal modulator with 60 dB reflectivity modulation [11]. They fabricated a photonic crystal on Si-on-insulator platform with three holes defect. The defect forms the optical nanocavity, which localizes the light inside the photonic crystal. After coating the surface of the cavity with graphene, the quality factor of the cavity is drastically reduced from 1500 to down to 500 due to additional optical losses. They observed narrowing in the cavity line width and an increase in the cavity reflectivity with increasing charge density on graphene. Both designs provide an electrically tunable photonic cavity operating in near-IR wavelengths. The physical footprint for the graphene-photonic crystal cavity (~a few μm^2) is significantly reduced when compared with graphene-waveguide optical modulator footprint (~25 μm^2). Very recently, Gao et al. demonstrated high speed photonic crystal modulator using graphene-boron-nitride heterostructure integrated with the photonic nanocavity [12]. Sandwiching graphene between two monalayer-BN layers, which operate as gate

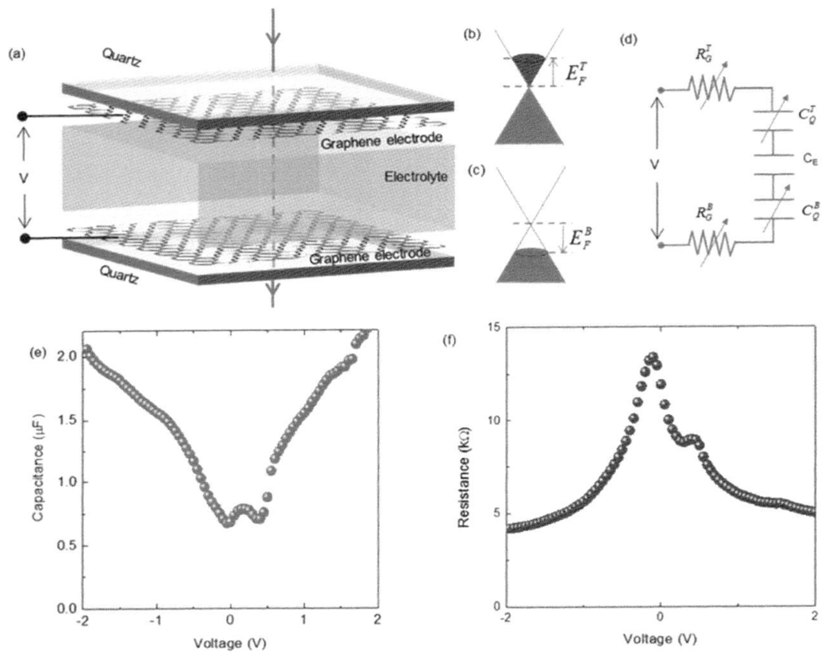

Fig. 13.4. Graphene supercapacitor operates as an optical modulator. (a) Schematic drawing of a parallel plate graphene supercapacitor formed by two large area graphene electrodes and electrolyte medium in between. (b,c) Band structure of graphene electrodes under a voltage bias. (d) Small signal circuit model of the device. (e,f) Variation of the resistance and capacitance of the device. Figure adapted with permission from: Ref. [14], American Chemical Society.

dielectric, yields very high mobility graphene due to the lower number of charged impurities on the substrate. The improvement of the quality of graphene yields a significant enhancement in the switching speed and reduction in the operation voltage. These initial works provide the foundation of graphene based electrically tunable integrated optical components that could yield more complex photonic integrated circuits.

13.2.3 *Graphene supercapacitor based optical modulators*

One unique advantage of graphene over other conventional 2D electronic systems (such as Si inversion layer and GaAs heterostructures) is that it can be grown over very large area [13]. Combining electrical tunability with the large area growth yields new perspective for large area electrically reconfigurable devices. Polat et al demonstrated a large area optical

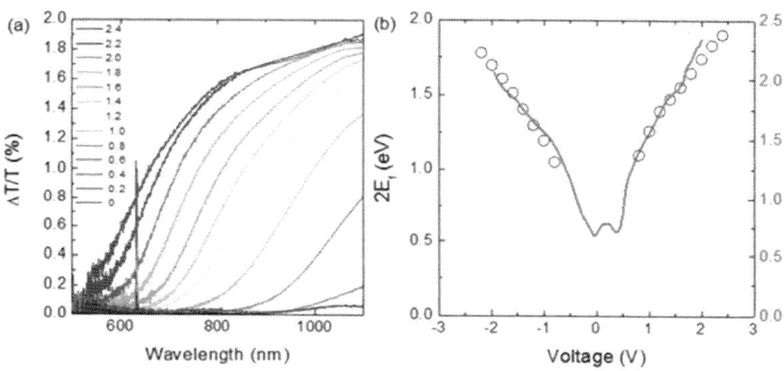

Fig. 13.5. (a) Variation of the optical transmittance spectra of graphene supercapacitor at different bias voltages. (b) The extracted Fermi energy and total capacitance of the device. Figure adapted with permission from: Ref. [14], American Chemical Society. (For color reference, go to page 497.)

modulators using graphene supercapacitor, which is a good example for graphene enabled devices [14]. Figure 13.4(a) shows their design which includes a parallel plate graphene supercapacitor filled with a liquid electrolyte. Two large area graphene electrodes are transferred onto optically transparent quartz substrates. Electrolyte is injected between the graphene electrodes. This device structure provides self-gating mechanism between the graphene electrodes without requirement of a metallic gate electrodes. Under a bias voltage, the electrolyte gets polarized and forms electrical double layers at the graphene-electrolyte interface. Since the electrical double layer is very thin, the accumulated ions yield very large electric field and increase the carrier concentration on the graphene electrodes. Holes and electrons accumulate on opposite electrodes. Figure 13.4(b) and 13.4(c) show the band structure of graphene electrodes under a bias voltage. In this design, the graphene electrodes have double role, they operate as (i) the conductive electrode and (ii) the active layer. As the charges accumulate on the electrodes, the Fermi energy shifts to higher energies resulting in variation of resistance and quantum capacitance of the electrodes. The small signal model of the device is shown in Fig. 13.4(d), where the graphene electrodes are modeled with a variable resistor and a capacitor and the electrolyte is modeled by a capacitor representing the double layer capacitance. Since the electrostatic capacitance of the double layers is much larger than the quantum capacitance of the

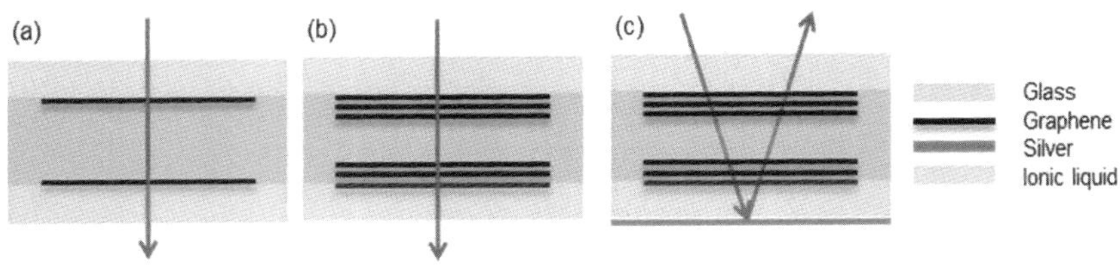

Fig. 13.6. Various modulator designs to improve the optical modulation. Figure adapted with permission from: ref. 14, American Chemical Society.

electrodes, total capacitance is limited by the quantum capacitance. The Fermi energy of the graphene defines the resistance (R) and quantum capacitance (C_Q) of the electrodes. Since the density of states of the atomic electrodes are very small, accumulation of charges results in large shifts of Fermi energy. Figure 13.4(e) and 13.4(f) show the variation of the total capacitance and resistance of the device. At charge neutrality point, the capacitance shows a minima and resistance reaches to a maxima. Two distinct maxima and minima are due to the differences between the Dirac point of the top and bottom electrodes.

Figure 13.5(a) shows the modulation of the transmittance at different bias voltages. Due to Pauli blocking, graphene becomes transparent to energies less than $2E_F$. The transmittance spectra shows a step-like function with cutoff energy at $2E_F$.

Fermi energy of the graphene electrodes can be extracted from the transmittance spectra. Figure 13.5(b) shows the variation of the Fermi energy with bias voltages and its comparison with the capacitance of the device which scales similarly with the charge density. The performance of the device is limited with the optical absorption of the graphene electrodes. Figure 13.6 shows various device configurations to improve the modulation depth of the graphene supercapacitor.

2.4 Graphene based electrochromic devices

Replacing single layer graphene with multilayer graphene in the supercapacitor modulators provides new perspective for optical modulators and electrochromic devices. Besides broad optical activity,

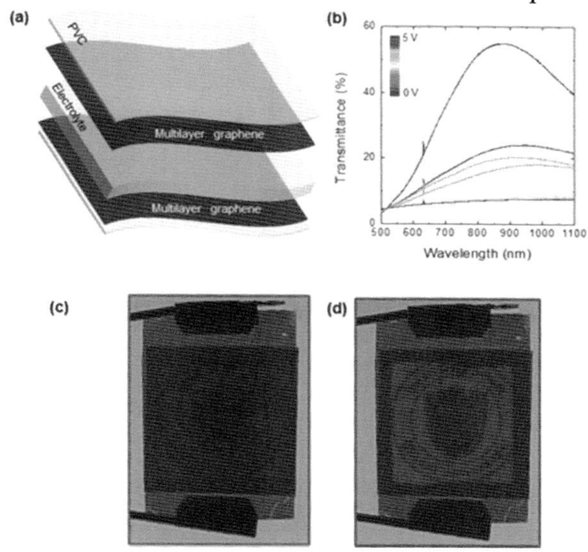

Fig. 13.7. (a) Schematic drawing of multilayer graphene based flexible electrochromic devices. (b) Optical transmittance of the device at different bias voltages. (c,d) Photograph of the device at 0 and 5 V bias voltage. Figure adapted with permission from: Ref. [15], Nature Publishing Group.

thickness of graphene enables fabrication of these devices on flexible substrates. Using CVD grown multilayer graphene, Polat et. al demonstrated new types of flexible electrochromic devices on polymer substrates [15]. Figure 13.7 depicts the schematic drawing of the flexible electrochromic device using multilayer graphene as an optically active layer. Multilayer graphene (MLG) simultaneously offers all key requirements for practical applications; high-contrast optical modulation over a broad spectrum, good electrical conductivity, and mechanical flexibility. This method relies on electro-modulation of interband transition of MLG via intercalation of ions into the graphene layers which yields significant modulation in optical transmittance.

Figure 13.8 shows the proposed mechanism for controlling optical properties of multilayer graphene supercapacitor. In single layer graphene case, formation of ionic double layer at graphene-electrolyte interface is sufficient to control doping on graphene electrodes. However, for multilayer graphene electrodes, accumulation of ions can only dopes first

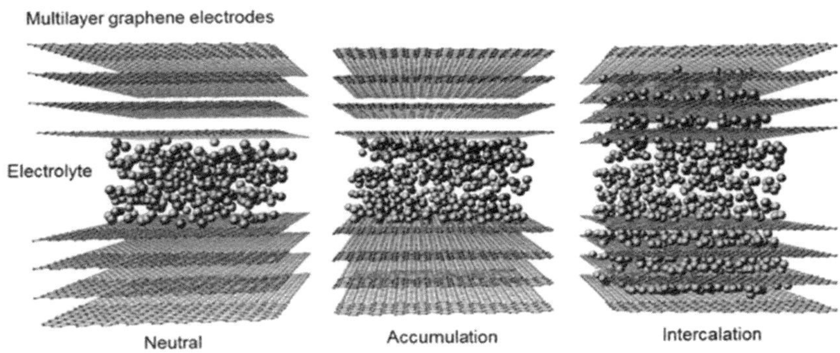

Fig. 13.8. Electrostatic doping of multilayer graphene via intercalation process. Unlike single layer graphene, formation of electrical double layer is not enough to gate the underneath layers due to screening. At large bias voltages, ions intercalate multilayer graphene which results in efficient gating of Fermi energies up to 1.5 eV. Figure adapted with permission from: Ref. [15], Nature Publishing Group. (For color reference, go to page 498.)

a few layers due to the electrostatic screening of the underneath layers. However, during ion intercalation process, ions can penetrate into thick film up to 100 graphene layers that yields optical modulation of 60% in the visible and near IR wavelengths, Fig. 13.9. There is a tradeoff between the modulation depth and thickness of the graphene film. As the thickness of the film increases, the modulation depth increases, however for very thick samples, the efficiency of intercalation process decreases and leads to less smaller

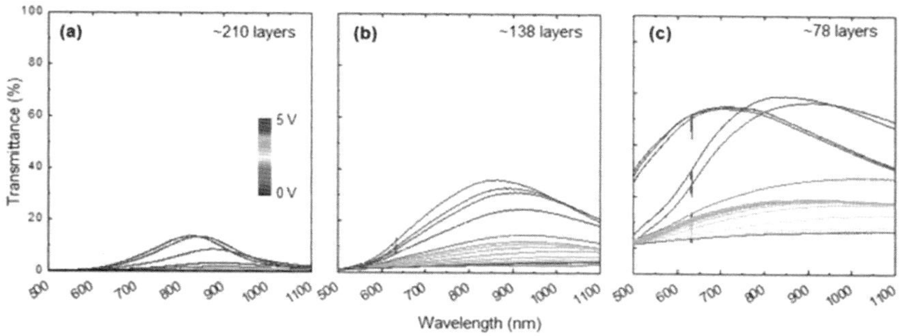

Fig. 13.9. Modulation of optical transmittance of multilayer graphene with varying multilayer graphene thicknesses. Figure adapted with permission from: ref. 15, Nature Publishing Group.

modulation and large insertion losses. Polat et al also demonstrated a reflection type electrochromic devices by replacing one graphene electrode with a reflecting gold electrode [15]. Furthermore, multi pixel display devices using pattered multilayer graphene has been demonstrated on flexible polymer substrates [15].

13.2.5 *Plasmonics based graphene optical modulators*

The third class of graphene based optical modulators is based on plasmonic based optical modulators where metallic nanostructures have been elegantly combined with single graphene layer [16-18]. Metallic nanostructures support surface plasmon polaritons (SPPs), which can concentrate incident light into a deep subwavelength volume, and thus SPPs lead to variety of appealing fundamental and practical applications such as spectroscopy, biosensing, nanophotonics, nanoelectronics, and solar cells [19]. To boost the performance of plasmonic devices at the nanoscale dimension and to enlarge the applicability of plasmonic nanostructures active control of the SPPs is always desirable and thus a large amount of efforts has been recently dedicated in this direction to actively control plasmon resonance in the visible and infrared part of the electromagnetic spectrum [20, 21]. Graphene provides a unique opportunity to modulate plasmon resonance at optical frequencies due to the presence of electrically controllable electronic interband transitions in graphene and the tunability of graphene conductivity by electronic gating as well. In addition, graphene can be easily integrated into the plasmonic nanostructures such as flat metal thin films, and metallic nanoparticles because of flexible nature of graphene sheets and graphene's compatibility with the nanofabrication processes. For example, Salihoglu et al placed graphene onto silver and gold surfaces and investigated SPPs on graphene-metal surfaces [22]. Ansel et al. fabricated and studied hybrid graphene plasmonic waveguide optical modulator, Fig. 13.10(a) [17]. The non-transparent grating on a gold surface couples incident light into SPPs, on the other hand, the transparent grating located at the other end of the device then decouples SPPs into light. The gated graphene placed on top of a boron

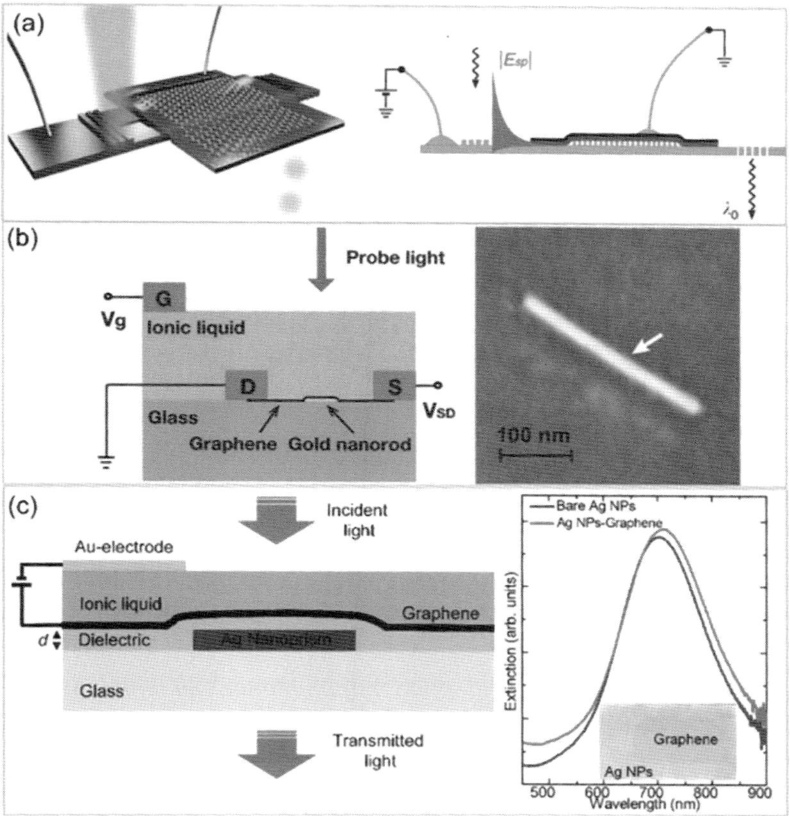

Fig. 13.10. Principles of plasmonic based optical modulators. (a) Hybrid plasmonic waveguide optic modulator. Uniform non-transparent grating couples incident light into SPPs which are modulated by graphene located on top of the dielectric spacer layer. Uniform transparent grating, on the other hand, decouples SPPs into light. (b) Graphene-gold nanowire hybrid nanostructure on a glass substrate. An electrolyte is used to control optical interband optical transitions in graphene. (c) Graphene-Ag nanoprism hybrid nanostructure placed on a glass substrate. The separation between graphene and Ag NPs are controlled with a spacer dielectric layer grown by ALD. Energy transfers from Ag NPs to graphene where graphene and Ag NPs behave as energy acceptors and donors, respectively. Figure adapted with permission from: a, Ref. [17], Nature Publishing Group, b, Ref. [16], American Chemical Society, and c, ref. 18, Optical Society of America. (For color reference, go to page 498.)

nitride flake affects SPPs propagation on the gold surface. In this way, the optical plasmonic modulator working at the telecom range was demonstrated. Furthermore, the demonstrated optical modulator had a modulation depth of >0.03dBμm^{-1} at low gating voltages. Graphene interaction with light has been greatly increased by fabricating a simple coplanar graphene configuration that involves SPPs propagation along a metal-dielectric interface adjacent to a graphene layer. At normal light incidence, graphene has relatively small absorption of 2.3% defined by the fine structure constant [4]. When light is incident on and transmitted through a graphene layer, the interaction between graphene and incident light has been enormously increased. Therefore, in this coplanar configuration a modulation depth of around 0.1 dB per graphene layer can be achieved.

In an another type of device structure, Kim et al demonstrated electrical controlled plasmonic resonance at infrared using gold nanorod-graphene hybrid nanostructure [16]. The principle of the plasmonic based optical modulator is shown in Fig. 13.10(b). It should be notified here that, in this study, the hybrid graphene-gold nanostructure was probed at the single-nanoparticle level using dark-field Rayleigh scattering spectroscopy of individual gold nanorods. Both the plasmon resonance and quality factor of gold nanorod plasmon is efficiently modulated. The results reveal that plasmon-graphene coupling is remarkably strong and they claim that even a single electron resting in graphene could affect the plasmon resonance in gold nanowire. Electrically tunable interband transitions in graphene at optical frequencies enable electrical modulation of plasmon resonance. For electrical gating of graphene a top electrolyte gating with an ionic liquid was used.

Recently, in the visible spectrum, Balci et al have actively modulated plasmon resonance of silver nanoprisms (Ag NPs) in graphene-Ag NPs hybrid nanostructure, Fig. 13.10(c) [18]. Ag NPs have tunable plasmon resonances ranging from 400 nm to 1100 nm and they can be uniformly coated on glass or silicon substrates [23]. The fabricated structure was in large area and the optical measurements were performed at the ensemble level. The interband transitions in graphene at optical frequencies can be easily controlled by ionic liquid top gating of graphene. As a result, the Fermi energy level of graphene can be tuned from around 0.2 eV to 1.2 eV which enables efficient control of graphene absorption in the visible spectrum because of the Pauli blocking. Both plasmon resonance

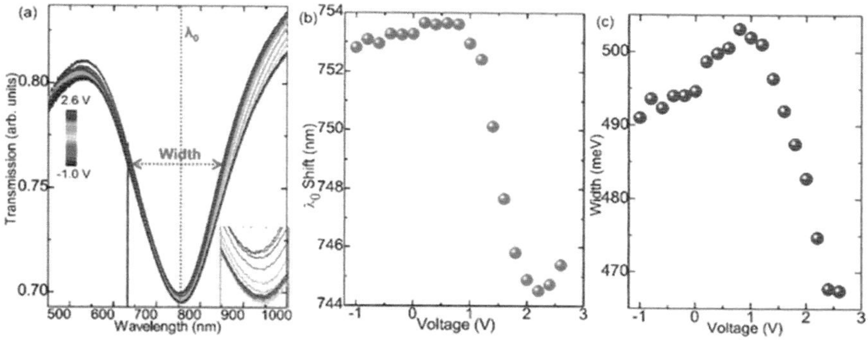

Fig. 13.11. Modulating plasmon resonance with graphene in the visible spectrum. (a) Optical transmission of the hybrid device for various applied bias voltages. (b) Plasmon resonance frequency (wavelength) increases (decreases) with the applied bias voltages. (c) Quality factor of the plasmon resonance at around 800 nm is significantly increased with the increase in the applied bias voltages. At low voltages, graphene induced loss in the medium is large due to the interband transitions in the graphene in the visible spectrum. At high voltages, the interband transitions are blocked and thus quality factor of the plasmon increases. Figure adapted with permission from: Ref. [18], Optical Society of America.

frequency and quality factor have been reversibly modulated by 20 and 35 meV, respectively. The separation between graphene and Ag NPs were controlled at the atomic scale by using atomic layer deposition (ALD) technique. The results indicate that the coupling between graphene and Ag NPs strongly depends on the dielectric layer thickness between them. Transmission spectra of the graphene-Ag NPs structure as a function of the applied bias voltage demonstrate that plasmon resonance of the Ag NPs can be electrically tunable, Fig. 13.11. The experimental results obtained in this work was explained by a damped harmonic oscillator (DHO) model [18]. In this model, a plasmon is considered as a damped harmonic oscillator in which an electron having a mass m attached to a spring having a spring constant k oscillates with a natural frequency $f = (k/m)^{1/2}/2\pi$. If there is a damping force represented as $F_{damping} = -bv$ where b and v are damping constant and velocity of oscillating object here an electron, respectively. Under damping condition, $f_{damped} = (k/m-(b^2/4m^2))^{1/2}/2\pi = (f^2-\gamma^2)^{1/2}$ where γ is the total damping rate. It is obvious here that under the damping condition, both the magnitude of the plasmon frequency and the quality factor of the plasmon resonance decrease. The DHO model successfully explains the observed modification in plasmon resonance as a function of the applied bias, Figure 13.11. At high bias voltages, the interband transitions in graphene are totally blocked and thus the loss in the medium because of the presence of graphene layer is blocked out and hence plasmon resonance frequency is shifted to high frequency. At low bias voltages, the interband transitions in graphene are not blocked out and the loss in the medium is large and thus plasmon resonance frequency is now at low frequency.

13.2.6 *Graphene-based long wavelength optical modulators*

In the previous sections, we summarize the graphene based optical modulators operating in the visible and near IR wavelengths. The operation principle of all of these devices are based on Pauli blocking principle that yields around 2.3% modulation per graphene layer. We now turn our attention to long wavelengths, i.e. terahertz and microwave. However, at long wavelengths, graphene operates as a tunable Drude metal, which yields much larger activity. The mechanism behind this large tunability is the intraband transitions whose rate is controlled by Fermi level and the associated density of charge carriers on the graphene. The activity in long wavelength makes graphene a unique material for active THz and microwave devices [24, 25]. The main challenge for long wavelength photonic devices is the physical dimensions of the device defined by the wavelength. For example, for an active microwave device operating at a frequency of 10 GHz (wavelength of 3 cm), the device dimension should be larger than 10 cm [26]. The conventional high mobility semiconducting materials are not suitable for large scale applications. High mobility of graphene, together with the ability to be synthesized over large areas, opens new opportunities for active long wavelength devices. Figure 13.12 summarizes some recent achievements in graphene enabled long wavelength active devices including switchable radar absorbing surfaces [25] and THz spatial light modulators [24]. Figure 13.12(a) shows an active microwave surface fabricated by graphene supercapacitors. Two very large area graphene sheets were synthesized by chemical vapor deposition and transferred on microwave transparent polymer substrates. Under a bias voltage, high mobility electrons accumulate on graphene electrodes and control the microwave

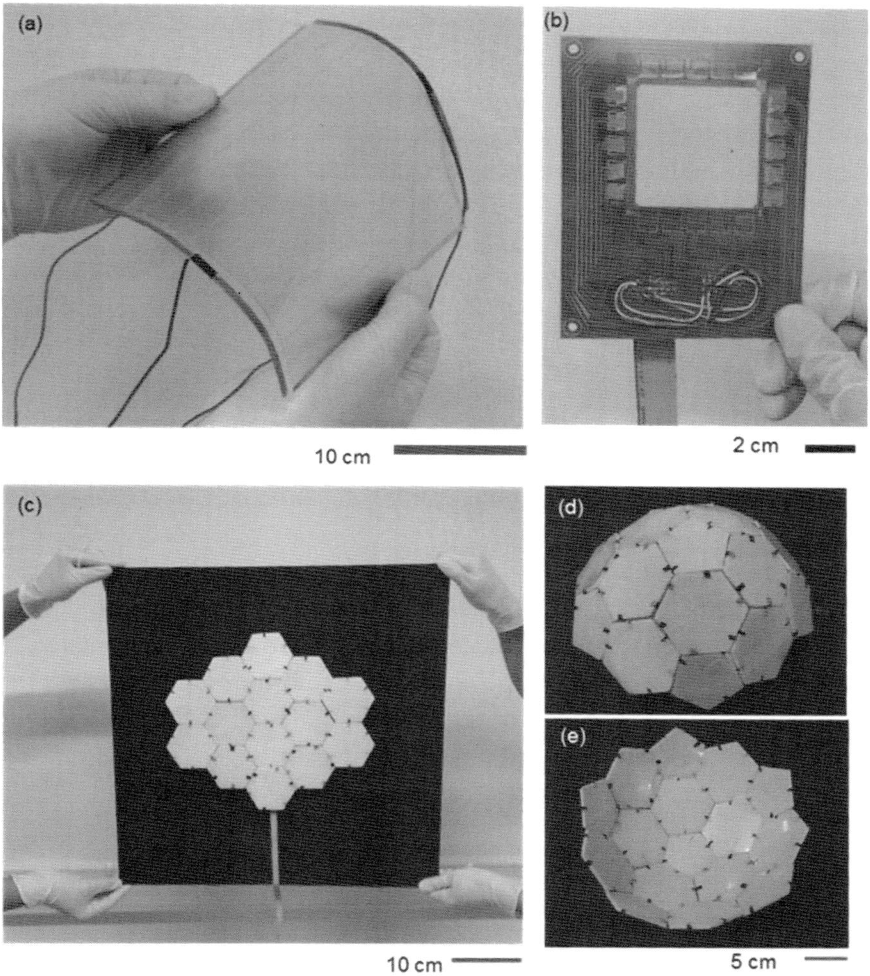

Fig. 13.12. Graphene enabled large scale active devices operating in THz and microwave frequencies. (a) Active microwave surface. (b) Multipixel spatial THz modulator. (c) Switchable radar absorbing surfaces. (d,e) Nonplanar active microwave surfaces. Figure adapted with permission from: a, c and d from Ref. [25], Nature Publishing Group. And b, from Ref. [24] Optical Society of America. (For color reference, go to page 499.)

reflectivity and transmittance of the surface. This device is a microwave analog of smart windows whose transmittance is modulated by electrical means. Combining these active microwave surfaces with different device designs enables electrically switchable radar absorbing surfaces with more than 50dB switchable suppression ratio. Kakenov et al. implemented similar device concept to control THz waves [26]. They were able to control THz transmittance through graphene capacitor with moderate bias voltages. They extended their work by integrated arrays of graphene device to form a spatial THz modulator (Fig. 13.12(b)) [24].

13.3. Conclusions

In conclusion, graphene provides a tunable optical platform to control light-matter interaction in a very broad spectrum ranging from visible to microwave frequencies. This profound tunability originates from high mobility free carriers on graphene surface whose density can be controlled by various gating schemes. Rich variety of active optoelectronic devices has been demonstrated. In this chapter, we summarize the recent progress on graphene based optical modulators fabricated by active graphene layer integrated on waveguides, photonic crystals, and plasmonic structures. In addition to the optoelectronic devices operating in the visible and infrared, there are recent examples for long wavelength modulators, which could find immediate applications, for example, in radar absorbing surfaces and THz communications.

Acknowledgements

Coskun Kocabas acknowledges the partial funding from TUBİTAK (The Scientific and Technological Research Council of Turkey) under the Project No. 113F278 and 114F052. He also acknowledges the support from the European Research Council (ERC) Consolidator Grant ERC – 682723 SmartGraphene.

References

[1] Novoselov, K.S., et al., *Electric field effect in atomically thin carbon films.* Science, 2004. **306**(5696): p. 666-669.

[2] Sun, Z., A. Martinez, and F. Wang, *Optical modulators with 2D layered materials.* Nat Photon, 2016. **10**(4): p. 227-238.

[3] Bonaccorso, F., et al., *Graphene photonics and optoelectronics.* Nature Photonics, 2010. **4**(9): p. 611-622.

[4] Nair, R.R., et al., *Fine structure constant defines visual transparency of graphene.* Science, 2008. **320**(5881): p. 1308-1308.

[5] Tunkara, E., et al., *Highly Proton Conductive Phosphoric Acid-Nonionic Surfactant Lyotropic Liquid Crystalline Mesophases and Application in Graphene Optical Modulators.* Acs Nano, 2014. **8**(10): p. 11007-11012.

[6] Lee, E.J., et al., *Active control of all-fibre graphene devices with electrical gating.* Nature Communications, 2015. **6**.

[7] Liu, M., et al., *A graphene-based broadband optical modulator.* Nature, 2011. **474**(7349): p. 64-67.

[8] Phare, C.T., et al., *Graphene electro-optic modulator with 30 GHz bandwidth.* Nature Photonics, 2015. **9**(8): p. 511-+.

[9] Liu, M., X.B. Yin, and X. Zhang, *Double-Layer Graphene Optical Modulator.* Nano Letters, 2012. **12**(3): p. 1482-1485.

[10] Gan, X.T., et al., *High-Contrast Electrooptic Modulation of a Photonic Crystal Nanocavity by Electrical Gating of Graphene.* Nano Letters, 2013. **13**(2): p. 691-696.

[11] Majumdar, A., et al., *Electrical Control of Silicon Photonic Crystal Cavity by Graphene.* Nano Letters, 2013. **13**(2): p. 515-518.

[12] Gao, Y.D., et al., *High-Speed Electro-Optic Modulator Integrated with Graphene-Boron Nitride Heterostructure and Photonic Crystal Nanocavity.* Nano Letters, 2015. **15**(3): p. 2001-2005.
[13] Polat, E.O., et al., *Synthesis of Large Area Graphene for High Performance in Flexible Optoelectronic Devices.* Scientific Reports, 2015. **5**.
[14] Polat, E.O. and C. Kocabas, *Broadband Optical Modulators Based on Graphene Supercapacitors.* Nano Letters, 2013. **13**(12): p. 5851-5857.
[15] Polat, E.O., O. Balci, and C. Kocabas, *Graphene based flexible electrochromic devices.* Scientific Reports, 2014. **4**.
[16] Kim, J., et al., *Electrical Control of Optical Plasmon Resonance with Graphene.* Nano Letters, 2012. **12**(11): p. 5598-5602.
[17] Ansell, D., et al., *Hybrid graphene plasmonic waveguide modulators.* Nature Communications, 2015. **6**.
[18] Balci, S., et al., *Dynamic tuning of plasmon resonance in the visible using graphene.* Optics Letters, 2016. **41**(6): p. 1241-1244.
[19] Ebbesen, T.W., et al., *Extraordinary optical transmission through sub-wavelength hole arrays.* Nature, 1998. **391**(6668): p. 667-669.
[20] Pala, R.A., et al., *A nonvolatile plasmonic switch employing photochromic molecules.* Nano Letters, 2008. **8**(5): p. 1506-1510.
[21] Sterl, F., et al., *Magnesium as Novel Material for Active Plasmonics in the Visible Wavelength Range.* Nano Letters, 2015. **15**(12): p. 7949-7955.
[22] Salihoglu, O., S. Balci, and C. Kocabas, *Plasmon-polaritons on graphene-metal surface and their use in biosensors.* Applied Physics Letters, 2012. **100**(21).
[23] Balci, S., *Ultrastrong plasmon-exciton coupling in metal nanoprisms with J-aggregates.* Optics Letters, 2013. **38**(21): p. 4498-4501.
[24] Kakenov, N., et al., *Graphene-enabled electrically controlled terahertz spatial light modulators.* Optics Letters, 2015. **40**(9): p. 1984-1987.
[25] Balci, O., et al., *Graphene-enabled electrically switchable radar-absorbing surfaces.* Nature Communications, 2015. **6**.
[26] Kakenov, N., et al., *Broadband terahertz modulators using self-gated graphene capacitors.* Journal of the Optical Society of America B-Optical Physics, 2015. **32**(9): p. 1861-1866.

Chapter 14

The Potential of Graphene as a Transparent Electrode

Wee Shing Koh[1,*], Wee Kee Phua[1] and Wei Peng Goh[2]
[1]*A*STAR Institute of High Performance Computing, Singapore*
[2]*A*STAR Institute of Materials Research and Engineering, Singapore*
[*]*kohws@ihpc.a-star.edu.sg*

This chapter evaluates the potential of current state-of-the-art doped graphene as a replacement of commercially available Transparent Conducting Oxides (TCOs) in display and photovoltaic applications. The requirements of transparent conductors, namely optical transmittance and sheet resistance, are elaborated. The optical properties of the doped graphene and two most commonly used TCOs, Indium Tin Oxide (ITO) and Fluorine-doped Tin oxide (FTO) are illustrated.

Two primary applications of TCOs in flat panel displays and photovoltaics are identified. In order to evaluate the feasibility in the use of doped graphene to replace ITO or FTO, the performance of the device for each application [i.e. Liquid Crystal Display (LCD) and Organic Photovoltaics (OPV)] is computed and compared. For flat panel displays, doped graphene has exhibited the potential to replace ITO or FTO as the transparent electrode with a relative 10% reduction in optical reflection for the reflective LCD when a four-layer graphene is used. Doped graphene has also demonstrated that it has the potential to replace ITO or FTO for the two organic solar cells simulated. In terms of the performance of the organic solar cell, the optical absorption in the polymer active layer of the P3HT:PCBM organic solar cell with the four-layer graphene as the transparent electrode is 5.1% less than that of ITO or FTO based devices. For the small-molecule ZnPc organic solar cell, the four-layer graphene device exhibit 4.4% and 3.3% less absorption in the active layer as compared to

ITO and FTO based devices respectively. It is also found that if substantial process improvements to the electrical characteristics or sheet resistance is achievable in the near future, monolayer graphene (instead of the four-layer graphene) exhibits optical performance that is better than FTO or marginally poorer as compared to ITO as the transparent conductor in both LCD display and organic solar cells.

14.1 Introduction

Graphene is a two-dimensional material made up of carbon atoms arranged in the planar honeycomb configuration. As graphene is just a single atom thick, the optical transparency of the material is confirmed to be very high. Although graphene is usually classified as a semiconductor, the absence of a bandgap enables graphene to exhibit metallic-like electrical conductivity [Geim and Novoselov (2007)] or low sheet resistance in the 200 $\Omega/\square$ range [Bonaccorso *et al.* (2010)]. In addition to good thermal and chemical stabilities, graphene can also be folded without drastic increase in the sheet resistance. For instance, Lee *et al.* has reported that the sheet resistance of the indium oxide sample increased by 321% under 2.14% strain, while graphene only increased by 10 to 15% under the same strain [Lee *et al.* (2012)].

Thus, graphene is potentially attractive for a huge variety of photonics and optoelectonics applications. One of the optoelectronic applications in which graphene has been touted as a potential replacement is the transparent conductor in photovoltaics, touch screens and various display technologies such as Liquid Crystal Display (LCD) and Organic Light Emitting Diode (OLED) based displays. In this chapter, we will focus on transparent conductors for two particular optoelectronic applications: (1) LCD and (2) Organic Photovoltaics.

14.2 Requirements of a Transparent Electrode

In order to to evaluate whether graphene can be used as a transparent electrode, it is necessary to understand the requirements of a transparent electrode for the LCD and the organic photovoltaic device. There are two primary requirements for transparent conductors for these two optoelectronic applications.

The first and most important requirement is optical transmittance. A transparent electrode candidate typically must be able to achieve at least

80% transmittance in the frequency band of interest for the corresponding application. For displays such as the LCD, the transmittance of visible light, in which the photon energy spanning 1.59eV to 3.26eV (380nm to 780nm), is important. Typical experimental optical characterization of transparent conductors for displays will evaluate the optical transmittance of the material at 2.254eV or 550nm (green light). This is because the photopic response of the human eye is most sensitive at that particular wavelength as illustrated by the CIE 1931 in Fig. 14.1. The CIE 1931 standard was developed by the International Commission on Illumination or the Commission Internationale de I'Eclairage (CIE).

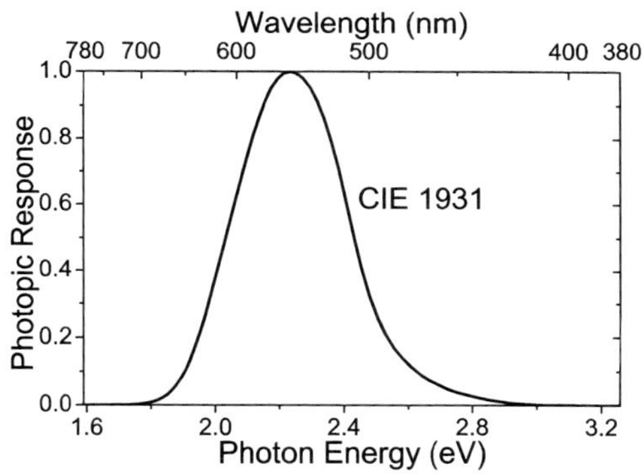

Fig. 14.1 Spectrum of CIE 1931 to replicate the photopic response of a human eye to evaluate visible light transmission.

For photovoltaic devices, the optical transmittance of $> 80\%$ has to cover from photon energies between 0.495eV to 4.42eV (280nm to 2500nm) where the bulk of the solar spectrum lies. The typical solar simulators available in the market mimics the reference solar spectral irradiance spectrum based on the Air Mass 1.5 Global (AM1.5G) or ASTM G-173 standard [ASTM G173-03 (2012)] developed by the American Society for Testing and Materials (ASTM) and their partners as shown in Fig. 14.2. The ASTM G-173 standard indicates that the intensity of the simulated sunlight is about $1000\mathrm{W/m^2}$. However, most photovoltaic absorbers are unable to absorb and generate electric current by utilizing all the light within the solar spectrum. For instance, the glass substrate used in a solar cell will screen out part of the solar spectrum in the ultraviolet (UV) region (i.e. photon energies > 3.54eV or $<\sim$350nm in terms of wavelength). In the case of organic photovoltaic devices (also categorized as third-generation solar cells), most organic media are only able to absorb incoming light from 1.55eV to 4.13eV (300nm to 800nm). Thus, it is only meaningful to study the optical characteristics of the transparent conductors within the frequency range of absorption of the photovoltaic active layers.

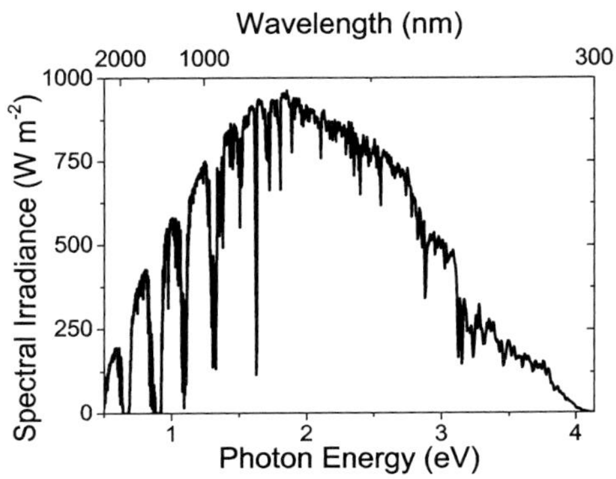

Fig. 14.2 Solar Spectrum Air Mass 1.5 Global (AM1.5G) to emulate outdoor illumination of solar cells.

The second requirement for a transparent electrode is the electrical conductivity. A good transparent conductor should have electrical conductivity close to that of a metal. However, as it is impossible for metal to achieve high transparency and yet retain its high conductivity. The only class of material that has both high conductivity and high transmittance in the visible light wavelengths (where the bulk of the solar energy is in) is the Transparent Conducting Oxide (TCO). The typical range of sheet resistance that are acceptable from TCOs or any transparent electrodes spans from 7Ω/□ to 30Ω/□. For instance, the commercial TEC series (TEC7, TEC8 & TEC15) of FTO glass (targeted at thin film photovoltaics) have sheet resistance ranging from 7-15Ω/□ and laboratory deposited TCOs on glass typically have higher sheet resistance of approximately 20-30Ω/□ [Wang *et al.* (2010); Zeng *et al.* (2010)].

14.3 Common Transparent Conducting Oxides Versus Graphene

The two most commonly available TCOs are Indium Tin Oxide (ITO) and Fluorine-doped Tin Oxide (FTO). Both ITO and FTO are commercially available as transparent conductors on substrates such as glass and polyethylene terephthalate (PET). ITO exhibits slightly better optical and electrical performance as compared to FTO. However, indium is a rare earth element that is limited in supply. Therefore, the cost of FTO is lower than of ITO and is used much more often commercially while ITO is used when the TCO's performance is important.

In order to evaluate if graphene is a suitable candidate to replace TCOs, it is essential to compare graphene to these two commercially available TCOs in terms of their optical and electrical characteristics. It is important to note that intrinsic graphene fabricated by current processing methods (such as Chemical Vapor Deposition (CVD) or roll-to-roll technology) has not been able to achieve the theoretical electrical conductivity or sheet resistance predicted [Bae *et al.* (2010)]. Therefore, chemical modification of graphene is essential to enhance the electrical properties of intrinsic graphene films. For instance, Bae *et al* has demonstrated that p-type doping of roll-to-roll fabricated monolayer graphene film by nitric acid (HNO_3) reduces the sheet resistance from $>200\Omega/\square$ to $\sim 100\Omega/\square$ [Bae *et al.* (2010)]. However, the calculated refractive indices (from [Koh *et al.* (2014)]) of intrinsic and doped graphene monolayers have negligible difference even though the sheet resistance (a function of carrier mobility in the graphene refractive index calculations) can be significantly different. The calculated real and imaginary refractive indices of graphene (assuming that graphene is 1nm thick) are illustrated in Fig. 14.3. The measured refractive indices of ITO and FTO are extracted from [Koh *et al.* (2011)] and [Ball *et al.* (2015)], respectively.

In terms of optical properties, the real part of the calculated refractive index of graphene is of the same magnitude as that of ITO and FTO as shown in Fig. 14.3. However, the imaginary part of the calculated refractive index of graphene is much much higher than both ITO and FTO. This implies that ITO and FTO have negligible absorption as compared to a graphene monolayer within the optical frequencies.

However, it is not straightforward to determine the actual optical transmittance of graphene or any of the TCOs from the refractive indices if the number of graphene layers or TCO thickness are unknown. Furthermore,

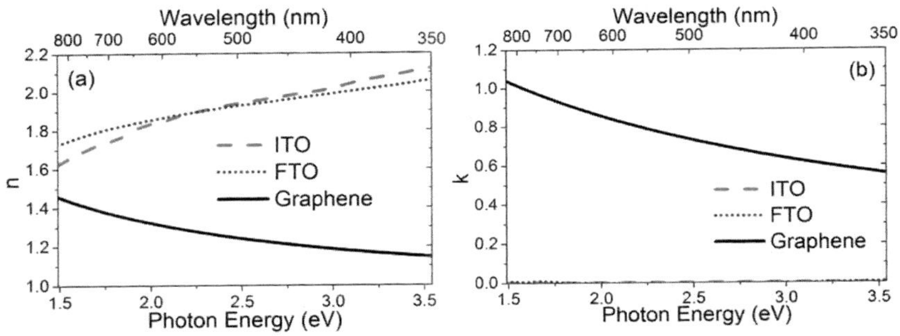

Fig. 14.3 (a) Real and (b) Imaginary refractive indices of ITO, FTO and monolayer graphene as a function of photon energy (eV). The refractive index of ITO and FTO are measured by ellipsometry by [Koh *et al.* (2011)] and [Ball *et al.* (2015)]. The refractive index of graphene is calculated based on the equations and assumptions specified in [Koh *et al.* (2011)] for doped graphene produced by roll-to-roll technology in [Bae *et al.* (2010)]

TCOs are usually deposited on a substrate such as glass, which is typically used in displays and photovoltaic devices. Therefore, it is instructive to assess the optical transmittance of graphene in comparison with ITO and FTO on a glass substrate.

For display applications, the integrated transmittance [$T_{int(CIE1931)}$] is weighted with the CIE 1931 frequency band of 1.59eV to 3.26eV (380nm to 780nm). For solar cells applications or organic solar cells, which will be covered in this chapter, the integrated transmittance is weighted with a subset of the AM1.5G solar spectrum spanning 1.55eV and 3.54eV (350nm to 800nm). Using the transfer matrix method (TMM), the graphene or TCO on glass structure is illuminated by monochromatic light from the glass side as illustrated in the inset of 14.4(b). The refractive index of glass is assumed to be 1.49 in the calculations. The complex refractive indices of ITO, FTO and doped graphene are plotted in Fig. 14.3. The normalized integrated transmittance weighting with the respective spectrum is given as

$$T_{\text{int(CIE1931)}} = \frac{\int_{1.59eV}^{3.26eV} T(E)S(E)dE}{\int_{1.59eV}^{3.26eV} S(E)dE} \tag{14.1}$$

$$T_{\text{int(AM1.5G)}} = \frac{\int_{1.55eV}^{3.54eV} T(E)S(E)dE}{\int_{1.55eV}^{3.54eV} S(E)dE} \tag{14.2}$$

where S(E) is the spectrum considered, T(E) = $P_{TCO/Air}(E)/I_0$ is the normalized frequency-dependent transmittance through the 1.1mm thick glass/TCO or graphene configuration at each photon energy or frequency E. The Poynting vector or power flowing out of the structure $P_{TCO/Air}(E)$ $=\vec{E} \times \vec{H}$ is computed from the solution of the Maxwell's Equations, where $\vec{E}$ and $\vec{H}$ are the electric and magnetic field vectors, respectively.

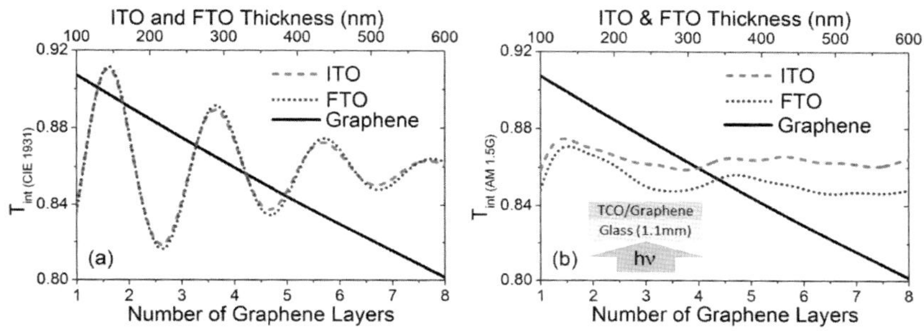

Fig. 14.4 Normalized Integrated transmittance (a) $T_{int(CIE1931)}$ of CIE1931 illumination (1.59eV - 3.26eV) and (b) $T_{int(AM1.5G)}$ of AM1.5G illumination (1.55eV - 3.54eV) illuminating 1.1mm thick glass / graphene (ITO or FTO) as a function of graphene layers (ITO or FTO thickness). The structure investigated is an inset in (b)

Figure 14.4 shows the normalized integrated transmittance (T_{int}) for different number of graphene layers and ITO or FTO thickness under (a) CIE 1931 and (b) AM1.5G spectrum illumination. It is interesting to note from both figures that the normalized integrated transmittance of graphene reduces from 90.7% to 80.2% almost linearly as the number of graphene layers increases from one to eight stacked layers. On the other hand, 140nm thick ITO and FTO on glass configuration both allows marginally more (91%) light to pass through the whole structure than a monolayer graphene under CIE 1931 illumination as shown in Fig. 14.4(a). Under AM1.5G illumination, 130nm ITO and FTO [denoted by the first peak from the left in Fig. 14.4(b)] exhibit lower peak transmittance as compared to monolayer graphene when illuminated by the partial AM1.5G spectrum spanning 1.55eV and 3.54eV (350nm to 800nm). The respective peak AM1.5G integrated transmittance of ITO and FTO are 87.5% and 87% respectively as observed in Fig. 14.4(b).

Figure 14.5 gives an overview of the variation of sheet resistance as a function of number of graphene layers and thickness of ITO or FTO. Undoped or intrinsic graphene film's sheet resistance relation for one to four layers are plotted as a dashed line with filled circular symbols, while the

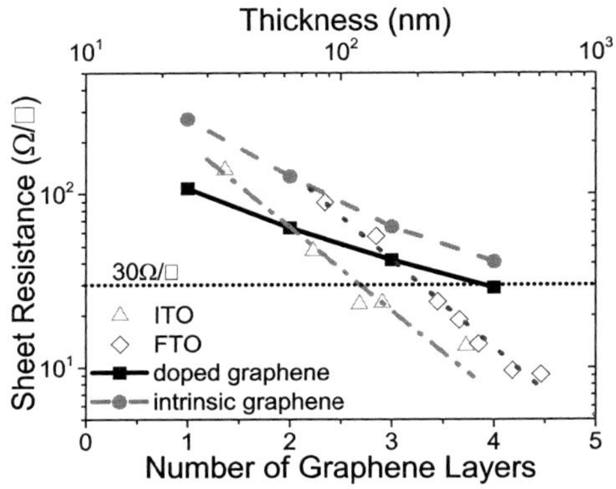

Fig. 14.5 Sheet Resistance versus Thickness for doped & intrinsic graphene, ITO and FTO. The data for doped & intrinsic graphene (or roll-to-roll doped graphene & roll-to-roll graphene) are extracted from [Bae *et al.* (2010)] and spray pyrolysis synthesized ITO and FTO data are extracted from [Aegerter (2016)]. A horizontal short-dotted line marks the 30Ω/□ requirement of a transparent conductor.

variation of sheet resistance with graphene layers for HNO_3 doped graphene is shown as a solid line with filled square symbols. In order for doped graphene to exhibit sheet resistance less than 30Ω/□, four layers of stacked doped graphene are required as four-layer intrinsic graphene exhibits sheet resistance of around 40Ω/□. At present, doped graphene has been demonstrated as the best candidate to compete with ITO or FTO based TCOs. **Thus, from this point forth, all subsequent mentions and comparisons made with graphene are assumed to refer to the HNO_3 doped graphene in Bae *et al*'s paper [Bae *et al.* (2010)].**

The measured sheet resistance of ITO are denoted as unfilled triangles and a dash-dot line indicates the approximate linear slope of the variation of ITO sheet resistance with thickness. Similarly, the measured sheet resistance of FTO are marked by unfilled diamonds and a dotted line shows the corresponding linear variation of FTO sheet resistance with thickness. Both ITO and FTO sheet resistance versus thickness data are extracted from [Aegerter (2016)] and the measured films are synthesized by spray pyrolysis. From Fig. 14.5, a minimum thickness of 100nm ITO and 200nm FTO are estimated to be necessary for the two TCOs to achieve at least 30Ω/□ sheet resistance. Note that even though the ITO and FTO data in Fig. 14.5 are based on spray pyrolysis, the trend and quantitative values are not significantly different from ITO or FTO deposited by other methods, such as sputtering and low temperature deposition [Wang *et al.* (2010)].

Knowing that four layers of doped graphene is required to meet both the optical transparency and sheet resistance requirement are not enough to conclude that graphene is a suitable candidate. This is because the incident light on the device with the transparent conductor embedded as part of the structure will also interact with the other layers of materials. Therefore, it is instructive to evaluate the optical performance of four-layer doped graphene and compare this with the same device with ITO or FTO.

Here, two common applications of transparent electrodes will be considered. They are the (1) Liquid Crystal Displays (LCDs) & LCD touchscreens and (2) thin film photovoltaics, in particular organic solar cells which are also known as the third generation photovoltaics.

14.4 Graphene for Displays and Touch Screens

For flat panel display, the liquid crystal display (LCD) technology is currently a widely used technology in both small mobile and large form factors. In this section, the comparison of graphene, ITO and FTO as a transparent electrode for a simple reflective LCD structure will be explored. A simplified structure of the reflective type LCD is presented in Fig. 14.6. The structure is deemed simplified as two thin polarization alignment layers sandwiching the nematic liquid crystal E7 are ignored due to the similarity in refractive indices with the thick E7 layer.

The corresponding real and imaginary part of the refractive index for nematic liquid crystal E7 (in ordinary and extraordinary wave modes) and back reflector aluminum are plotted as Fig. 14.7. The refractive index of E7 is extrapolated from the numerical fitting by Li [Li (2005)] and the refractive index of aluminum is obtained from Palik's handbook [Palik (1997)]. Here, the refractive index of glass in the simplified structure is assumed to be 1.49 and is used in the calculations in this section. Note that E7 has zero imaginary refractive index. This means that E7 is non-absorbing in nature.

In terms of display applications, the light source entering from the glass side is assumed to be a constant or of unit intensity, across the visible spectrum with frequencies between 1.59eV and 3.26eV (380nm to 780nm). Light enters the structure from the air to glass as indicated by the dotted arrow in Fig. 14.6. The equivalent illuminant standard defined by CIE is Illuminant E. As the LCD type of interest is the reflective type, the performance benchmark of the LCD is the normalized reflectance of light (with respect to the intensity of the incident light) by the LCD structure.

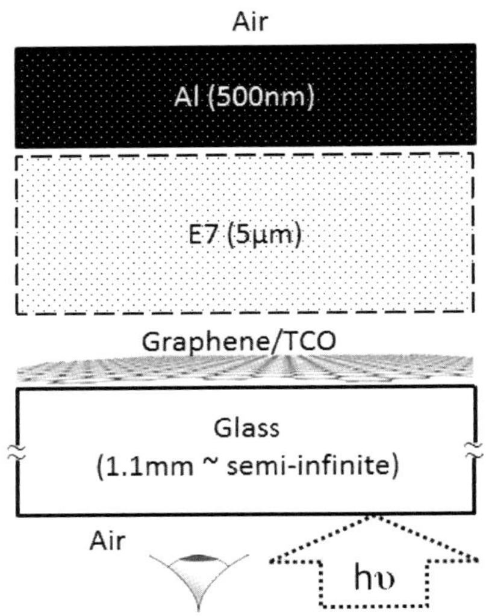

Fig. 14.6 Simplified structure of reflective type LCD with one TCO layer.

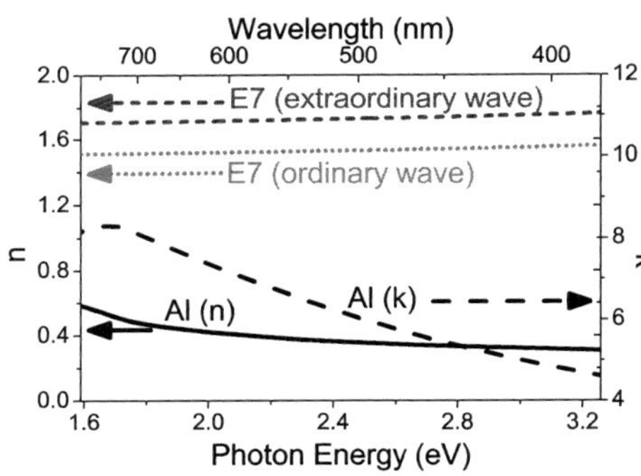

Fig. 14.7 Refractive indices of aluminum and nematic liquid crystal E7 in a LCD. The solid line and dashed line denote the real (n) and imaginary (k) refractive index of aluminum, while the short-dotted and short-dashed lines depict the real refractive index (n) of E7 for ordinary wave and extraordinary wave modes, respectively.

The reflected light from the LCD will then enter the eye of the viewer and therefore, it is necessary to weigh the frequency-dependent normalized reflection with the photopic response of the eye as described by CIE 1931 standard shown by Fig. 14.1 in Section 14.2.

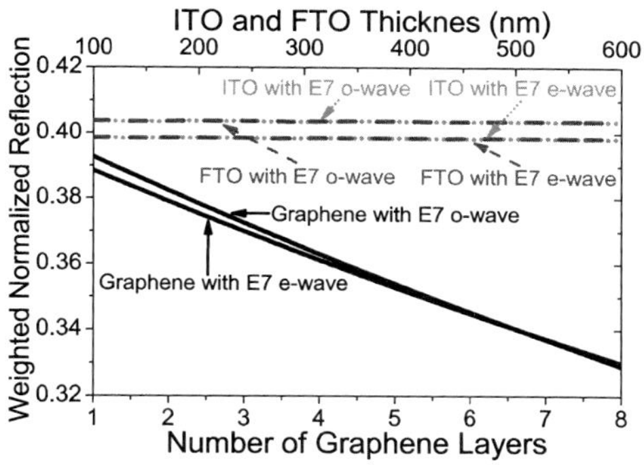

Fig. 14.8 Weighted normalized reflection of the simplified reflective type LCD structure as a function of the number of graphene layers and ITO & FTO thickness.

Figure 14.8 illustrates the CIE 1931 weighted normalized reflection of the simplified reflective type LCD structure as a function of the number of graphene layers and ITO & FTO thickness. The LCDs with graphene as the transparent electrode are denoted in solid black lines; LCDs with FTO are denoted in gray dashed lines and LCDs with ITO are indicated in light gray dotted lines. When graphene is used as the transparent electrode, it is observed that the weighted normalized reflection (when the LC layer E7 is in either ordinary-wave or extraordinary-wave mode) decreases from ~0.39 to 0.33 as the number of layers of graphene increases from one to eight. On the other hand, when ITO or FTO is the transparent electrode, the weighted normalized reflection characteristics are similar for both TCOs. The weighted normalized reflection of the reflective LCD in ordinary-wave mode and extraordinary-wave mode with ITO and FTO is 0.404% and 0.399% respectively. It is apparent that both ITO and FTO are able to outperform monolayer graphene as a transparent electrode. If the sheet resistance requirement is to be met ($< 30\Omega/\square$), four-layer graphene is required and the CIE 1931 weighted normalized reflection is reduced to ~0.362 (or ~10% less than the TCOs based LCD device) as observed in Fig. 14.8. Therefore, it is obvious that the increase in the number of graphene layers will increase the corresponding absorption of light in graphene, which leads to less light reflected back to the eye of the observer since the incident light makes at least two passes through the graphene transparent electrode. Since the weighted normalized reflection is similar for both ITO and FTO for TCO thickness ranging from 100nm to 600nm, the thickness of subsequent spectral evaluations will be determined primarily based on sheet resistance considerations. In Fig. 14.5 of Section 14.3, it was shown that the sheet resistance of ITO is smaller than that of FTO for the same thickness. Here, the thickness of FTO was chosen to be the same as that of a commercially available TEC15 FTO glass [Ball *et al.* (2015)] at 340nm with sheet resistance of <15Ω/□. Since the optical response is independent of the thickness of the ITO as observed in Fig. 14.8, the thickness of 160nm ITO was selected to meet the minimum sheet resistance requirement as illustrated in Fig. 14.5.

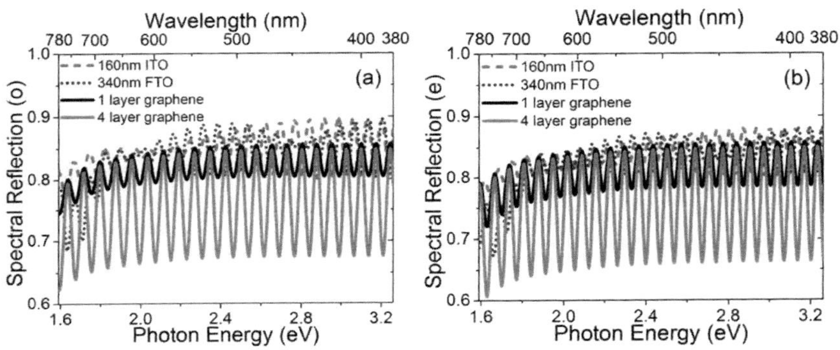

Fig. 14.9 Normalized reflection of reflective type LCD for (a) ordinary- and (b) extraordinary-wave modes.

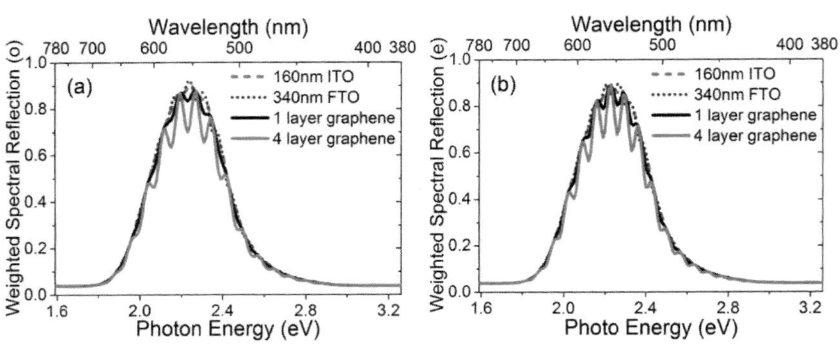

Fig. 14.10 Weighted normalized reflection of reflective type LCD with the CIE1931 spectrum for (a) ordinary- and (b) extraordinary-wave modes.

Figures 14.9 and 14.10 illustrate the spectral reflection of the LCD structure without and with CIE1931 photopic response weighting for 160nm ITO, 340nm ITO, monolayer and four-layer graphene. The oscillatory features for both ordinary-wave and extraordinary-wave modes are induced due to multiple reflections in the thick ($\gg$ incident wavelength) E7 liquid crystal layer. It is interesting to note from Fig. 14.9 that the amplitude of oscillations for the spectral reflection is largest when four-layer graphene is used as the transparent conductor for both ordinary-wave and extraordinary-wave modes. Both monolayer and four-layer graphene are able to maintain their respective amplitude of oscillations for photon energies in the visible light range, while the amplitude of oscillations in the spectral reflection for 160nm ITO and 340nm FTO reflective LCD fluctuates. This fluctuation in amplitude is due to the additional reflections induced by the thick (>100nm) TCOs. On the other hand, the variation in amplitudes due to spectral reflections is absent in the graphene cases as monolayer or four-layer graphene is too thin ($\ll$ than the wavelength of visible light) to induce additional reflections that may interfere constructively or destructively with the reflections induced by 5μm thick E7 liquid crystal. Figs. 14.9 illustrate that the TCOs based LCD devices are expected to exhibit better spectral reflection performance for both ordinary-wave and extraordinary-wave modes (i.e. more visible light will to be reflected back to the eye of the observer) for frequencies between 1.55eV to 3.26eV (380nm to 780nm).

However, after weighting with the observer's photopic response (as described by the CIE 1931 standard), the weighted spectral reflection in Figs. 14.10 show that the performance of monolayer graphene as a transparent electode is just slightly poorer than both ITO and FTO. This performance difference is primarily due to less reflection arising from thin graphene. In terms of the amount of light captured by the observer, a four-layer graphene based reflective LCD reflects ~10% less light relative to ITO or FTO. Hence, a four-layer graphene can potentially replace the two popular TCOs.

Figure 14.11 shows the CIE1931 spectrally weighted normalized angular reflection of light from the reflective LCD structure with 160nm thick ITO, 340nm thick FTO, monolayer and four-layer graphene as the transparent electrode. The weighted normalized reflection curves for both Figs. 14.11(a) and (b) are computed based on the average of the TE and TM reflection at each incident angle of illuminating plane waves. For both ordinary- and extraordinary-wave modes, monolayer and four-layer graphene exhibit poorer angular reflectance than both ITO and FTO when used as the trans-

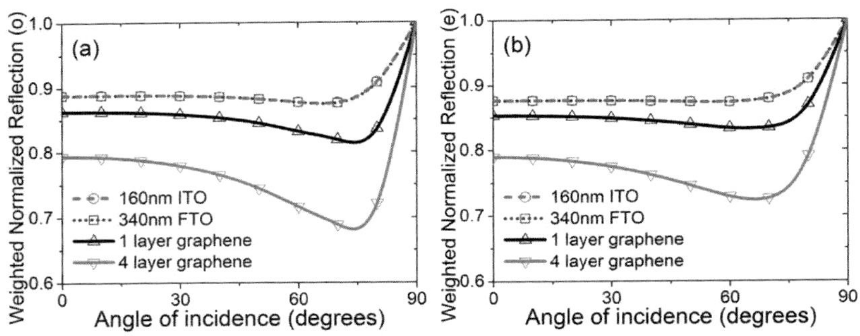

Fig. 14.11 Angular weighted normalized reflection of reflective type LCD with the CIE1931 spectrum for (a) ordinary and (b) extraordinary-wave modes.

parent electrode. It is even more obvious at larger angles of incident light of ~45-80 degrees, where the percentage of reflected light weighted with the CIE 1931 spectrum of monolayer and four-layer graphene reduces from 0.86 to 0.81 and 0.79 to 0.69, respectively, for ordinary-wave mode as shown in Fig 14.11(a). A minima at an incident angle ~75 degrees is observed for both monolayer and four-layer graphene. In contrast, the weighted normalized reflection of ITO and FTO based LCD do not change much for angles of incident light up to 75 degrees as observed in the same figure.

For extraordinary-wave mode, the magnitude of reduction in reflection as the incident light angle increases (up to 70 degrees) for monolayer and four-layer graphene based device is smaller than that of ordinary-wave mode. However, it is obvious that both mono-layer and four-layer graphene also exhibit a minimum at ~70 degrees incident angle. Similar to the ordinary-wave mode, the weighted normalized reflection of the reflective LCD with ITO or FTO as the transparent conductor in extraordinary-wave mode are very similar. The weighted normalized reflection is above 85% at incident angles of illumination up to 70 degrees.

14.5 Graphene for organic solar cells

For organic solar cells, the comparison of graphene, ITO and FTO as transparent electrodes for the well studied polymer-based poly(3-hexylthiophene-2,5-diyl):phenyl-C_{61}-butyric acid methyl ester or P3HT: PCBM bulk-heterojunction and the small molecule zinc phthalocyanine:carbon-60 or ZnPc:C_{60} organic solar cells will be shown. These two types of organic solar cell are chosen as they absorb in different frequency ranges. Sometimes, both material systems are used together in a tandem solar cell as their absorption range are complementary. In the next two sub-sections, the optical performance of the two different organic solar cells will be evaluated with doped graphene, ITO or FTO as the transparent conductor.

To benchmark the performance of the OPV device, the normalized absorbed power in the active layer and also the transparent electrode (i.e. graphene and ITO) are computed with TMM. The normalized absorbed power of the layer of interest is given by

$$A = \frac{\int_{1.55eV}^{3.54eV} [P_n(E) - P_{n+1}(E)]\ S(E)\ dE}{\int_{1.55eV}^{3.54eV} S(E)\ dE} \tag{14.3}$$

where $P_n(E) = \vec{E}_n \times \vec{H}_n$ and $P_{n+1}(E) == \vec{E}_{n+1} \times \vec{H}_{n+1}$ are the Poynting vectors of the two interfaces (n and $n+1$) bounding the layer of interest, such as the active layer or the TCO layer in an organic solar cell. $\vec{E}$ and $\vec{H}$ are the electric field and magnetic field vectors, respectively.

14.5.1 *Graphene in P3HT:PCBM organic solar cell*

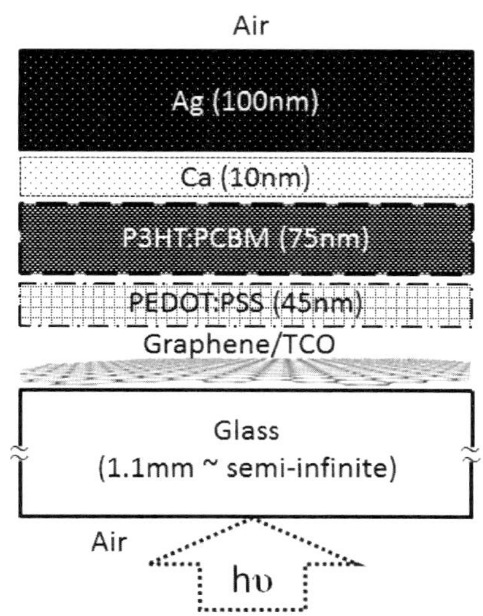

Fig. 14.12 Structure of polymer based P3HT:PCBM bulk heterojunction solar cell.

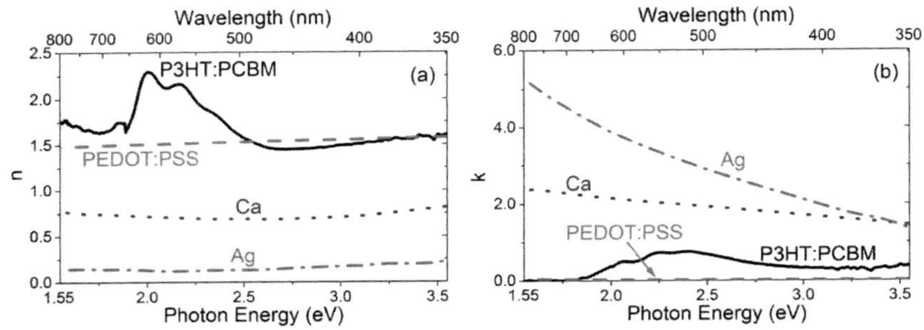

Fig. 14.13 (a) Real and (b) imaginary refractive indices of P3HT:PCBM, PEDOT:PSS, Ca and Ag in a polymer based P3HT:PCBM bulk heterojunction solar cell.

A conventional structure of the P3HT:PCBM organic solar cell and the refractive indices of each material are shown in Fig. 14.12 and Fig. 14.13 respectively. Note that the primary active layer of interest in this solar cell is the P3HT:PCBM bulk heterojunction which absorbs photons with energies greater than 1.8eV (up to 690nm), as observed from the imaginary refractive index k in Fig. 14.13(b). For photon energies greater than 3.54eV (350nm onwards), the absorption of the glass substrate dominates. As a result, no photons with energy greater than 3.54eV will be able to pass through the glass substrate and be absorbed by the P3HT:PCBM bulk heterojunction.

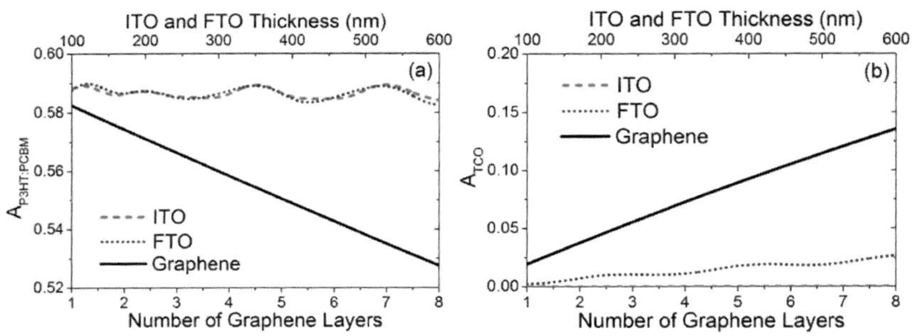

Fig. 14.14 AM1.5G weighted normalized absorptance of (a) P3HT:PCBM and (b) graphene/TCOs as a function of the number of graphene layers or TCO thickness.

As mentioned earlier in Section 14.2, the AM1.5G solar spectrum with an integrated intensity of approximately 1000W/m^2 is the standard spectrum used to emulate outdoor illumination of solar cells. In order to evaluate the performance of the P3HT:PCBM bulk heterojunction solar cell with different transparent conductors (on top of the glass substrate), the proportion of incident solar intensity or the AM1.5G weighted normalized absorptance in P3HT:PCBM are plotted in Fig. 14.14(a) for comparison. As the number of layers of graphene is varied from one to eight stacked layers, the weighted absorptance in P3HT:PCBM exhibits a decreasing trend that is in line with the increased absorption in graphene as illustrated in Fig. 14.14(b). On the other hand, the behavior of the weighted absorptance in P3HT:PCBM with ITO and FTO are very similar and varies between 0.58 and 0.6 as the thickness of the TCOs are increased from 100nm to 600nm. However, it is apparent that the weighted absorptance of FTO rises as the thickness of FTO increases, while ITO has negligible absorption. The optimal ITO thickness for the P3HT:PCBM bulk heterojunction solar cells is ~130nm. However, as FTO typically has higher sheet resistance compared to ITO of the same thickness, the choice of less than 200nm is undesirable from an electrical point of view. Thus, the next best optimal thickness of FTO is ~340nm to ensure that the sheet resistance of FTO is comparable to that of ITO at 130nm.

Figure 14.14(a) shows that the the four-layer graphene (with sheet resistance $<30\Omega/\square$) device allows the polymer active layer (P3HT:PCBM) to absorb 55.8% of the incoming light. Both ITO and FTO based devices allow P3HT:PCBM to absorb 58.8% of the light. This implies a relative 5.1% less absorption when the practical four-layer graphene is used as the transparent electrode as compared to ITO and FTO based P3HT:PCBM bulk heterojunction solar cell. Observations derived from Fig. 14.14(b) indicate that absorption of graphene is significant as compared to both ITO and FTO. Based on the weighted integrated absorptance of graphene and the TCOs, the absorptance in graphene and the two TCOs do not account for the same magnitude of reduction in absorptance in P3HT:PCBM itself.

To gain more insight into the reduction of P3HT:PCBM absorptance with graphene as the transparent electrode, it would be informative to compare the spectral absorption of P3HT:PCBM with ITO, FTO, and four-layer graphene. Figure 14.15 illustrates that the spectral response of ITO and FTO based devices are pretty much determined by interference effects as multiple peaks and troughs can be observed in the P3HT:PCBM spectral absorption. Graphene based P3HT:PCBM solar cell, on the other hand, do not exhibit any enhancement (peak) around the frequencies around $3.1eV$ as compared to ITO and FTO based devices. This can be explained by

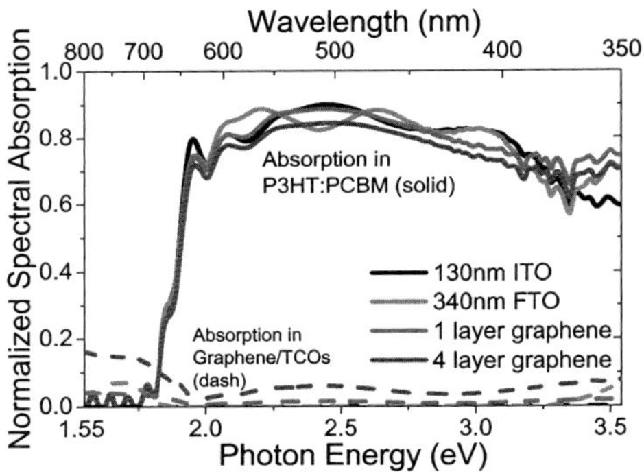

Fig. 14.15 Normalized spectral absorption of P3HT:PCBM and graphene/TCO as a function of photon energy (eV). (For color reference, go to page 500.)

the small thickness of monolayer (∼1nm) and four-layer (∼4nm) graphene. However, the absence of interference based enhancement does not necessarily mean that graphene based electrodes will always perform poorly as compared to ITO or FTO. It is also observed that graphene enables higher absorption in P3HT:PCBM in the UV frequencies (>3.3eV) with both mono-layer and four-layer graphene as compared to the solar cell with ITO and FTO where the dip in P3HT:PCBM absorption is most probably due to destructive interference induced by ITO and FTO layers. Another observation that can be drawn from Fig. 14.15 is that the addition of graphene layers from one to four shows a decrease of the absorption in P3HT:PCBM by approximately 3-4%. This reduction in P3HT:PCBM absorption can be explained by a similar increase in absorption in four-layer graphene as compared to monolayer graphene.

Figure 14.16 shows the AM1.5G spectrum weighted normalized angular absorption of the active layer P3HT:PCBM with monolayer and four-layer graphene, compared with 190nm ITO and 340nm FTO as transparent electrodes for the P3HT:PCBM polymer organic solar cell. It is clearly shown in the figure that four-layer graphene has a poorer angular performance in the P3HT:PCBM organic solar cell as compared to ITO or FTO. However, it is interesting to note that if the sheet resistance of monolayer graphene can be reduced to that of the four-layer graphene, the angular absorption of P3HT:PCBM with monolayer graphene as the transparent electrode is marginally poorer as compared to ITO or FTO for incident illumination up to 40 degrees. For incident light entering the monolayer graphene based

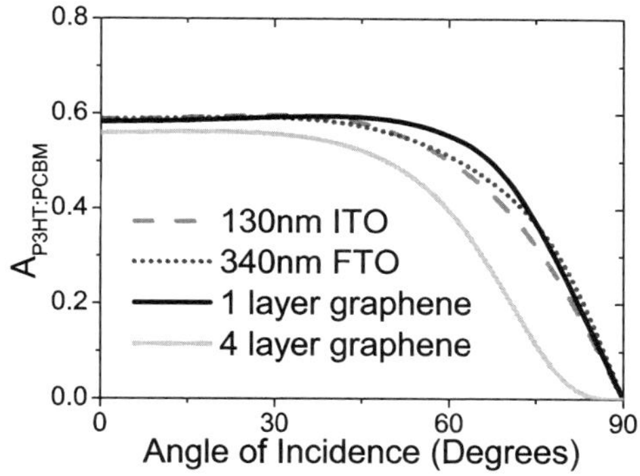

Fig. 14.16 Normalized angular absorption of P3HT:PCBM with Graphene, ITO and FTO.

P3HT:PCBM solar cell from 40 to 75 degrees, monolayer graphene allows more light to be absorbed by the active layer than the same device with ITO or FTO as the transparent electrode.

14.5.2 *Graphene in ZnPc:C_{60} organic solar cell*

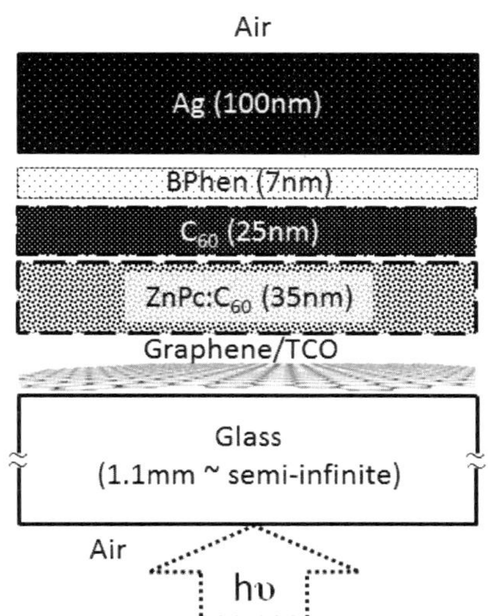

Fig. 14.17 Structure of small molecule based ZnPc:C_{60} bulk heterojunction solar cell.

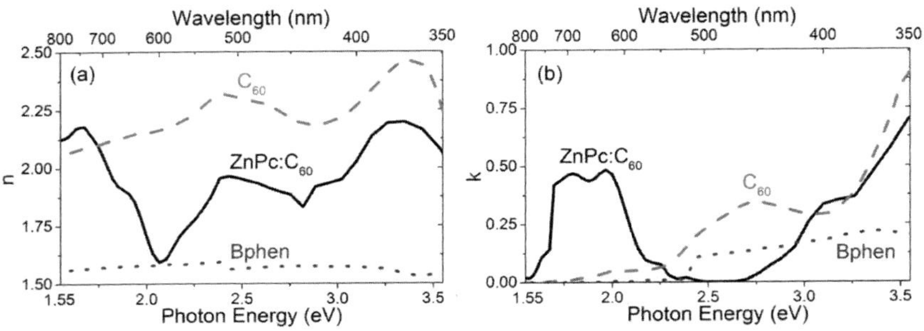

Fig. 14.18 (a) Real and (b) Imaginary refractive index of ZnPc:C_{60}, C_{60}, and BPhen layers in the small molecule based ZnPc:C_{60} + C_{60} bulk heterojunction solar cell.

In addition to the polymer based P3HT:PCBM bulk heterojunction solar cell, the small molecule based zinc phthalocyanine:carbon-60 (ZnPc:C_{60}) with graphene, ITO and FTO transparent electrodes are also considered. The conventional structure of the ZnPC:C_{60} solar cell is illustrated in Fig 14.17. The real and imaginary refractive indices of the ZnPc:C_{60} bulk heterojunction, C_{60} and 4,7-diphenyl-1,10-phenanthroline (BPhen) exciton-blocking layer are obtained from ellipsometric measurements and are plotted in Fig. 14.18. For the ZnPc:C_{60} bulk heterojunction solar cell, the primary organic active layers are both the ZnPc:C_{60} bulk heterojunction and the C_{60} layer deposited on top of the ZnPc:C_{60}. The C_{60} layer on top of ZnPc:C_{60} forms a separate planar heterojunction to facilitate both charge separation and exciton dissociation. It is important to note that the ZnPc:C_{60}+C_{60} active layers have a wider range of absorption for frequencies >1.55eV (up to 800nm) as compared to >1.8eV (up to 690nm) for P3HT:PCBM bulk heterojunction.

Figure 14.19(a) shows the AM1.5G weighted normalized absorptance of the small molecule organic absorber ZnPc:C_{60}+C_{60} as a function of the number of graphene layers, ITO and FTO thickness for illumination photons with energies ranging from 1.55eV to 3.54eV (350nm to 800nm). The four-layer graphene device exhibits 4.4% and 3.3% less absorption in the active layer as compared to ITO and FTO based devices respectively, where the four-layer graphene enables the ZnPc:C_{60}+C_{60} active layer to absorb 47.3% of the incident solar intensity as compared to 49.5% for 190nm ITO and 49% for 340nm FTO. Similar to the weighted normalized absoprtance of P3HT:PCBM bulk heterojunction in Fig. 14.14, there are multiple peaks and troughs observed as the ITO and FTO thickness increase from 100nm

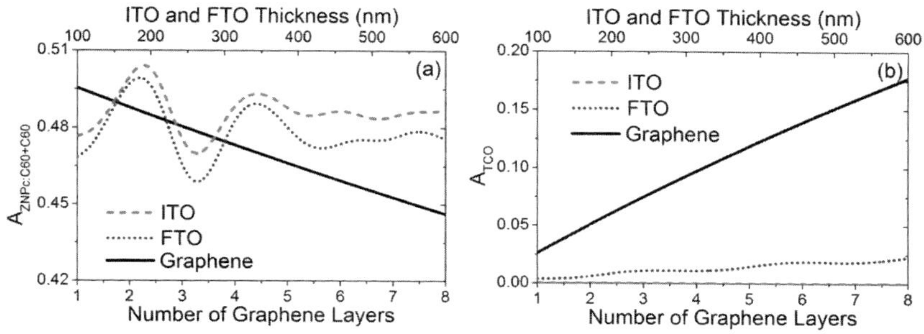

Fig. 14.19 AM1.5G weighted normalized absorptance of (a) ZnPc:C_{60}+C_{60} and (b) graphene/TCOs as a function of the number of graphene layers and TCO thickness.

to 600nm. Therefore, interference effect is the the most important consideration to determine the optimal thickness of TCOs. On the other hand, the weighted normalized absorptance of ZnPc:C_{60}+C_{60} decreases when the number of graphene layers increases. Thus, this further confirms that absorption in graphene is limiting optical transmission of incident light to the ZnPc:C_{60}+C_{60} active layers as shown by the substantial absorptance by graphene in Fig. 14.19(b). It is also interesting to note that the use of monolayer graphene as the transparent electrode in ZnPc organic solar cell is able to facilitate higher absorption in ZnPc:C_{60}+C_{60} than the case when 340nm FTO is used. Thus, this further reinforces the idea that if the sheet resistance of monolayer graphene can be reduced to 20-30$\Omega/\square$, just a single layer of graphene will be able to exhibit better optical performance for ZnPc:C_{60} organic solar cells as compared to FTO.

The normalized spectral absorption characteristics of ZnPc:C_{60}+C_{60} with graphene and the two TCOs (ITO and FTO) are plotted in Fig. 14.20 in solid lines. The corresponding normalized spectral absorption of the transparent electrodes are plotted in the same figure with dash lines. The choice of 190nm thick ITO and 340nm thick FTO are the optimal thicknesses obtained from Fig. 14.19(a). In Fig. 14.20, the effect of constructive interference for ITO and FTO can be observed for frequencies >2.5eV where the presence of the 190nm ITO electrode induces significantly higher absorption for ZnPc:C_{60}+C_{60} at around 2.8eV & 3.5eV (~354nm–443nm), and the 340nm FTO induces two peaks in ZnPc:C_{60}+C_{60} absorption between 2.5eV & 3.3eV (~376nm–496nm) as compared to the spectral absorption curves of either monolayer and four-layer graphene based ZnPc:C_{60} device. Otherwise, monolayer graphene is actually able to perform on par or

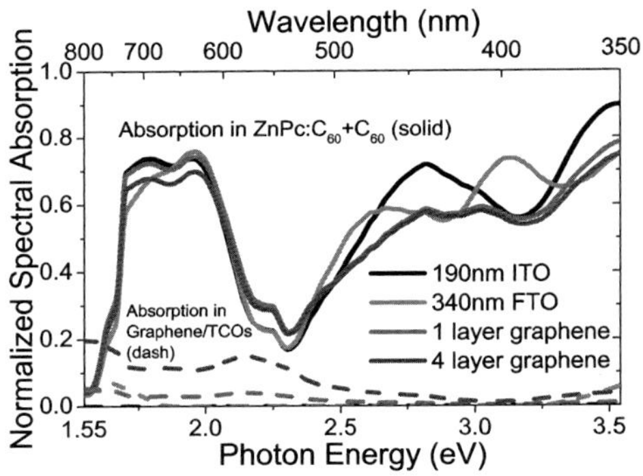

Fig. 14.20 Normalized spectral absorption of ZnPc:C_{60}+C_{60} and graphene/TCO as a function of photon energy (eV). (For color reference, go to page 500.)

even better than 190nm ITO and 340nm FTO for frequencies <2.4eV (up to 517nm). However, with the current state-of-the-art four-layer graphene required to achieve sheet resistance $\sim 30\Omega/\Box$, the spectral absorption in ZnPc:C_{60}+C_{60} is poorer than that of ITO and FTO for most of the frequencies in the band of interest between 1.55eV to 3.54eV (350nm to 800nm).

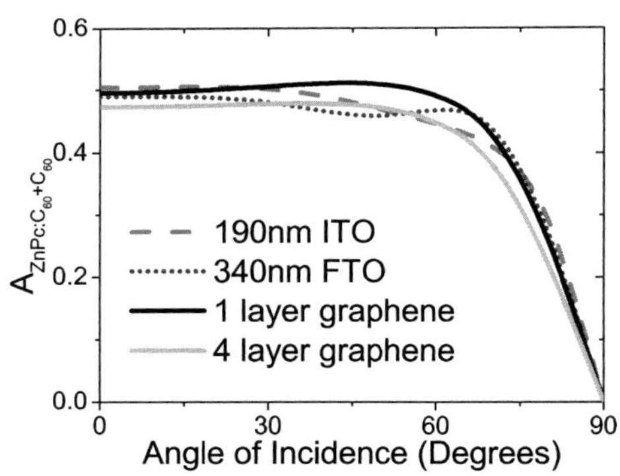

Fig. 14.21 Normalized angular absorption of ZnPc:C_{60}+C_{60} with graphene, ITO and FTO.

Figure 14.21 shows the normalized angular absorption of the active layer ZnPc:C_{60}+C_{60} with monolayer and four-layer graphene, compared with 190nm ITO and 340nm FTO. Similar to the angular absoprtion of P3HT:PCBM, the four-layer graphene generally allows less solar energy to be absorbed by the ZnPc:C_{60}+C_{60} active layer as compared to ITO or FTO transparent electrodes. There is an exception for angles between 40-50 degrees, where four-layer graphene is marginally superior to FTO. However, the angular absorption of the ZnPc:C_{60} organic solar cell's active layer, with monolayer graphene as the transparent electrode, can surpass that of a similar ITO or FTO device for incident illumination up to ~65 degrees. Thus, if the sheet resistance of monolayer graphene can be reduced to that of its four-layer stacked graphene, graphene can definitely be a potential replacement for ITO or FTO in ZnPc:C_{60} organic solar cell.

14.5.3 *Discussion*

Based on the spectral absorption characteristics of the active layer, where P3HT:PCBM has a shorter spectral absorption (1.8eV to 3.54eV or 350nm to 690nm) as compared to ZnPc:C_{60} (1.55eV to 3.54eV or 350nm to 800nm), four-layer graphene performs 0.7% to 1.8% better in the ZnPc:C_{60} bulk heterojunction than in the P3HT:PCBM bulk-heterojunction solar cell when compared to the ITO or FTO equivalent. This is probably because interference enhances over a narrow frequency band and graphene has the advantage of wide band transmission due to the negligible thickness. Thus, it is expected that graphene will be able to outperform ITO and FTO as the transparent electrode for the perovskite solar cell which has a wider absorption frequency band that can span from 1.38eV to 3.54eV (350nm to 900nm) [Cortecchia *et al.* (2016)].

14.6 Summary

In summary, this chapter introduces the requirements of a transparent electrode that should enable (1) at least 80% transmittance of visible light for its use in displays or the solar spectrum for solar cells applications, and (2) less than 30$\Omega/\square$ sheet resistance to function as a reasonable conductor of electrical current. The optical and electrical characteristics of the two most common transparent conducting oxides, ITO and FTO, are also illustrated and compared with that of the current state-of-the-art HNO_3-doped graphene produced by roll-to-roll technology.

The optical performance of graphene as a transparent electrode in the reflective type LCD is then compared to that of the same device with ITO and FTO. Using the reflection of incident light spectrally scaled with the photopic response of the eye with the CIE 1931 standard as a gauge of the LCD's performance, it is determined that the four-layer graphene LCD device reflects >10% less light due to substantial absorption of graphene relative to ITO or FTO's performance for incident angles of light up to 70 degrees. Even if the sheet resistance of monolayer graphene can be reduced to that of the four-layer stacked graphene ($\sim$28 $\Omega/\square$), a graphene LCD device will not be able to outperform the ITO or FTO as the transparent conductor in the reflective type LCD investigated here.

Two different organic solar cells are evaluated with graphene as the transparent electrode. The optical absorption of the active layers are computed and compared with the same device with ITO and FTO. For the polymer based P3HT:PCBM bulk heterojunction solar cell which absorbs photons with energies from 1.8eV to 3.54eV (350nm to 690nm), the four-layer graphene is unable to outperform both ITO and FTO. However, if the monolayer graphene is used as the transparent electrode, the AM1.5G weighted absorption in the P3HT:PCBM active layer is marginally higher than that of the same ITO and FTO photovoltaic device at oblique illumination from angles 40 to 70 degrees. For other angles of illumination (especially normally incident light), monolayer graphene based P3HT:PCBM device is limited by its intrinsic absorption even though it is much thinner than ITO or FTO when applied to the same device.

In contrast to P3HT:PCBM bulk heterojunction solar cell, the small molecule ZnPc:C_{60} bulk heterojunction has a broader absorption band spanning photon energies from 1.55eV to 3.54eV (350nm to 800nm). At normal incidence, both ITO and FTO outperform four-layer graphene as the transparent electrode in the ZnPc:C_{60} organic solar cell, but when compared to monolayer graphene, only the ITO based ZnPc:C_{60} organic solar records slightly marginally higher absorption in the active layer. In terms of angular absorption, both monolayer and four-layer ITO allows more light to be absorbed by the active layer of the ZnPc:C_{60} organic solar cell at angles of illumination between 40 to 50 degrees. It is interesting to note that the monolayer graphene has a more desirable angular response which allows higher absorption in the active layers of both P3HT:PCBM and ZnPc:C_{60} bulk heterojunctions solar cell.

Bibliography

[1] Aegerter, M. A. (2016). http://www.solgel.com/labfoc/inm/spray_e.html.

[2] ASTM G173-03 (2012). Standard tables for reference solar spectral irradiances: Direct normal and hemispherical on 37° tilted surface, http://www.astm.org/Standards/G173.htm.

[3] Bae, S., Kim, H., Lee, Y., Xu, X., Park, J.-S., Zheng, Y., Balakrishnan, J., Lei, T., Kim, H. R., Song, Y. I., Kim, Y.-J., Kim, K. S., Ozyilmaz, B., Ahn, J.-H., Hong, B. H., and Iijima, S. (2010). Roll-to-roll production of 30-inch graphene films for transparent electrodes, *Nat. Nano.* **5**, pp. 574–578.

[4] Ball, J. M., Stranks, S. D., Hörantner, M. T., Hüttner, S., and Zhang, W. (2015). Optical properties and limiting photocurrent of thin-film perovskite solar cells, *Energy Environ. Sci.* **8**, pp. 602–609, doi:10.1039/C4EE03224A, suppliementary Information.

[5] Bonaccorso, F., Sun, Z., Hasan, T., and Ferrari, A. C. (2010). Graphene photonics and optoelectronics, *Nat. Photonics* **4**, pp. 611–622.

[6] Cortecchia, D., Dewi, H. A., Yin, J., Bruno, A., Chen, S., Baike, T., Boix, P. P., Grätzel, M., Mhaisalkar, S., Soci, C., and Mathews, N. (2016). Lead-free ma2cuclxbr4x hybrid perovskites, *Inorg. Chem.* **55**, 3, pp. 1044–1052.

[7] Geim, A. K. and Novoselov, K. S. (2007). The rise of graphene, *Nat. Mat.* **6**, pp. 183–191.

[8] Koh, W. S., Gan, C. H., Phua, W. K., Akimov, Y. A., and Bai, P. (2014). The potential of graphene as an ito replacement in organic solars: An optical perspective, *IEEE J. Sel. Top. Quantum Electron.* **20**, 1, pp. 36–42.

[9] Koh, W. S., Pant, M., Akimov, Y. A., Goh, W. P., and Li, Y. (2011). Three-dimensional optoelectronic model for organic bulk heterojunction solar cells, *IEEE J. Photovolt.* **1**, 1, pp. 84–92.

[10] Lee, S., Lee, K., Liu, C.-H., and Zhong, Z. (2012). Homogeneous bilayer graphene film based flexible transparent conductor, *Nanoscale* **4**, pp. 639–644.

[11] Li, J. (2005). *Refractive Indices of Liquid Crystals and their Applicaions in Display and Photonic Devices*, Ph.D. thesis, University of Central Florida, Orlando, Florida, USA.

[12] Palik, E. D. (1997). *Handbook of Optical Constants of Solids*, Vol. V, 2nd edn. (Elsevier, Reading, MA), ISBN 978-0-12-544415-6.

[13] Wang, X. Z., Ng, G. M., Ho, J. W., Tam, H. L., and Zhu, F. (2010). Efficient semitransparent bulk-heterojunction organic photovoltaic cells with high-performance low processing temperature indium tin oxide top electrode, *IEEE J. Sel. Top. Quantum Electron.* **16**, 6, pp. 1685–1689.

[14] Zeng, W., Yong, K. S., Kam, Z. M., Zhu, F., and Li, Y. (2010). Effect of blend layer morphology on performance of znpc:c60-based photovoltaic cells, *Appl. Phys. lett.* **97**, 13, p. 133304.

Color Plates

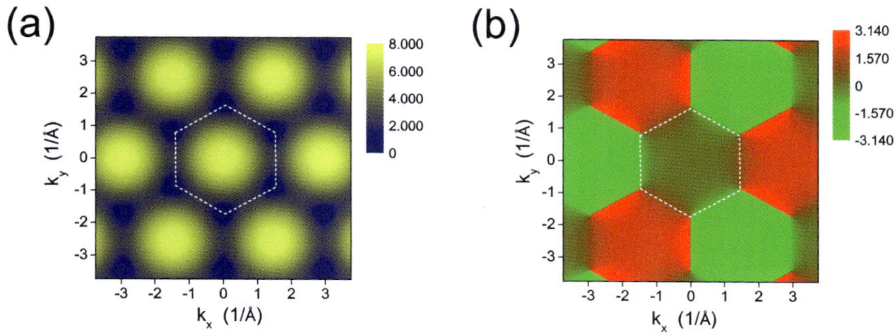

Fig. 1.7: First appeared on page 14.

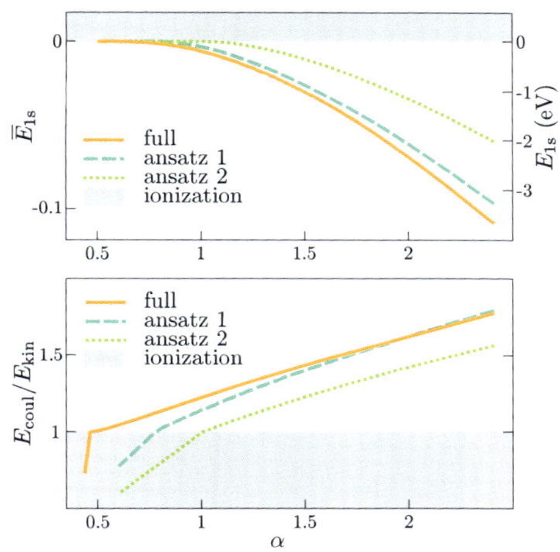

Fig. 2.5: First appeared on page 68.

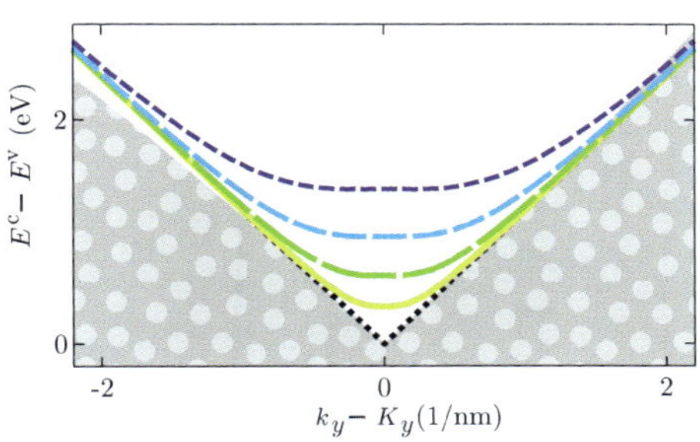

Fig. 2.6: First appeared on page 70.

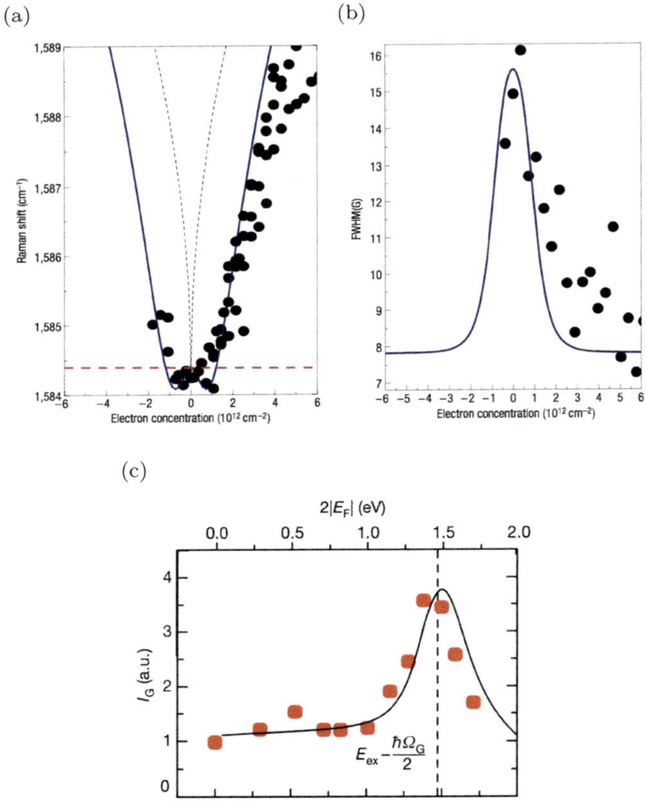

Fig. 3.10: First appeared on page 121.

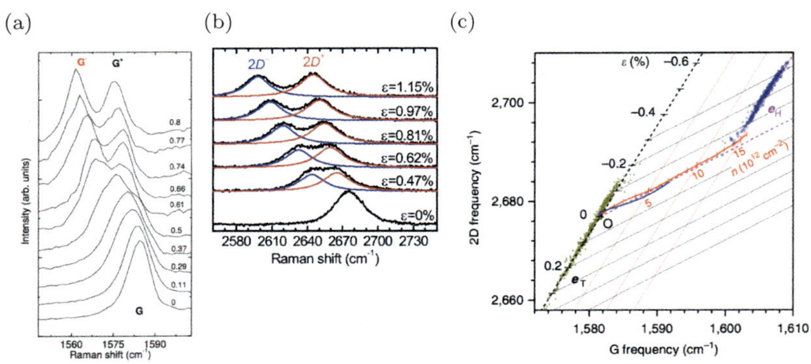

Fig. 3.11: First appeared on page 127.

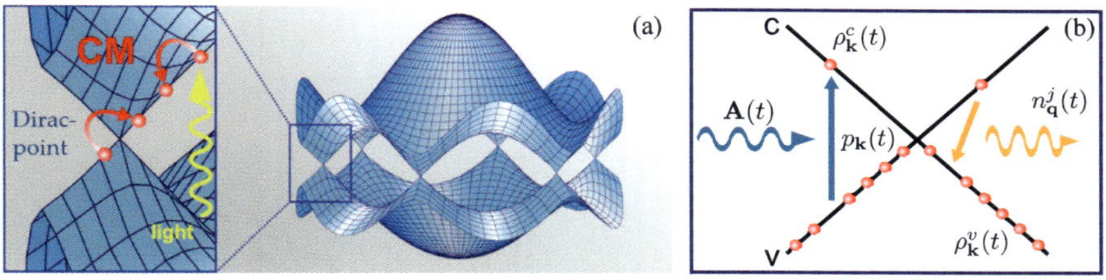

Fig. 4.1: First appeared on page 136.

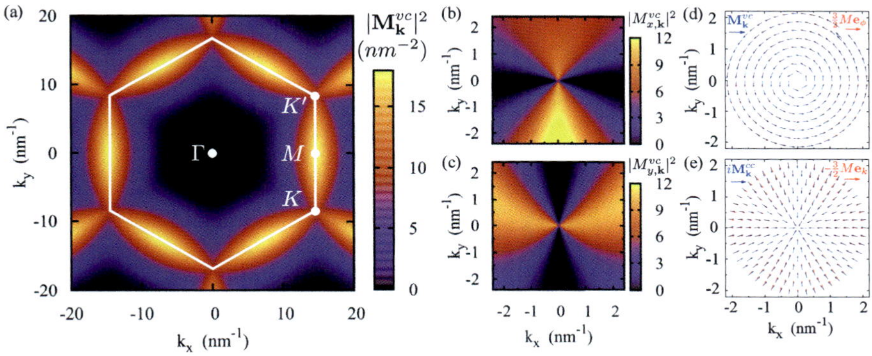

Fig. 4.2: First appeared on page 140.

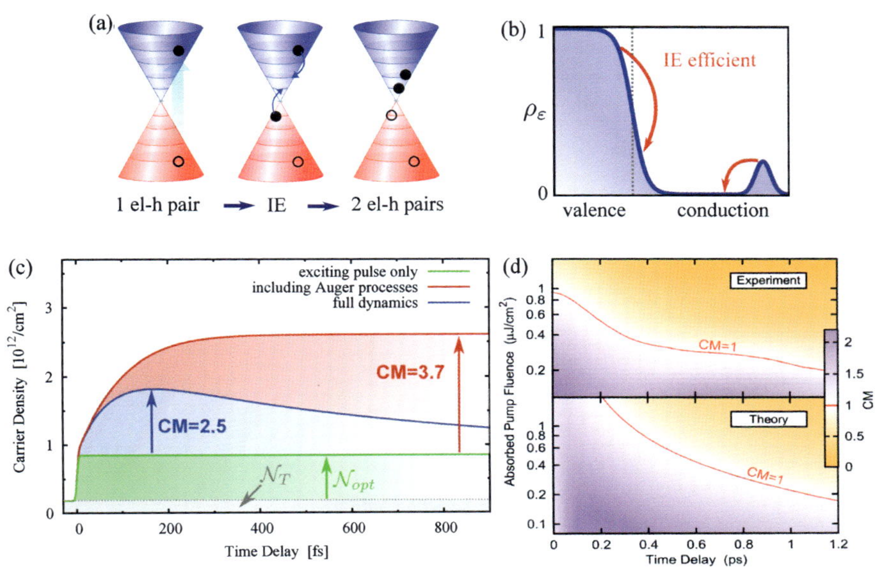

Fig. 4.19: First appeared on page 168.

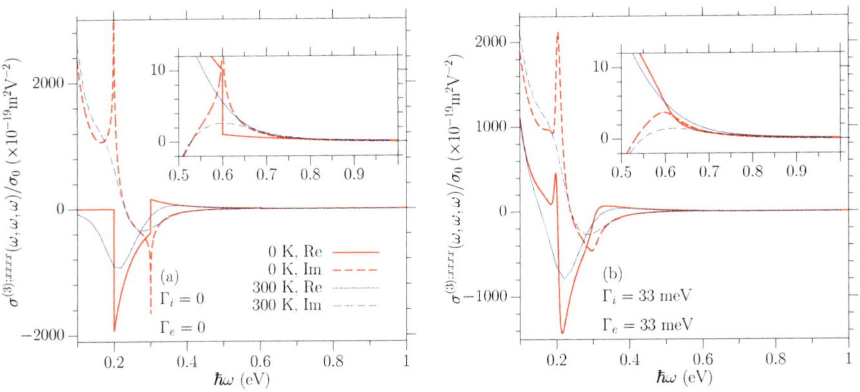

Fig. 5.4: First appeared on page 202.

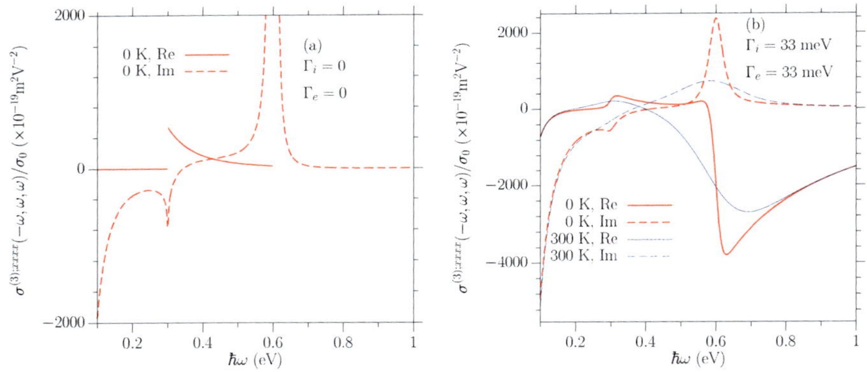

Fig. 5.5: First appeared on page 205.

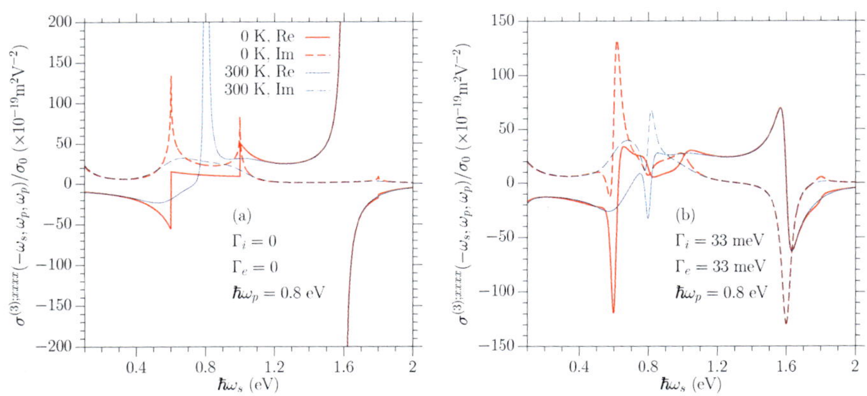

Fig. 5.6: First appeared on page 207.

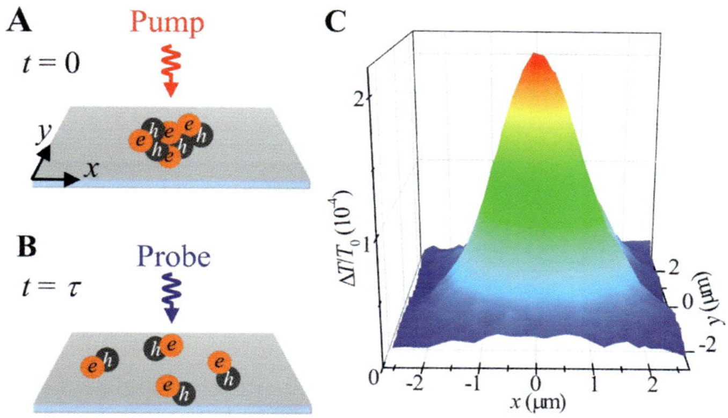

Fig. 6.9: First appeared on page 236.

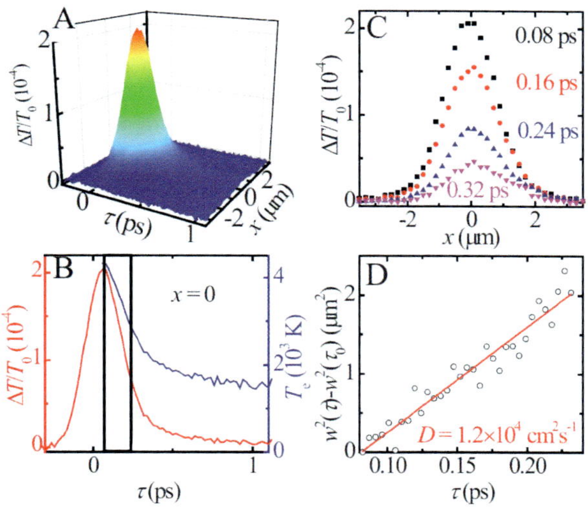

Fig. 6.10: First appeared on page 237.

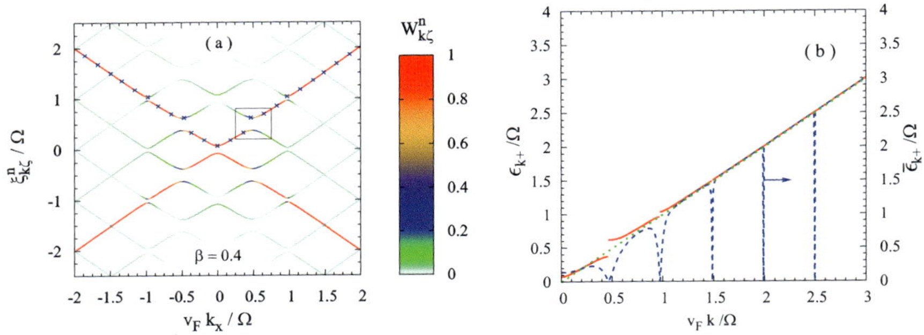

Fig. 7.2: First appeared on page 253.

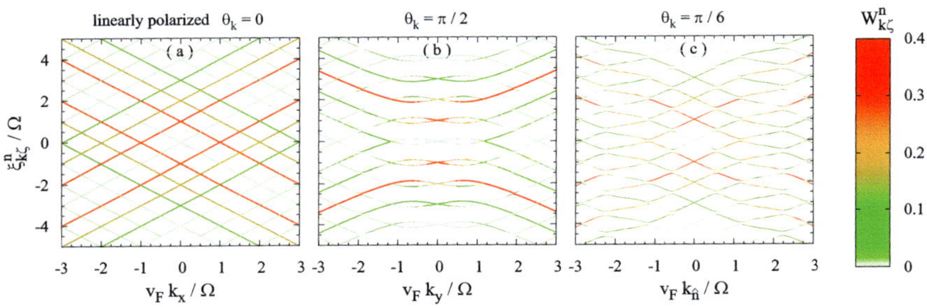

Fig. 7.3: First appeared on page 253.

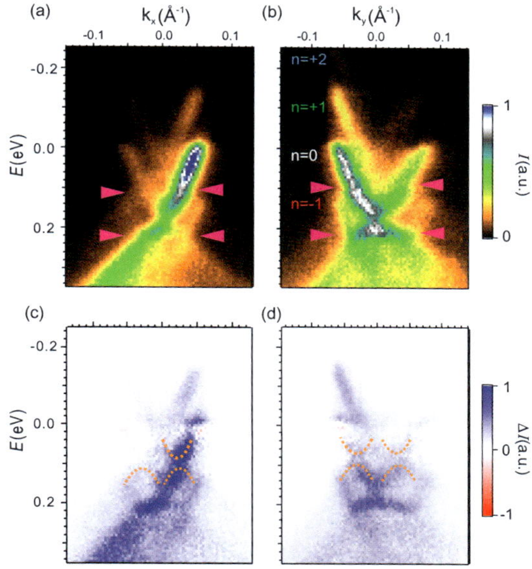

Fig. 7.6: First appeared on page 257.

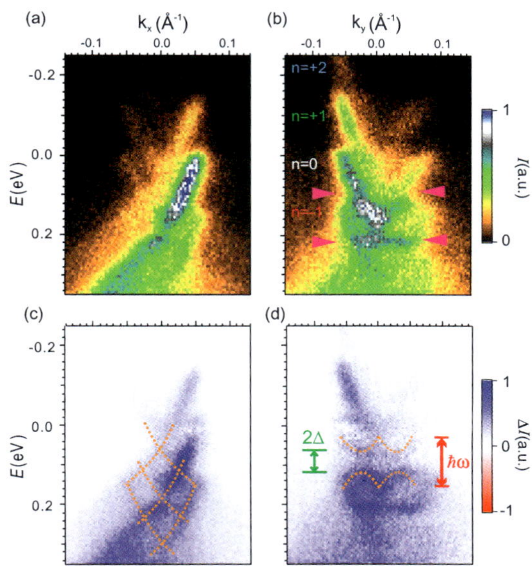

Fig. 7.7: First appeared on page 258.

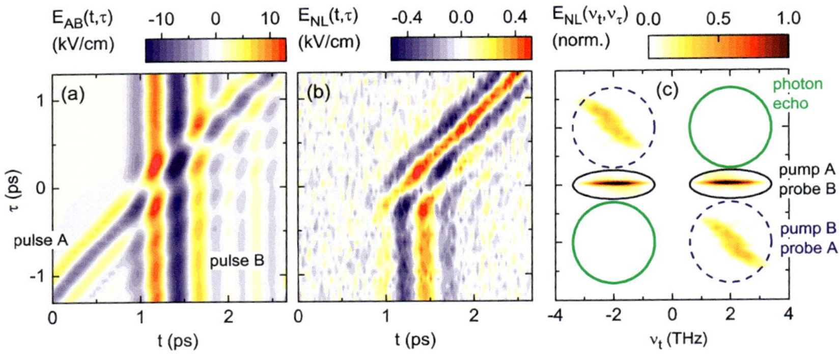

Fig. 8.6: First appeared on page 279.

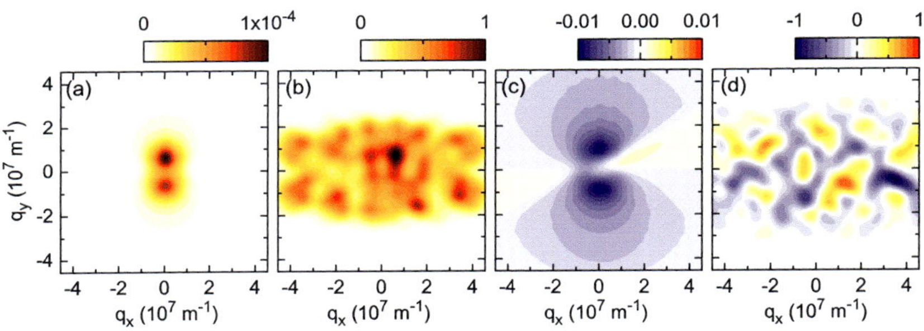

Fig. 8.9: First appeared on page 283.

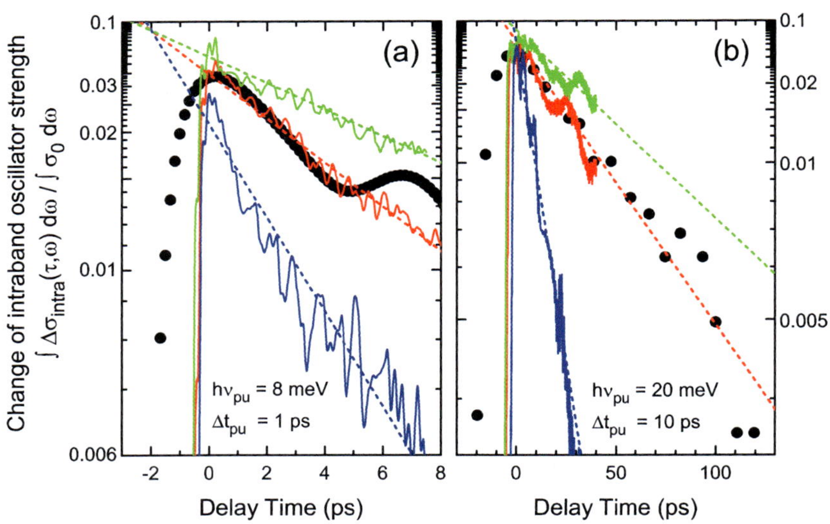

Fig. 8.10: First appeared on page 285.

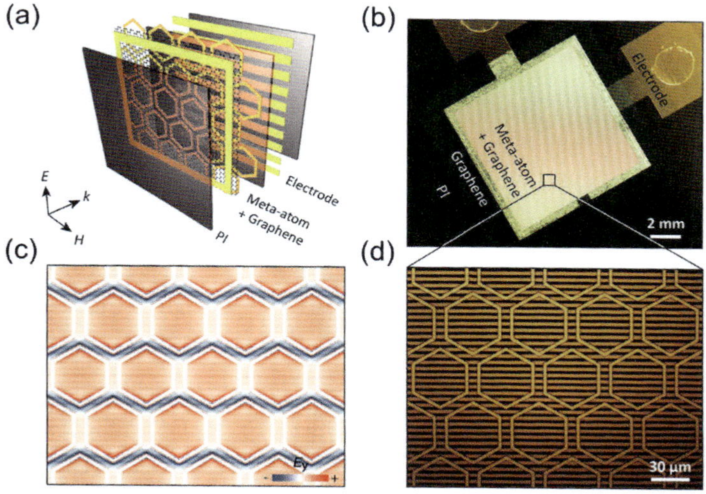

Fig. 9.9: First appeared on page 310.

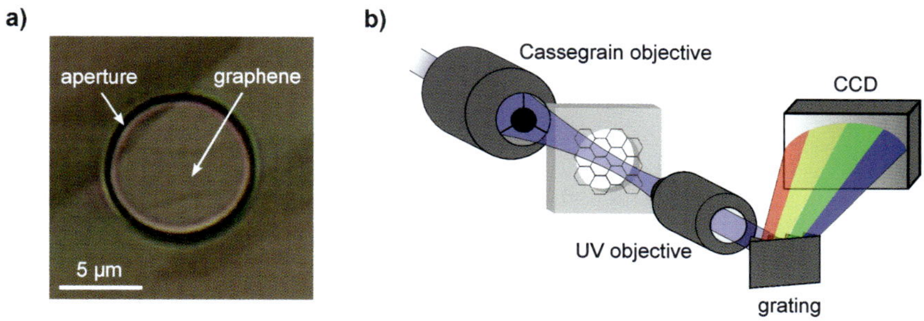

Fig. 10.2: First appeared on page 329.

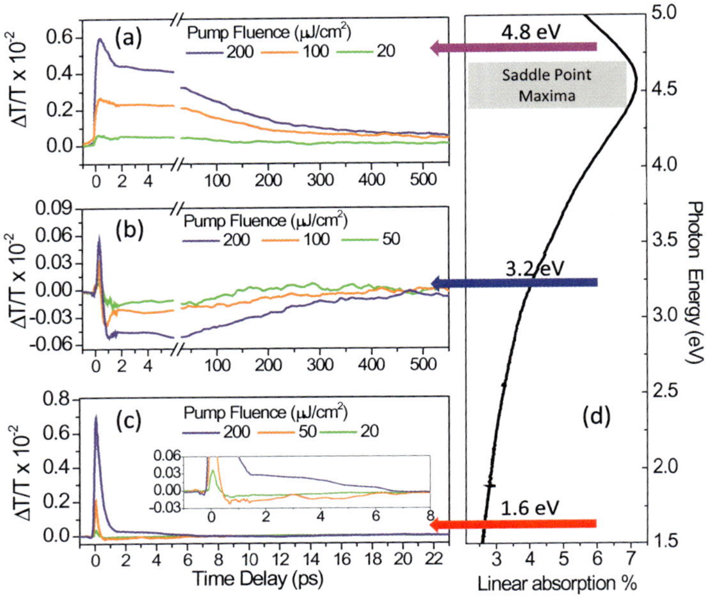

Fig. 11.4: First appeared on page 355.

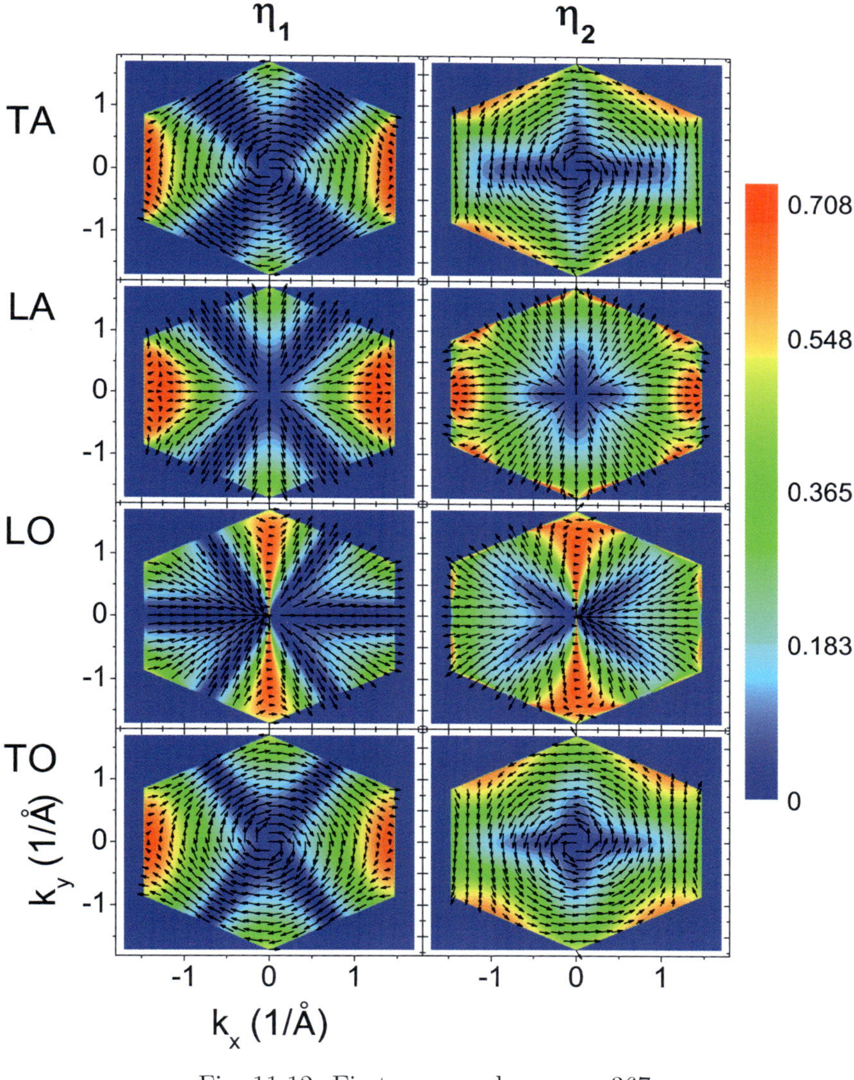

Fig. 11.12: First appeared on page 367.

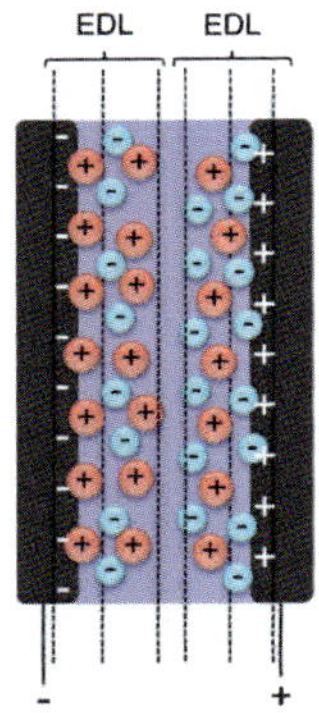

Fig. 12.15: First appeared on page 406.

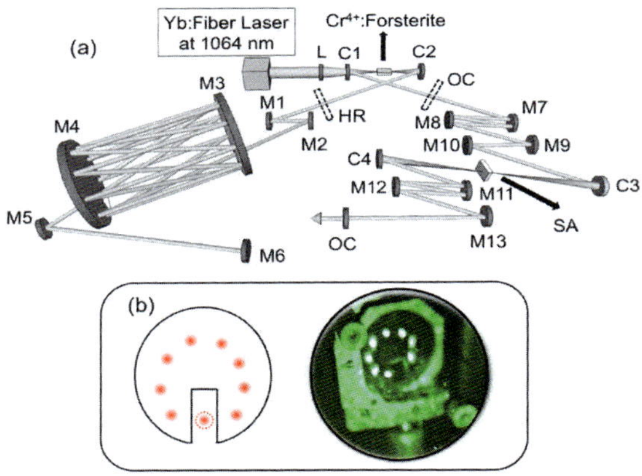

Fig. 12.14: First appeared on page 419.

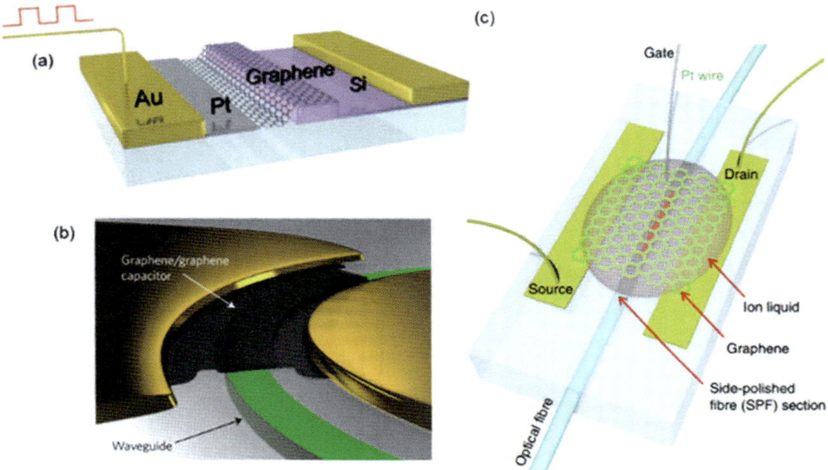

Fig. 13.3: First appeared on page 440.

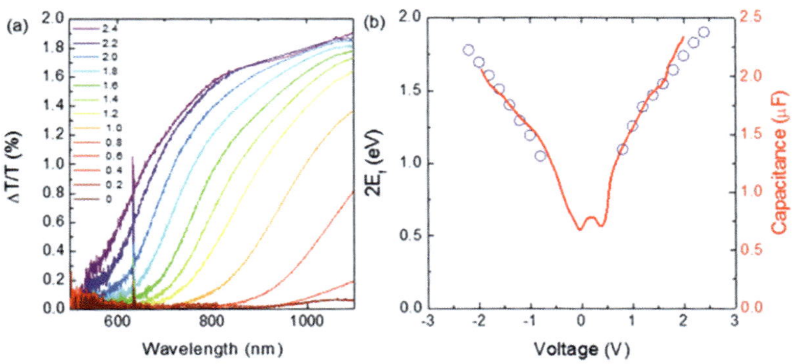

Fig. 13.5: First appeared on page 443.

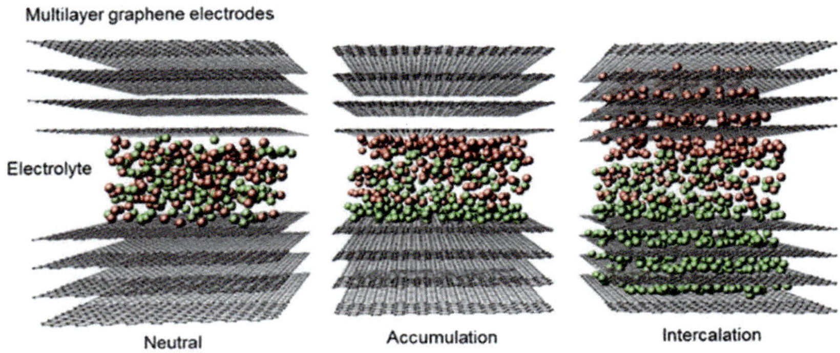

Fig. 13.8: First appeared on page 446.

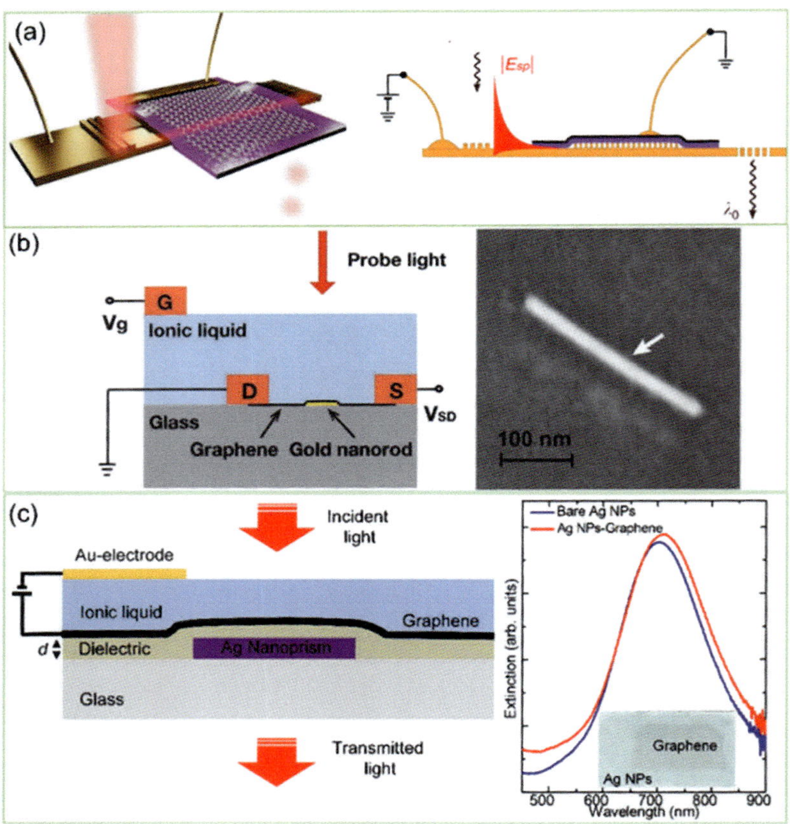

Fig. 13.10: First appeared on page 449.

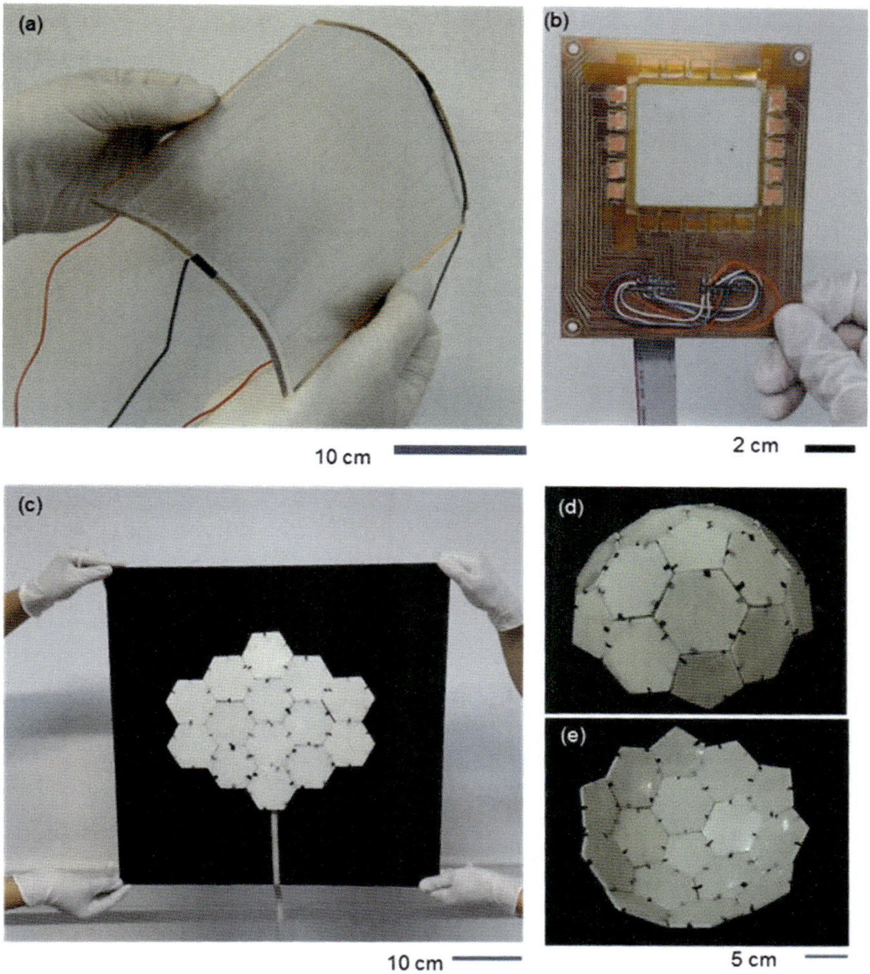

Fig. 13.12: First appeared on page 453.

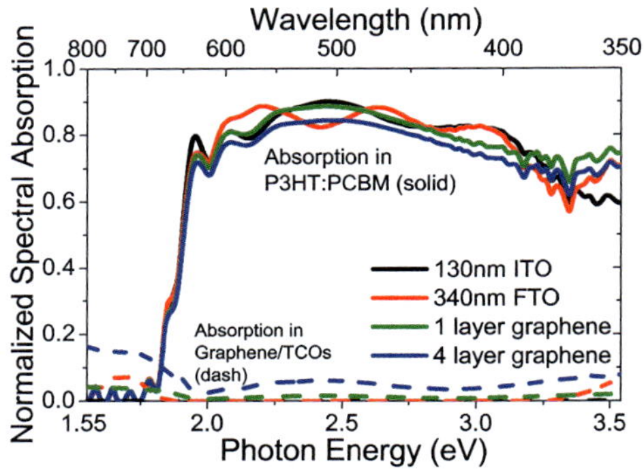

Fig. 14.15: First appeared on page 474.

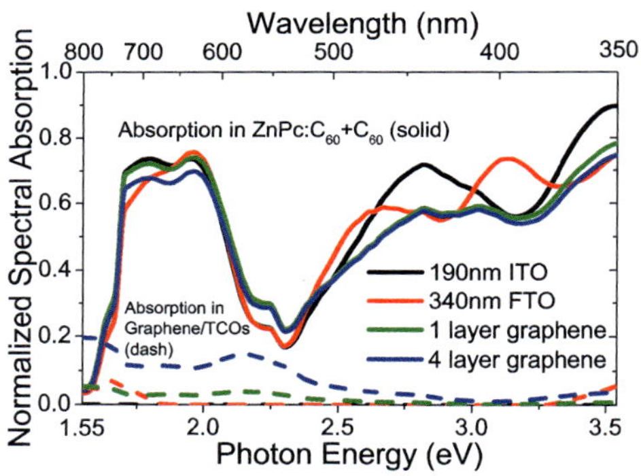

Fig. 14.20: First appeared on page 478.

Index